Vol. I.   A.D. 1900.

THE OFFICIAL MANUAL

OF THE

CRIPPLE CREEK DISTRICT

COLORADO, U.S.A.

PUBLISHED BY

FRED HILLS, E.M.

Colorado Springs, Colo.

PRICE **5** DOLLARS

Introduction

Sylvanite Publishing and Miningbooks.com are proud to put back into print this long out of print publication. The information in this publication is still valid and informative for historians, researchers, prospectors, miners, geologists and more. Much of this information is becoming lost as more and more books are discarded from libraries or schools, as older material disintegrates, and as publications are thrown out because someone does not know their value.

Our goal is to put hundreds of these publications back into print at a reasonable cost. Many of these originals can cost hundreds of dollars which is out of the reach of many who just want the information for study. In the process of this we try and clean up the books and text best we can to produce a quality product. There are many publishers reprinting books these days but they are essentially scan shops that put them into print errors, smudges, writing in the books, and all. Very occasionally we will do the same if we cannot obtain a copy in decent condition.

If you would like to see a particular book or are and author that has written a book in the past that is now out of print in our subject field, feel free to contact us and we will see what we can do to get that publication back in print.

THE CITY OF CRIPPLE CREEK—1900.

SHOWING MOUNT PISGAH IN THE BACKGROUND.

INTRODUCTION

IN submitting this, the first volume of the Official Manual of the Cripple Creek District, the publisher desires to thank the officers of the various companies and others who have assisted him in its compilation, and to state that it has entailed eight months of arduous labor and minute research to attain the results now presented.

It has been his aim to not only make the work correct and complete, but also the best publication ever issued of the greatest mining camp in the world—which Cripple Creek justly deserves—this being evidenced by the typographical part produced by The Smith-Brooks Printing Company, and the etchings and half tones by The Williamson-Haffner Engraving Company, both of Denver.

All those who appreciate these efforts are particularly requested to forward data relative to any changes in the companies represented herein, and to submit, for the next volume, all information regarding new corporations. The observance of this request will greatly facilitate the compilation and correctness of subsequent issues.

FRED HILLS, *Mining Engineer,*
PUBLISHER,
COLORADO SPRINGS, COLORADO, U. S. A.

June, 1900.

MISCELLANEOUS INDEX.

INDEX TO ILLUSTRATIONS.

INDEX TO COMPANIES.

THE COLORED MAP

*Which is folded in the back of this work,
shows in color the surface holdings of each
company described herein. These holdings
are designated by the company name, and
not by that of the claim, as the latter is
given in the individual plat of the property.*

THE CRIPPLE CREEK DISTRICT
ITS PAST AND FUTURE

WRITTEN SPECIALLY FOR THE *OFFICIAL MANUAL*

By T. A. RICKARD, DENVER
State Geologist of Colorado

I. HISTORICAL AND GENERAL.

The name of Cripple Creek is said to have originated from the fact that at a certain point along the course of the little stream flowing through the ranch on which the town was subsequently built there was a morass in which straying cattle wandered and. in their efforts to extricate themselves, were occasionally lamed. So says one of the survivors of the band of men who once tended the herds that grazed on the hills now pierced with many shafts. However, the name needs no apology today; the magic baptism of golden discovery has made it sound as alliterative and impressive as the most exacting historian could demand.

Cripple Creek has had its vicissitudes. In the early days of Colorado's history the country around Pike's Peak was searched for gold by the pioneers who were attracted from afar by the snowy crest of the old beacon mountain. They found no gold worthy of mention either in the cañons or amid the hills that lie at the foot of the peak. The interval of a whole generation elapsed before the expectations of the first prospectors became justified, and during the intervening period of over thirty years the district was the scene of a peculiar incident which passed into local history under the name of the Mount Pisgah fiasco.

Among the hills which, like a flock of sheep, cluster around the southern base of Pike's Peak there is a dark cone standing in solitude among its smaller brethren. This is Mt. Pisgah. In April, 1884, the rumor of rich discoveries of gold caused a horde of prospectors to hurry from Leadville and the adjoining camps to the country south of Pike's Peak. Mt. Pisgah was mentioned as the spot from which the miner might hope to see the promised land of wealth and plenty. The dawn of the next day found an excited crowd of four thousand men camping upon the shoulder of the hill. They found no gold in workable quantity save in the prospect holes made by the original locators. Salting was suspected; the man who had instigated the rush was conspicuous by absence; an accomplice was caught with a bottle of yellow stuff in his pocket. It was not whisky, but its modern antidote, the chloride of gold. Man had endeavored to remedy nature's seeming niggardliness and the rock had been artificially enriched. Angry feelings found vent in threats of lynching, but in the failure to lay hands on the real perpetrators of the fraud, the affair broke up in a big picnic and a general drunk. A little digging had been done, one or two veins were uncovered, but the comparative poverty of the ore only added bitterness to the general disappointment. The crowd disappeared as quickly as it had come. The hillsides resumed the quiet aspect of the cattle range for which they seemed best fitted. The incident was over.

Not until 1893 did Cripple Creek make its mark; two years before, the name had begun to be mentioned in mining circles, but the Mt. Pisgah excitement had discredited the district, and the contemporaneous discoveries of large silver lodes at Creede, at the head waters of the Rio Grande, diverted attention for a time.

For several years preceding 1891, prospectors had wandered over the hills between Colorado Springs and Florissant, unconsciously treading in the footsteps of the men of 1858. Among the earliest of the gold-seekers was Robert Womack, who once owned a small ranch in the district. He sold it to Bennett & Myers, the proprietors, at that time, of the cattle range which covered a large part of the area now forming the environs of the town of Cripple Creek. For ten years, from 1880 to 1890, Bob Womack had been living in the district, doing occasional work for Bennett & Myers. He spent his spare time in prospecting, for he had previously had some experience in Gilpin County, and knew gold ore when he saw it. In the course of desultory digging, he found several veins, and when seen, at intervals, at Colorado Springs, he would exhibit pieces of float (surface ore) as evidence of his discoveries; but having a reputation for honesty rather than shrewdness, his statements made little impression. For many years he worked a hole in Poverty Gulch without staking a claim in proper form. There seemed no need to do so; no one came to disturb him. The whole hill country was at that time fenced in so as to serve as a summer range for cattle. The cowboys and herdsmen looked good-naturedly at Bob's digging, and considered it of no moment. In December, 1890, E. M. De la Vergne and F. F. Frisbee came up from Colorado Springs to prospect. The hills were under snow and only a few bare spots permitted of any investigation. On Guyot Hill, in Eclipse and Poverty Gulches, they found evidence of gold veins, from which they took away samples. Encouraged by their first visit, De la Vergne and Frisbee returned early in February, 1891. They found Bob Womack at work in Poverty Gulch. He had sunk a shaft to a depth of forty-eight feet, and encountered good ore. The claim he had pegged out was called the Chance, and a number of stakes indicated that he had re-located it six years in succession without recording the fact or complying with the conditions of the mining law in regard to the amount of assessment work annually required. When he found that the new-comers were making inquiries, he re-located the claim as the El Paso, and De la Vergne, finding another lode, heavy in iron pyrites, to the west of Womack's vein, located a claim which he called the El Dorado. It was recorded a few days later, and in the certificate the district was called, for the first time, by the name which it now bears.

II. THE ROMANCE OF THE INDEPENDENCE AND PORTLAND MINES.

In May, Frisbee and De la Vergne happened to be at Colorado Springs and met W. S. Stratton, to whom they showed certain assays of ores brought down by them from Cripple Creek. Stratton, in the intervals of his occupation as builder, had been prospecting in different parts of Colorado for fully twenty years previous to this date. After the meeting with De la Vergne and Frisbee he went to Cripple Creek and began to look over the district. Toward the middle of June he was induced by William Fernay to do some prospecting on what is now called Battle Mountain. Next day he descended the hillside and came upon a big outcrop of granite. It was the Independence vein, which had been seen by many, miners as well as cattlemen, to be condemned by all of them as worthless granite. Stratton did as the others had done. On examining the outcrop he remarked the absence of any metallic sulphides or of vein-quartz such as he had been accustomed to in his previous experience, and he therefore concluded that it was unlikely looking rock. He disregarded it, but not irretrievably. Several days later he went to Colorado Springs with several samples which he thought might indicate the position of the lode, the existence of which he had been led to suspect from the fact that the loose material all over that part of Battle Mountain yielded gold on panning. The samples gave low results. It suddenly occurred to him that the granite outcrop, which stood out so boldly upon the grass-covered hill-slope, must be the lode he had been searching for. Acting on the impulse he took a horse forthwith and rode to the spot. On arrival he pegged out two claims, the Washington and the Independence. Thus did he celebrate the Fourth of July, 1891. I doubt if ever a man celebrated a national holiday to better purpose. On the 27th of April, 1899, he sold the mine for $10,000,000. Such is the romance of mining!

The story of the Independence mine is well known. That of its neighbor, the Portland, has not been told so often, but it is equally interesting. The dramatis personæ are John Harnan, James Doyle and James F. Burns, and to them must be added, later on in the story, W. S. Stratton, also. John Harnan, a native of Pennsylvania, was an experienced coal miner when he first went prospecting in Idaho and Montana. From the northwest he worked his way, subsequently, to Colorado. Later on he was employed in digging trenches at Woodland Park, from which place he walked into Cripple Creek. Fortune was adverse for a time, he could not get employment for several weeks and was about to leave in disappointment when he met Stratton, who had just begun to open up the prospect on Battle Mountain which became the Independence mine. He remained with Stratton for several months, doing the cooking for ten men, besides his work as a miner. When he had accumulated $300 he decided to start prospecting on his own account. On his way up the hill behind Stratton's mine he stopped at a log cabin where he found Burns, who, just at that moment, was in a very discouraged mood. Doyle was absent at

Colorado Springs. He and Burns owned a fractional claim, covering one-sixth of an acre, which Doyle had located on January 22, 1892. It had been christened the Portland, and upon it they had been working without success for several months. Harnan, in a joking way, asked Burns why he did not make a mine of it. The latter retorted that he and Doyle would take him in as a partner, giving him one-third of their holding, if he succeeded in finding pay ore. He went to work forthwith. These three men were typical of the gold-seekers then scattering over the Cripple Creek hills. Harnan, as has been stated, was an experienced prospector and miner, with an inquiring turn of mind, and had some idea of the lode-structure on Battle Mountain. Doyle had been carpenter's apprentice, and, later, superintendent of irrigation at Colorado Springs. Burns had served as engineer in the sugar plantations of Louisiana, previous to working as pipefitter, at Colorado Springs, also.

Ten days after the above partnership was inaugurated Harnan broke into the top of the bonanza of the Portland mine. The shaft was already down a little over twenty feet, but it had been sunk nearly alongside the vein. Harnan put in a cross-cut at six feet below the surface and struck it rich. The ore showed free gold, but, at this time, the claim was not patented and, on account of the claim-jumping which was general in the early days of the district, the three owners were afraid to let the fact of the discovery become known. So they removed their ore by night, two of them sacking it underground, while the third stood on guard at the surface to warn them of the approach of a passer-by. The ore was carried by stealth to the cabin and then, at intervals, one of them would strap a sack of it to his shoulders and trudge away by night to Colorado Springs, thirty-five miles distant, where, after sufficient had been accumulated, it was consigned to the smelters. In this manner they made enough money to pay for supplies and tools while pushing ahead with the development of the mine.

By and by they decided to do business on a larger scale and arranged to make a big shipment of ore. A wagon was brought up beside the dump and loaded, but rain had lately fallen, and they got into difficulties. The wagon, with the first lot of ore, broke down a few hundred yards from the mine; they covered it with sacks so that no one was any the wiser, but in making the second shipment they were more careful and used rough-locks on the wagon, with the result that tell-tale tracks could be seen all the way down the hill from the shaft. The prospectors on adjacent claims saw the marks and came to their own conclusions. Straightway the Portland became besieged by adverse litigation with those who were working adjoining mines.

Harnan, Doyle and Burns had pegged out several other claims at a distance from the original Portland. When they became involved in litigation they found that the richer their ore the more they became harassed by their neighbors. They decided to consult Stratton, who was going ahead successfully with the development of his own mine, lower down on the hill. Stratton had begun to believe that Battle Mountain was rich ground and he agreed to back them. He did. There were at that time 27 different law-suits filed against the Portland claims. Stratton helped them to buy out some of the contending interests, to warn off others, and so, by the use of money and diplomacy, enabled them to effect the first consolidation which comprised over a hundred acres of ground. Stratton became the first president of the Portland Gold Mining Company.

In 1894 an option was given on the property for $200,000. It was not taken up. A dividend of $70,000 was declared. The ore was discovered to be coming from disputed ground. An injunction was obtained and further extraction was prevented. The owners of the Portland found rich ore in another part of their property and were thus enabled to pay the dividend. Every time an option was given, a part of their property came into conflict with the territory of those around them. This was slowly overcome by their acquiring more acreage by purchase. The great richness of the mine gave them the necessary funds. In the meantime Stratton was buying up ground north of the Independence group, and the owners of the Portland were extending their territory southward, so that the two properties finally adjoined, and upon one side, at least, they became clear of litigation. Then other troubles began. They quarreled with Stratton. Lawsuits were started. About that time the Exploration Company, of London, commenced negotiations for the Portland, and Mr. Hamilton Smith came to examine. He found that there were conflicts between Stratton and the Portland Company. There were nearly thirty suits and counter-suits. A further consolidation was the only safe recourse. The option was made to include the disputed claims. The price was $2,250,000, of which $1,000,000 covered the seventeen disputed claims, fifteen of them belonging to Stratton. The deal fell through because the engineer who examined the property failed to find ore enough in sight to justify the completion of the transaction.

The shares of the Portland were then selling for 54 cents, and the dividends were at the rate of 1 per cent. per month on an issued capital of $2,250,000, this being equivalent to $270,000 per year. However, although the sale to the Exploration Company did not mature, the consolidation was effected, and within a month of the event the value of the group, as quoted on the share market, rose to $4,000,000. Since then there has been no litigation worthy of mention, and the property has been judiciously enlarged until it covers 206 acres. The shares of the Portland sell to-day for $3.00, equivalent to a value of $9,000,000 for the mine. The dividends to date aggregate $2,977,080. The story speaks well for the grit of the three Irish-Americans who made a claim of less than an acre the stepping stone to a magnificent property covering 200 acres. It is one of the many romances which have characterized the growth of Cripple Creek.

III. SITUATION AND APPEARANCE OF THE REGION.

The known gold-bearing portion of the district covers an area of about ten square miles, occupying a group of hills which rise from 300 feet to 1,000 feet above the general surface, and attain an average altitude of 10,500 to 11,000 feet above the sea. The drainage of the district flows into the Arkansas river, whose gateway into the plains is at Cañon City. The general slope is southward, and the sunny aspect incident to this configuration of the surface has caused the hillsides to be clad with sufficient grass, and enabled them, at one time, despite the high altitude, to yield good pasturage.

Few mining camps have so picturesque a situation, and Cripple Creek is further notable because the picturesque is not obtained at any sacrifice of accessibility. The beauty of the panoramic view to be obtained from most of the mines is not due to mere ruggedness or to the ordinary grandeur of a mountainous country; it is traceable to a position upon the slopes flanking Pike's Peak, which permits of an uninterrupted view of snow-clad ranges a hundred miles away. It is a panorama rather than a picture. In front are hills like giants tumbled in troubled sleep, whose feet touch the plateau of the South Park. To the left are the Arkansas hills that confine the river of the same name to its tumultous gorge; further south is the Wet Mountain valley, and beyond that the long, magnificent, serrated range of the Sangre de Cristo, telling of the shattered dream of Spanish conquest, of which no trace now survives, save in the occasional name of a mountain or a stream. These remind a practical age of the priestly warriors and the warlike priests who once sought to win the golden treasures of a land, the aboriginal people of which have almost passed away. The wind blows the snow of the Sangre de Cristo into streaming banners, and the clouds, like marshalled armies in retreat, fade across the far horizon. Turning northward, the valley of the Arkansas can be seen dividing the mountains which overlook Leadville. Further to the right are the beautiful Kenosha hills, at the head waters of the Platte, and beyond them are further peaks ennobled with coronets of snow. The details of the view are lost in the vastness of it, which impresses the observer no less because he may be surrounded by a noisy murmur of trains, steam whistles, wagons and machinery, which tell of the activity going on about him. Nor should he feel annoyance with his surroundings, for there is a nobility of human endeavor and successful achievement no less impressive than the beauty of the snowclad peak and sunlit plain.

The physical condition of the surface had much to do with the checkered history of the district. Owing to a southern exposure and the comparative absence of a protecting growth of trees, the rocks, which mostly possess a fissile structure, have been shattered by frost so as to overspread the solid formation with a thickness of debris, to which the tufted grass has given a further covering. Water, owing to its expansion between 4° C. and the freezing point, is a ceaselessly destructive agent. When it penetrates the cracks and crannies of the rocks it serves as a wedge, shattering their stony substance with resistless power. The heat of the day and the cold of night, the warmth of summer and the snows of winter, alike aid this disintegrating process. A high altitude and a southerly slope afford the conditions most favorable to such action. Thus it came about that the district of Cripple Creek is largely covered with the shattered rock which the miners call "wash," incorrectly however, because it is not composed of rounded water-worn material, but of angular fragments which, if not in place, are not far from their original position, having slid down the hill-slope in obedience to the laws of gravity. This shattering of the rock surface has caused one very important and, in Cripple Creek's case, far-reaching result. There are no outcrops. Ordinarily, the veins of gold ore stand above the surface with that boldness which caused the Australian miner to term them "reefs," and the Californian to call them "ledges." The ore, as will be seen when discussing the geology of the gold-field, is essentially altered and enriched rock, comparatively devoid of the quartz composing the typical lodes of other districts in America and Australia, and consequently it shares with the rock the tendency to undergo easy shattering. Solid vein-stone, therefore, rarely survives amid the general disintegration, the outcrop of the Independence being a very notable exception.

The first discoveries in mining are usually due to the finding of outcrops; in the absence of them, deep explorations are seldom undertaken. Deep ravines often afford good natural sections of the rock formation. The Cripple Creek district was as deficient in the one feature as in the other. The absence of steep declivities and abrupt rock-faces was characteristic of the pastoral landscape, and the angular debris covering the rounded hillsides made digging difficult. For these reasons, although the district was traversed by many thousands of prospectors at successive epochs, the existence of rich lodes was not surmised until a very recent date, and many experienced miners failed of success at first because they encountered conditions unfamiliar to them. Among the early arrivals, in 1891 and 1892, were the miners from Gilpin, Leadville and Aspen, men of knowledge in their own habitat, but unable to understand the peculiar vein-structure which they saw at Cripple Creek. It was the adverse opinion of these men, rather than the views of geologists or scientific observers, which injured the reputation of the gold-field in the beginning of its development.

IV. THE GEOLOGY OF THE GOLD FIELD.

The Cripple Creek district occupies the ground-floor of a volcano, the superstructure of which has been removed by erosion. This interesting fact is responsible for many of the peculiar features of the region. The mines are situated amid a complex of volcanic rocks lying upon the southern slope of the mass of granite whose culminating point is Pike's Peak. These volcanic rocks found a passage through the underlying granite during the comparatively recent period known to science as the Miocene, an early part of the last three great subdivisions of geological time. The granite was formed in the very beginning; out of the substance of it the foundations of Pike's Peak were upbuilt and the crest of the mountain was chiseled. It is the basal rock of the region and at one time probably formed the bed of the ancient seas which received the sediments now composing the sandstones and limestones flanking the Front Range. The granite is of a particular type, known, because of its prevalence in this locality, as the Pike's Peak granite. It is coarsely crystalline, and its three ingredients, the minerals quartz, mica and feldspar, are easily distinguishable by the unaided eye. A beautiful red tint, mainly due to the color of the feldspar, characterizes it and renders it recognizable by the least observant.

Long subsequent to the formation of the granite, and also that of the sedimentary rocks which were laid down upon it, there began an elevatory movement supposed to be traceable to the readjustment of the earth's exterior to its cooling and shrinking interior. Accompanying this movement there occurred a general fracturing of the rocks thus affected, so as to permit volcanic matter to force a way upward, after the manner of water rising through cracks in the overlying ice. The volcanic matter thus brought to the surface of the granite slowly filled the hollows of its uneven surface, and spread over a large area since then diminished by the patient forces of atmospheric erosion, which, during the long period of time separating the Miocene from the present day, have slowly sculptured the hills and valleys of the district.

A glance at the geological map of the gold field exhibits a great variety of volcanic rocks. The principal of these is andesite breccia.* The very nature of the breccia suggests the violence of the volcanic action which brought it to the surface of the granite. The miners call the breccia "porphyry"** from its apparent resemblance to the rocks of that class with which they were previously familiar in the Leadville and in the Gilpin county mines. The porphyry of Leadville is quartz-felsite; that of Gilpin is quartz-andesite. Porphyry is an adjective-noun and refers to the structure rather than to the com-

---

* "Andesite" is derived from Andes, the mountain range where this rock is especially prevalent. "Breccia" is a word of Italian origin, and means "broken." It is a term applied to rocks which are made up of fragmentary material.

** "Porphyry" comes to us through the Greek word "porphyra," signifying purple. It was first used to designate a beautiful rock of this type which the Romans obtained from the quarries of Gebel Dokhan, on the shores of the Red sea. This original "porphyry," called by the Italians "porfido rosso antico," had, according to Zirkel, a beautiful blood-red ground-mass speckled with small snow-white and rose-red crystals of feldspar. But the first meaning of the term, which depended on the color, has long been lost in another meaning, which refers to the structure. A rock is a "porphyry," or, more correctly, is "porphyritic," when some peculiar constituent mineral, very often feldspar, stands out well defined from the general ground-mass, as in the Western miners' familiar "bird's-eye porphyry."

position of a rock, so that there is "granite-porphyry," "diorite-porphyry," "andesite-porphyry," etc., the term being applied to rocks of igneous origin in which particular minerals are distinguishable amid the ground-mass of the rock so as to give it a speckled appearance. The Cripple Creek breccia has this appearance, but it is due to the fact that it is made up of a heterogeneous mass of rock particles of every size, from the most minute powder to fragments as large as a man's head. These consist mainly of andesite, but the other rocks are also included, especially near the edges of the volcanic vent. Some of this material is mere volcanic dust, called "tuff,"† which, when consolidated under pressure and cemented by silicious waters, becomes compacted into a dense, hard substance difficult to distinguish from a true crystalline rock; so that it is not to be wondered that the miners often label it with an incorrect name.

The breccia lies in the uneven hollows of the granite and probably fills a large part of the crater formed by the energies of the volcano. The thickness of the breccia has not been proved, nor has the exact position of the original vent of the volcano been discovered. There is evidence, of a general kind, chiefly in the composition of certain masses of eruptive rock, which indicates that there were two vents, at least—one near Goldfield and the other near Guyot Hill. The mine workings which happen to be near the rim of the central basin have penetrated through the breccia into the underlying granite. A depth of over 1,000 feet has thus been proved by the sinking of shafts. There is, however, evidence to indicate that the maximum depth of the breccia formation, in the vicinity of the vent, must be several times 1,000 feet.

Fig. 1.

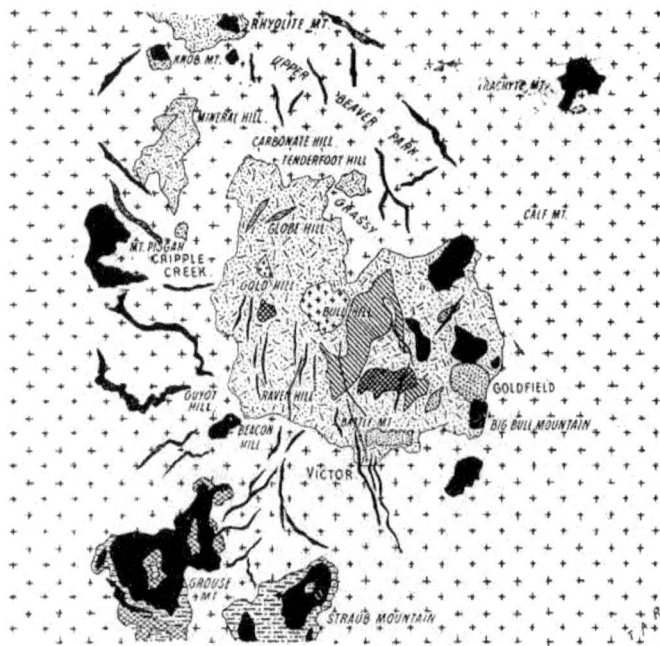

GEOLOGICAL MAP OF THE CRIPPLE CREEK DISTRICT.
(After U S Geological Survey)

The geological map and the ideal section of the volcano which are presented herewith, will illustrate the geological structure of the district. See Figs. 1 and 2.

The breccia is penetrated and traversed by later volcanic rocks of which phonolite‡ is the most important in its relation to the occurrence of ore. Until recent years phonolite was not known as a rock species save as forming the Wolf rock in Cornwall, and, therefore, its association with great mineral wealth at Cripple Creek has been one of the most interesting features of the development of that district. The phonolite occurs for the most part in dykes; that is to say, in approximately vertical sheets which traverse the older formations, the granite and the breccia, in various directions, and are probably united, at depths far beyond the reach of human exploration, to larger masses of rock having a similar composition, just as the cracks in ice are filled with a liquid similar to that beneath.

These dykes follow such lines of weakness in the older rocks as were developed into fractures at a time when the rocks underwent strains, the latter being considered to

---

† "Tuff" comes from the Italian "tufa." Vesuvius is responsible for the Italian nomenclature of many volcanic products.

‡ "Phonolite" is derived from two Greek words, "phone," signifying sound, and "lithos," meaning stone. It owes this name to the fact that it rings when struck by a hammer. This is due to its hardness and close texture. It is also called "clinkstone." The essential constituents of phonolite are nepheline and the glassy variety of feldspar, called sanidine.

be the result of the slow wrinkling of the earth's crust due to its readjustment over a cooling and shrinking interior.

The phonolite rose in a mobile, if not molten, condition through the fractures thus formed, after the manner of water rising through the cracks in the overlying ice. The structural conditions thus created gave a direction to the subsequent circulation of underground waters. The deposition of ore is the result of such circulation, the underground waters being the vehicle by which the metals are leached out of the rocks at one place and laid down at another in such a concentrated form and within such a distance of the surface as to render them valuable to man. The place of ultimate origin is surmised, but vaguely, as being deeper than our deepest mines, and the place of deposition of the ore is not always the place where the miner finds it. Lines of weakness, healed and strengthened by the cementing effects of hot igneous rock, in the form of dykes, afford new lines of lesser resistance, parallel to the old ones and along the contact of the two rocks of unlike hardness and texture. For this reason ore-bearing veins so often accompany dykes. They do so at Cripple Creek.

It is well to begin the discussion of a subject by defining the terms employed. A "lode" is something which leads a miner, the words "lode" and "lead" having an identical Saxon origin. Australian miners designate a small, continuous vein connecting larger ore bodies as a "leader." "Lode" is therefore a comprehensive term covering many diverse forms of ore occurrence. The word "vein" has a more restricted usage, and describes those lodes in which the ore is supposed to occur in a tabular form, occupying continuous planes which are approximately vertical, and traversing the rocks like interminable sheets of paper set on edge; that is, they are supposed to fill simple fractures made in a perfectly homogeneous material. This term was originally borrowed from the human anatomy, and the oldest writers have used the simile of the rock veined with the precious metal. Nature does not recognize the definitions of the technical dictionary, and in mining practice it has been found that regularity of structure is the exception rather than the rule. The geologist of fifty years ago, when the science of geology was more the product of the library and the laboratory than of actual observation underground, conceived the ore as having filled gaping fissures in the rock, comparable to the crevasses of a glacier; and when he had noted the dissimilarity between the ore and the encasing rock, he imagined the former to have been due to an upwelling of molten metallic matter. The ideas of the present day are still slightly tainted by the imaginations of the past, and the terms of an obsolete philosophy continue to cling to our nomenclature.

Modern investigations, based on accurate chemical knowledge, as well as geological observation, have all gone to prove that gold ores are not the product of direct volcanic action, but that they have been conveyed to the place where the miner finds them through the medium of water, the metals having been dissolved, in various chemical combinations, by underground solutions, and precipitated along those fractures in the rocks which have been first lines of least resistance, and then lines of maximum circulation. The mineral solutions can not have come from indefinite depths, because the increasing pressure encountered would finally tend to close the channels of circulation, and also because the increase of heat (1° F. for every forty-eight feet of descent), observable in the sinking of shafts and bore-holes, indicates that at a horizon of about 20,000 feet below the present surface a temperature (the critical point, 773° F.) would be attained at which water, in spite of the pressure to which it would be subject, becomes dissociated into its constituent gases. It is considered probable, from

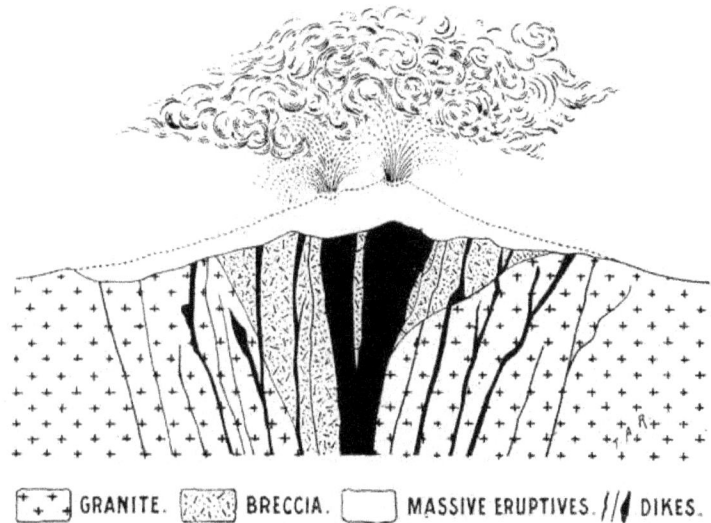

Fig 2.

GRANITE.   BRECCIA.   MASSIVE ERUPTIVES.   DIKES.

IDEAL SECTION OF THE CRIPPLE CREEK VOLCANO.

the evidence yielded by certain classes of lodes, particularly those of nickel ores, that volcanic action serves to bring the metals from these great depths to that zone of the earth's exterior wherein solvent waters can circulate. The experience of gold mining corroborates this view, the association of volcanic rocks with bodies of valuable ore having become almost proverbial.

It is not surprising, therefore, that this very fact has tended to cause a confusion of ideas between volcanic action and lode formation.

In a railway cutting between the towns of Cripple Creek and Anaconda there is a bit of nature's testimony which will be of service in getting a clear idea of the essential characteristics of gold-bearing veins as compared with dykes of volcanic rock. The accompanying drawings will help the description. In Fig. 3 there is afforded an

Fig. 3.   Fig. 4.

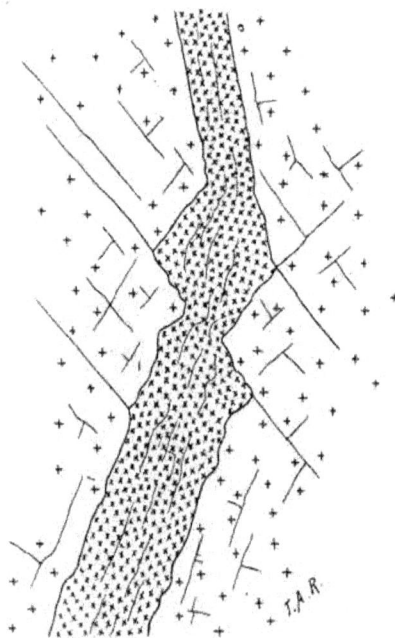
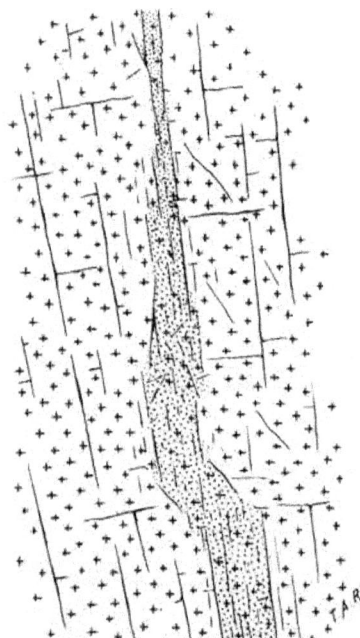

[░░░] BASALT.   [+ + +] GRANITE.   [+ + +] GRANITE.   [░░░] ORE.

excellent illustration of simple dyke structure. The dyke in this case is composed of basalt; it is from nine to fifteen inches in width, and can be easily traced as an irregular dark band traversing the coarse-grained pink granite. The dyke is very well defined, exhibiting clear-cut lines of demarcation from the enclosing granite, and it is evident from the contour of the walls that it occupies a fault-fissure. The outline of the east wall corresponds exactly to that of the western one, the movement of the latter having been upward, causing a displacement equal to about fourteen inches. It is a clean-cut fissure in the granite, filled with foreign material, a basic volcanic rock, which probably welled upward in a mobile condition, filling the fissure as it was formed, so as at no time to permit of a vacuity. Compare this with Fig. 4, which is a sketch of a gold-bearing vein, situated at a distance of a few yards from the dyke illustrated in Fig. 3. The country is the same, viz., granite, but in this instance the vein-filling is not foreign matter, but essentially rock in place; it is granite, altered indeed, but easily recognizable, in spite of the kaolinization of the feldspar and the partial removal of the mica. There are no clearly-defined boundaries between the decomposed vein-matter and the enclosing country, nor is there any evidence of faulting. The lines of fracture shown in the granite are the joints of that rock, and those which are observable in the vein itself are not continuous, but rather a closely-knit series of little breaks, which have afforded a passage for a liquid more subtle than the basalt. The vein occupies a line of maximum porosity along which water, more searching than any molten lava, has found a way, decomposing the soluble ingredients of the rock, and depositing a minute quantity of gold, insufficient to make the decomposed granite of the vein differ essentially from the outer country, but rendering one gold-bearing ore, and leaving the other barren rock.

Here we have a dyke compared with a vein and volcanic agencies brought into strong contrast with aqueous action. The faulting along the fissure followed by the dyke is easily seen, but no evidences of such movement can be discerned along the seam of altered granite, which forms the gold vein. Nevertheless there must have been some movement, however slight, because a crack or break, not made evident, can be considered as only latent, until the two faces of it are caused, by that very shifting, so to disagree as to produce the irregularities which, when linked together, form the visible line of fracture. Even the joints in the solid granite require such an explanation, and however insignificant the shifting may be, it marks the adjustment of the rock to the effects of stresses, traceable in this case, probably, to the volcanic energies which extruded the large masses of breccia forming the characteristic feature of the geology of Cripple Creek. Permit me to repeat, however insignificant this shifting may have been, it made the rocks pervious to underground mineral-bearing solutions, and where it occurred it developed a series of united passages, which afforded a line of maximum porosity, permitting the circulation of gold-bearing waters.

The mines afford illustrations of a great diversity of lode-structure. This diversity is traceable to the complexity of the enclosing rocks. The variations in ore-occurrence due to this fact explain the vicissitudes which marked the early history of the district, and the recognition of them should contribute toward the success of future exploratory work.

Many of the lodes are essentially dykes which have undergone fracturing, thereby affording an opportunity for their impregnation with gold through the agency of circulating solutions. The Moose vein will exemplify this type. The accompanying sketch, Fig. 5, was made at the 350-foot level. From D to F is the width of the dyke, which is

Fig. 5

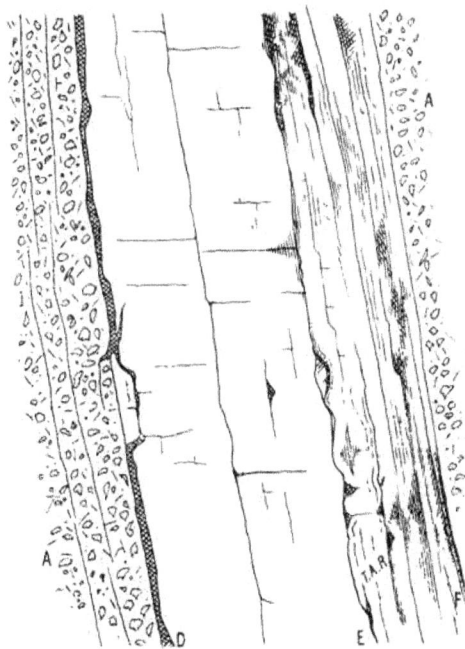

Fig. 6.

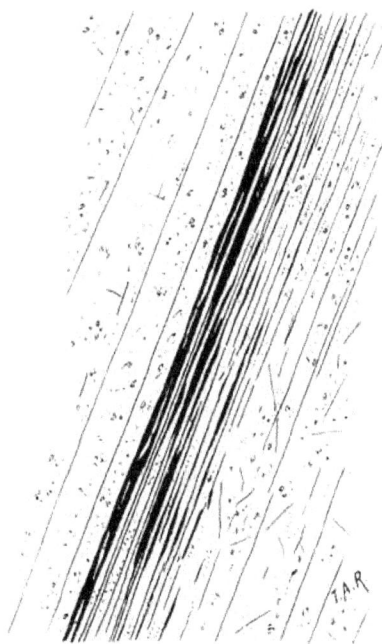

ANDESITE BRECCIA   NEPHELINE BASALT.   VEIN MATTER.   ORE ALONG SHEETED ZONE.

composed of nepheline basalt. It traverses the andesite breccia, which is indicated at AA. The pay-ore extends from E to F, with a width of ten inches, and is distinguished from the remaining and comparatively barren portion of the dyke, ED, a dark bluish grey rock, by being iron-stained and seamed with brown threads in which free gold and tellurides occur. The multiple fracturing, parallel to the walls of the dyke, is a characteristic feature of this and similar lodes, experience having also shown that there is reason to expect the lode to consist of rich ore when it becomes threaded with minute seams following these lines of fracture. This feature can be described as a sheeting of the rock; it is a very important factor in ore deposition.

An example of the latter is exhibited in Fig. 6, which represents a lode in the Moon-Anchor mine. This type of ore occurrence is thoroughly characteristic of the mines in that part of the district known as Gold Hill. The breccia is fine-grained. The partings are about a quarter of an inch apart. They are followed by minute seams of red, gritty clay in which the tellurides can be distinguished. The individual seams are united by transverse impregnations which collectively make a pocket or small body of ore, in which it is not unusual to encounter patches consisting of an almost solid aggregate of crystalline sylvanite. The sheeted structure dies out into the enclosing country by the process of a gradual widening of the space intervening between each successive parting.

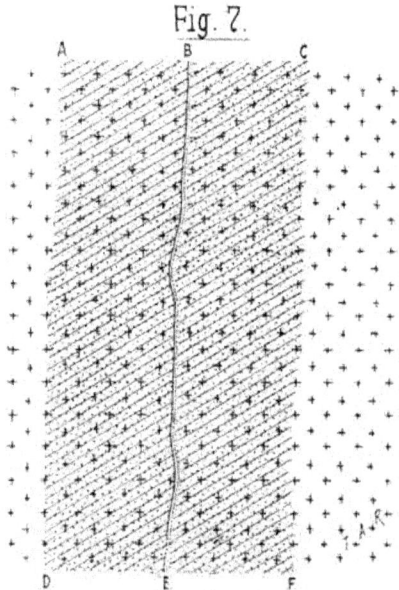

Fig. 7.

SIMPLE LODE IN GRANITE.

Independence Vein in the Washington Claim.

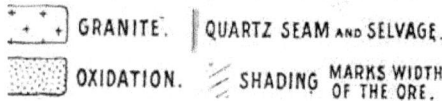

GRANITE.   QUARTZ SEAM and SELVAGE.

OXIDATION.   SHADING MARKS WIDTH OF THE ORE.

The occurrence represented in Fig. 7 is a peculiarly suggestive illustration of the manner in which a gold-vein may be formed. It exhibits the Independence lode in the Washington claim, which is a portion of Stratton's Independence property. The lode, which further north traverses the breccia and is closely associated with a phonolite dyke, is seen here as a band of decomposed granite sub-divided equally by a central thread of quartz. The ore is essentially granite. The enclosing rock is also granite. The portion AC, DF is four feet wide and carries a little over three ounces of gold per ton. The width of four feet which carries gold, and is therefore ore, has no parting or wall separating it from the outer rock which carries none, and is therefore regarded as waste, but it is distinguished from the latter in many ways. The outer granite is fresh and unaltered, exhibiting its constituent minerals, white quartz, black biotite mica and pink orthoclase feldspar, with great clearness. The inner gold-bearing rock is much altered by decomposition and replacement; the orthoclase alone appears to have survived the general metamorphism; the mica has been removed and, in its stead, chlorite can be seen in green patches; the original crystalline quartz is largely gone and the presence of purple fluorite suggests that hydrofluoric acid may have been a primary agent in that removal; secondary hydrous quartz fills many of the interstices between the crystalline constituents of the rock; in iron-stained cavities free gold can be seen by the aid of a pocket lens and the gold is observed to have the dark, lusterless appearance which characterizes it when derived from the oxidation of tellurides. The entire width of this gold-bearing, decomposed granite is heavily iron-stained by the oxides resulting from the disintegration of the small crystals of iron pyrites which can be seen in an unaltered condition disseminated throughout the same lode at lower levels. In the centre of the band of ore there is a distinct parting, BE, which is separated from a persistent thread of white quartz, only about a quarter of an inch in width, by a slight selvage of red clay. At a distance the lode appears as a distinct broad band of iron-stained granite, and it is only by closer examination that the boundaries of it are seen to consist, not of "walls" or of any such evident demarcation, but merely of a transition from decomposed into undecomposed granite.

The lodes which have been described will exemplify the extraordinary variety of structure to be observed in the district. Much evidence has been accumulated in support of the most recent theories of ore deposition which are based upon a study of the underground water-circulation. An idea has been voiced by those not having authority, such as the scribes of the daily press, that Cripple Creek, both in its earlier stages and now, has set at naught the accepted teachings of geological science. It is a fantasy, born of the necessity of filling newspaper space. On the contrary, it has always been insisted, by the writer among others, that no mining region yields so much corroborative evidence in support of those views on ore deposition which have won general acceptance, namely: that the formation of ore is due to the precipitation of the matter carried in solution by underground waters which have circulated in obedience to the conditions created by the structural relations of the rocks.

V. THE ORES AND MINERALS.

The gold, the search for which is the basis of all the mining activity, occurs either in a native condition or as a telluride. It is found distributed among the interstices of the rock, lining the fractures or penetrating the substances of it in threads of varying minuteness. In lodes traversing the granite, the gold, or the tellurides containing it, will be scattered amid the porous cavities due to the removal of certain more soluble portions of the rock; in phonolite, the values will be found more frequently along fractures than in the heart of it. This renders the last mentioned class of ore very difficult of estimation. In the andesite breccia, the component fragments of which are so heterogeneous, the physical character of the rock varies considerably, and the gold values will partake of an irregular sporadic distribution.

In the early days of the district many blunders were caused by the fact that the ores are, for the most part, essentially altered rock, in place, impregnated with gold-bearing minerals to a slight extent as regards percentage, but to a notable degree in respect of commercial value. The comparative scarcity of quartz and other minerals usually characteristic of gold veins puzzled the early prospectors, and remains to this day a feature of peculiar interest. In 1893, when W. S. Stratton, the owner of the Independence, sent several carloads of rich ore from his mine to one of the Denver smelters, the officials at the works thought a blunder had been made, and that loads of ballast had been inadvertently consigned to them. The ore was obviously granite and it required a trained eye, such as that of Dr. Richard Pearce, manager of the Boston and Colorado Smelting Company, to detect the fact that the mica of the granite had been largely removed, leaving small, iron-stained spots amid which were disseminated dull yellow specks of gold. The ore found in the breccia would also deceive the unwary. A glance into the ore-bins of the chlorination establishments will exhibit a mixture of broken rock, which looks more like the spoil of a barren cross-cut than the yield of a rich stope. The petrographer, in looking over this material, would easily label the rock, but if not initiated, he would wonder whether it could be gold-bearing. On being assured that it was valuable as such, he would take a few pieces and break them open so as to examine a fresh surface, and it would not be long before he would see, on the planes of fracture, evidences of richness.

The occurrence of tellurides characterizes the district. Whatever free gold is found in the oxidized ores near the surface has invariably that peculiar appearance which characterizes the precious metal when it has originated from the disintegration of tellurides. It is dull-looking and brown, resembling gold which has been precipitated from solution, and is often found in splashes that look like reddish-brown clay, but, on being burnished, by scratching, exhibit the unmistakable glint of gold.

Tellurides are compounds which tellurium forms with the metals, in a manner similar to the combinations formed by sulphur and selenium. The first determination of this interesting mineral species is due to Klaproth, who, in 1802, recognized them in the ores of Zalathna, in Transylvania. Tellurium is a non-metallic element with a metallic lustre. In its chemical combinations it acts in a manner analogous to sulphur, which, it appears, at times, to replace. Native tellurium is a tin-white brittle substance with a bright metallic lustre. Its commercial value is $3.50 per ounce, but the demand for it in the arts is very slight and a few shipments demoralize the market, as is the case with most of the rare earths. It is extremely uncommon at Cripple Creek, but in Boulder county it is frequently encountered. A mass, twenty-five pounds in weight, was found, in 1877, in the John Jay mine, near Jimtown. In Gunnison county, in southern Colorado, it has lately been found in the Vulcan mine, in mica schist, associated with a lode of gold-bearing iron-pyrites, which, in the oxidized zone, includes masses of native sulphur.

Telluride ores have become an important source of gold during the past five years on account of the discoveries made in Colorado and West Australia. They have been treated as something quite new and phenomenal. As the result of the success of the two new districts there has grown up an idea that this mode of gold occurrence indicates ore bodies of an especially persistent nature, a fallacy akin to the older one which assumes an enrichment with depth as a general characteristic of gold veins. Tellurides have been mined in Transylvania for a century; they have been known in Colorado, as important ores of gold and silver, since 1872, two districts, the La Plata mountains and Boulder county, yielding them in commercial quantities. In none of those regions have they been characterized by any especial continuity in depth. Quite the contrary. Until Cripple Creek, and then Kalgoorlie, commenced to make a record, it was generally held, among those who were aware of the facts, that telluride ores were erratic in behavior and difficult to treat. The former is not more true of them now than of gold deposits in general, while the latter has been largely overcome by the advance of metallurgical practice.

The following is a simple test for tellurides: Remove a small bit of the suspected mineral with the point of an old knife, and put it in a porcelain dish or a white saucer. Add three or four drops of strong sulphuric acid, and heat over a lamp. Should tellurium enter into the composition of the suspected mineral, a beautiful purple will suffuse the colorless acid. The miner's time-honored test is to put the ore in the fire of a blacksmith's forge and roast it. Tellurium fuses at a comparatively low temperature, and becomes volatalized, passing off in white fumes of telluric oxide. If the telluride mineral contains gold, the latter will remain in the form of globules. Even the precious telluride hidden in the seams of the piece of ore will be exuded as a yellow perspiration. The miner calls this process "sweating," and the reason for it becomes obvious when the results are observed.

The principal telluride minerals found in the Cripple Creek ores are sylvanite, calaverite, and petzite. Sylvanite, named after the place of its discovery, the historic gold field of Transylvania, is the most characteristic of Cripple Creek ores. It is a double telluride, containing gold and silver, an average composition being 28 per cent. gold, 16 per cent. silver, and 56 per cent. tellurium. Sylvanite is a brilliant silvery-white mineral, having a characteristic crystalline habit, to which it owes its other name, "graphic tellurium," the arrangement of the crystals resembling Arabic lettering.

On account of the absence of silver in the composition of the richest ores of Cripple Creek, it has been concluded that the gold occurs, for the most part, combined with tellurium alone, in the form of the mineral calaverite. Calaverite is the simple telluride of gold; pure specimens contain 44.5 per cent. gold and 55.5 per cent. tellurium. It is named after the county of Calaveras, in California, where it was first found, at the Stanislaus mine. It usually carries from 2 to 3 per cent. of silver, which then must be considered as an impurity. The purest varieties have a bronze-yellow color. It is the characteristic mineral of the rich ores of Kalgoorlie, in West Australia. Calaverite is rather difficult to distinguish from iron pyrites; the difference in color is slight, but the former is easily cut by a knife, while the latter will not permit it. The cubic crystalline habit of pyrites will usually help to make it known, because calaverite rarely occurs in any other than a massive form, and has a distinctly conchoidal fracture.

Petzite, named after the German chemist Petz, is, like sylvanite, a double telluride of gold and silver, its average composition being 25 per cent. gold, 42 per cent. silver, and 33 per cent. tellurium. It is much darker than sylvanite, being steel-gray to iron-black; it is also slightly harder and more brittle.

All these tellurides are distinguished from the baser minerals, with which they may be occasionally confounded, by their peculiarly rich lustre.

The lodes of Cripple Creek are further characterized by the presence of fluorite or fluorspar (the fluoride of calcium), a beautiful purple mineral which is so notably associated with the ores of the district as to have led to the idea that it could be accepted as an indication of the richness of the veins in which it was found. But, like similar attempts at short cuts to knowledge of this kind, the generalization has proved fallacious. There are several large lodes of very low-grade ore in the district which are purple with fluorite, and there are some very rich ones almost devoid of it. The association of the gold and the fluorspar points to a similarity, and possibly a contemporaneity, of origin; but this fact does not, and could not, predicate whether the quantity of gold present will give the ore an average value of $2 or of $200 per ton. As a matter of science, both kinds of ore may be considered gold-bearing; as a matter of business, one spells losses and the other dividends.

Below the depth, which ranges from 100 to 400 feet, reached by surface waters, the unaltered tellurides appear in all their untarnished beauty. At a further depth, from 500 to 700 feet, the ores become more complex, because of the increasing percentage of baser minerals, chiefly iron pyrites, but including also galena (the sulphide of lead) and stibnite (the sulphide of antimony). This change is important chiefly from a metallurgical standpoint, because the increase of sulphur renders the roasting of the ores more expensive. The general question of the probable changes to be encountered as the mines become deeper will be discussed under another heading at the close of this description of the district.

VI. THE FUTURE OF THE DISTRICT.

In spite of the proverbial difficulty of attempts to prophesy the future, one can not consider the condition of a mining district so productive, as that of which Cripple Creek is the center, without a glance at the possibilities created by the vigorous development which is now in progress. In the ninth year of its existence Cripple Creek produced an amount of gold valued at $15,500,000, of which 24 per cent. was distributed in the form of profits upon the operations of the mines. Is this to be the height of achievement?

The present boundaries of the productive portion of the district have already overlapped the area occupied by the breccia, which, not long ago, was supposed to limit the territory open to successful mining. The central mass of andesite-breccia covers about eight square miles. It is undoubtedly the proper territory for further exploration. Recent developments have proved that no part of it affords greater promise of affording large masses of ore than that which lies along the border separating the breccia from the surrounding granite. However, an important feature of the active exploratory work of the past two years has been the discovery of pay ore outside the breccia, away from the contact referred to, and far out in the body of the granite which, like the sea around an island, encloses the central volcanic area of Cripple Creek. It was surmised by the writer, many years ago, that such discoveries would be made, because of the known fact that the phonolite dykes pass out of the breccia into the granite, both in depth and in strike; and wherever the dykes extend, there also the facilities for the circulation of mineral-bearing solutions must in all probability have been afforded sufficiently to permit of the deposition of gold ores.

The feature which characterizes the occurrence of ore in the Cripple Creek district is the comparative absence of any dependence upon a particular rock encasement. Rich ore bodies have been discovered in both the granite and the breccia, and also as impregnations in syenite, phonolite and basalt. Wherever passageways have been afforded by the fracturing of the rocks, there the ore-bearing solutions appear to have circulated and precipitated their golden freight. Such passageways have varied in their nature according to the physical condition of the rocks they traversed, so that a line of linked breaks in a rock of fragmentary composition, like breccia, has become a depository of ore fully as much as the cleaner-cut fissuring in a homogeneous crystalline rock, such as the granite and syenite. The direction of the fractures followed by the ore has been determined by the lines of weakness previously developed through the formation of dykes and other bodies of eruptive rock. As a consequence the lodes have a course which is generally sympathetic to the structural relations of their geological environment, although they do not slavishly follow the exact lines of the dykes. The latter are often tortuous in their strike and dip, while the lodes, for the formation of which they afforded the necessary facilities, tend to maintain a straighter line, which may be seen, as, for example, on Battle Mountain, to cut across bends and irregular sinuosities of the especial dyke with which any particular lode is associated. The results, as expressed on a map, may be likened to a road which follows a river, the windings of which it generally, but not particularly, follows.

The map published by the United States Geological Survey marks the existence of several dykes and cores of phonolite outside the central area. Mining explorations have uncovered many others, unknown at the time the government geologists were investigating the district. Pay ore—that is, material which under existing conditions yields a profit—has been discovered in several directions outside the boundaries formerly supposed to limit the gold-bearing territory. The principal finds have been made to the northeast, on the slopes rising from both sides of the valley known as Grassy; to the southwest, on Beacon Hill; and to the south, at Victor. The last is in many respects the most important, because a very large ore body has been opened up in ground which is fully 2,000 feet south of the contact, and therefore far out in the granite. This is an event of great importance to the future of the district. Further favorable evidence is afforded by the extension of the Independence vein, which carries ore in the breccia, at the breccia-granite contact, and for several hundred feet southward into the granite along a line which keeps in touch with a dyke of phonolite. The occurrence of very rich, although restricted, ore bodies in the flat sheets of phonolite which traverse the granite of Beacon Hill, is another discovery, made during recent years, of much suggestiveness.

The ore-bearing area has been thus satisfactory extended in strike. Next comes the question, what of the deep? Has increasing depth coincided with any evidence of impoverishment? No. At least to no greater extent than in other gold-mining regions. That is, some mines have got out of ore in depth, some have penetrated into better ore, others appear to be just holding their own. It is undoubtedly true that a few years ago there was a good deal of timidity regarding the future of the mines when they should become deeper. It was known that the lodes would eventually penetrate into the underlying granite, and it was obvious that those near the edges of the depression occupied by the breccia would be the first to do so. The workings of the Portland and Independence were among the earliest to pass into the granite which dips under them on the south and west. Other mines on the adjacent hill have had a similar experience. But the discouraging results anticipated by the timorous have not supervened. The ore bodies have been found to continue their course amid this changed geological environment, and they are as rich as they were in the breccia. On the other hand, one or two of the deepest shafts have been unsuccessful in developing productive mines, but in these instances the upper ground gave no particular promise of any betterment in depth. Mines, like men, must be exhausted at last; some men lose their vigor long before old age, many die as infants. It is so with mines. Every mine must have a horizon at which it is at its maximum productiveness. What that horizon is depends upon the particular conditions which obtain in each instance. Some ore bodies are like new-born infants and die out as soon as they are chronicled. Many last long enough to enrich one owner and then to delude his successor. Others persist with masterful strength to a great depth and threaten to test the utmost limits of human ingenuity in the pursuit of them underground.

The last year has seen a notable introduction of capital from the outside. There is an evident increase of interest in the opportunities, for both speculation and investment, offered by the mines. The sale of several properties has not led to the retirement of their previous owners, but to a revival of activity on their part in new directions. The mining men of Colorado have been bred in an atmosphere of gold-seeking, and the acquirement of riches due to a successful operation only leads to the inauguration of fresh enterprises. The result of these transfers of property will cause an increase of energy in prospecting, and in the wake of such new explorations there will come discoveries, one or more of which may prove far-reaching. Whatever development is undertaken is likely to be better directed than formerly; an increased knowledge of the local geological conditions and the introduction of a larger measure of technical skill in the underground prospecting will doubtless become evident, both in the lessened costs of operation and in a diminution of merely haphazard explorations.

The continued improvement in economic conditions tends to increase the tonnage of ore capable of yielding a profit. The two existing railroads will probably be supplemented by others in the immediate future, and the further competition thus incited must decrease the rates of transport to the mills and smelters. The latter are in the valley, nor is there any reason to believe that it will at any time be found advantageous to erect reduction plants at the mines. Lack of fuel, a restricted water supply, and the high price of labor all operate heavily against such a plan. Meanwhile the mills of Florence and Colorado City are increasing in capacity and in number, the demand for ores keeping pace with the growth of the ore supply. Although the rates of treatment do, at rare intervals, rise to a level which the mine owners consider extortionate, yet in the long run the producer of ore is on the side of the big battalions, because the addition to the number of mills always tends to catch up with the supply of ore and then to exceed it. All the economic factors favor a gradual decrease in the cost of realizing the value of the gold, and this renders available a larger amount of lower grade ore each year, giving an enhanced value to mines which not long ago could not meet the working charges.

All the signs are favorable; the brilliant discoveries of the early days and the successful development of the present promise a future which should long maintain Cripple Creek in the proud position which it now occupies, namely, that of the largest and richest gold mining district on the American continent.

CRIPPLE CREEK FROM THE STAND-POINT OF STATISTICS

By GEORGE REX BUCKMAN

WRITTEN SPECIALLY FOR THE *OFFICIAL MANUAL*

Mr. Rickard, in his comprehensive, luminous and exceedingly valuable article, has related the story, in most interesting detail, of the conversion in less than a decade of a mountain wilderness into one of the greatest of the world's gold-bearing districts. He has told the whole story; nothing remains for the writer except to present a few figures which in tabular form may serve to further impress the fact of Cripple Creek's rapid growth, as well as conduce to a better understanding of the pages which follow.

The spring of 1891 marks the real beginning of mining, though in a small and crude way, in what later came to be known as the Cripple Creek district. The production for this year is a very uncertain quantity, but it surely did not exceed $200,000, and may have been very much less. During the following year comparatively little progress was made in actual mining, prospectors being busy with the location of the available ground in what was believed to be the mineral zone; and the gold product barely reached $600,000. In 1893 a new impetus to gold mining was given by the tremendous fall in the price of silver; many experienced miners from other parts of the state turned to the new gold fields; Cripple Creek underwent rapid development, and production rose to about $2,000,-000. The great labor strike of 1894 gave the district a temporary setback, the principal mines being closed down for several months, and as a consequence the ratio of increase was considerably reduced, the production amounting to but $3,250,000. But from that time until the present, development has gone forward by leaps and bounds, until in 1899 the production reached the magnificent total of $19,500,000. The following table shows the production of the Cripple Creek district for the nine years of its history:

GOLD PRODUCTION OF CRIPPLE CREEK, BY YEARS.

1891	$ 200,000
1892	587,310
1893	2,010,400
1894	3,250,000
1895	6,100,000
1896	8,750,000
1897	12,000,000
1898	15,000,000
1899	19,500,000

According to the estimate of the Director of the Mint, the gold production of the United States for 1899 was $70,500,000, and of Colorado, $26,000,000. Cripple Creek's production will doubtless be placed by this authority at about $17,500,000, these figures being based on returns from smelters, reduction plants, refining works, and on mint receipts, no account being taken of ore mined and unshipped, in transit, or at reduction works, the amount of which was at the close of 1899 far greater than at any previous time. From these figures, which may fairly be compared, it will be seen that during the past year Cripple Creek furnished over two-thirds of the gold production of Colorado, and nearly one-fourth of the entire production of the United States.

Assuming that the world production for the same period was not far from $300,-000,000, it is further seen that Cripple Creek, during 1899, contributed about one-eighteenth of the tremendous total. These figures are sufficiently surprising; but, unless the promise of the future should prove illusory, Cripple Creek is destined to play an even more prominent and commanding part in the production of the world's gold. On this point I quote from a statement made by United States Mint Director Roberts, whose expectations of the future of Cripple Creek's gold production are certainly fully justified:

"The world does not depend on the Transvaal for its stock of gold. Colorado will produce more gold in 1900 than all North America yielded ten years ago, and this continent will yield more next year than the whole world did ten years ago. Cripple Creek is up to the record of the Transvaal five years ago, and Cripple Creek, with Alaska and the Klondike, will next year exceed the Transvaal production of 1897."

The profits of Cripple Creek mining, as shown by the dividends paid by the companies operating in the district, are found to have fully kept pace with the increase in production, as will be seen from the following table:

CRIPPLE CREEK DIVIDENDS, BY YEARS.

1893	$ 88,940.00
1894	853,450.21
1895	1,231,000.00
1896	1,176,744.37
1897	1,269,395.00
1898	2,596,145.00
1899	4,354,402.85
Total	$11,570,077.43

These totals do not by any means represent the profits of the several years or during the camp's productive life. Though nearly all the mining property in the district is owned by companies, a portion is in the hands of lessees, while a further portion is owned by companies whose stocks are closely held, and of whose operations the public is not informed. In addition, a number of properties included among the profitable producers are owned privately.

Has the climax of Cripple Creek's achievement been reached? The answer by those whose opinions are of the greatest value is an unqualified negative. Marvelous as has been the development during recent years, there are no signs of the limit of production being neared; on the contrary, each succeeding year seems to project this point further into the future. But a fraction of the assured gold-bearing area has been as yet exploited even on the surface. There are scarcely a half dozen shafts in the district that have reached a depth of 1,000 feet; the workings of a considerable proportion of the foremost mines are still above the 500-ft. level. Development is still in its beginnings, even within the limits of the producing area, while the outlying districts which compass this central zone offer magnificent possibilities. For a decade, at least, Cripple Creek promises to increase. Whether it will finally distance its great South African rival can only be conjectured, but the expectation is a reasonable one that within a few years its output will equal that of the United States at the present time.

Cripple Creek is unique among the great gold-producing districts of the world in that it has been developed almost wholly out of itself. It was discovered and prospected by poor men, who located the claims and determined the course of the great dykes and mineral-bearing veins; the foundations of practically all the great mining companies were laid by men whose only capital was their muscle. The shafts and tunnels which penetrate the golden hills, and the great plants of machinery which dot their slopes have, in the great majority of cases, been paid for with the gold obtained from the identical ground. Few of the companies which are to-day furnishing the bulk of the district's output originally possessed any cash working capital. The leasing system was naturally evolved from the necessities of the case, and through its operations practically all the great producing properties of the district received their initial development. These conditions have favored the multiplication of companies; so that, when properly understood, the fact that but a small fraction of the five hundred properties shown in this volume are developed mines, does not argue against the general success of mining in the district. A great majority of these are owned by companies without working capital, which must depend for at least the beginnings of development on the desultory operations of lessees. But as time goes on the list of "mines" gradually lengthens, as the result of the successful labors of these lessees, while the "prospects" are in turn recruited as the debatable borderland of the district is pushed still farther away. Accumulated royalties from leases enable the companies on their expiration to prosecute systematic development. This is the manner in which the average Cripple Creek mine is evolved; and while the process generally requires time, and is attended with many uncertainties, it is perhaps, on the whole, the most satisfactory that could be devised.

A history of the development of Cripple Creek would be incomplete that did not include a reference to the growth of the mining stock business of Colorado Springs. Since it was Colorado Springs men who discovered and developed the new gold fields— but twenty-five miles distant, as the bird flies—it was but natural that the business of trading in the shares of the companies organized to operate therein should center in that city. It very soon became evident that an exchange was necessary to systematize the business and to afford protection to both broker and client. Of the several exchanges which have been organized, the original as well as the most important is The Colorado Springs Mining Stock Association, established in May, 1894. During the six years of its existence it has grown steadily in importance and in the confidence and respect of the investing public, until to-day it is recognized as the leading institution of its kind in the United States. It is safe to say that it occupies a unique place among the exchanges of the country. It was organized by the foremost business men of Colorado Springs—bankers, merchants and capitalists; and according to its constitution these non-board members must compose the majority upon its governing and principal committees. These men, among whom are the presidents of the four banks of Colorado Springs, had seen the incalculable injury which has been inflicted upon the business interests of many other cities through the questionable practice of their mining exchanges; and hence they lend their names and give their time and the benefit of their experience to the direction of this most important business interest of the city. It is not surprising, therefore, that as a consequence of these exceptional conditions of its organization and management, the Colorado Springs Mining Stock Exchange has from the beginning been accorded the respect and confidence of the public; and it is largely because of the reputation made by this exchange that Colorado Springs, though a comparatively small city, has become the foremost mining stock market in the United States, upon which business is daily transacted not only throughout this country, but with most of the European capitals. The growth of this business has been very rapid, particularly during the last three years, as is shown by the following table:

SALES ON THE COLORADO SPRINGS MINING STOCK EXCHANGE:

	Shares.	Cash Value.
1897	49,723,857	$ 7,573,629.00
1898	66,575,999	10,253,146.00
1899	236,165,443	34,446,956.54

The price of seats on this exchange has also advanced rapidly and is now over $3,000. Its quarters, which were considered ambitious three years ago, are now entirely inadequate, and a proposal has just been accepted for the erection by Mr. W. S. Stratton of a building for the exclusive occupancy of the institution which will cost upwards of $250,000.

The Colorado Springs Board of Brokers' Association is a newer organization, having begun business in September, 1899. It has a large membership and occupies commodious and admirably arranged quarters. That it is already doing a considerable business will be seen from the following:

SALES ON THE COLORADO SPRINGS BOARD OF BROKERS' ASSOCIATION:

	Shares.	Cash Value.
September to December 31, 1899	34,540,000	$2,498,900
January 1 to April 1, 1900	31,993,500	2,679,200

At the present time (June, 1900) the development of the Cripple Creek district is proceeding at a more rapid rate than ever before in its history. Mining activity, both in the district proper and in the outlying sections, is unprecedented. A third railroad is under active construction, connecting Colorado Springs directly with the gold camp, the completion of which in October next will inaugurate important reductions in freight rates. New plants for ore treatment are building which will increase the daily capacity fully 33 per cent. As reflecting this growth and expansion, it may be stated that the bank deposits of Colorado Springs, the business and financial center of the Cripple Creek district, have increased 500 per cent. in the past six years.

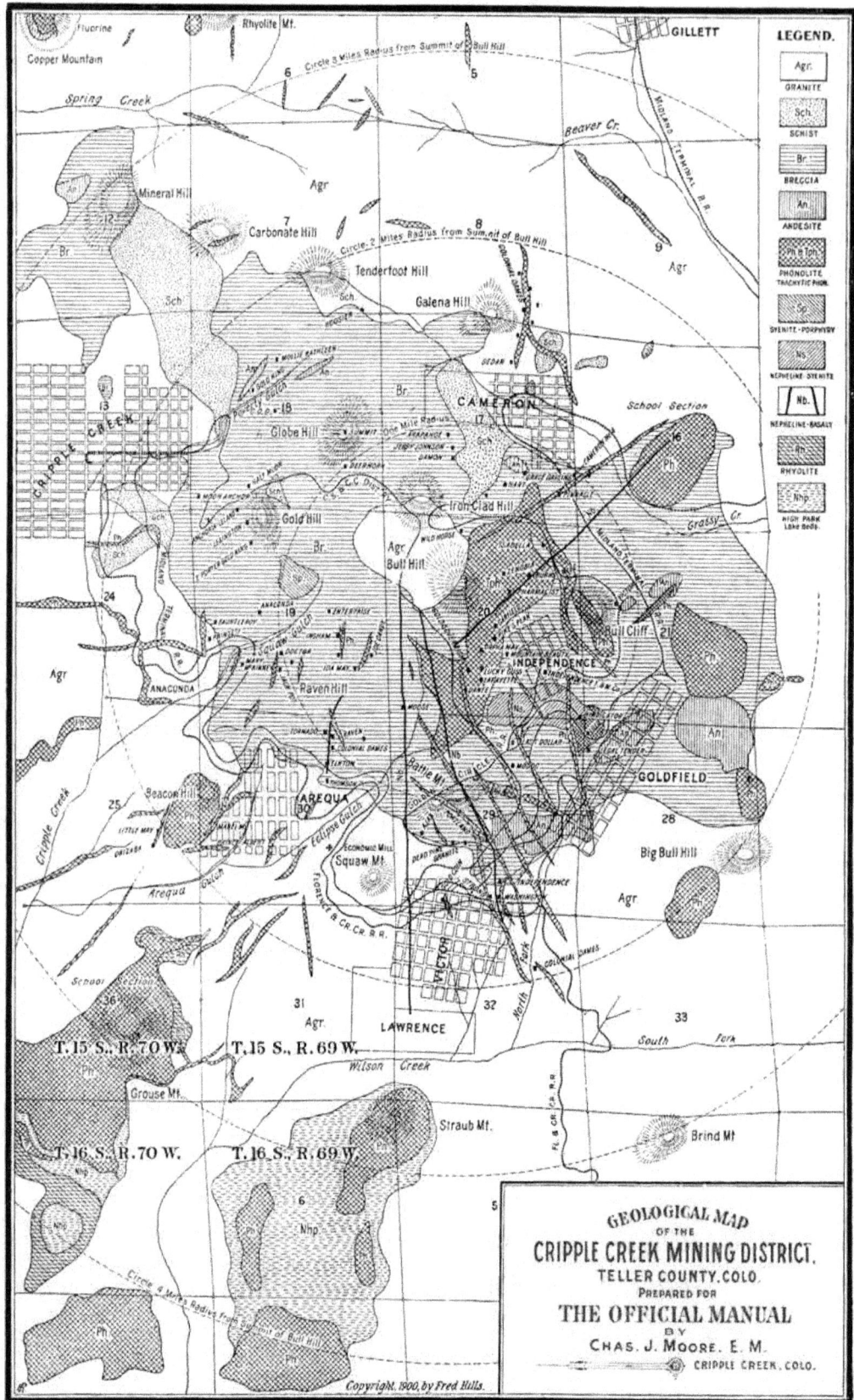

GEOLOGICAL MAP
OF THE
CRIPPLE CREEK MINING DISTRICT,
TELLER COUNTY, COLO.
PREPARED FOR
THE OFFICIAL MANUAL
BY
CHAS. J. MOORE, E. M.
CRIPPLE CREEK, COLO.

Copyright, 1900, by Fred Hills.

LEGEND.

Agr. GRANITE
Sch. SCHIST
Br. BRECCIA
An. ANDESITE
Ph. & Tph. PHONOLITE TRACHYTIC PHON.
Sp. SYENITE-PORPHYRY
Ns. NEPHELINE-SYENITE
Nb. NEPHELINE-BASALT
Rh. RHYOLITE
Nhp. HIGH PARK Lake Beds.

Copper Mountain
Fluorine
Rhyolite Mt.
GILLETT
Spring Creek
Beaver Cr.
Midland Terminal R.R.
Mineral Hill
Carbonate Hill
Circle 3 Miles Radius from Summit of Bull Hill
Circle 2 Miles Radius from Summit of Bull Hill
Tenderfoot Hill
Galena Hill
School Section
CAMERON
CRIPPLE CREEK
Globe Hill
Iron Clad Hill
Grassy Cr.
Gold Hill
Agr. Bull Hill
Bull Cliff
INDEPENDENCE
ANACONDA
Raven Hill
GOLDFIELD
Beacon Hill
Squaw Mt.
AREQUA
Big Bull Hill
Eclipse Gulch
Arequa Gulch
Cripple Creek
VICTOR
COLONIAL DAMES
Florence & Cr. Cr. R.R.
LAWRENCE
School Section
T.15 S., R.70 W.
T.15 S., R. 69 W.
Grouse Mt.
Wilson Creek
North Fork
South Fork
T. 16 S., R. 70 W.
T. 16 S., R. 69 W.
Straub Mt.
Brind Mt.
F. & Cr. Cr. R.R.

The Abdallah Gold Mining Company.

Incorporated April 18, 1896.

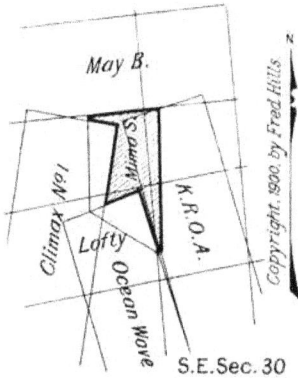

D. N. Heizer.....................President Directors
Sidney R. Bartlett..........Vice-President
C. E. Heizer....................Secretary

Main Office—No. 16 N. Nevada avenue, Colorado Springs, Colorado.

1,500,000 shares. Par value, $1.00.   Capitalization
In treasury January 1, 1900, 196,500 shares.

Owns the Mima S., adjoining the May B.,   Property of the Woods Investment Company, and the Climax No. 1 of the Little Puck Company. The Mima S. has contracted rights to follow and mine its veins through and across the May B. and the M. K. & T. claims, thus giving it about 780 feet additional vein rights. This property contains 1.23 acres, and is located in section 30, on Squaw mountain.

This property has a main shaft 200 feet deep, and 300 feet of drifting. Development A steam hoist has been ordered and the property is being developed by lessees.

Gross production to January 1, 1900, $12,000.00.   Production

The Acme Consolidated Gold Mining Company.

Incorporated December, 1895.

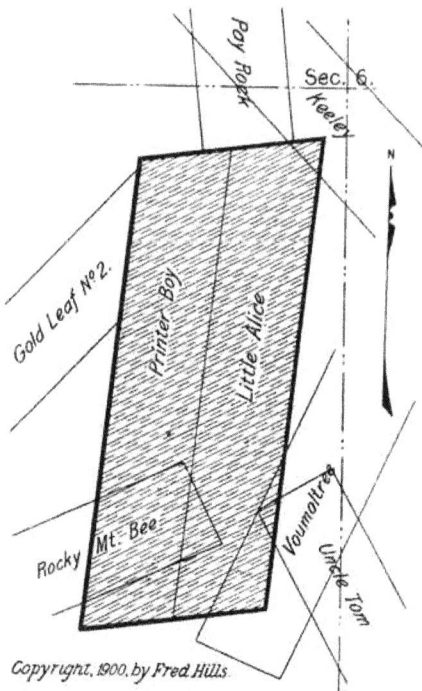

J. K. Vanatta...Pres. and Treasurer Directors
W. K. Jewett.......Vice-President
M. C. Meek.............Secretary
   H. C. Hollister. Walter Scott.

Main Office—16 N. Nevada avenue, Colorado Springs, Colo.

1,250,000 shares. Par value, $1.00. Capitalization
In treasury January 1, 1900, 85,000 shares.

Owns the Printer Boy and Little Property Alice, consisting of 20.165 acres, in the S. W. quarter section 6 on the east side of Rhyolite mountain; patented; claims sideline with each other, making all ground in a body.

There is a 100-foot shaft on one Development claim and a 65-foot shaft on the other. No ore has been shipped; company is now considering leasing its property.

Highest price for stock during 1899, 1 cent; lowest price for stock during 1899, 1 cent.

The Acacia Gold Mining Company.

Incorporated 1895.

Directors
John E. Hundley........President
W. J. Chambers.....Vice-President
P. C. Dockstater........Secretary
Bi-Metallic Bank........Treasurer
F. E. Robinson.   D. R. McArthur.
A. B. Shilling.

Main Office—Bank block, Colorado Springs. Transfer Office—International Trust Company, Colorado Springs.

Capitalization
1,500,000 shares. Par value, $1.00.
In treasury January 1, 1900, 86,-900 shares; in treasury January 1, 1900, $10,000.00 cash.

Property
Owns the Burns and the Morning Star claims, both patented and containing 20-2-3 acres, situated in sections 17 and 20 on Bull hill; the property being formerly owned by the Calumet M. and M. Co.

Development
On the main works of the Burns is a shaft house, 40x85 feet, and a three-drill compressor and steam hoist. Two lessees, operating on the north end of the Burns, are equipped with steam hoist and power drills. The main shaft is 350 feet deep, with a 400-foot incline from the 350-foot level. There is also about 1,000 feet of levels and cross-cuts. On this claim there are two other shafts, one on the north end, 400 feet deep, with 300 feet of drifting, the other, 200 feet deep, with 100 feet of drifting. The Morning Star has a shaft 350 feet deep and 200 feet of levels, and has a compressor plant (3 drills) and steam hoist.

Production
Gross production from July, 1899, to March 15, 1900, $81,000; net royalties to the company during 1899, $13,000.00.

History
In 1895 the Acacia Company absorbed the Calumet Company. Highest price for stock during 1899, 54 cents; lowest price for stock during 1899, 4 cents.

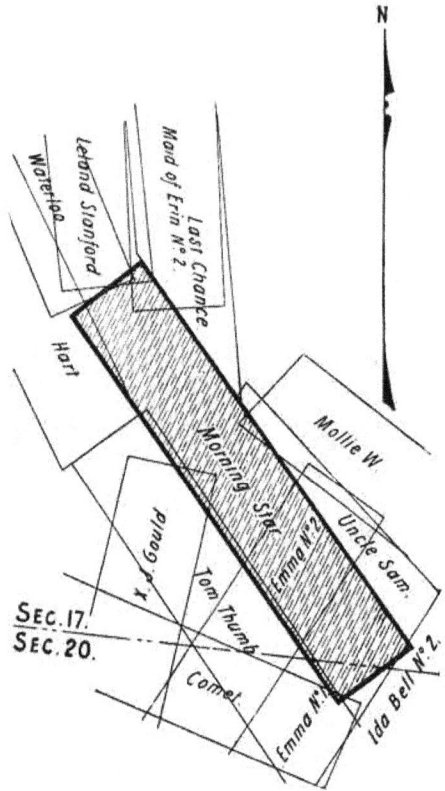

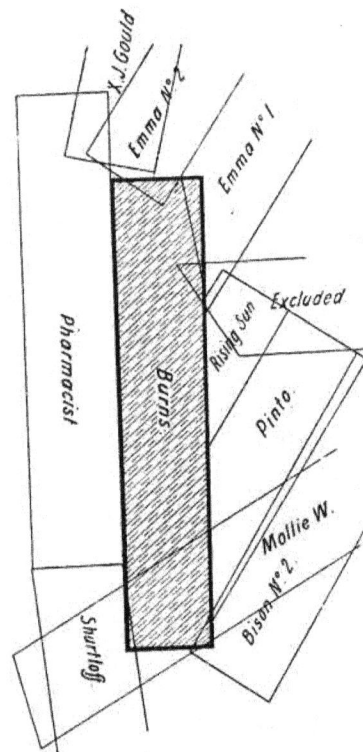

N

Copyright, 1900, by Fred Hills

The Acorn Gold Mining Company.

Incorporated December 28, 1895.

G. C. Hemenway........................President

L. E. Hawkins......................Vice-President

S. R. Bartlett.............Secretary and Treasurer

    C. E. Maclan.   A. W. Singer.

Main Office—64 Postoffice building, Colorado Springs, Colorado.

Copyright, 1900, by Fred Hills

1,250,000 shares. Par value, $1.00.   Capitalization

In treasury January 1, 1900, 250,000 shares.

Owns the Last Chance Nos. 1 and 2, and Sure Pay Nos. 1 and 2, con- Property taining 41 acres, in the N. W. ¼ Sec. 3, on Trachyte Mountain. Patented.

The Adams Express Gold Mining Company.

Incorporated 1894.

W. G. Rice................President and Treasurer
H. A. Lees......................Vice-President
L. H. Harding.......................Secretary
Frank Smith.   J. R. Adams.

Main Office—No. 15 North Tejon street, Colorado Springs, Colorado.

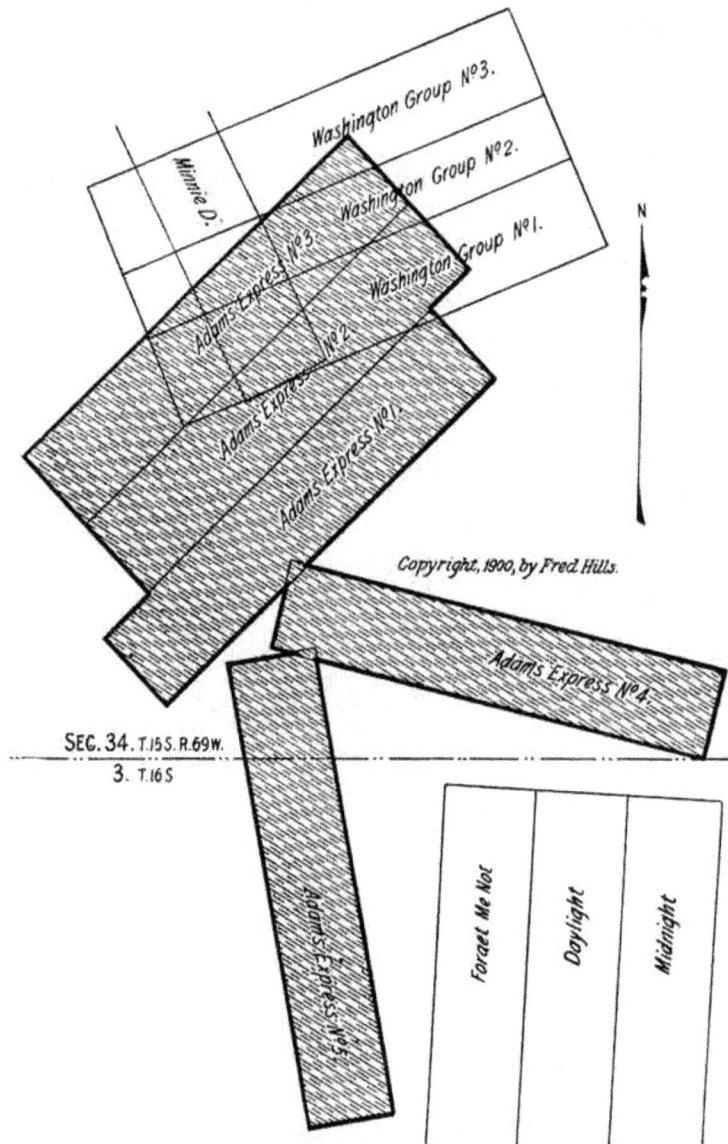

Copyright, 1900, by Fred Hills.

1,500,000 shares. Par value, $1.00.
In treasury January 1, 1900, 120,000 shares.

Owns the Adams Express Nos. 1, 2, 3, 4 and 5, in all 50 acres, in the S. E. ¼ of Sec. 34, on Little Bull Hill.
The property is patented.

Highest price for stock during 1899, $4.50 per M. Lowest price for stock during 1899, $3.00 per M.

The Addie C. Mining Company.

Incorporated February 25, 1892.

<table>
<tr><td>L. C. De Morse.........................President</td><td rowspan="3">Directors</td></tr>
<tr><td>D. Le Duc..........................Vice-President</td></tr>
<tr><td>C. C. Lunt.................Secretary and Treasurer</td></tr>
</table>

Main Office—No. 206 Mining Exchange building, Denver, Colorado.

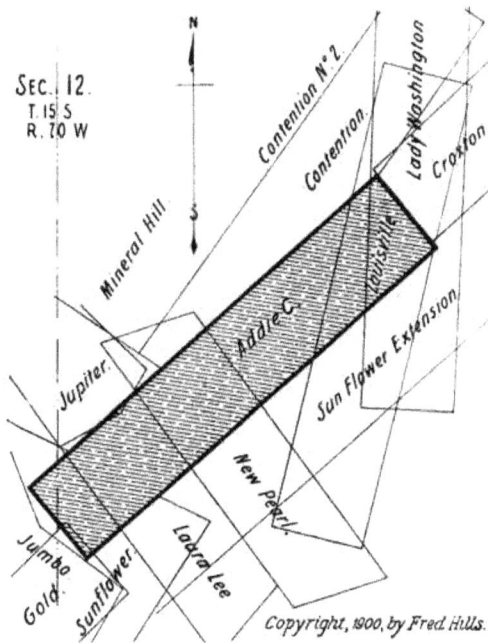

Copyright, 1900, by Fred Hills.

500,000 shares. Par value, $1.00.   Capitalization

Owns the Addie C. mining claim, in the S. E. quarter of section 12, Property
on Mineral hill, containing 10 1-3 acres. Patented.

There is a shaft 209 feet deep, and cross-cuts and drifts of about 50 Development
to 75 feet. The property was leased for 20 months, from November 18,
1899, with bond for $30,000. This is simply a prospect and is being
worked in a small way.

Highest price for stock during 1899, $11.25 per thousand; lowest
price for stock during 1899, $5.00.

The Addie E. Mining and Milling Company.

Incorporated April, 1892.

Directors

E. J. Eaton...........................President

J. E. Rockwell.............Secretary and Treasurer

C. H. Mallon.   E. S. Woolley.

Main Office—The El Paso County Abstract Company, No. 113 East Kiowa street, Colorado Springs, Colorado.

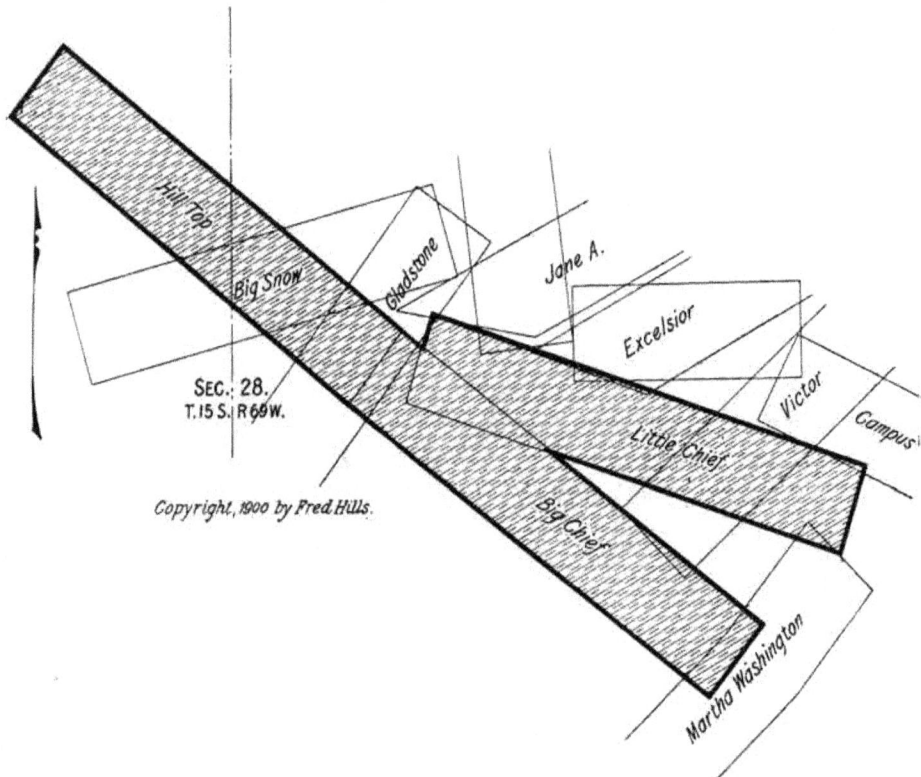

Copyright, 1900 by Fred Hills.

Capitalization

1,250,000 shares.   Par value, $1.00.

In treasury January 1, 1900, 250,000 shares.

Property

Owns the Hill Top, the Big Chief and the Little Chief, a total of 29.13 acres, situated in the S. E. ¼ of Sec. 28, on Big Bull Hill.   All patented.

Development

There is a 200-foot shaft on the property.   Greater part of the development work has been done on the Hill Top.

Highest price for stock during 1899, $7.50 per M.   Lowest price for stock during 1899, $3.50 per M.

The Ader Bell Mining and Tunnel Company.

Incorporated November 25, 1895.

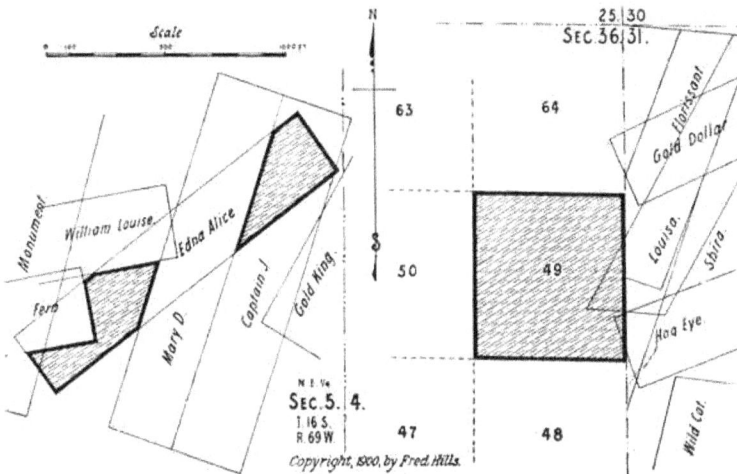

Directors

Main Office—
Pueblo, Colorado.

Branch Office—
Colorado Springs,
Colorado.

2,000,000 shares. Par value, $1.00. **Capitalization**

In treasury January 1, 1900, 200,000 shares; in treasury January 1, 1900, $100.00 cash.

Owns the Gold Coin, 2 1-2 acres, in the N. E. 1-4 of section 19, on Gold hill; the Hazel Kirk, 5.84 acres, in the N. E. 1-4 of section 32, on Big Bull hill; the Todos Santos, 7.77 acres, in the E. 1-2 of section 32, on Little Bull hill; the Hi-Ki, 4.61 acres, in the W. 1-2 of section 33, on Little Bull hill; and the Edna Alice, 4.90 acres, in the N. E. 1-4 of section 5, township 16 south, range 69 west. Also a 20-year lease on lot No. 49, in school section 36, Grouse mountain, 10 acres. All the above claims are patented. **Property**

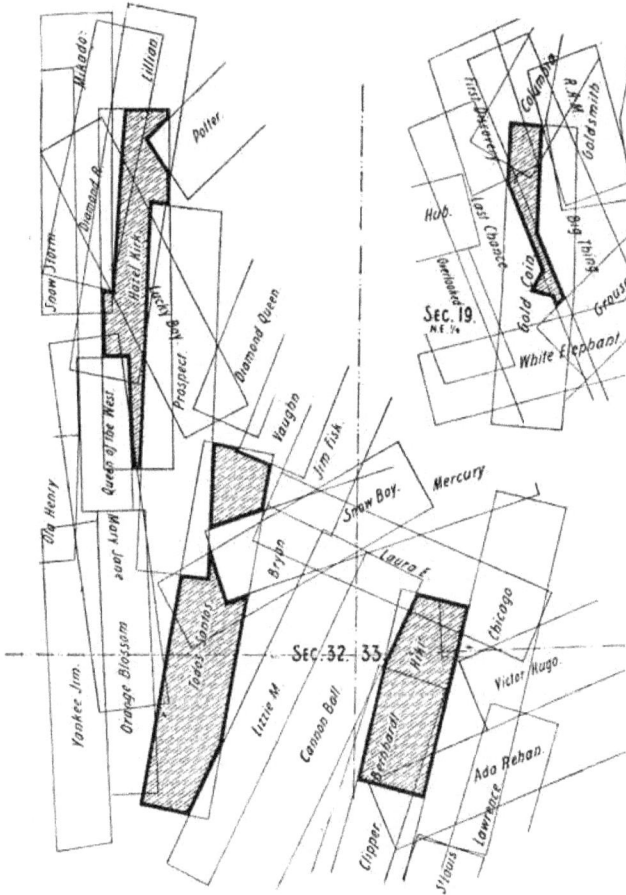

Considerable development work has been done on all the claims owned by this company. The Gold Coin has been leased for two years. **Development**

Highest price for stock during 1899, 1 1-2 cents; lowest price for stock during 1899, 1 cent.

The Advance Gold Mining, Bonding and Leasing Company.

Incorporated December 13, 1896.

Directors

A. R. McIntyre............President and Treasurer
Josiah Winchester ....Vice-President
A. C. Labrie.............................Secretary
    A. B. Atwater.   W. L. Quigley.

Main Office—P. O. box No. 681, Cripple Creek, Colorado.

Capitalization

1,500,000 shares. Par value, $1.00.

In treasury January 1, 1900, 200,000 shares; in treasury January 1, 1900, $1,000.00 cash.

SEC. 14.
T. 15 S. R. 70 W.

Little Violet

Copyright, 1900 by Fred Hills.

Property

Owns the Mary Jane, 1-2 acre, in the S. 1-2 section 19, on Raven hill; the Bohemian, 6 1-4 acres, in the S. E. 1-4 section 14, on Gold Quartz hill; the Golden Wedge, 4 1-2 acres (owns 7-16 interest only), in the S. 1-2 section 19, on Raven hill; also the Gold Bell, in the S. W. 1-4 section 28, on Big Bull hill. The Mary Jane, Golden Wedge and Bohemian lodes are patented. The Gold Bell (not shown on plat) is in process. The company also holds, by lease, the Julia E., located on Raven hill, adjoining shaft No. 1 of the Advance G. M. B. & L. Co.

Development

The Mary Jane shaft No. 1 is down 110 feet, with drift of 145 feet from the 100-foot level. This shaft is being worked with whim. A suitable shaft house has been erected over this shaft. Shaft No. 2 is down 475 feet; stations at the 200, 300, 350, 400 and 440-foot levels. About December 1, 1899, the vein was cut at the 440-foot level. Machinery of this shaft includes a 15-horse power electric hoist, and suitable buildings, including ore house of 30 tons capacity. On the Golden Wedge there is a good shaft house. The shaft on this claim has been sunk 400 feet. The greater part of the development work is being done on the Mary Jane lode.

Production

Gross production to January 1, 1900, 127 tons.

Highest price for stock during 1899, 10 3-4 cents; lowest price for stock during 1899, 3 3-4 cents.

40

The Ajax Gold Mining Company.

Incorporated 1894.

E. A. Colburn..............................President   Directors
J. E. McRay...........................Vice-President
C. H. Dudley..............Secretary and Treasurer
J. C. Helm.

Main Office—14 N. Nevada avenue, Colorado Springs, Colo.

100,000 shares. Par value, $1.00.   Capitalization

Owns the Apex, Monarch, Mammoth Pearl, Champion, June Blizzard   Property
and a controlling interest in Hallett & Hamburg G. M. Co.; all patented

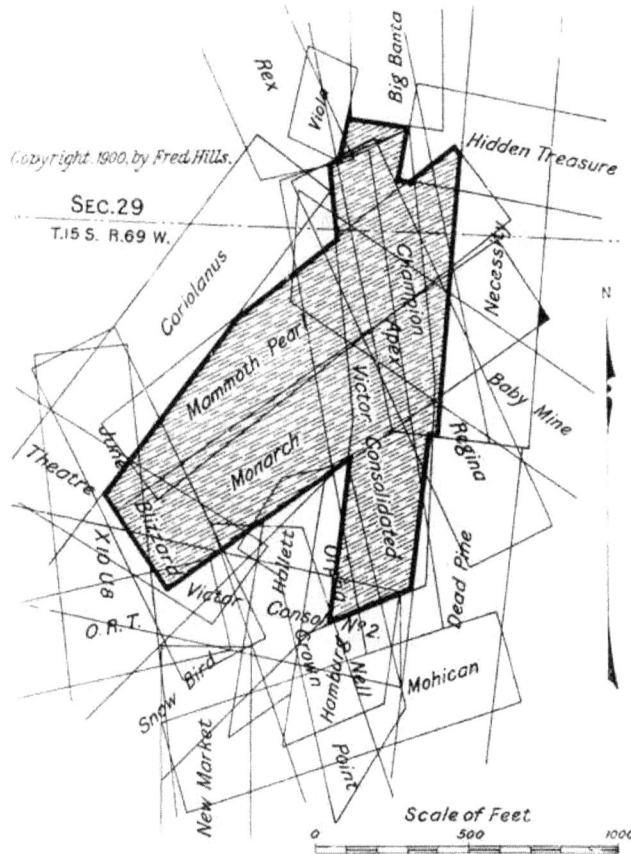

except June Blizzard and the Orpha Nell (receiver's receipt); about 28
acres in all, in the S. W. 1-4 section 29, on Battle mountain, near the Dead
Pine. The Hallett & Hamburg and the Orpha Nell are a late purchase
and comprise the holdings of the Hallett & Hamburg G. M. Co. Refer
to latter for plat.

There is on the property a shaft house, engine house, blacksmith   Development
shop and ore house combined; two 100-horse power boilers; two Norwalk
compressors; one double reel first motion hoist and a triple compartment
shaft. The main shaft is 600 feet deep, with about 5,000 feet of drifts and
cross-cuts, also a few small shafts. The property has been worked by the
company mostly since 1895. The company is now adding one of the
largest hoisting plants in the district.

Owing to this being a very close corporation, it is impossible to ascer-
tain the production, which has been considerable—in fact very large.

41

The Alamo Mining Company.

Incorporated 1892.

Directors

John W. Proudfit.........................President

C. H. Bryan............ ....Secretary and Treasurer

K. R. Babbitt.   J. J. Hughes.

J. McK. Ferriday.

Main Office—10 Hagerman building, Colorado Springs, Colorado.

Capitalization

1,000,000 shares.   Par value, $1.00.

In treasury January 1, 1900, $50,000 cash.

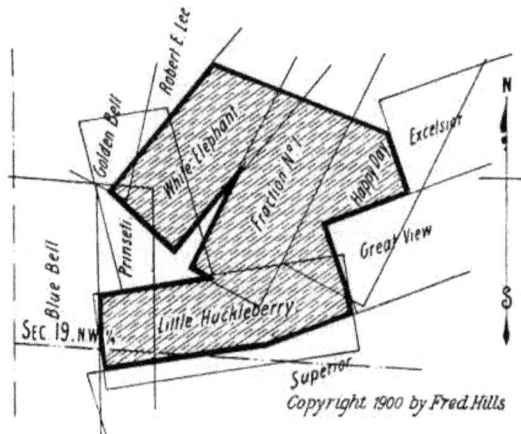

Copyright 1900 by Fred Hills

Property

Owns Little Huckleberry and parts of White Elephant, Happy Day and Fraction No. 1, in the N. W. 1-4 section 19, on Gold hill, near the Mary McKinney, Anaconda, Lexington and Anchoria-Leland properties. All the property is patented.   In 1899 the northern portion of the original holdings of this company's property was sold to the Rittenhouse G. M. Co.

Development

A shaft has been sunk 200 feet deep and 50 feet of drifting on the Happy Day claim.

Highest price for stock during 1899, 12 1-2 cents; lowest price for stock during 1899, 4 3-4 cents.

The Albemarle Gold Mining and Milling Company.

Incorporated November 19, 1895.

W. H. Metz..............................President.
B. P. Anderson.....................Vice-President
A. C. Hart............................Secretary
I. S. Harris..........................Treasurer
J. K. Vanatta.

Directors

Main Office—No. 112 East Cucharras street, Colorado Springs, Colorado.

1,000,000 shares. Par value $1.00.
In treasury January 1, 1900, 69,000 shares.

Capitalization

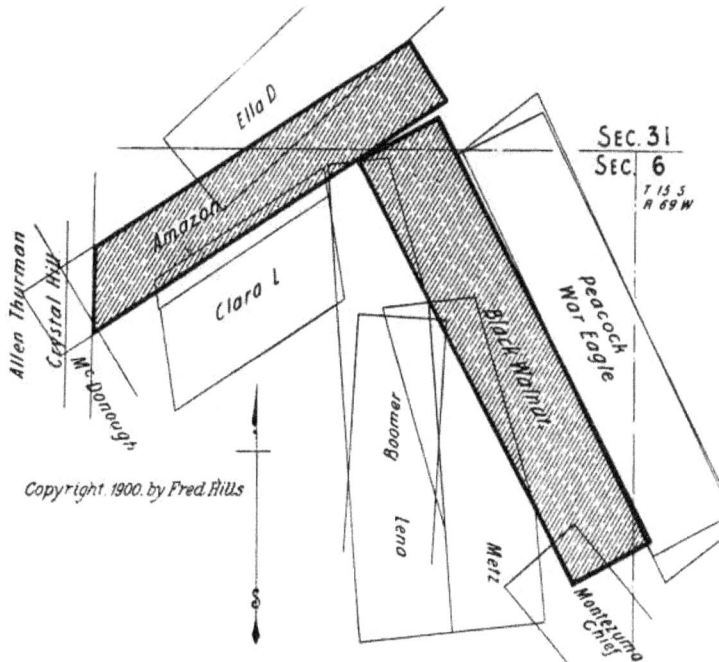

Copyright, 1900, by Fred Hills.

Owns the Black Walnut, 10 1-3 acres, in the N. W. 1-4 section 6, on Rhyolite mountain; the Amazon, about 8 1-2 acres, in the N. W. 1-4 section 6, on Rhyolite mountain; and the Clover Leaf, 9.45 acres, in the N. E. 1-4 section 19, township 14 south, range 69 west, on Aspen mountain. All patented. The Clover Leaf, being outside the Cripple Creek district, is not shown on plat.

Property

One log cabin. There are several shafts, none of them over 100 feet deep; also one tunnel of 100 feet. Active development work has ceased for the present.

Development

The company discontinued work when funds were exhausted, although the indications were very favorable. The company has had several opportunities to lease, refusing them because bond was required. The property shows strong mineralized leads and has furnished some high assays, the best being $324.00.

History

The Alcyone Gold Mining Company.

Incorporated October 1, 1896.

Directors

R. W. Griswold....................Vice-President

Spencer Penrose......................Secretary

C. M. MacNeill.......................Treasurer

Geo. F. Fry.

Main Office—Room No. 11, El Paso Bank building, Colorado Springs, Colorado.

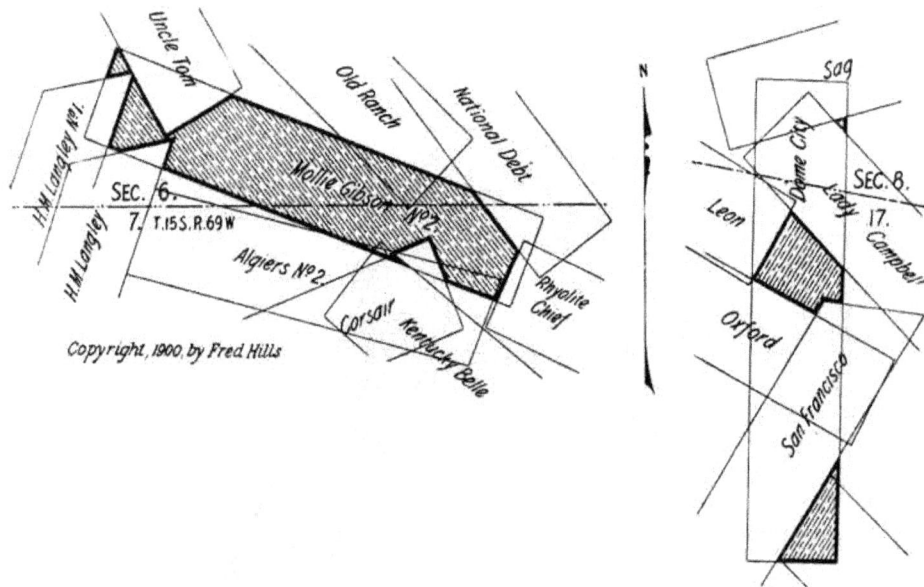

Copyright, 1900, by Fred Hills

Capitalization

1,250,000 shares.   Par value, $1.00.

In treasury January 1, 1900, 112,700 shares.

Property

Owns the Dome City, about 2 acres, in the N. E. 1-4 section 17; the Bull King, about 3 1-2 acres, adjoining the Dome City, not shown on map; the Mollie Gibson No. 2, 9.24 acres, in the N. E. 1-4 section 7; the Toga, Alcyone and Atlas, about 27 acres, on the east slope of Big Bull mountain, not shown on map; the Inez M., about 3 1-2 acres, on the west slope of Squaw mountain, not shown on map. The Mollie Gibson No. 2 is patented. Dome City and Bull King in process. Remainder held by location. The Inez M. is in litigation.

Development

Work sufficient to obtain a patent has been done on all the claims. The Inez M. has a shaft 150 feet deep. The greater part of the development work is being done on the Inez M. and the Mollie Gibson No. 2.

The Alert Gold Mining Company.

Incorporated January 14, 1896.

Wm. H. Reynolds.........................President
W. S. Harwood......................Vice-President
J. W. D. Stovell........................Secretary
O. W. Pitcher...........................Treasurer
John H. Whyte.

Directors

Main Office—Rooms 21 and 22, Postoffice block, Colorado Springs, Colorado.

Capitalization

1,500,000 shares. Par value, $1.00.

In treasury January 1, 1900, 50,000 shares; in treasury January 1, 1900, about $5,000.00 cash.

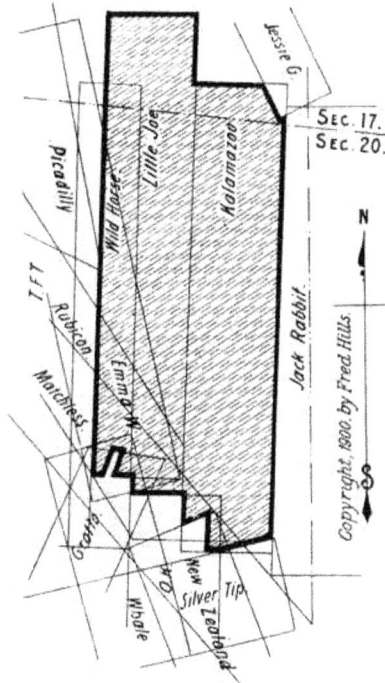

Property

Owns the Kalamazoo and Little Joe, containing 19 5-8 acres, situated in the N. W. 1-4 of section 20, on Bull hill; the Cozad No. 2, containing about 3 1-2 acres, in the N. W. 1-4 of section 24, on Signal hill, which is not shown on plat; also owns lease on a portion of the Diamond and Hardwood claims of the Damon company, to run 14 months from date, or until May 1, 1901. Kalamazoo and Little Joe are patented; the Cozad No. 2 in process.

Development

30 H. P. electric hoist, 45 H. P. steam hoist, 4 drill air compressor, 80 H. P. boiler on the Damon lease; a large shaft house on the Kalamazoo. The Kalamazoo has a shaft 500 feet deep and from 1,000 to 1,200 feet of cross-cutting. On this claim and on the Little Joe there are several shafts of from 20 to 65 feet. Greater part of the development work has been done on the Kalamazoo, where there were several veins cut in cross-cutting. The work on this property was done before the company owned it.

Production

Gross production to January 1, 1900 (on lease), about $125,000.

Stock selling for 30 cents in April, 1900.

The Alice Gray Gold Mining Company.

Incorporated November, 1895.

Directors

R. T. Fahey.............................President
Joseph Parnell ....Vice-President
John Bridge..............Secretary and Treasurer
I. S. Harris.   J. D. Gavitt.

Main Office—The Out West Printing and Stationery Company, Colorado Springs, Colorado.

Capitalization   1,000,000 shares. Par value, $1.00.

Property   Owns the Alice Gray, the Queen Isabelle, the Helen E., the Ben Bolt and the Sam Gale, about 28 acres in all, situated in the N. E. 1-4 of section 6, township 16 south, range 69 west. In one group on Straub mountain. All patented.

Development   On Alice Gray a 30-foot shaft has been sunk and a 225-foot tunnel, 112 feet of which is timbered. The Queen Isabella has a 23-foot shaft, a 49-foot shaft, a 45-foot shaft, and 90 feet of trenching. The Sam Gale has a 23-foot shaft and 25 feet of trenching. The Helen E. has a 50-foot shaft. The Ben Bolt has a 23-foot shaft and 36 feet of trenching. All of the above shafts are timbered.

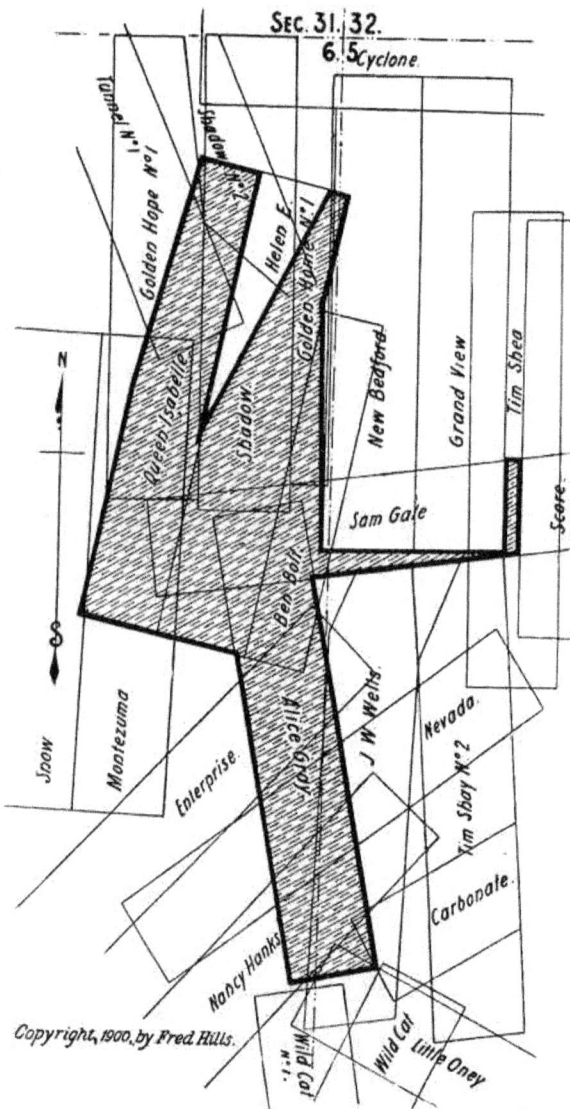

History   The property was located January, 1892. No shipments have been made, but all claims show good prospective values.

46

The Alice M. Mining Company.

Incorporated December 14, 1895.

P. W. Middagh..........................President
Walter Kennedy ....Vice-President
J. L. Middagh..............Secretary and Treasurer

Directors

Main Office—Room 35, P. O. building, Colorado Springs, Colorado.

1,250,000 shares.   Par value, $1.00.
In treasury January 1, 1900, 42,000 shares.

Capitalization

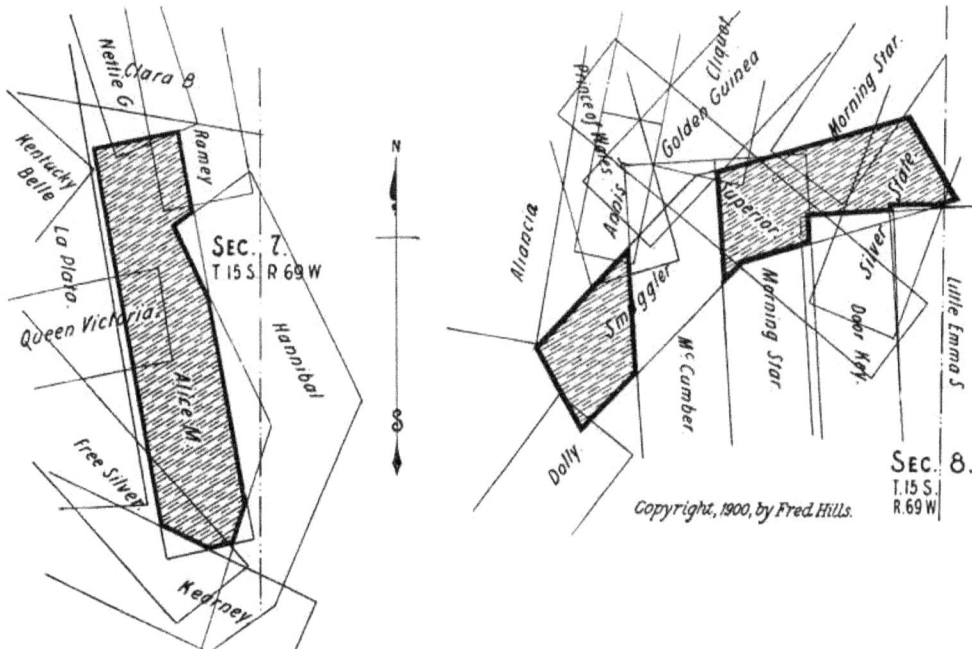

Owns the Alice M., situated in the N. E. 1-4 of section 7, contain-Property ing 8 acres; also the Smuggler, in the N. W. 1-4 of section 8, containing 7 acres.   Both claims patented.   These properties are located on the north and east slope of Tenderfoot hill.

The Alice M. claim has a 90-foot shaft and the Smuggler one ofDevelopment 60 feet.

Highest price for stock during 1899, 1 1-2 cents; lowest price for stock during 1899, $3.00 per thousand.

The Alpha Gold Mining Company.

Incorporated December, 1895.

Directors

A. Reynolds . . . . President
E. P. Shove . . . . Secretary
Sherwood Aldrich.   C. A. Steyn.
S. S. Morris.

Main Office—9 S. Tejon street, Colorado Springs, Colorado.

Capitalization

1,000,000 shares. Par value, $1.00.

In treasury January 1, 1900, 375,000 shares.

Copyright, 1900, by Fred Hills.

Property

    Owns the Alpha, containing 8.677 acres, situated in the S. W. 1-4 of section 31, township 15, south of range 69 west; also the Noonday, containing 4.747 acres, same location, both on Squaw mountain. Both claims are patented.

Development

    On the Alpha is a shaft 160 feet deep. The Noonday has a shaft 100 feet deep. The greater part of the development work is being done on the Alpha.

The Alton Gold Mining Company.

Incorporated December 5, 1895.

Thos. Gough...........................President
F. A. Pine........................Vice-President
E. S. Cohen..............Secretary and Treasurer
H. E. Keenan.   C. L. Wells.

Directors

Main Office—Colorado Springs, Colorado.

1,000,000 shares.   Par value, $1.00.
In treasury January 1, 1900, 60,000 shares.

Capitalization

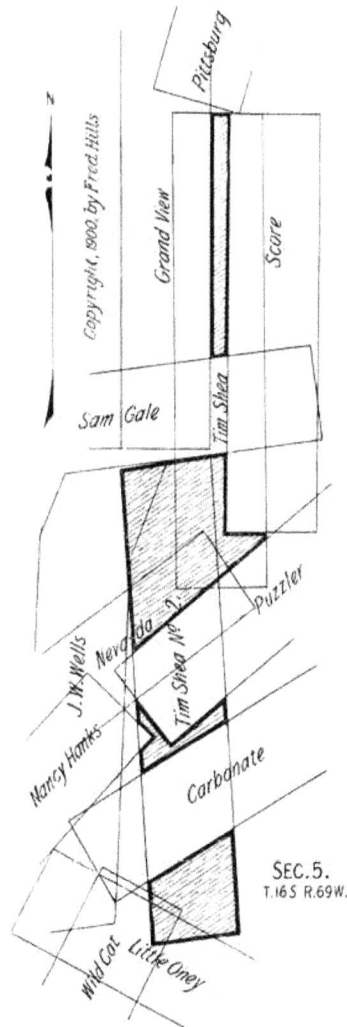

Owns the Tim Shea and Tim Shea No. 2, containing 7.17 acres, situated in the N. W. 1-4 of section 5, township 16 south of range 70 west; the Ida Taylor, containing 1 1-2 acres, on Gold hill. This latter is not shown on map. Both in process of patenting.

Property

Highest price for stock during 1899, $5.00 per thousand; lowest price for stock during 1899, $2.00 per thousand.

The Amarillo Gold Mining Company.

Incorporated September 23, 1895.

Directors

John Lennox..........................President
W. W. Bryan.....................Vice-President
K. MacMillan..........................Secretary
F. D. Fox.

Main Office—No. 24 Midland block, Colorado Springs, Colorado.

Capitalization   1,000,000 shares. Par value, $1.00. In treasury January 1, 1900, 106,000 shares.

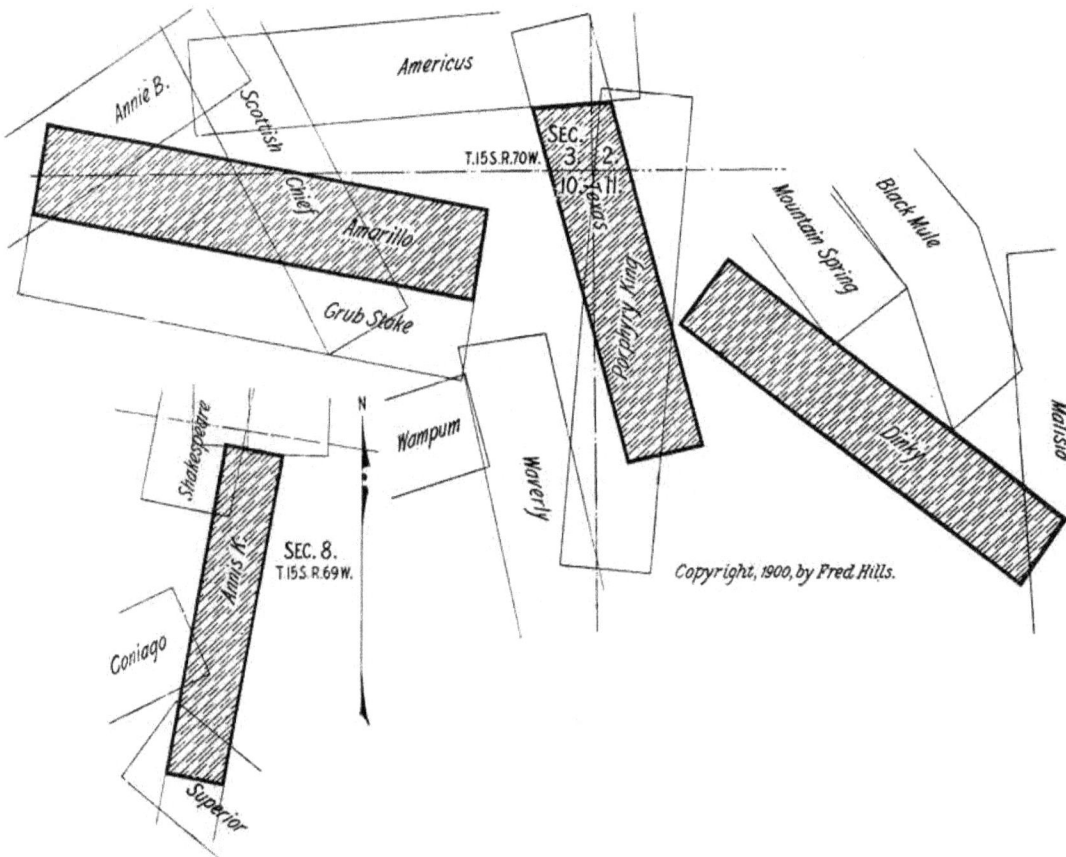

Copyright, 1900, by Fred Hills.

Property   Owns the Amarillo, Dinky, Texas, containing respectively 10.331 acres, 7.511 acres, 7.313 acres, situated on Iron mountain, in the N. E. 1-4 of section 10, township 15 south, range 70 west, and in the N. W. 1-4 of section 11, township 15 south, range 70 west; also the Annis K., 6.409 acres, in the N. W. 1-4 of section 8, on Tenderfoot hill. The Amarillo is patented; receiver's receipts held for the other claims.

Development   On this property there is a shaft and tunnel workings of about 300 feet. The greater part of the work is being done on the Amarillo, which is leased for eighteen months from September 25, 1899.

The Amazon Mining Company.

Incorporated.

James F. Burns........................President

Wm. Lennox.......................Vice-President

Frank G. Peck............Secretary and Treasurer

John Harnan.   Jas. J. Ducey.

Directors

Main Office—Colorado Springs, Colorado.

600,000 shares. Par value, $1.00.

Capitalization

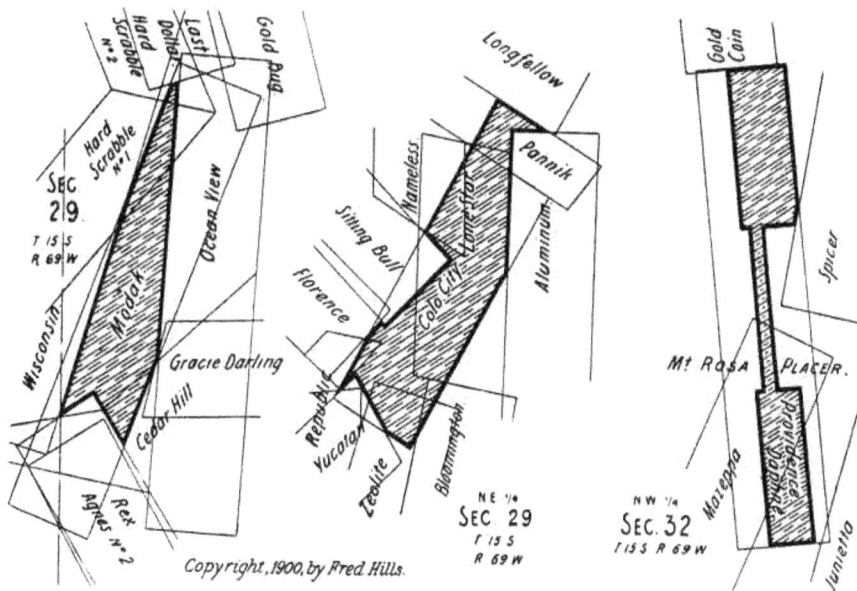

Copyright, 1900, by Fred Hills.

Owns the Providence, 6 acres, in the N. W. 1-4 section 32; the Colorado City, 6 acres, in the N. E. 1-4 section 29; and a part of the Modak, 4 acres, in the N. 1-2 section 29. All patented. The Colorado City claim is under bond for $100,000.00.

All the above property is leased.

Dividends up to March 15, 1900, $62,000.00, the last being paid March 15, 1900. As the Official Manual goes to press, May, 1900, it is reported that the Providence claim has been sold to James Doyle of Victor for $40,000.00, and also the Modoc claim to the Portland Gold Mining Company for $50,000.00. The stock of this company has never been placed on the market.

Property

Dividends

The American Consolidated Mining and Milling Company.

Incorporated 1893.

Directors

Max Straus ....President
W. E. Hannon.........................Vice-President
A. B. Hays...........................Secretary
N. Leipheimer........................Treasurer
Francis Wright.   J. A. Wright.

Main Office—31 P. O. building, Colorado Springs, Colorado.

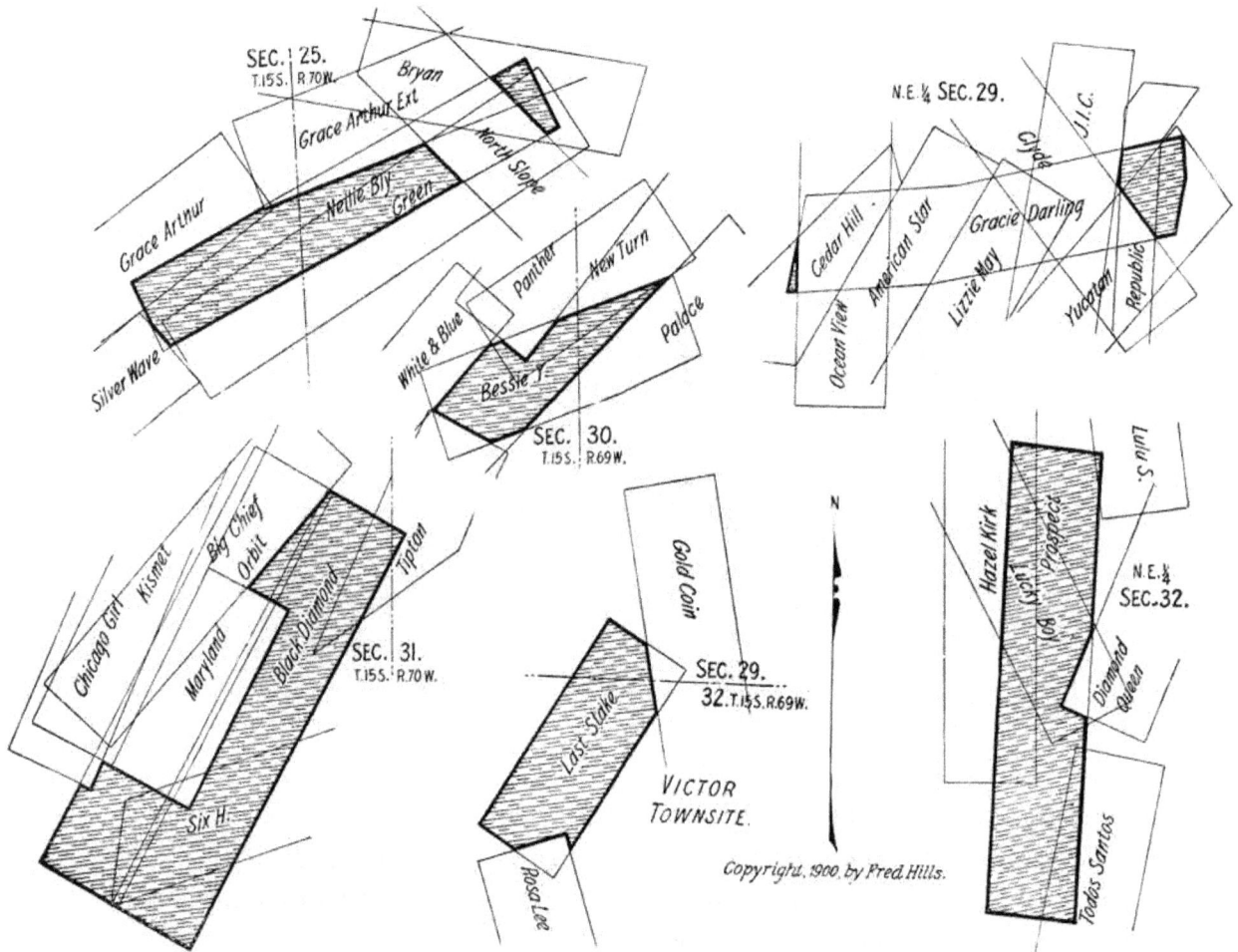

Capitalization

2,000,000 shares. Par value, $1.00.
In treasury January 1, 1900, 500,000 shares.

Property

Owns the Nellie Bly, 7 acres, in the S. 1-2 of section 25, on Beacon hill; the Bessie Y., 3 acres, in the S. 1-2 of section 30; the Orbit and the Black Diamond, 11 acres, in the N. W. 1-4 of section 31, on Squaw mountain; the Gracie Darling, 2 acres, in the N. W. 1-4 of section 29, on Battle mountain; the Last Stake, 5 acres, in the N. W. 1-4 of section 32; also Prospect, 10 acres, in the N. E. 1-4 of section 32. All patented except the Prospect, which is held by location.

Development

The greater part of the development work is being done on the Nellie Bly.

Highest price for stock during 1899, 8 3-4 cents; lowest price for stock during 1899, 4 cents.

The American Eagles.

Not Incorporated.

This group, which is owned by Mr. W. S. Stratton, of Colorado Springs, is situated in section 20, on Bull hill, and comprises the following claims: The John A. Logan, Lottie, Lucy, Brooklyn, American Eagle, American Eagle No. 2 and American Eagle No. 3, containing, in all, about 32 acres.

Copyright, 1900, by Fred Hills.

Scale of Feet
0   500   1000

The American Gold Mining and Milling Company.

Incorporated November 1, 1898.

O. L. Linch, president; S. F. Keith, vice-president; Chas. H. Peters, secretary; F. A. Bailey, treasurer; H. S. Shaw.   *Directors*

Main Office—307 Mining Exchange building, Denver, Colorado.

1,500,000 shares. Par value, $1.00. In treasury January   *Capitalization* 1, 1900, 400,000 shares.

Owns the Yankee Jim, situated in the N. E. 1-4 of section   *Property* 32, containing 7 1-2 acres, on the south slope of Battle mountain; also the Gold Magnet, containing 9 acres, on the north slope of Tenderfoot hill, and the Waldorf, situated southwest of Cripple Creek. The last two are not shown on plat. The Yankee Jim is patented. The Waldorf and Gold Magnet in process. The company also owns the American and North American on Gunnell hill, Central City, patented, and the Edna May, which adjoins the Clay County mine.

On the various claims about $7,000 of development work has been   *Development* done up to date. Ore has been shipped from the Central City property.

Highest price for stock during 1899, 1 3-4 cents; lowest price for stock during 1899, 1 cent.

Copyright, 1900, by Fred Hills.

The Anaconda Gold Mining Company.

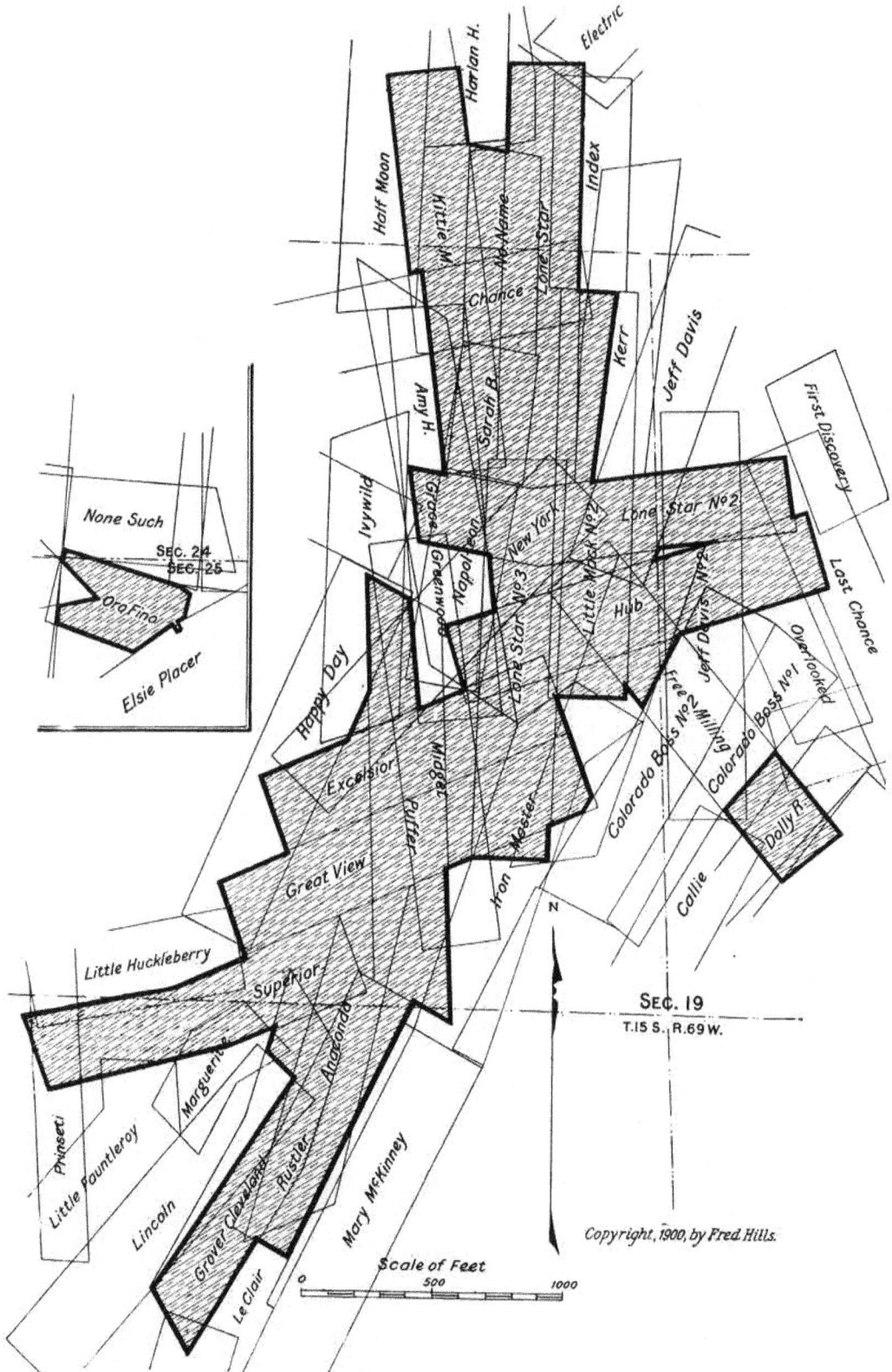

Electric

Harlan H.

Index

Half Moon

Kittie M.

No Name

Lone Star

Chance

Kerr

Jeff Davis

First Discovery

Amy H.

Sarah B.

Ivywild

Grace

Greenwood

Napoleon

New York

Little Miner No 2

Lone Star No 2

Lost Chance

None Such

SEC. 24
SEC. 25

Ora Fino

Elsie Placer

Happy Day

Lone Star No 3

Hub

Colorado Boss No 2

Jeff Davis No 2

Free No 2 Milling

Overlooked

Colorado Boss No 1

Dolly R.

Excelsior

Midget

Puffer

Iron

Master

Great View

Collie

N

Little Huckleberry

Superior

SEC. 19
T.15 S. R.69 W.

Anaconda

Prince T.

Little Fauntleroy

Marguerite

Lincoln

Grover Cleveland

Rustler

Le Clair

Mary McKinney

Copyright, 1900, by Fred Hills.

Scale of Feet
0   500   1000

54

The Anaconda Gold Mining Company.

Incorporated June 21, 1892.

D. H. Moffat...........................President
David Rubidge....................Vice-President
R. H. Reid.............................Secretary
G. E. Ross-Lewin......................Treasurer
    Eben Smith.   Thos. Keely.
        D. L. Webb.

Directors

Main Office—827 Exchange building, Denver, Colorado.

1,000,000 shares.   Par value $5.00.

Capitalization

In treasury January 1, 1900, 15,000 shares; in treasury January 1, 1900, $505.30 cash.

Owns the Lone Star, Lone Star No. 2, Lone Star No. 3, Rustler, *Property* Puffer, Anaconda, Grover Cleveland, Superior, Excelsior, Great View. Hub, Little Mack, Free Milling, Sarah B., Napoleon, Kittie M., No Name, Oro Fino and Ivy Wild, containing 150 acres, in the N. W. 1-4 of section 19, on Gold hill. Patented.

The equipment on the property is complete. This company reports *Development* on January 1, 1900, that there is opened up on this property 31,483 feet of tunnels, drifts and winzes; in its last report it claims ore sales during 1899 of 9,325.58 tons of mill ore of average net value of $7.35 per ton, making $58,598.33, and 21.37 tons smelter ore, average value $112.86, making $2,411.00, or total of $71,010.24.

Production to January 1, 1900, $1,001,476.82.

Production

Highest price for stock during 1899, 59 cents; lowest price for stock during 1899, 43 cents.

In May, 1900, an effort was made to voluntarily assess every stockholder at the rate of 15 cents per share to liquidate an indebtedness on the property of about $112,000, and to provide funds for further operations. The stockholders failed to respond, and as this Manual goes to press (June) a reorganization is being perfected by Mr. J. T. Milliken and associates, who have assumed the liabilities.

The Anchoria-Leland Mining and Milling Company.

Incorporated May 13, 1892.

Irving Howbert.........................President
C. W. Howbert.....................Vice-President
F. H. Gay.............................Secretary
J. C. Plumb............................Treasurer
    S. N. Nye.   Edgar Howbert.
    John Potter.   Theop. Harrison.
        W. F. Anderson.

Directors

Main Office—22 First National Bank block, Colorado Springs, Colorado.

(CONTINUED ON PAGE 56.)

Capitalization   600,000 shares.   Par value, $1.00.

Property   Owns Anchor, Anchor No. 2, Conundrum, Midland, Lillian Leland, Chance, City View and Cottontail.   Property is located in N. W. 1-4 of section 19, N. E. 1-4 of section 24 and the S. E. 1-4 of section 13, Gold hill, Cripple Creek district, Colorado.   All property patented.   Parts of Conundrum, Anchor and Chance are leased.

Development   Surface improvements consist of shaft house, machinery, assay building and office building; boiler house and water tank.   Machinery comprises two 100-horse power boilers, one eight-drill air compressor, one four-drill air compressor, and hoist good for 1,000-foot shaft, all on the Chance claim.   On the Lillian Leland, shaft house and small plant of machinery.   The Chance claim has shaft 950 feet deep and over 6,000 feet in drifts.   The Lillian Leland shaft is 200 feet deep and a small amount

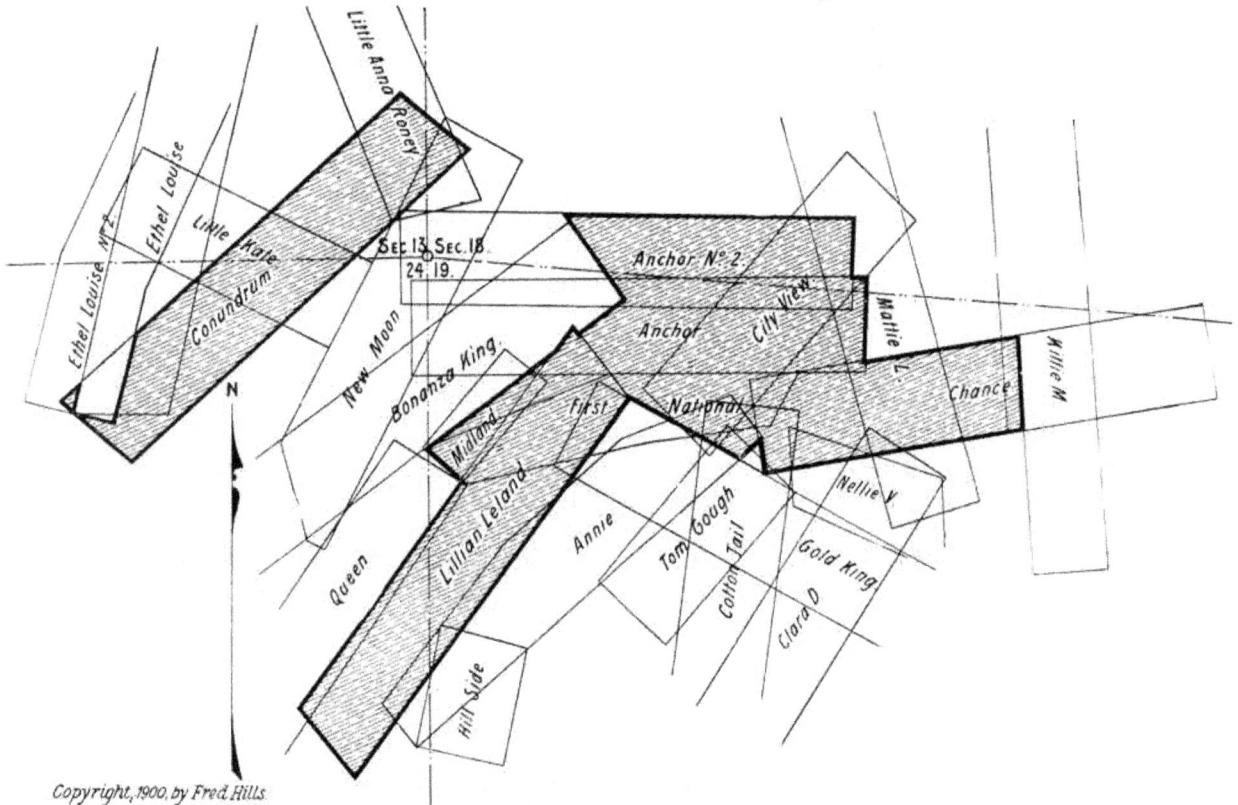

Copyright, 1900, by Fred Hills.

of drifts.   On the Conundrum, Anchor, Anchor No. 2, Midland and City View, surface prospecting only has been done, most of the development work having been done on the Anchor and Chance claims.   This company being a close corporation, nearly all of the stock is held by the directors and their personal friends, there being only about 10,000 shares for general sale.   The company has been very successful in its operations.

Some time ago the workings of this company were somewhat retarded by the heavy flow of water, but the various tunnels that are working under Gold hill have now drained the mine so that a much heavier output is expected.

Production   Production to January 1, 1900, over $1,000,000.00.   Dividends up to January 1, 1900, $198,000.00.   Last dividend paid was April 15, 1899, $18,000.00.

Highest price for stock during 1899, 95 cents; lowest price for stock during 1899, 60 cents.

The Anchor Gold Mining and Milling Company.

Incorporated May, 1894.

L. C. Weyand.....President
M. S. Herring.....Vice-Pres
E. M. Purdy..Sec. and Treas
D. Weyand.   D. P. Cathcart.

Directors

Main Office—Freeman building, Colorado Springs, Colorado.

1,250,000 shares. Par value, $1.00.

Capitalization

In treasury January 1, 1900, 219,250 shares; in treasury January 1, 1900, $1,400 cash.

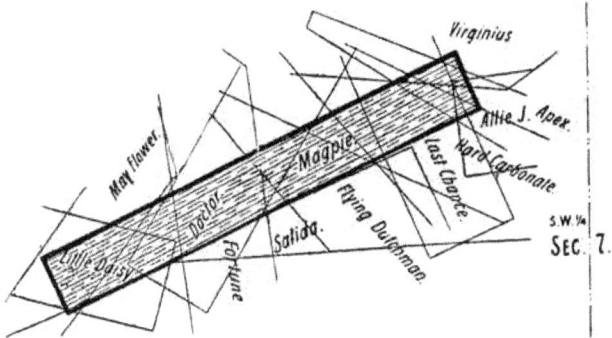

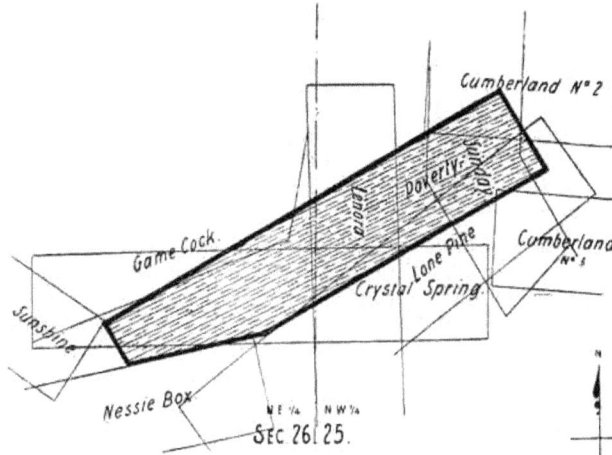

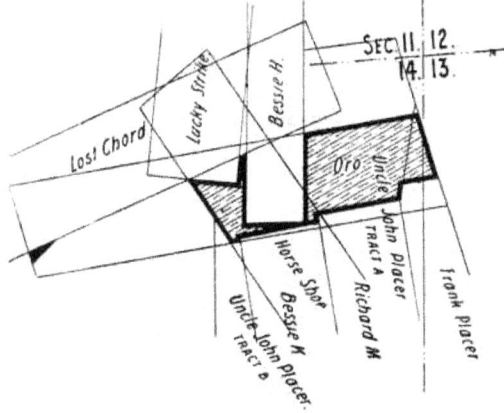

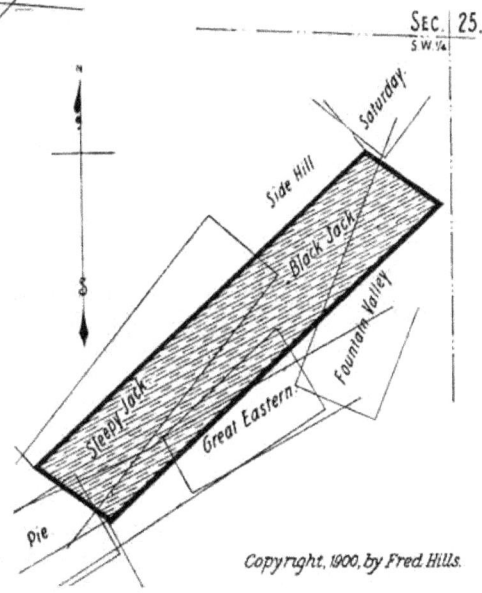

Copyright, 1900, by Fred Hills.

Owns the Magpie, 7 acres, in the S. W. 1-4 of section 7, Tenderfoot hill; Black Jack, 10 acres, S. W. 1-4 of section 25, on Beacon hill; Poverty, 7 acres, in the N. W. 1-4 of section 25, on Beacon hill, and the Oro, 3 acres, in the N. E. 1-4 of section 14, range 70 west, on Mineral hill. All patented. On April 27, 1900, this company purchased, for $10,000.00, the Flying Dutchman claim of 7 acres, on Tenderfoot hill, crossing the Magpie claim of this company, as shown on plat.

Property

The Magpie has two shafts, one about 75 feet deep and one about 80 feet deep; the Black Jack has one 125-foot shaft; Poverty has a 75-foot shaft; Oro has a 100-foot shaft.

Development

Highest price for stock during 1899, 3 1-2 cents; lowest price for stock during 1899, 1 1-4 cents.

The Anna May Gold Mining Company.

Incorporated December 20, 1895.

Directors

D. N. Heizer..............President
A. Freeman..........Vice-President
C. E. Heizer..............Secretary
Dr. T. G. Horn...........Treasurer

Main Office—No. 16 N. Nevada avenue, Colorado Springs, Colorado.

Capitalization

1,000,000 shares. Par value, $1.00.

In treasury January 1, 1900, 52,626 shares.

Property

Owns Anna May, containing 5.105 acres, situated in E. 1-2 of section 30, on Raven hill; patented; adjoining the Elkton and Gould properties.

Development

There is a lease on the Anna May for two years on graded royalties; it has a shaft 165 feet deep; a depth of 300 feet is probably necessary to catch the Elkton, Walter vein.

No shipments have been made.

Highest price for stock during 1899, 6 cents; lowest price for stock during 1899, $7.50 per 1,000.

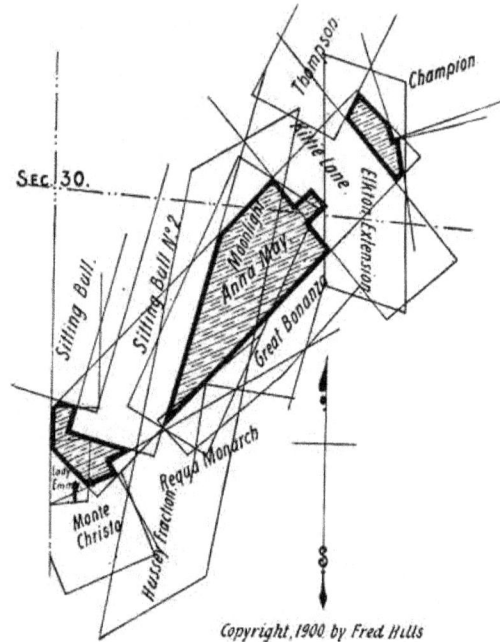

Copyright, 1900 by Fred Hills

The Annie Gold Mining Company.

Incorporated.

Directors

Charles L. Tutt.....President

Charles S. Hebard...Vice-Pres

Spencer Penrose.........

...Secretary and Treasurer

Main Office—Room 11 El Paso bldg., Colorado Springs, Colorado.

Capitalization

500,000 shares. Par value, $1.00.

In treasury January 1, 1900, $2,500 cash.

Copyright, 1900 by Fred Hills

Property

Owns the Annie, in the N. W. 1-4 section 19, containing 2 acres, on Gold hill. Patented.

Development

Two shafts have been sunk, one of 200 feet and one of 100 feet. Stock is closely held and has not been placed on the market.

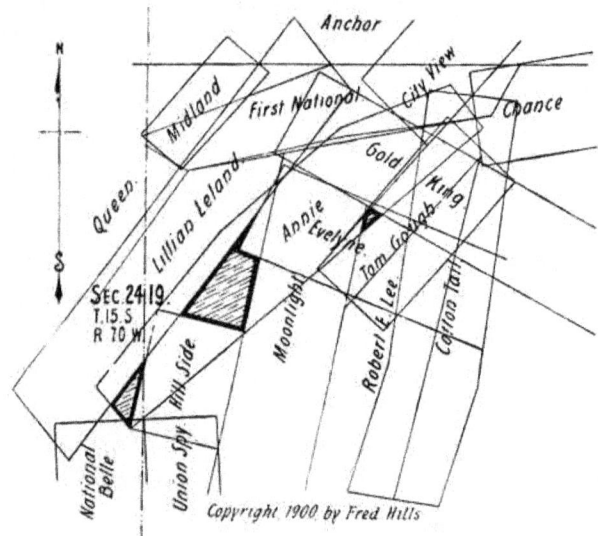

The Antelope Gold Mining Company.

Incorporated August, 1899.

Clarence Edsall.........................President   Directors
John J. Key...........................Vice-President
B. N. Beal..................Secretary and Treasurer
   W. E. Jones.   John H. Hobbs.

Main Office—Hagerman building, Colorado Springs, Colorado.

1,250,000 shares. Par value, $1.00.   Capitalization

Shares in treasury January 1, 1900, 150,000; cash in treasury January 1, 1900, about $1,000.00.

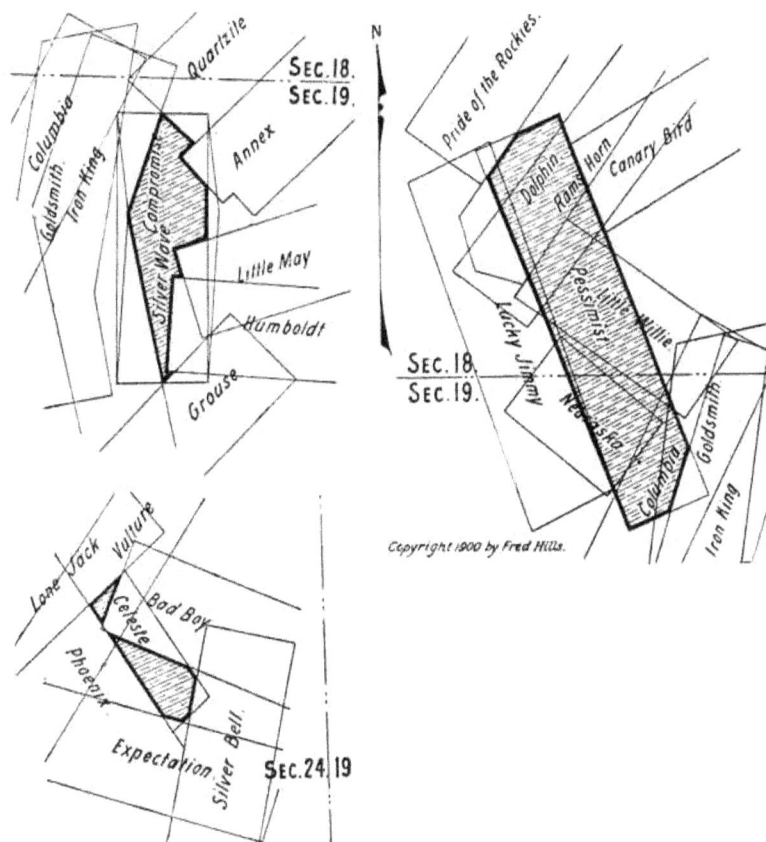

Copyright 1900 by Fred Hills.

Owns Pessimist, Compromise and Celeste, containing 12 1-2 acres Property in all, which are patented; Pessimist, 9 acres, is located in section 18 on Globe hill; Compromise, 2 acres, on Ironclad, in section 19; Celeste, 1 1-2 acres, section 24, on Gold hill.

There is an electric hoist on the Pessimist claim; one shaft down Development 150 feet, from which there are 150 feet of drifts; one shaft 100 feet and one 80 feet. On the Compromise there is a shaft 100 feet deep, with 100 feet of cross-cutting.

This is a new company, with good prospects for mineral on account of its proximity to paying mines.

Highest price for stock during 1899, 4 1-2 cents; lowest price for stock during 1899, 2 1-2 cents.

The Antlers Gold Mining Company.

Incorporated 1896.

Directors

B. B. Grover..........President
W. K. Sinton.....Vice-President
R. B. Taylor....Sec. and Treas.
Robert Gale.   J. K. Brunner.

Main Office—No. 25 Midland block, Colorado Springs, Colorado.

Capitalization

1,500,000 shares. Par value, $1.00.

In treasury January 1, 1900, 500,000 shares.

Property

Owns the Rufus, containing 10.096 acres, situated in the S. W. 1-4 section 6, township 15 south, range 69 west, on Tenderfoot hill; also the Leonard and the Cumberland claims, containing about 4 acres, situated in the N. E. 1-4 section 24. The Cumberland and the Rufus are patented. The Leonard is in process of patenting.

Development

On the Rufus there is a tunnel from 60 to 75 feet long. The Leonard has a shaft about 40 feet deep, and also some drifting. These two claims are receiving the greater part of the development work.

Highest price for stock during 1899, 3 cents; lowest price for stock during 1899, $2.00 per M.

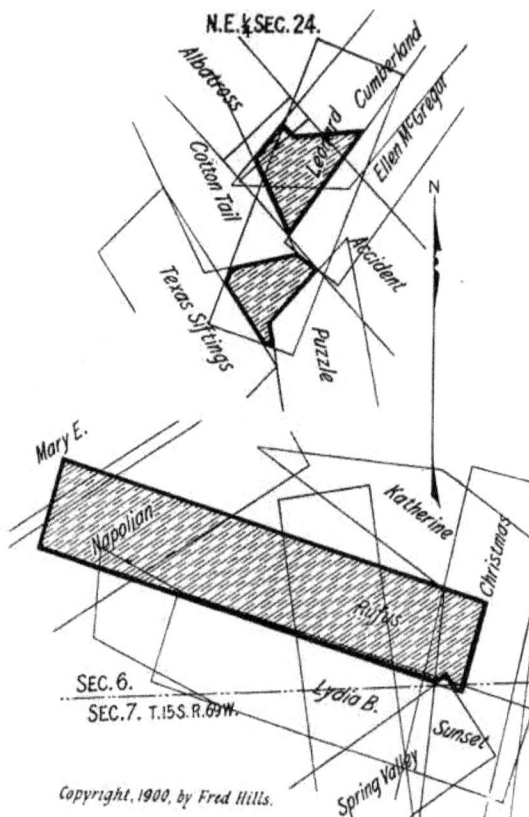

The Aola Gold Mining Company.

Incorporated 1893.

Directors

D. Weyand................ ....President
H. V. Wandell..Secretary and Treasurer
F. W. Laurie.   W. O. McFarlan.

Main Office—No. 112 East Pike's Peak avenue, Colorado Springs, Colo.

Capitalization

1,000,000 shares. Par value, $1.00.

In treasury January 1, 1900, 15,000 shares; in treasury January 1, 1900, $225.00 cash.

Property

Owns the Gold Cup and Something Good, 6 acres in all, situated in the W. 1-2 of section 19, on the north slope of Raven hill. All patented.

Development

There is a small shaft house and a horse whim, owned by the lessees. A shaft has been sunk 240 feet. Greater part of development work has been done on Gold Cup.

Highest price for stock during 1899, 7 1-4 cents; lowest price for stock during 1899, 4 cents.

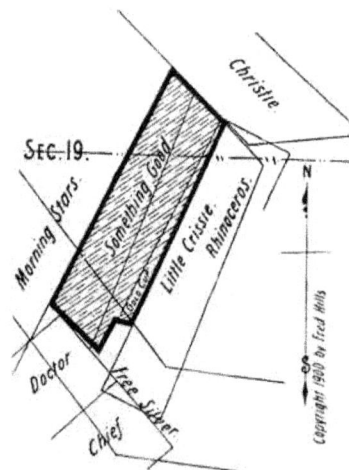

The Apothecaries Gold Mining Company.

Incorporated December 14, 1895.

S. Ben Smith, president; J. D. Allaire, vice-president; Horace Granfield, secretary; Adolph Fehringer, treasurer; J. K. Miller, B. B. Grover, M. D., P. A. Primeau. Directors

Main Office—Colorado Springs, Colorado.

Transfer Offices—Room No. 1, B. & M. block, Denver, Colorado; suite A and B, St. Paul building, New York City.

2,250,000 shares. Par value, $1.00. Capitalization

In treasury January 1, 1900, about 100,000 shares.

Owns the Chicago, 3.562 acres, in the S. 1-2 of Property section 2, on Red mountain; the Silent Friend, Hidden Treasure, Monarch and Bonanza King, containing 25.942 acres, in the S. 1-2 of section 2, on Red mountain; the Bonita, 7.771 acres, in the S. 1-2 of section 2 and N. 1-2 of section 11; the Grand Republic, Frenchman and Don Pedro, containing 9.305 acres, in sections 11 and 12, on the north slope of Mineral hill and on Spring creek; the Mammoth, 3.162 acres, in the N. W. 1-4 of section 12, on the north slope of Mineral hill; and the Ecce Oro, containing about 7 acres, in the N. W. 1-4 section 12 and S. W. 1-4 section 1. The total holdings of this company comprise 58 acres. The Chicago is patented. Receiver's receipts are held for the following: Silent Friend, Hidden Treasure, Monarch and Mammoth. All the others are in process of patenting.

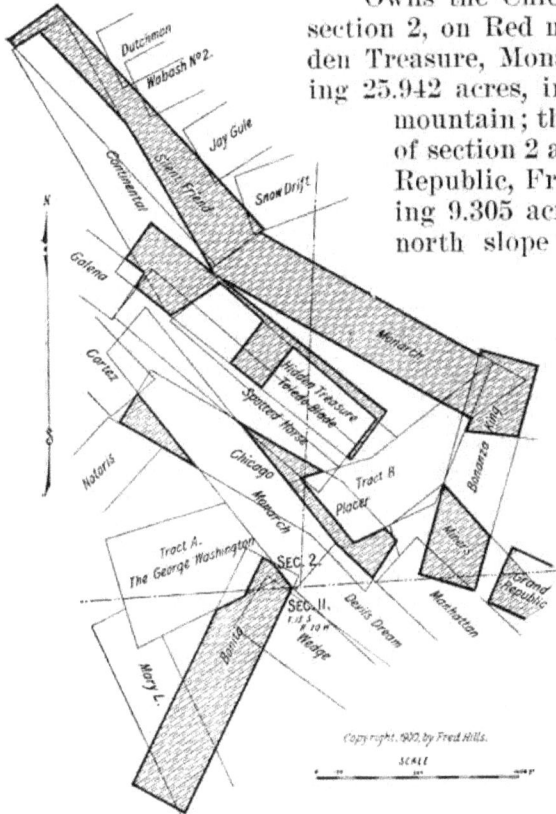

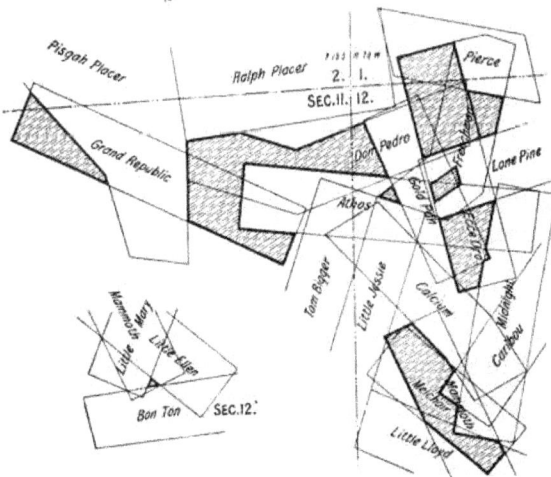

The Grand Republic has a Development cross-cut tunnel 155 feet in length. Values assay from $12 to $30. The Chicago has a 50-foot tunnel. Assays from $10 to $22. Monarch has a shaft of 68 feet and drift of 42 feet. Assays average from $20 to $25. The Silent Friend has two shafts, one of 35 feet, one of 50 feet. Don Pedro has a 30-foot shaft; also trenches. On the Hidden Treasure is a well-timbered shaft, about 70 feet deep. Bonita has a shaft 50 feet deep; Frenchman a shaft of 60 feet depth. Assays range from $22 to $35. Mammoth has a 50-foot shaft. Assays from $8 to $30. On the numerous claims there are also various other prospecting shafts. Already between $15,000 and $20,000 have been expended by the company on these properties. A few small shipments have been made by the lessees. Stock not listed.

The Arapahoe Gold Mining Company.

Incorporated January 16, 1896.

Directors

J. A. Valentine.........................President
G. E. Page.............................Secretary
Edw. Rollandet.........................Treasurer

Main Office—Denver, Colorado.

Capitalization

1,500,000 shares.   Par value, $1.00.

In treasury January 1, 1900, 500,000 shares.

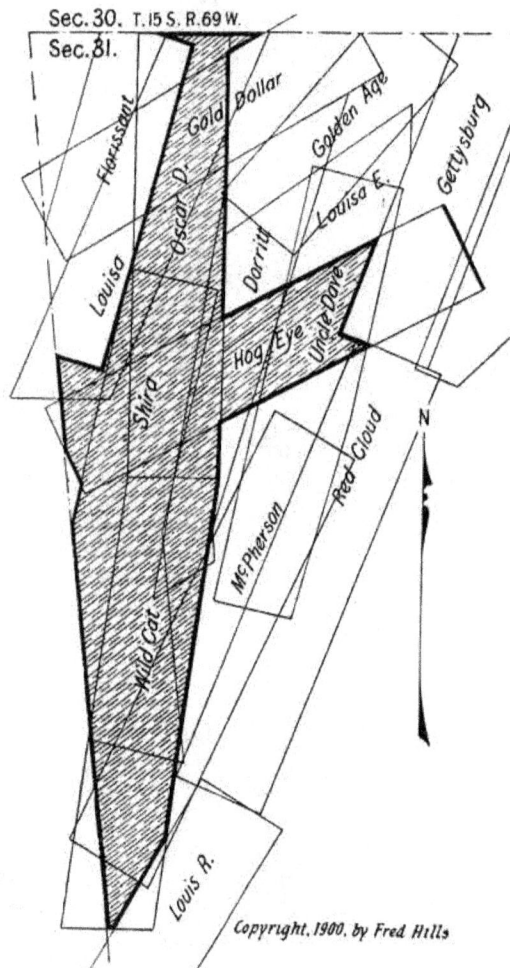

Copyright. 1900. by Fred Hills

Property

Owns the Hog Eye, Oscar D., Shira, Wild Cat and Louisa, 24.378
acres, in the N. W. 1-4 section 31, on Grouse mountain, 20 acres of which
are patented, and the balance is in process of patenting.

Development

Shaft house, with several shafts for prospecting purposes running
from 50 to 150 feet deep. Assays of $35 and upwards have been obtained.

62

The Arequa Gold Mining Company.

Incorporated January 3, 1896.

Copyright, 1900, by Fred Hills.

Wm. A. Otis.................President
E. W. Giddings, Jr........Vice-President
C. S. Wilson......Secretary and Treasurer
C. E. Titus...Asst. Secretary and Treasurer
J. C. Connor.   Geo. E. Lindley.

Main Office—Wm. A. Otis & Co., Giddings building, Colorado Springs, Colorado.

1,250,000 shares.   Par value, $1.00.

Capitalization

In treasury January 1, 1900, 265,195 shares.

Property

Owns lots 1-13, block No. 6; lots 7-10, 38-41, in block No. 3; lots 3-10, 13-39, in block No. 4; also the Homestead claim, in the N. E. 1-4 section 31. The company also own vein rights on block No. 5. All the property, an estimated 9 1-2 acres, is patented.

Development

One shaft 67 feet deep, with 50 feet of drifts, on block 6, on the same dyke as the Mabel M. Shaft 27 feet deep on block 4, on a large dyke, running northwest towards the Chicken Hawk. Shaft house on this claim. These two dykes intersect near the contact, or granite rim. Lessees are contemplating spending $10,000 in development work on the crossing of the two dykes mentioned and have incorporated a leasing company to begin this work at once, under the name of the Big Five Leasing Company.

History

The above lots were purchased at different times from Messrs. Bennett & Meyers, and the Homestead lode was located by Mr. Leavenworth. The dyke lode, or vein, running northwest on Guyot hill, is over 100 feet in width; and the other, running northeast from Beacon hill, extends in width from the Homestead west to the St. Thomas lode, crossing both blocks No. 5 and No. 6.

The Arcadia Consolidated Mining Company.

Incorporated.

Property

Owns a part of the Lone Star No. 1 and a part of the Abe Lincoln, containing about 5 acres, in the S. E. 1-4 section 13. This property is owned by Mr. W. S. Stratton, of Colorado Springs, and map of same can be seen by referring to "Stratton's group" under "S."

The Argon Gold Mining and Milling Company.

Incorporated January, 1896.

Directors

Chas. E. Cherrington........President
Jno. F. Bishop........Vice-President
Wm. Barber......Sec. and Treasurer
Augustine Fromm.   W. K. Dudley.

Main Office—235 N. Union avenue, Pueblo, Colorado.

Capitalization

1,200,000 shares. Par value, $1.00.
In treasury January 1, 1900, 270,000 shares.

Property

Owns the Cripple and Little Florence claims, situated in the N. W. 1-4 of section 12, containing 7 acres, on Mineral hill. Receiver's receipt for both.

Development

Deepest shaft, 65 feet.
Highest price for stock during 1899. 1 cent per share; lowest price for stock during 1899, $3.00 per thousand.

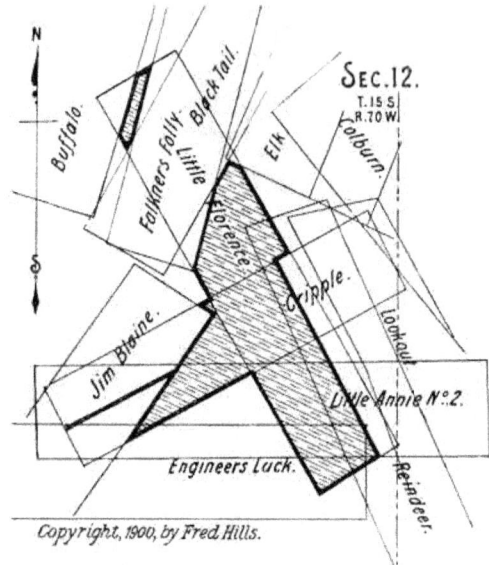

Copyright, 1900, by Fred. Hills.

The Arrow Gold Mining Company.

Incorporated July 20, 1899.

Directors

Chas. Farnsworth......President
Franklin E. Brooks....Vice-Pres
Amos S. Anderson.....Secretary
R. P. Davie..........Treasurer
    H. L. Shepherd.

Main Office—50 Bank building, Colorado Springs. Transfer Office —International Trust Company, Colorado Springs.

Capitalization

1,250,000 shares. Par value, $1.00.

In treasury January 1, 1900, 150,000 shares.

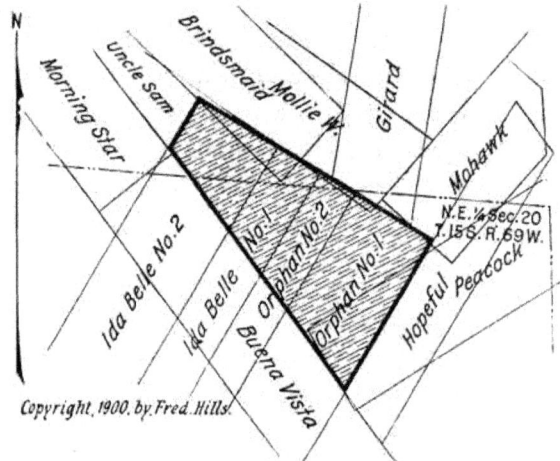

Copyright, 1900, by Fred. Hills.

Property

Owns a portion of the Ida Bell Nos. 1 and 2; also a portion of the Orphan Nos. 1 and 2, containing 7.6 acres. Patented. These claims adjoin the Buena Vista lode of the Isabella company, in the N. E. 1-4 section 20, on Bull Hill.

The property is being developed by lessees. On the property is a hoist, horizontal boiler and air compressor. Between July 1, 1899, and January 1, 1900, $15,000.00 has been expended in development work. The present outlook seems very favorable. Prior to the organization of this company this property, together with the property of the Orphan Gold Mining Company, belonged to the Orphan Bell M. & M. Company. The latter shipped considerable ore under lease.

Highest price for stock during 1899, 9 1-4 cents; lowest price for stock during 1899, 7 1-2 cents.

The Arvilla Tunnel and Mining Company.

Incorporated October, 1892.

W. F. Kendrick.........................President
C. D. Wood.........................Vice-President
C. B. Lowther.........................Secretary
P. M. Kendrick.........................Treasurer

M. B. Carpenter.	Earl B. Coe.
James A. McClurg.	D. L. Webb.
W. R. Owen.	M. J. McNamara.

Directors

Main Office—No. 720 Mining Exchange, Denver, Colorado.

1,250,000 shares.   Par value, $1.00.   Capitalization

Present indebtedness of the company, about $10,000.00.

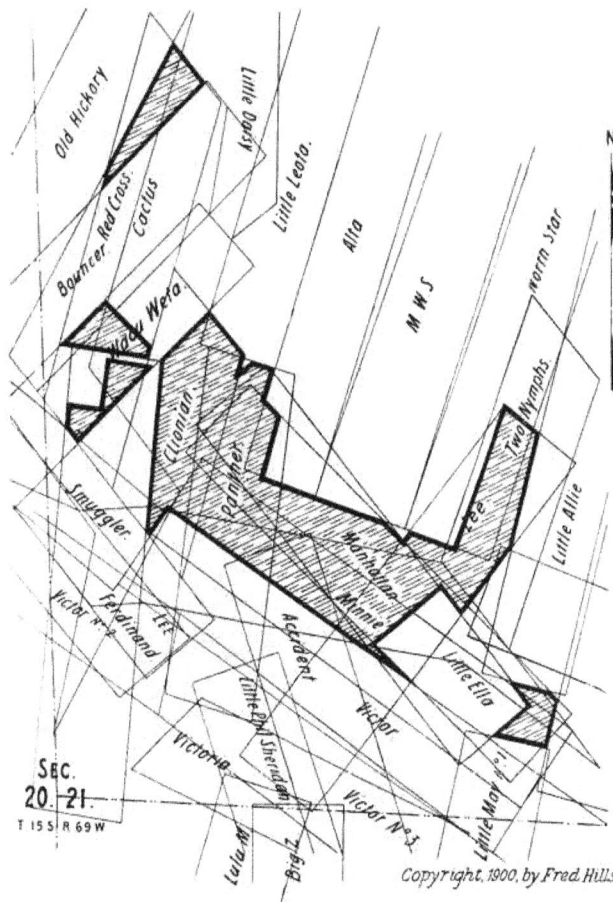

Copyright, 1900, by Fred Hills.

Owns the Little Ella, Minnie, Lee, Cactus and the Clionian, containing in all 26 acres, in the N. W. 1-4 section 21.   Survey No. 8,407. All patented, and adjoining the properties of the Victor Gold Mining Company and the Isabella Gold Mining Company.   Property

The company have under consideration a lease and bond of $750,000 on this property.

Highest price for stock during 1899, 20 cents; lowest price for stock during 1899, 12 cents.

The Astor Gold Mining Company.

Incorporated.

Directors

F. H. Arcularius.........................President

H. S. Sommers........................Vice-President

C. P. Bently......................Assistant Secretary

L. C. Hall...........................Treasurer

F. P. Buck.   Chas. Neuer.

Main Office—No. 53 First National Bank building, Colorado Springs, Colorado.

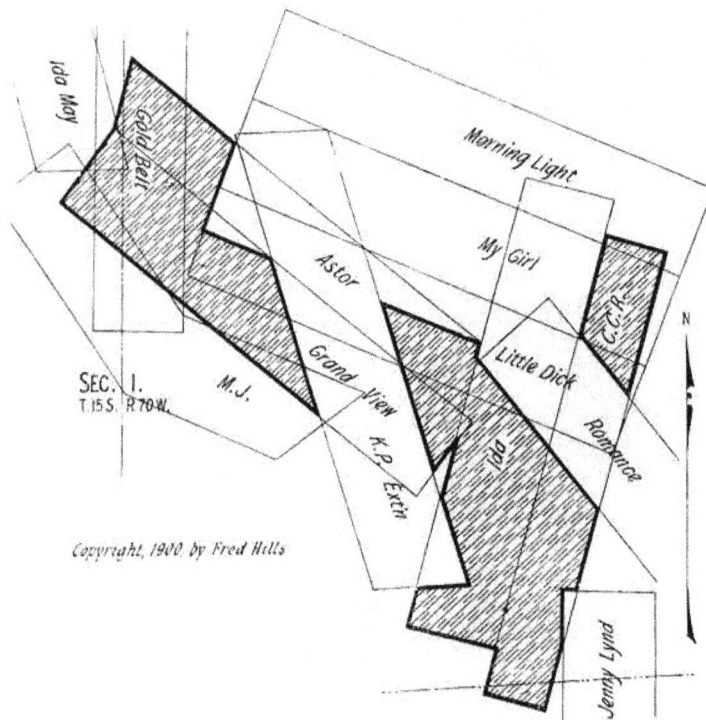

Copyright, 1900, by Fred Hills

Capitalization   1,500,000 shares.   Par value, $1.00.

In treasury January 1, 1900, 140,000 shares.

Property   Owns the Astor, Grand View, Ida and the C. C. R., all in a group on the southeast slope of Copper mountain, in the S. E. 1-4 section 1, containing about 16 acres. Receiver's receipt held for the Astor, Grand View and the Ida. C. C. R. in process of patenting.

Development   Sufficient work has been done on each of these claims to obtain a patent.

The Atlanta, Cripple Creek and Creede Mining Company.

Reorganized December 30. 1895.

James F. Smith......................President
H. L. Shepherd....................Vice-President
Albert Wagner.......................Secretary
W. B. Pullen........................Treasurer

T. J. Dalzell.   H. A. Pettit.
W. Gus Smith.

Directors

Main Office—No. 367 Bennett avenue, Cripple Creek, Colorado.

1,000,000 shares.   Par value, $1.00.

Capitalization

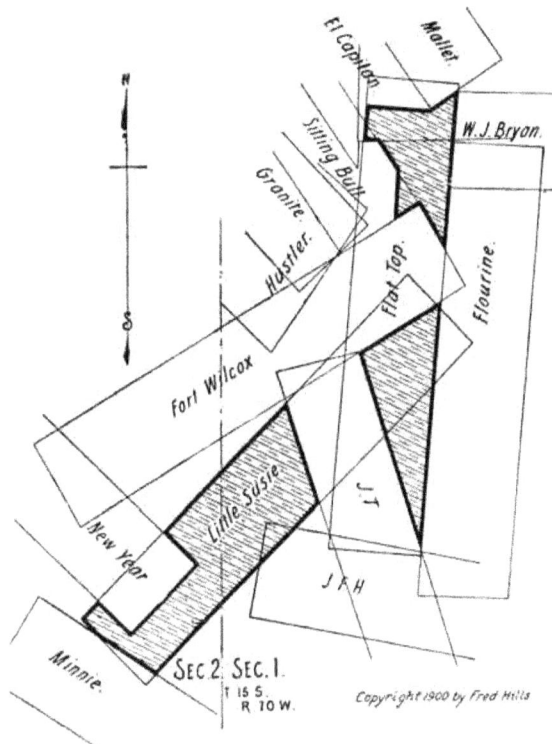

Copyright 1900 by Fred Hills

Owns the Flat Top and the Little Susie, containing 11 acres, situated in the S. W. 1-4 section 1, on Copper mountain. Both claims are patented.

Property

Plant of machinery; air drills. There is a shaft 80 feet deep on the Little Susie; also three other shafts from 20 to 30 feet deep and some trenching. The greater part of the development work is being done on the Flat Top, which is leased to November, 1902. This claim has a shaft 80 feet deep and 300 feet of cross-cuts and drifting. It has also a tunnel 400 feet in length. Work on this tunnel is progressing at the rate of 125 feet per month. This is being driven from the south end of the claim to the north. The lease compels the lessees to continue driving the tunnel during life of lease, until the north end line is reached.

Development

The property has been worked by various lessees. Good assays have been obtained. This property side-lines with the Fluorine of the Montreal Company, which has produced $150,000. There is no indebtedness.

History

Highest price for stock during 1899, 5 3-4 cents; lowest price for stock during 1899, $4.00 per M.

The Atlantic and Pacific Gold Mining and Milling Company.

Incorporated November 1, 1895.

Directors

H. G. Laing..............................President
Frank Cotten..............Secretary and Treasurer
Walter Burlew.   W. J. Rice.
E. D. Lowe.

Main Office—No. 23 1-2 North Tejon street, Colorado Springs, Colorado.

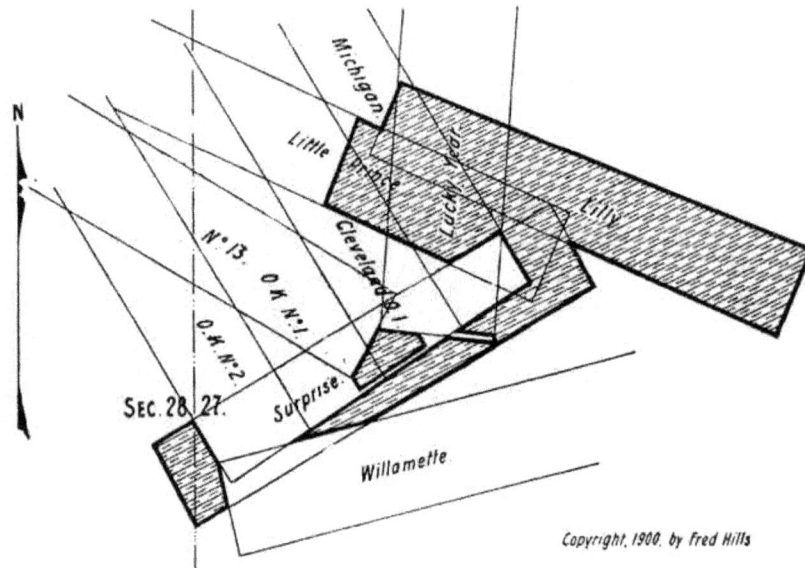

Copyright, 1900, by Fred Hills

Capitalization

1,000,000 shares.   Par value, $1.00.

Property

Owns Lilly, Little Prince and Surprise, located in the S. W. 1-4 section 27, Big Bull hill.   About 16 acres in all.   Lilly claim, containing about 10 acres, is patented, and receiver's receipt is held for Surprise and Little Prince claims.

Development

This property is being developed mostly on the Lilly claim by lessees, and driving is being continued on a tunnel, which is about 200 feet long.

No shipment has been made.

Highest price for stock during 1899, $15.00 per 1,000 shares; lowest price for stock during 1899, $3.50 per 1,000 shares.

The Atlantis Mines Corporation.

Incorporated 1899.

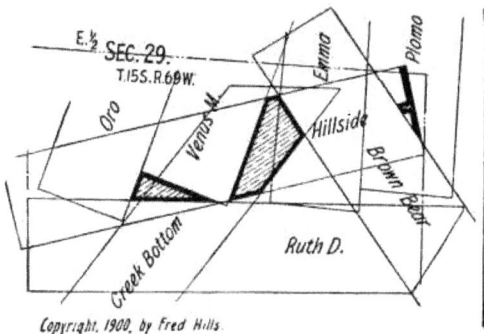

Copyright. 1900. by Fred Hills.

Directors

A. B. Heath............President
E. F. Conant........Vice-President
C. F. Potter....Sec'y and Treasurer

Main Office—No. 619 Ernest & Cranmer building, Denver, Colorado.

Capitalization

1,250,000 shares. Par value, $1.00.

In treasury April 1, 1900, 250,000 shares.

Property

Owns the Hillside, 2 acres, in the S. E. 1-4 section 29, in the town of Goldfield. The company also holds lease and bond for $100,000 expiring October 1, 1901, on the Santa Rita, from which about $70,000 worth of ore had been shipped prior to the lease.

Development

The company has expended $20,000 on the Santa Rita. A main working shaft has been sunk to a depth of 500 feet, where a large and valuable body of ore has been opened up, from which shipments are being made. The officers of the company state that the bond will be taken up.

No stock has ever been placed on the market, an offer being made and refused on April 1 last for 20,000 shares of treasury stock at 40 cents.

The Atlas Consolidated Gold Mining Company.

Incorporated.

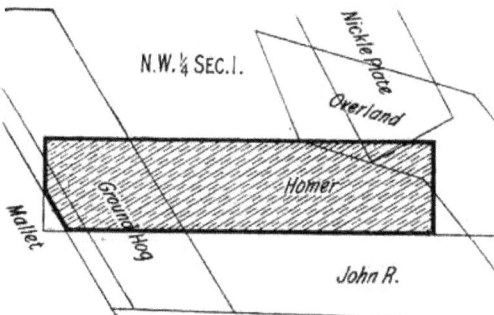

Copyright, 1900, by Fred Hills.

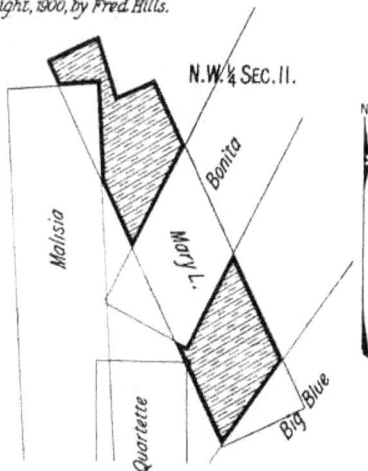

Directors

D. N. Heizer...........President
C. E. Heizer............Secretary

Main Office—No. 16 North Nevada avenue, Colorado Springs, Colorado.

Capitalization

1,250,000 shares. Par value, $1.00.

In treasury January 1, 1900, 1,500 shares.

Property

Owns the Homer, 8.761 acres, in the N. W. 1-4 section 1, on Copper mountain; and the Mary L., 5.214 acres, in the N. 1-2 section 11, on Red mountain. The Homer is patented. Receiver's receipt held for the Mary L.

Development

The Homer is leased on graded royalties. Sub-leases are also in operation. The greater part of the development work has been done on this claim. Lessees on the Homer claim have opened up a vein of ore which assays from $8.20 to $136 a ton.

Highest price for stock during 1899, 1 3-4 cents; lowest price for stock during 1899, $2.00 per M.

The Avalon Gold Mining Company.

Incorporated December 16, 1895.

Directors D. N. Heizer, president; J. C. Helm, vice-president; C. E. Heizer, secretary; E. J. Eaton, treasurer.

Main Office—No. 16 North Nevada avenue, Colorado Springs, Colorado.

Capitalization 2,000,000 shares. Par value, $1.00.

In treasury January 1, 1900, 200,000 shares.

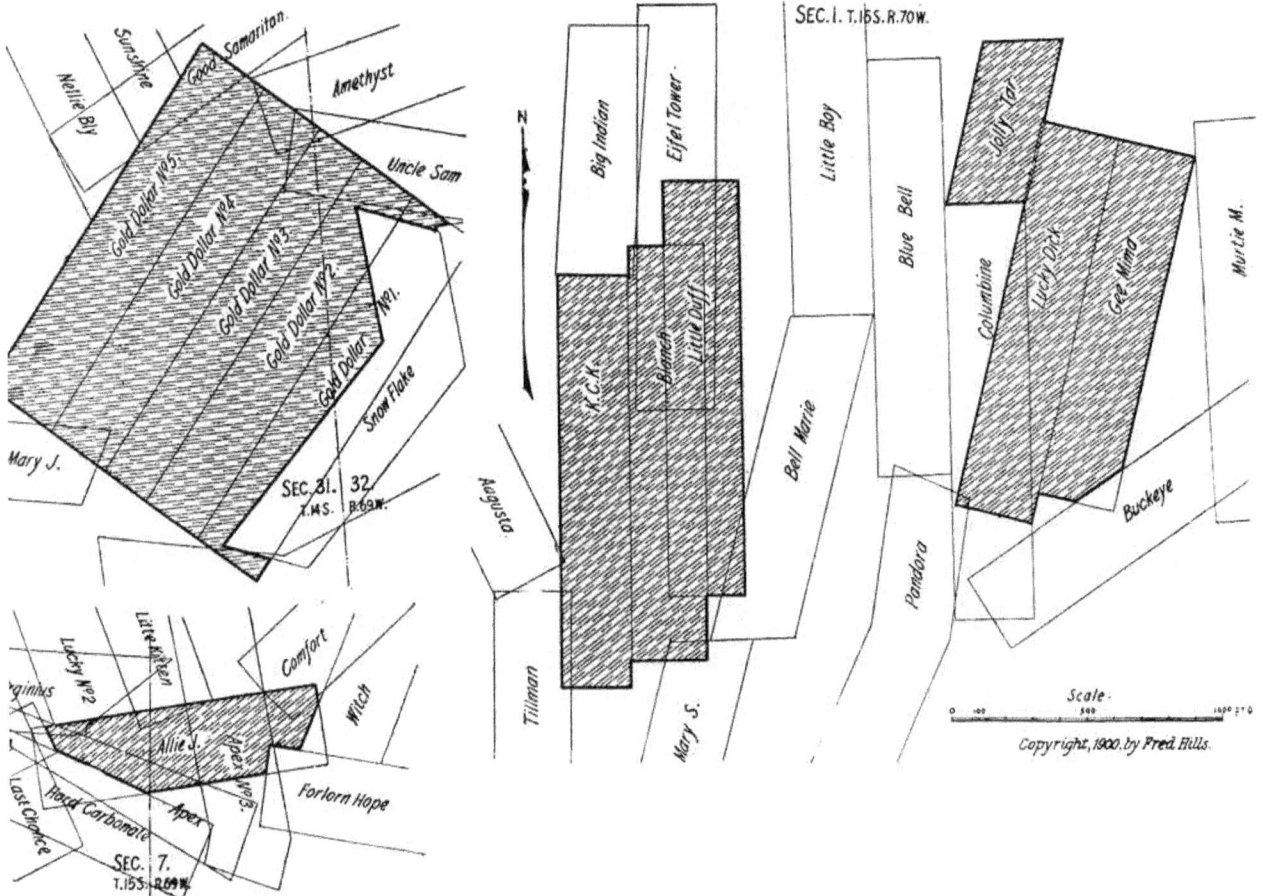

Copyright, 1900, by Fred Hills.

Property Owns the Allie J. claim, in the S. 1-2 of section 7, containing 6.031 acres, on Tenderfoot hill, adjoining the Forlorn Hope claim of the Chicolo company, and the Hard Carbonate claim of the Hard Carbonate company; the K. C. K., Blanche and Little Duff claims, in the E. 1-2 of section 1, township 16 south, range 70 west, containing in all 24.907 acres, on the northwest slope of Straub mountain; the Lucky Dick, Geemina and the Jolly Tar claims, in the E. 1-2 of section 1, township 16 south, range 70 west, containing 22.23 acres, on Straub mountain; and the Gold Dollar Nos. 1, 2, 3, 4 and 5, in the S. E. 1-4 of section 31, township 14 south, range 69 west, containing in all 37.301 acres, on the N. E. slope of Rhyolite mountain. The company owns a total acreage of 90.469 acres. All patented.

Development The Lucky Dick is leased for three years on a graded royalty.

The stock is listed on the Board of Brokers' Association of Colorado Springs, but is largely held as an investment on the growth of the property.

The Autumn Bell Gold Mining Company.

Incorporated January, 1895.

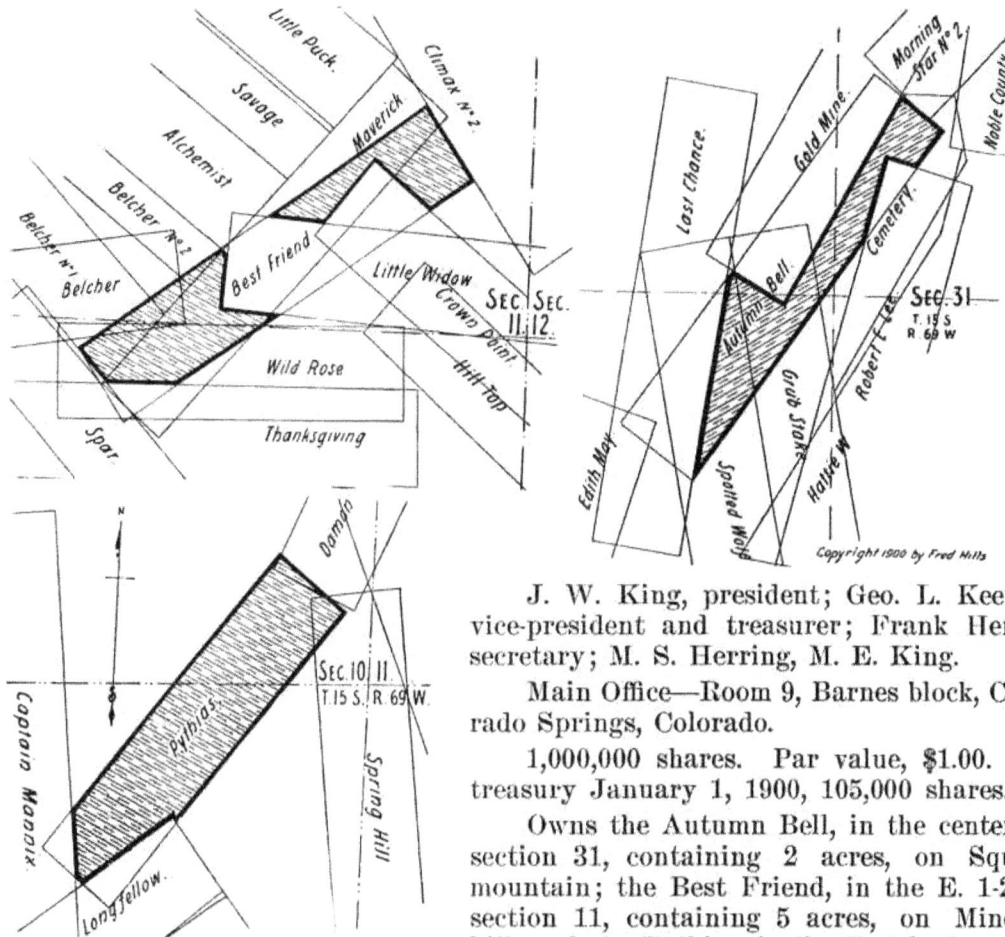

Copyright 1900 by Fred Hills

J. W. King, president; Geo. L. Keener, **Directors** vice-president and treasurer; Frank Heron, secretary; M. S. Herring, M. E. King.

Main Office—Room 9, Barnes block, Colorado Springs, Colorado.

1,000,000 shares. Par value, $1.00. In **Capitalization** treasury January 1, 1900, 105,000 shares.

Owns the Autumn Bell, in the center of **Property** section 31, containing 2 acres, on Squaw mountain; the Best Friend, in the E. 1-2 of section 11, containing 5 acres, on Mineral hill, and the Pythias, in the E. 1-2 of section 10, containing 7 1-2 acres, on Cow mountain. The Pythias is patented; the other two in process.

The Autumn Bell, on which the greater part of the development **Development** work has been done, has a 60-foot shaft. On the Best Friend and the Pythias simply enough work has been done to patent.

Highest price for stock during 1899, 1 cent; lowest price for stock during 1899, $2.50 per M.

The Avondale Gold Mining Company.

Incorporated January 12, 1896.

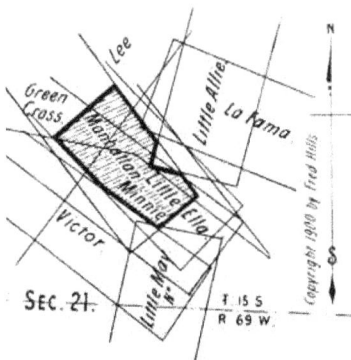

M. A. Leddy, vice-president; W. H. Clotworthy, **Directors** secretary; H. H. Grafton, treasurer; J. Fischal.

Main Office—Strang's Cigar Store, 119 N. Tejon street, Colorado Springs, Colorado.

1,250,000 shares. Par value, $1.00. In treasury **Capitalization** January 1, 1900, 125,000 shares. In treasury January 1, 1900, $1,000.00 cash.

Owns the Manhattan, containing 2 1-2 acres, in **Property** the N. W. 1-4 of section 21, on Bull hill, adjoining the Victor mine. Patented. The property is bonded and leased for two years from February 15, 1900.

A shaft has been sunk 90 feet deep and some drifting has been prose- **Development** cuted. Highest price for stock during 1899, 2 1-2 cents; lowest price for stock during 1899, $8.00 per M.

The Bankers Gold Mining and Milling Company.

Incorporated 1892.

Directors

W. H. Chittenden........................President

J. A. Swarthout....................Vice-President

H. S. Morgan..........................Secretary

H. K. Chittenden......................Treasurer

G. G. Newcomb.   Albert Smith.

G. J. Chittenden.

Main Office—Ernest & Cranmer building, Denver, Colorado.

Capitalization

1,250,000 shares. Par value, $1.00.

In treasury January 1, 1900, shares, none; in treasury January 1, 1900, cash, none.

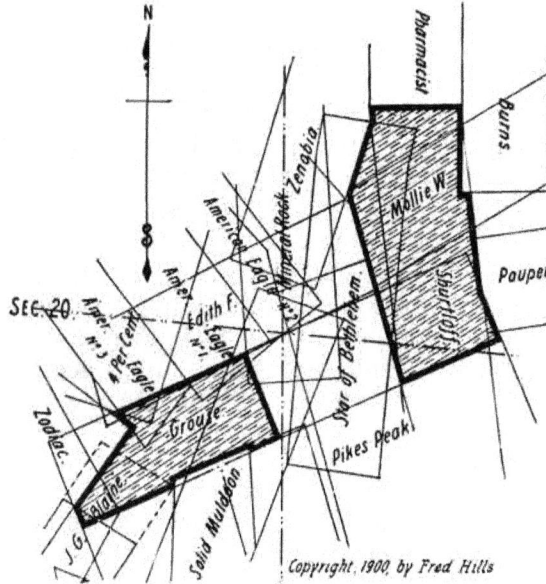

Copyright, 1900, by Fred Hills

Property

Owns the Grouse, Star of Bethlehem, Shurtloff, Mollie W. and New Discovery, about 11 acres in all, in the center of section 20, on Bull hill. Of the Mollie W., .061 of an acre is patented. The rest is held by location and patent will not be issued until pending litigation is settled in favor of the company. There has been considerable talk of consolidation between the Bankers and Garfield companies, but nothing definite has been given out.

Development

Shaft house, ore bin and engine room; 6x8 hoist; 25-horse power boiler; 500 feet of shaft on the Grouse, with about 2,500 feet of drifting and cross-cutting. Considerable ore was produced from this property some time ago.

Highest price for stock during 1899, 15 1-2 cents; lowest price for stock during 1899, 7 1-2 cents.

The Banner Gold Mining Company.

Incorporated 1895.

L. E. Sherman, president; J. R. McKinnie, vice-president; E. C. Directors Sharer, secretary; L. L. Aitken, treasurer; J. M. Auld.

Main Office—No. 25 East Pike's Peak avenue, Colorado Springs, Colorado.

1,500,000 shares. Par value, $1.00.

In treasury January 1, 1900, 48,925 shares; in treasury January 1, 1900, $12,000.00 cash.

Capitalization

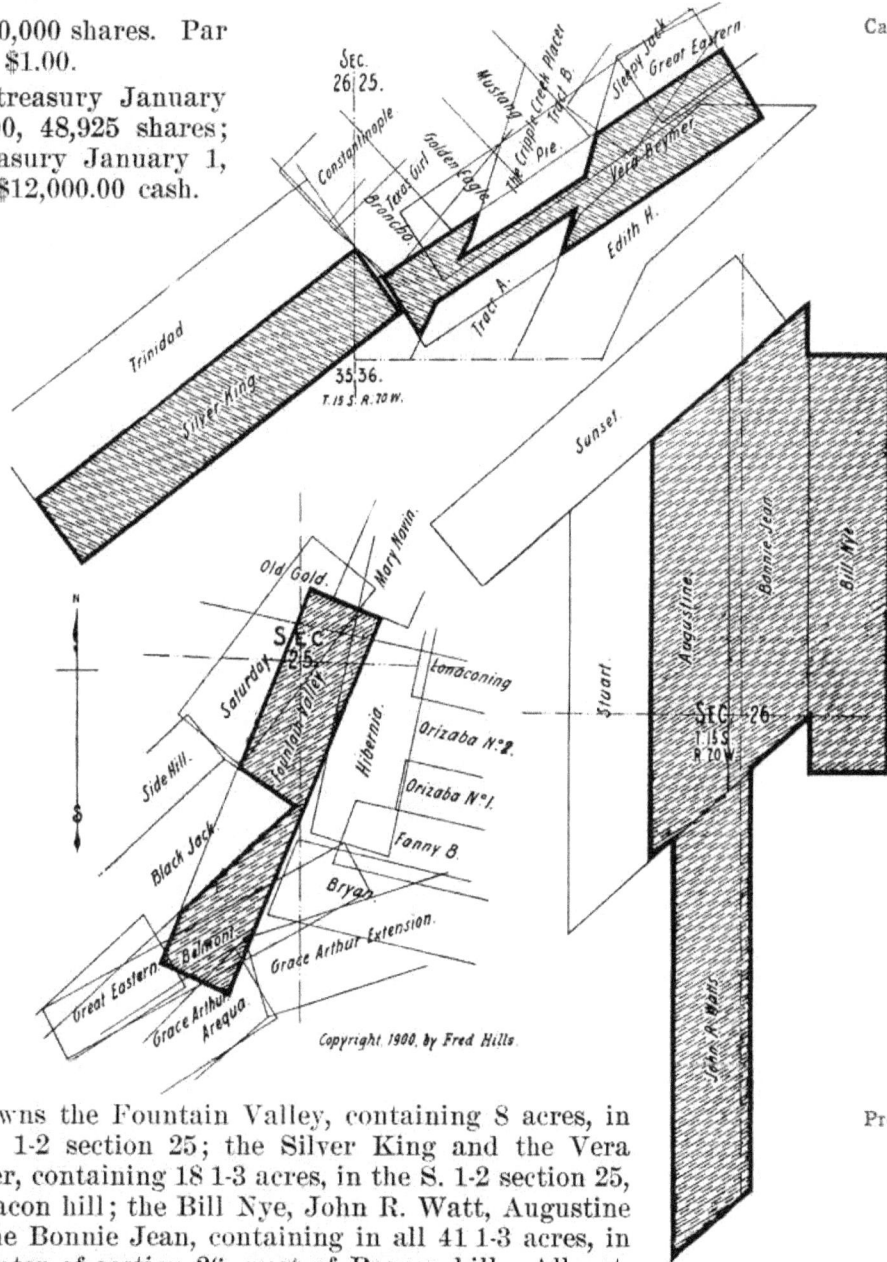

Copyright 1900, by Fred Hills.

Owns the Fountain Valley, containing 8 acres, in the S. 1-2 section 25; the Silver King and the Vera Beymer, containing 18 1-3 acres, in the S. 1-2 section 25, on Beacon hill; the Bill Nye, John R. Watt, Augustine and the Bonnie Jean, containing in all 41 1-3 acres, in the center of section 26, west of Beacon hill. All patented.

Property

Whim, buckets, etc. The Silver King now down 140 feet and cross- Development cutting is in progress. This shaft is sunk jointly by the Banner G. M. Co. and the Texas Girl G. M. Co. for mutual benefit. The shaft on the Fountain Valley lode is now down 110 feet. These two claims are receiving the greater part of the development work. Both the Silver King and the Fountain Valley shafts are timbered.

Highest price for stock during 1899, 5 7-8 cents; lowest price for stock during 1899, 1 3-4 cents.

The Battle Mountain Consolidated Gold Mines Company.

Copyright, 1900, by Fred Hills

The Battle Mountain Consolidated Gold Mines Company.

Incorporated February 14, 1896.

Warren Woods......................President
H. E. Woods......................Vice-President
F. M. Woods..............Secretary and Treasurer
    J. M. Allen.   C. L. Arzeno.

Directors

Main Office—Giddings building, Colorado Springs, Colorado.

Branch Office—Victor, Colorado.

2,500,000 shares.   Par value, $1.00.

Capitalization

In treasury January 1, 1900, 250,000 shares.

Owns the Viola, 1 1-4 acres; Rex, Regina, Duchess and the Big Theatre, 5.6 acres; the Junietta, 6 acres; Hypatia, Conejos and the Eldorado placer, 4.234 acres; and undivided interests in the Justice, Kodak, May and the Babey June; the company also controls the property of the Requa Gold and Silver Mining and Milling Company. The Uinta Tunnel, Mining and Transportation Company has been absorbed by them. They also own the Uinta Tunnel No. 1, 2.5 acres; the Pitkin, 0.95 acre; a three-fourths interest in the Blue Stocking, 4 acres; the Black Jasper, the Scorpion, 4.6 acres; the Trail, the Big Banta, Old Ironsides and Lost Fraction, 27.6 acres, in sections 20, 29, 30 and 32.

Property

On the Trail group there are ten shafts, aggregating 1,200 feet, and four tunnels having a total length of 2,500 feet, together with various stopes, levels, drifts and cross-cuts. There are 2,300 feet of lateral work on the Uinta tunnel, and 900 feet of drifts, stopes, cross-cuts, etc. There are 400 feet of shafts and 300 feet of drifts on the Viola. The Big Theatre has a 375-foot tunnel.

Development

Highest price for stock during 1899, 42 cents. Lowest price for stock during 1899, 22 cents.

Beacon Hill-Ajax Gold Mining Company.
Incorporated October 18, 1899.

Directors

J. R. McKinnie, president; W. P. Wagy, vice-president; R. P. Davie, secretary; J. T. Burkholder, treasurer; J. W. Graham.

Main Office—25 East Pike's Peak avenue, Colorado Springs, Colorado.

Capitalization

1,250,000 shares. Par value, $1.00.

In treasury January 1, 1900, 200,000 shares; in treasury January 1, 1900, over $1,000.

Property

Owns Ajax, 8 acres, situated on the N. W. 1-4 section 30, south slope of Beacon hill, Cripple Creek district. All the property is patented. This property is surrounded by several paying mines. The indications for pay ore are favorable.

Development

The company has sunk a shaft of about 150 feet. Two sets of lessees are at work, each to sink a shaft of at least 15 feet per month. Shaft 4x7.

Highest price paid for stock during 1899, 4 3-4 cents; lowest price paid for stock during 1899, 3 7-8 cents.

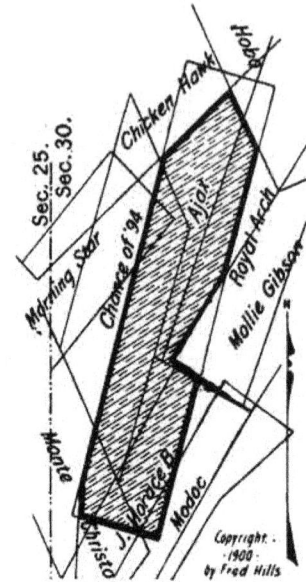

The Beacon Hill Gold Mining Company.
Incorporated November, 1894.

Directors

W. H. Anderson, president; F. W. Ford, vice-president; A. A. Ford, secretary; M. E. Anderson, treasurer; A. Ford.

Main Office—Nos. 1 and 2, Safe Deposit building, Cripple Creek, Colorado.

Capitalization

1,000,000 shares. Par value, $1.00.

In treasury January 1, 1900, $2,000.00 cash.

Property

Owns the Little May lode in the S. E. 1-4 section 25, on the west slope of Beacon hill, in process of patenting. Full claim adversed by the Hiawatha for ground in conflict. Adverse claim still pending.

Development

One shaft has been sunk 250 feet; three other shafts are from 50 to 100 feet deep. A large amount of drifting and cross-cutting has been done; over 1,000 feet of work in all. The main shaft has one plant of steam machinery.

Production

Gross production to January 1, 1900, about $35,000.00. Net profit on ore mined during 1899, about $500.00.

History

The property has been in litigation for three years. The company have won out in all matters in the lower courts, and final hearing in the Supreme Court will be had about April, 1900. The suit grows out of the claim of Burris et al. to try to enforce forfeited contract to purchase at $75,000.00. The stock is not listed, all being held by a few individuals.

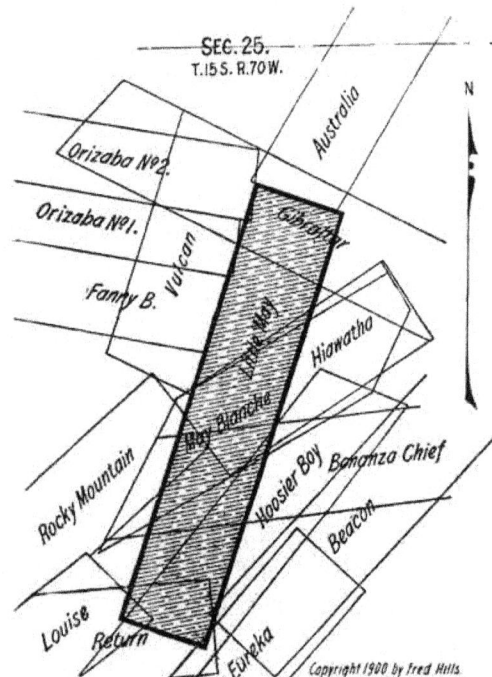

The Beacon Light Gold Mining Company.

Incorporated 1896.

Directors

E. J. Eaton..........................President
A. M. Ripley.....................Vice-President
W. R. Barnes.............Secretary and Treasurer
A. F. Woodward.   E. F. Wright.

Main Office—The El Paso County Abstract Company, No. 113 East Kiowa street, Colorado Springs, Colorado.

Capitalization

1,250,000 shares. Par value, $1.00.

In treasury January 1, 1900, 250,000 shares; in treasury January 1, 1900, $25.00 cash.

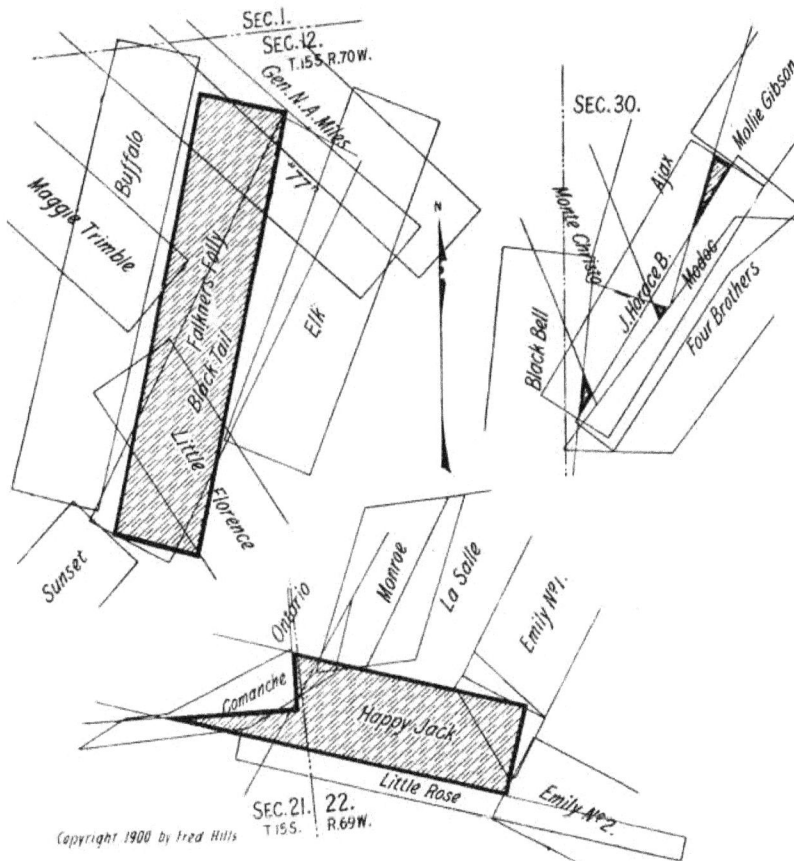

Copyright 1900 by Fred Hills

Owns Falkner's Folly, 9.070 acres, in the N. W. 1-4 section 12, township 15 south, range 70 west; the Happy Jack, 5.678 acres, in the N. W. 1-4 section 22, township 15 south, range 69 west; and the J. Horace B., 0.558 acre, in the N. W. 1-4 section 30. All patented. Total holdings of this company comprise 15.036 acres.

Property

On the J. Horace B. there is a shaft 175 feet deep. The Happy Jack has a shaft of 150 feet depth. Falkner's Folly has one 50 feet deep. At present no work of importance is being done on this company's property.

Development

Highest price for stock during 1899, 1 cent; lowest price for stock during 1899, 1-2 cent.

The Beatrice Gold Mining and Milling Company.

Incorporated 1896.

Directors

M. S. Herring, president; C. J. Watson, vice-president and treasurer; J. T. Sanderson, secretary; D. C. Sanderson. Main Office—No. 1 East Huerfano street, Colorado Springs, Colorado.

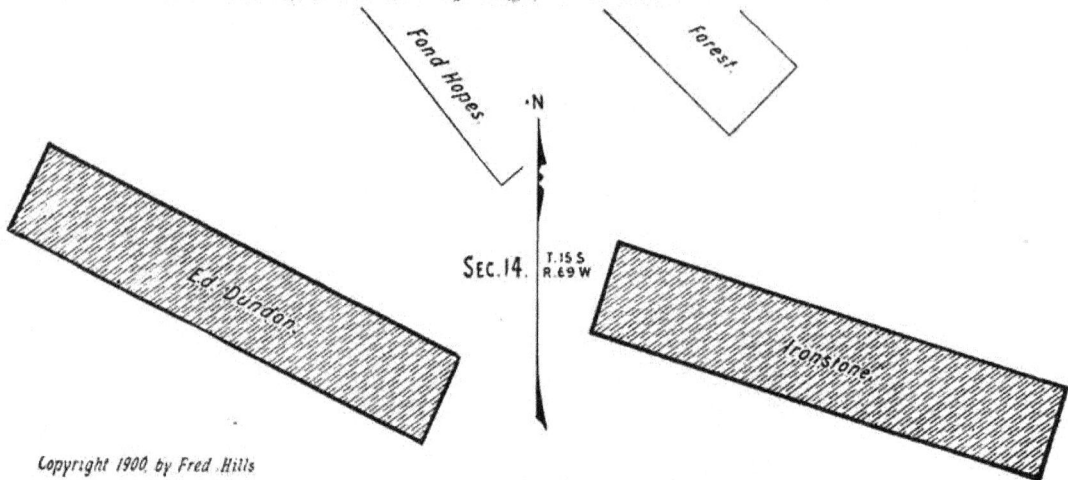

Copyright 1900, by Fred Hills

Capitalization

1,000,000 shares. Par value, $1.00.

In treasury January 1, 1900, 32,000 shares; in treasury January 1, 1900, nominal amount of cash.

Property

Owns the Ed. Dundon and the Ironstone, containing 20 1-2 acres, in the center of section 14, township 15 south, range 69 west, on the south slope of Cow mountain. Patented.

Development

Only patent work has been prosecuted. The prospects for this becoming a shipping mine seem favorable. The greater part of the development work is being done on the Ironstone, which is held by lease.

The stock is closely held.

The Beaver Bell Gold Mining Company.

Incorporated March, 1896.

Directors

G. P. Robinson, president; L. C. Nickerson, vice-president; A. O. Downs, secretary and treasurer; R. R. Latta, W. C. Grafton.

Main Office—129 North Tejon street, Colorado Springs, Colorado.

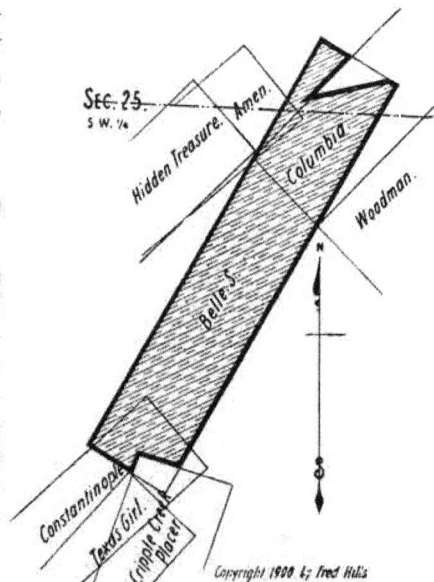

Copyright 1900 by Fred Hills

Capitalization

1,000,000 shares. Par value, $1.00.

In treasury January 1, 1900, 117,000 shares.

Property

Owns the Belle S. claim, 10 acres, in the S. W. 1-4 section 25, on the west slope of Beacon hill, adjoining the Texas Girl Company, and is near Banner Gold Company's properties. About 2 acres of the Columbia claim acquired in settlement with the Merrimac Gold Mining Company, and lying next to the Belle S. claim; 20 acres of mining land, patented, in Woodland Park district, is also owned by the company.

Development

Just sufficient work has been done to patent the claim. The Belle S. was patented by the company; the two acres from the Columbia claim are to be deeded when the owners of said claim secure patent. Highest price for stock during 1899, $6.00 per 1,000; lowest price for stock during 1899, $3.00 per 1,000.

THE TOWN OF GOLDFIELD—1900.

SHOWING BULL HILL AND TOWN OF INDEPENDENCE IN BACKGROUND.

The Bedford Gold Mining Company.
Incorporated 1896.

Directors

James P. Merriden.............President
Chas. M. Thayer..Secretary and Treasurer
Joseph Masse.   Henry A. Rideout.
E. J. Moxley.   F. H. Dunnington.
Frank Baergalupo.

Main Office—No. 7 Exchange Place, Boston, Mass.

Capitalization

1,250,000 shares. Par value, $1.00.

In treasury January 1, 1900, 500,000 shares.

Property

Owns the Minnie, the Minnie No. 2 and the Maggie A., comprising 28 acres, situated in the N. E. 1-4 section 11, township 15 south, range 69 west, on Cow mountain. All the property is patented.

Sec. II.
T. 15 S. R. 69 W.

Development

The company have sunk three shafts of from 30 to 50 feet, also a tunnel of 150 feet. Work on the property is being pushed vigorously, assays running from $6.00 to $54.00 per ton.

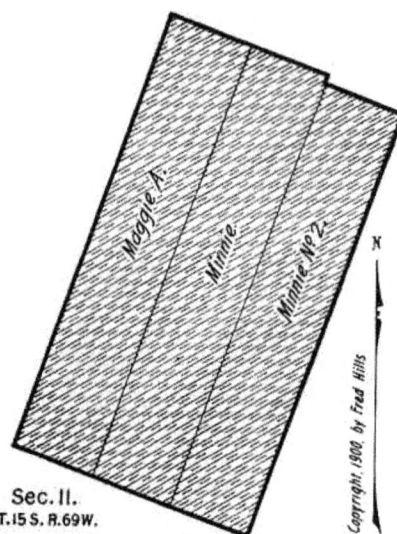

The Ben Hur Mining and Milling Company.
Incorporated June 9, 1892.

Directors

F. H. Pettingell........................President
Theoph. Harrison.....................Vice-President
L. A. Civill..............................Secretary
F. F. Schreiber.......................Treasurer
J. M. Roseberry.   W. O. Wirt.
J. McK. Ferriday.

Main Office—No. 11 Bank building, Colorado Springs, Colorado.

Capitalization

900,000 shares. Par value, $1.00.

In treasury January 1, 1900, about $100.00 cash. Present indebtedness of the company, $500.00.

Property

Owns the Little King and Queen, 10.39 acres, in the N. E. 1-4 section 24, on Gold hill; the Optimus, 3.14 acres, in the N. 1-2 section 19, on Gold hill; the Minnie H. and the Moss Back, in the center of section 18, containing 3.45 acres, on Globe hill; the Bon Ton, 7.289 acres, in the S. E. 1-4 section 31, on Squaw mountain, and the Tejon, 8.28 acres, in the S. E. 1-4 section 31, on Squaw mountain.

Development

The greater part of the development work is being done on the Little King and Queen and the Optimus.

Highest price for stock during 1899, 9 1-2 cents; lowest price for stock during 1899, 3 7-8 cents.

The Ben Hur Mining and Milling Company.

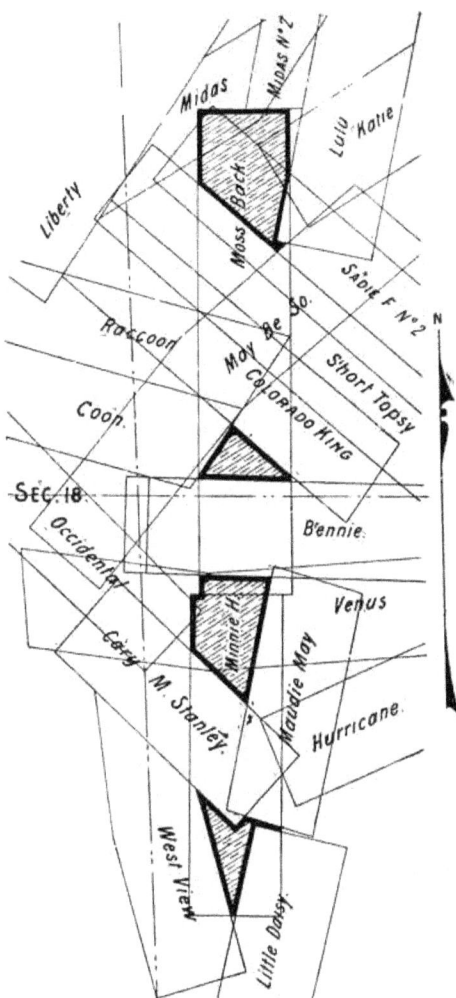

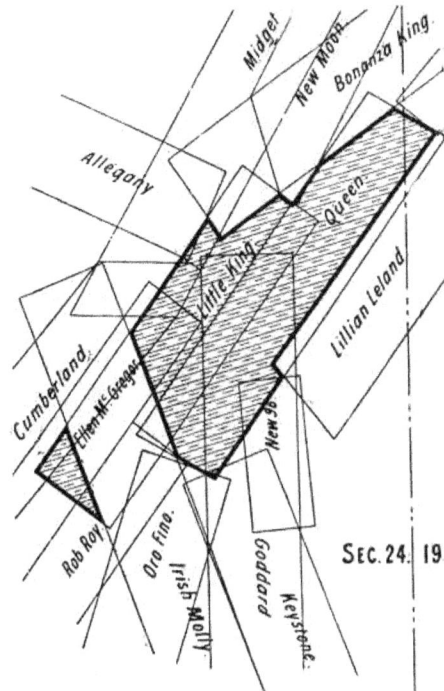

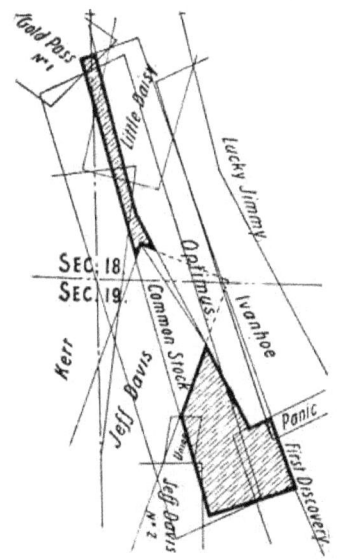

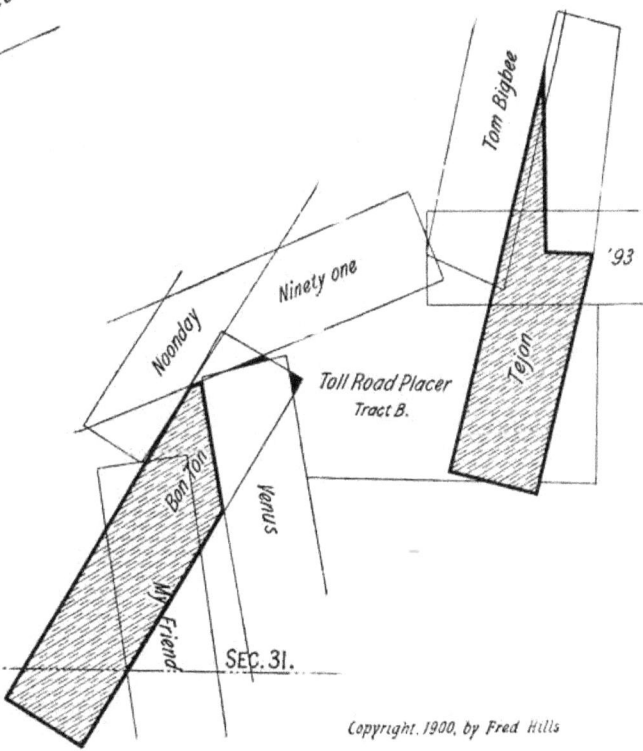

The Benny and Reform Gold Mining Company.

Incorporated 1899.

Directors

Jas. Doyle....................President
Scott Ashton..........Vice-President
Chas. Walden.................Secretary
A. A. Rollestone..............Treasurer
J. B. Cunningham.

Main Office—Victor, Colorado.

Capitalization

600,000 shares. Par value, $1.00.

In treasury January 1, 1900, 100,000 shares.

Property

Owns the Benny and Reform, containing 3 acres, situated in the N. 1-2 of section 29, in the Cripple Creek mining district. Property will soon be patented. This company also owns the Shoshone and Florence claims, adjoining the celebrated Caribou mine in Boulder county, which property is not shown on plat in this Manual. The Benny and Reform are situated between the Last Dollar and the Los Angeles; they have been in litigation for some time past, but everything has been settled up and the patent will soon be received.

Development

On the Benny and Reform is a fine shaft and an excellent plant of machinery, ore house, etc. Shaft 330 feet deep; 300 feet of drifting; large bodies of low grade ore in sight. On the Shoshone and Florence, a plant of machinery, shaft house, shaft 150 feet deep; two tons of ore shipped just before shutting down that will run $84.00 in gold.

The stock of the company is all owned by James Doyle, Chas. Walden, J. B. Cunningham and Scott Ashton. No stock has ever been sold or offered for sale, the stockholders paying for the development of the properties, in proportion to their stock, as needed.

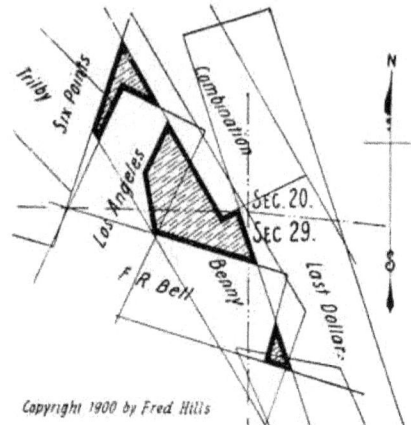

The Big Dick Mining Company.

Incorporated September 21, 1899.

Directors

J. T. Milliken............................President
A. P. Mackey.......................Vice-President
Wm. A. Otis..............Secretary and Treasurer
C. E. Titus.....................Assistant Secretary
F. H. Morley.   J. C. Connor.

Main Office—Wm. A. Otis & Co., Giddings block, Colorado Springs, Colorado.

Capitalization

1,000,000 shares. Par value, $1.00.
In treasury January 1, 1900, 200,000 shares.

Property

Owns the Big Dick, containing 8 acres, situated in the N. W. 1-4 section 17, survey No. 7,409, on Tenderfoot hill. The property is patented.

History

The property is leased. As the company has only recently been organized, there has been no production thus far.

Highest price for stock during 1899, 5 cents; lowest price for stock during 1899, 2 1-2 cents.

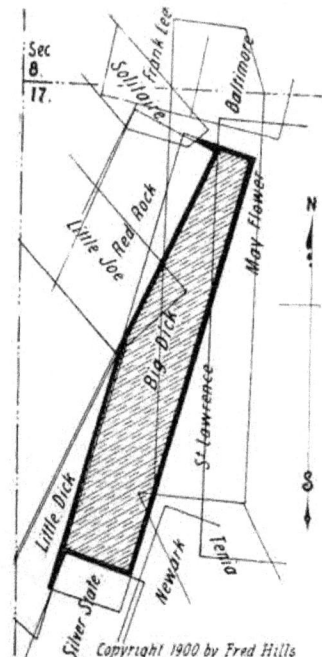

The Big Four Gold Mining Company.

Incorporated December 23, 1895.

T. B. Burbridge.....................President   Directors

Chas. N. Miller...................Vice-President

B. F. N. Macrorie.........Secretary and Treasurer

A. G. Young.   J. E. Hunter.

Main Office—271 Bennett avenue, Cripple Creek, Colorado.

1,000,000 shares. Par value, $1.00.   Capitalization

In treasury January 1, 1900, 29,120 shares; in treasury January 1, 1900, $13.00 cash.

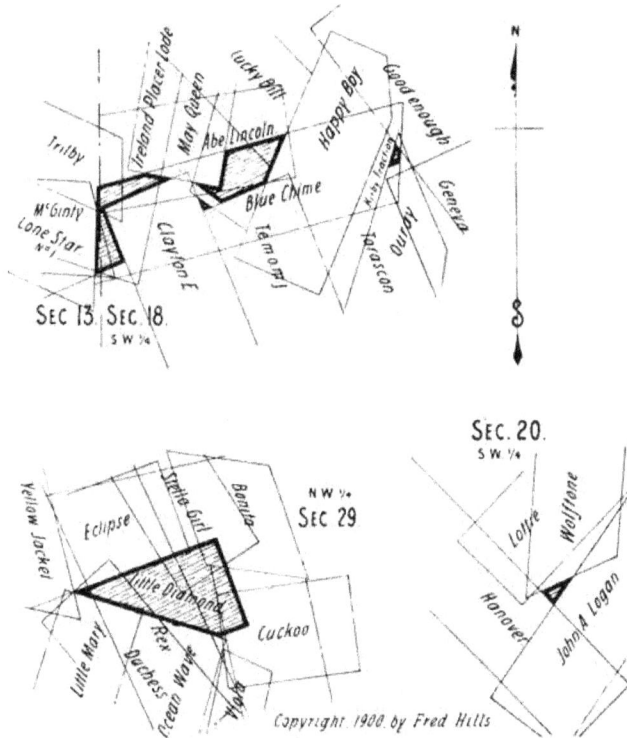

Copyright 1900 by Fred Hills

Owns the Blue Chime, 1 3-4 acres, in the S. W. 1-4 section 18, in   Property
Poverty gulch; Lonely, 1-2 acre, in the S. W. 1-4 section 20, on Bull hill, and the Little Diamond, 2 1-4 acres, in the N. W. 1-4 section 29, on Battle mountain. The Negropontus claim, a fraction, in the N. W. 1-4 section 20. The Little Diamond is patented, while the Lonely, Negropontus and Blue Chime are held by location. The Negropontus does not show on map.

About 500 feet of development work done on various properties;   Development
the deepest shaft being about 120 feet on the Lonely. There are three sets of lessees working, and they are all talking of putting up a steam hoist. The present officers and stockholders only acquired their holdings in September, 1899, and know little or nothing of the previous history of the company's properties.

Production to January 1, 1900, about $5,000.00.   Production

Highest price for stock during 1899, 4 cents; lowest price for stock during 1899, 2 1-4 cents.

The Black Belle Gold Mining Company.

Incorporated January, 1896.

Directors

D. B. Fairley.........................President

G. M. Carter........................Vice-President

J. T. Burkholder.......................Secretary

C. W. Fairley........................Treasurer

J. A. Himebaugh.   W. H. Gandy.

J. B. Carnes.

Main Office—Fairley Bros., No. 23 South Tejon street, Colorado Springs, Colorado.

Capitalization

1,250,000 shares.   Par value, $1.00.

In treasury January 1, 1900, 250,000 shares; in treasury January 1, 1900, $6,000 cash.

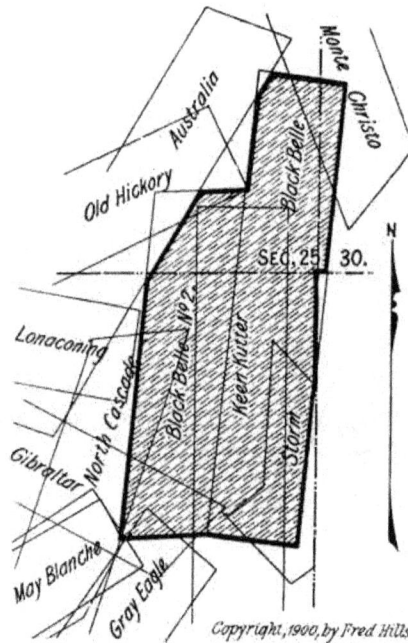

Copyright, 1900, by Fred Hills

Property

Owns the Black Belle and Black Belle No. 2, comprising 17.08 acres, situated in the S. E. 1-4 of section 25, on Beacon hill.   Property is patented.

Development

The company have two steam hoisting plants.   One shaft has been sunk 265 feet, and one 165 feet, with about 700 feet of levels.   The south 600 feet of the Black Belle is leased.   The greater part of the development work is being done on the Black Belle.

Production and Dividend

Production to January 1, 1900, about $100,000.00.   Dividends up to January 1, 1900, about $5,000.00.

Highest price for stock during 1899, 23 cents; lowest price for stock during 1899, 13 cents.

The Blackstone Gold Mining Company.

Incorporated July, 1895.

H. Hutchinson........................President Directors

E. R. Whitmarsh.........Secretary and Treasurer

W. C. Crane.   R. C. Day.

Main Office—Colorado Springs, Colorado.

800,000 shares. Par value, $1.00. Capitalization

In treasury January 1, 1900, 50,000 shares.

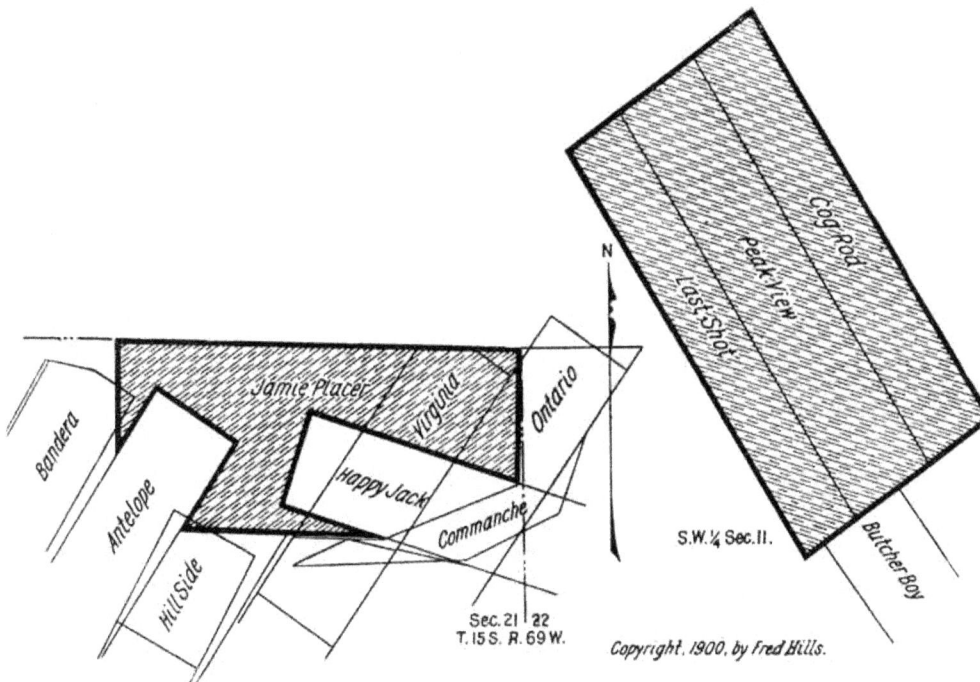

S.W. ¼ Sec. 11.

Sec. 21 | 22
T. 15 S. R. 69 W.

Copyright. 1900. by Fred Hills.

Owns the Cog Road, the Peak View and the Last Shot, containing Property
in all 26 acres, in the W. 1-2 of section 11, township 15 south, range 69
west, on Cow mountain. The company also own the Jamie placer, containing 11 1-2 acres, in the N. E. 1-4 section 21, township 15 south, range
69 west, adjoining the S. E. 1-4 of school section 16. All the above property is patented.

Only enough development work to obtain patent has been done. Development

Highest price for stock during 1899, $8.00 per M; lowest price for
stock during 1899, $2.50 per M.

The Black Wonder Gold Mining Company.

Incorporated December 26, 1894.

Directors

J. N. Kimzey...........................President
C. J. Cover......................Vice-President
J. R. Talpey..............Secretary and Treasurer
    D. P. Cathcart.   J. K. Brunner.

    Main Office—8 Bank block, Colorado Springs, Colorado.

Capitalization

    500,000 shares. Par value, $1.00.
    In treasury January 1, 1900, $50.00 cash.

Property

    Owns the Black Wonder and the M. C. lode claims, containing 11 acres, situated in the S. W. 1-4 of section 21, on the southwest slope of Bull hill. Property is patented.

Development

    There are two shafts on the property, each of about 200 feet depth; also about 100 feet of drifting. The greater part of the development work is being done on the Black Wonder lode.

    No stock has been sold, as it is closely held.

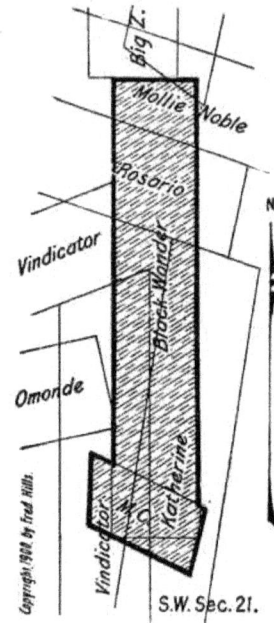

The Blanche Gold Mining Company.

Incorporated October, 1899.

Directors

A. E. Carlton....................President
Frank Gilpin................Vice-President
F. G. Whipp.....................Secretary
E. C. Newcomb...................Treasurer
    R. G. Withers.

    Main Office—Cripple Creek, Colorado.

Capitalization

    1,250,000 shares. Par value, $1.00.
    In treasury January 1, 1900, 150,000 shares; in treasury January 1, 1900, $10,000 cash.

Property

    Owns Uncle Sam, 2 1-2 acres, in the S. E. 1-4 section 17, on Bull hill. Patented.

Development

    There is a plant of machinery on the Uncle Sam, valued at $5,000.00. One main working shaft 125 feet deep, and a prospect shaft 90 feet deep.

Production

    Production to January 1, 1900, $30,000.00.

    Highest price for stock during 1899, 21 cents; lowest price for stock during 1899, 10 cents.

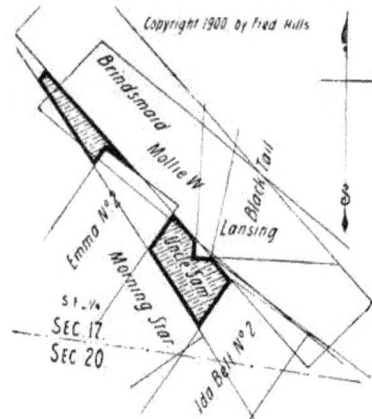

The Blue Bell Mining, Milling and Prospecting Company.

Incorporated 1892.

Wm. A. Otis............President Directors
Wm. Barber........Vice-President
Wm. P. Sargeant....Sec. and Treas
 Chas. Walden.
  John McConaghy, Manager.

Main Office—Wm. A. Otis & Co., Giddings block, Colorado Springs, Colorado.

1,000,000 shares. Par value, $1.00. Capitalization
In treasury January 1, 1900, 200,-000 shares; in treasury January 1, 1900, $500.00 cash.

Owns the Blanche, the Blue Bell, Property
the Blue Bell Mill Site, containing in all 25 acres, on line between section 24 and section 19; also owns the Robt. E. Lee, containing 7 acres, situated in the N. W. 1-4 of section 19. All on Gold hill. All patented. The company also owns 318,325 shares of the capital stock of the Katinka G. M. Co., for which 30 cents per share is bid. The Blanche, Blue Bell and the north 1-2 of the Robt. E. Lee, also the Blue Bell Mill Site, are leased.

There is an 80-foot shaft on the Development
Robt. E. Lee. Through the Blanche is a tunnel at a depth of 600 feet now being worked.

Highest price for stock during 1899, 16 3-4 cents; lowest price for stock during 1899, 4 1-2 cents.

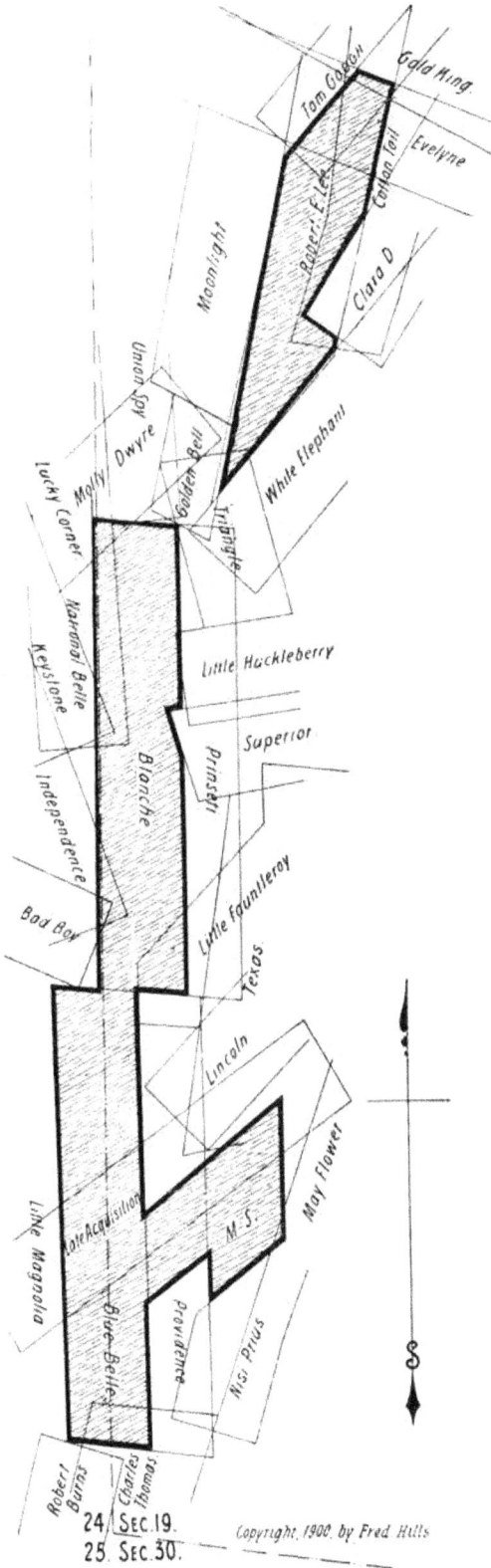

Copyright. 1900. by Fred Hills

The Blue Bird Gold Mining and Milling Company.

Incorporated December 26, 1895.

Directors
E. S. Johnson.........President and Treasurer
W. S. Jackson.................Vice-President
A. T. Gunnell....................Secretary
Della A. Johnson.   Luther Lee Johnson. .

Main Office—P. O. building, Colorado Springs, Colorado.

Capitalization
1,000,000 shares. Par value, $1.00.

In treasury January 1, 1900, $5,000.00 cash.

Property
Owns the Blue Bird, in the S. W. 1-4 of section 20, containing 10 1-3 acres, on Bull hill; patented. The company also have bond and lease on the Whippoorwill, running two years, area 3 acres. This property adjoins the Blue Bird and is patented.

Development
Four-story shaft and ore house. Electric hoist suitable for working 1,000 feet. Equipped with air plant and air drills. Also an electric hoist for sinking purposes. A 4 1-2x9 shaft has been sunk to a depth of 500 feet, timbered with square sets 8x8. The greater part of the development work is being done on the Blue Bird.

Production
Gross production to January 1, 1900, $300,000.00.

This is a close corporation. Stock not listed.

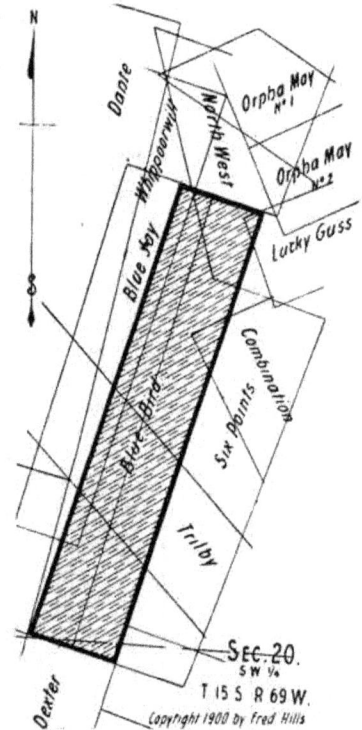

The Bob Lee Gold Mining Company.

Incorporated January 9, 1894.

E. D. Marr............................President
P. W. Middagh....................Vice-President
Jas. W. Avery............Secretary and Treasurer
F. J. Brown.   J. L. Middagh.

Directors

Main Office—34, 35 and 36 P. O. building, Colorado Springs, Colorado.

1,500,000 shares. Par value, $1.00.

In treasury January 1, 1900, 3,000 shares.

Capitalization

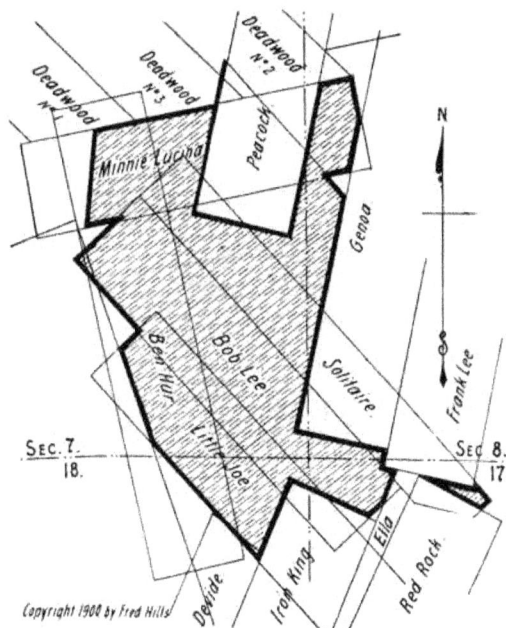

Owns the Solitaire, Bob Lee, Little Joe, Minnie Lucina and Lucky Diamond, consisting of 21 acres, in the S. E. 1-4 section 7, on Tenderfoot hill. All patented.

Property

There is a shaft on the Little Joe claim 60 feet deep, with about 50 feet of drifting. The property is being worked by lessees, and adjoins the Hoosier of the Grafton Company, which has produced about $150,-000.00 worth of ore, and the lessees are working on a phonolite dyke which comes from the Hoosier, and from which assays have been had running about $100.00 per ton.

Development

Highest price for stock during 1899, 10 cents; lowest price for stock during 1899, 1 1-4 cents.

The Bonanza King Gold Mining and Milling Company.

Incorporated February, 1894.

Directors

J. J. Grier.............................President

J. C. Manchester....................Vice-President

T. F. McCarthy........................Secretary

R. G. Miller...........................Treasurer

Bruce Glidden.

Main Office—Depot Hotel, Colorado Springs, Colorado.

Capitalization

700,000 shares. Par value, $1.00.

In treasury January 1, 1900, 55,000 shares.

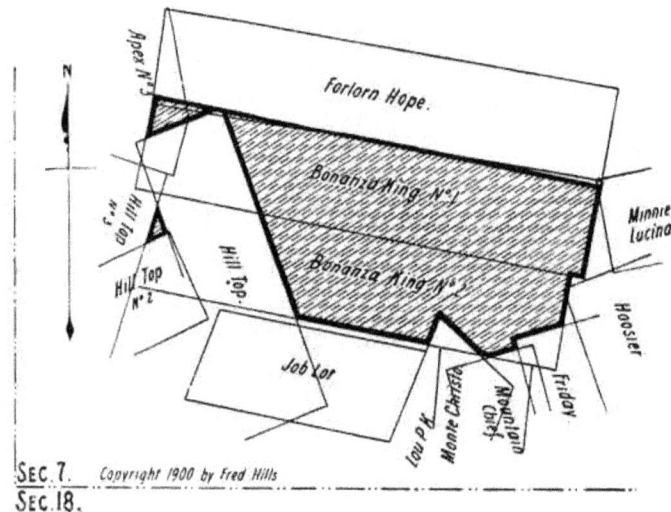

Property

Owns the Bonanza King No. 1 and Bonanza King No. 2, containing 14.71 acres, in the S. E. 1-4 section 7, township 15 south, range 69 west, on Tenderfoot hill. All patented.

Development

There is one shaft on the property, containing small vein, which is 85 feet deep, and one 60 feet deep, with a fine vein but no values yet; one 45 feet deep, with good vein but no value. The property is all leased. Work only began July 12, 1899. The showing is very good, as there are two good veins opened. Owing to its proximity to the Hoosier lode claim, the prospects are very flattering.

Highest price for stock during 1899, 8 cents; lowest price for stock during 1899, 1 1-2 cents.

The Bonnie Nell Gold Mining Company.

Incorporated October 5, 1899.

J. W. Pring..............................President   Directors

H. H. Barbee.......................Vice-President

T. C. McDonald.........................Secretary

R. J. Gwillim..........................Treasurer

A. F. Woodward.

Main Office—Room No. 8, Freeman building, Colorado Springs, Colorado.

1,500,000 shares. Par value, $1.00.   Capitalization

In treasury January 1, 1900, 179,000 shares; in treasury January 1, 1900, about $4,000.00 cash.

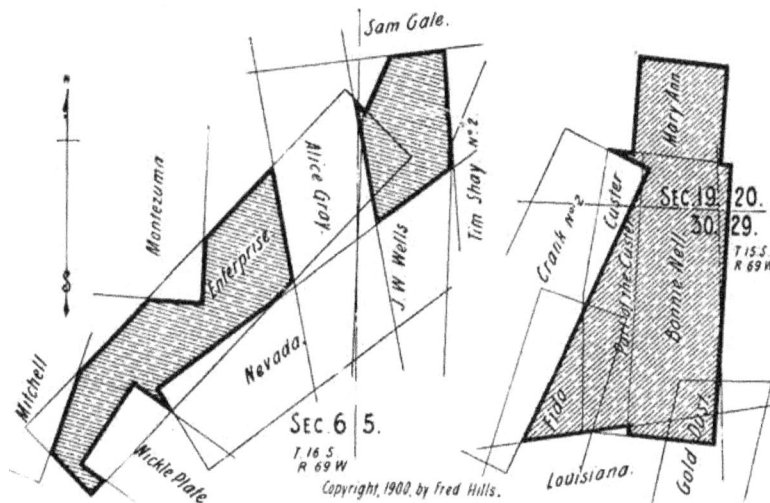

Copyright, 1900, by Fred Hills.

Owns the Fido, Custer, Bonnie Nell, Mary Ann, containing in all   Property
12 acres, in the N. W. 1-4 section 30, on Raven hill; also the Enterprise, a part of the Alice Gray and a part of the J. W. Wells, containing 8 acres, in the E. 1-2 section 6, township 16 south, range 69 west, on Straub mountain. All patented.

Two good steam hoists; one electric hoist. The Bonnie Nell has a   Development
shaft 125 feet deep; the Custer one of 150 feet. There is also another shaft 120 feet deep, on the north end of the Bonnie Nell. The greater part of the development work has been done on this claim.

Highest price for stock during 1899, 9 cents; lowest price for stock during 1899, 5 cents.

The Bourse Mining Company.

Incorporated 1896.

Directors

Joseph Milner..........................President

J. L. Brown.......................Vice-President

F. C. Matthews............Secretary and Treasurer

J. C. Freeman.   C. R. Lawrence.

Main Office—92 State street, Boston, Massachusetts.

Capitalization

1,250,000 shares. Par value, $1.00.

In treasury January 1, 1900, $2,000 cash.

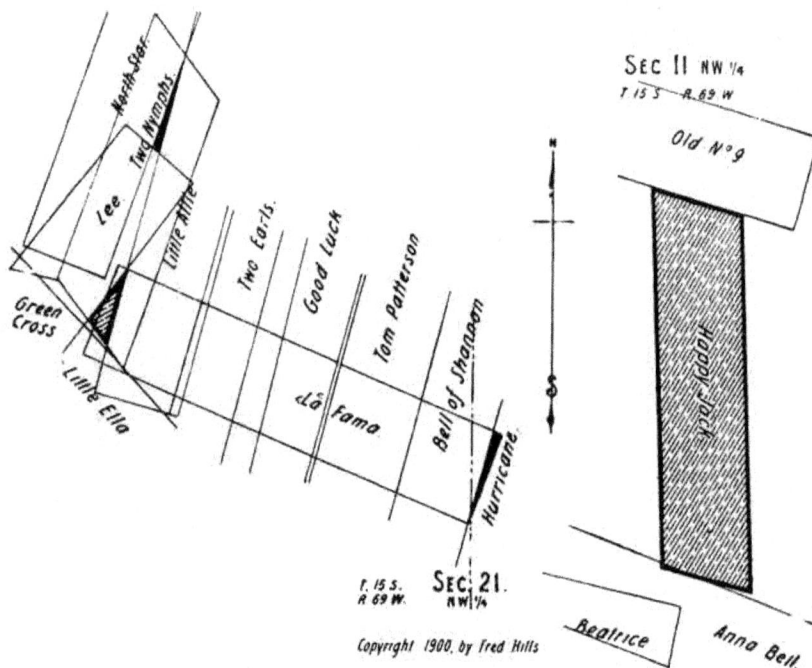

Copyright 1900, by Fred Hills

Property

Owns the La Fama and the Two Nymphs, containing 0.461 acre, in the N. W. 1-4 section 21, on Bull hill; the Happy Jack, 8.108 acres, in the N. W. 1-4 section 11, township 15 south, range 69 west, on Cow mountain; also the Fitzsimmons, in Poverty gulch, a location, and, as such, not shown on plat. All patented except the Fitzsimmons.

Development

The La Fama has a shaft 55 feet deep, and the ore at this depth assays $6.00 per ton. The greater part of the development work is being done on this claim. Both the La Fama and the Two Nymphs are surrounded by such heavy producers as the Victor, Isabella, Lillie, etc.

The B. P. O. E. Gold Mining Company.

Incorporated.

A. P. Mackey..........................President
Phil. Starr.......................Vice-President
W. J. Matthews........................Secretary
H. M. Mason................Assistant Secretary
Geo. W. Lloyd.

Main Office—No. 53 De Graff building, Colorado Springs, Colorado.
2,000,000 shares. Par value, $1.00.
In treasury January 1, 1900, 250,000 shares.

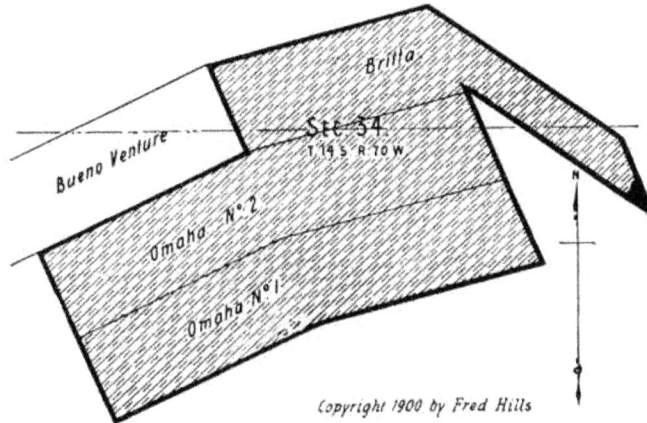

Copyright 1900 by Fred Hills

Owns the Britta and the Omaha No. 1 and No. 2, containing about 26 acres, situated in the W. 1-2 section 34, township 14 south, range 70 west, west of Copper mountain. Receiver's receipt for the Omaha No. 1. The Omaha No. 2 and the Britta in process of patenting. Only sufficient development work to obtain a patent has been done.

Highest price for stock during 1899, $11.00 per M; lowest price for stock during 1899, $2.50 per M.

The Bradford Gold Mining Company.

Incorporated 1896.

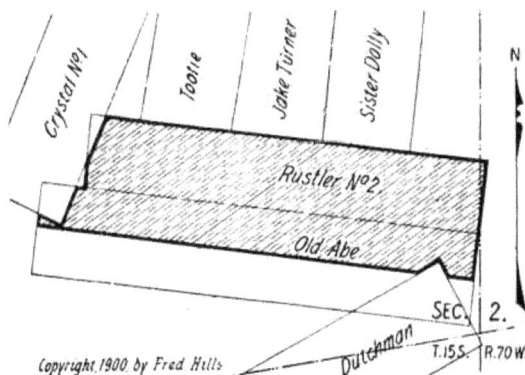

Copyright 1900 by Fred Hills

J. C. Manchester......President
C. A. McLain...V-Pres and Treas
R. G. Miller..........Secretary
W. H. Spurgeon.
Lucinda Manchester.

Main Office—719 N. Nevada avenue, Colorado Springs, Colorado.

1,000,000 shares. Par value, $1.00.

In treasury January 1, 1900, 125,000 shares; in debt January 1, 1900, $1,000 cash.

Owns the Rustler No. 2, containing 7 1-2 acres, and the Old Abe, containing 4 1-2 acres, both situated in the N. W. 1-4 of section 2, on Copper mountain. Both patented.

No development work has been done as yet, excepting what has been necessary for patent.

The Broken Hill Gold Mining Company.

Incorporated June 15, 1894.

Directors

Horace M. Cutler, president; Walter Ames, vice-president; Wm. C. Robinson, secretary; Samuel Musso, treasurer. Chas. Lawrence. Albert C. Litchfield.

Main Office—Bank building, Colorado Springs, Colorado. Branch Office—Lynn, Massachusetts.

Capitalization

1,000,000 shares. Par value, $1.00.

Property

Owns the Maggie, containing 10 acres, in the center of section 32, Survey No. 9,020, south of Battle mountain; patented.

Development

There is a shaft house about 20x70 feet, boiler, hoist, etc. A shaft has been sunk about 425 feet deep and there has been some drifting and cross-cutting.

Highest price for stock during 1899, 2 cents; lowest price for stock during 1899, 1 cent.

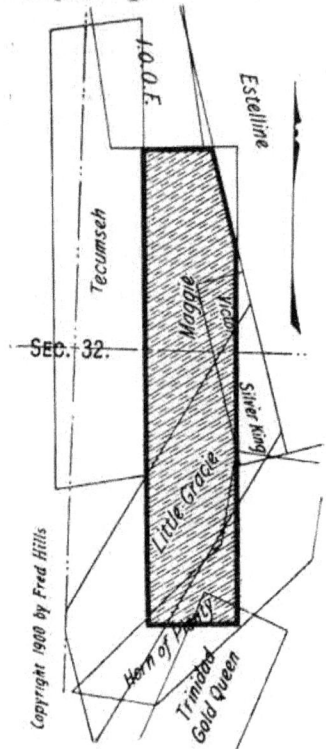

The Buffalo and Cripple Creek Gold Mining Company.

Incorporated February, 1897.

Directors

Frank S. Oakes, president; Willis Tew, vice-president; John E. Lundstrom, secretary and treasurer; Thos. J. Farrar; Wm. E. Newell; Olin G. Rich; John M. Wilson.

Main Office—Colorado Springs, Colorado.

500,000 shares. Par value, $1.00.

Capitalization

In treasury January 1, 1900, 100,000 shares.

Property

Owns lease on lots Nos. 53, 60 and 61, containing 30 acres, in school section 36; the Winner lode claim, containing 7 acres, located July 2, 1892; and the Winner tunnel site. Patented. As the company could furnish no data in regard to the Winner claim, it is not shown on map.

Development

The company owns and is operating a first-class plant of machinery, including an air compressor. Fine office and living rooms for the general manager. The buildings containing the boiler, engine and compressor are complete, and, being situated at the mouth of the Newell tunnel, the company are thus enabled to develop the mine economically. Underground machinery: Air pipes and air drills, with electric bells from the power house to the remotest workings. The Newell tunnel enters Grouse mountain from the northwest side of school lot No. 60, striking 19 degrees south of east. About 1,000 feet from the portal a vein is crossed 27 feet wide. There has been about 600 feet of drifting on this vein north and south. The Winner tunnel has also been run some 170 feet into the mountain. The greater part of the development work is being done on school lot No. 61.

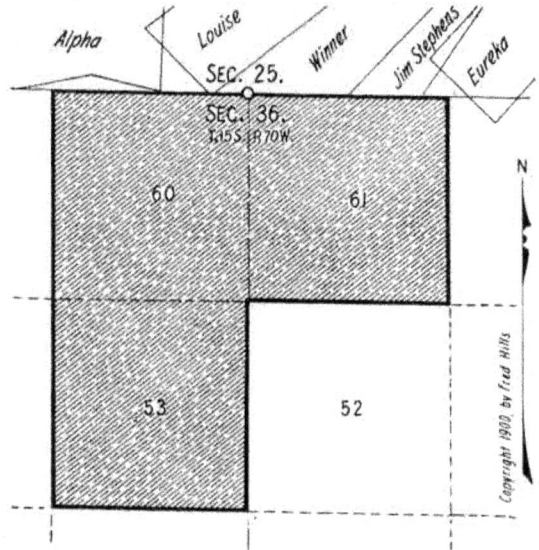

The Buckeye Gold Mining Company.

Incorporated October 16, 1895.

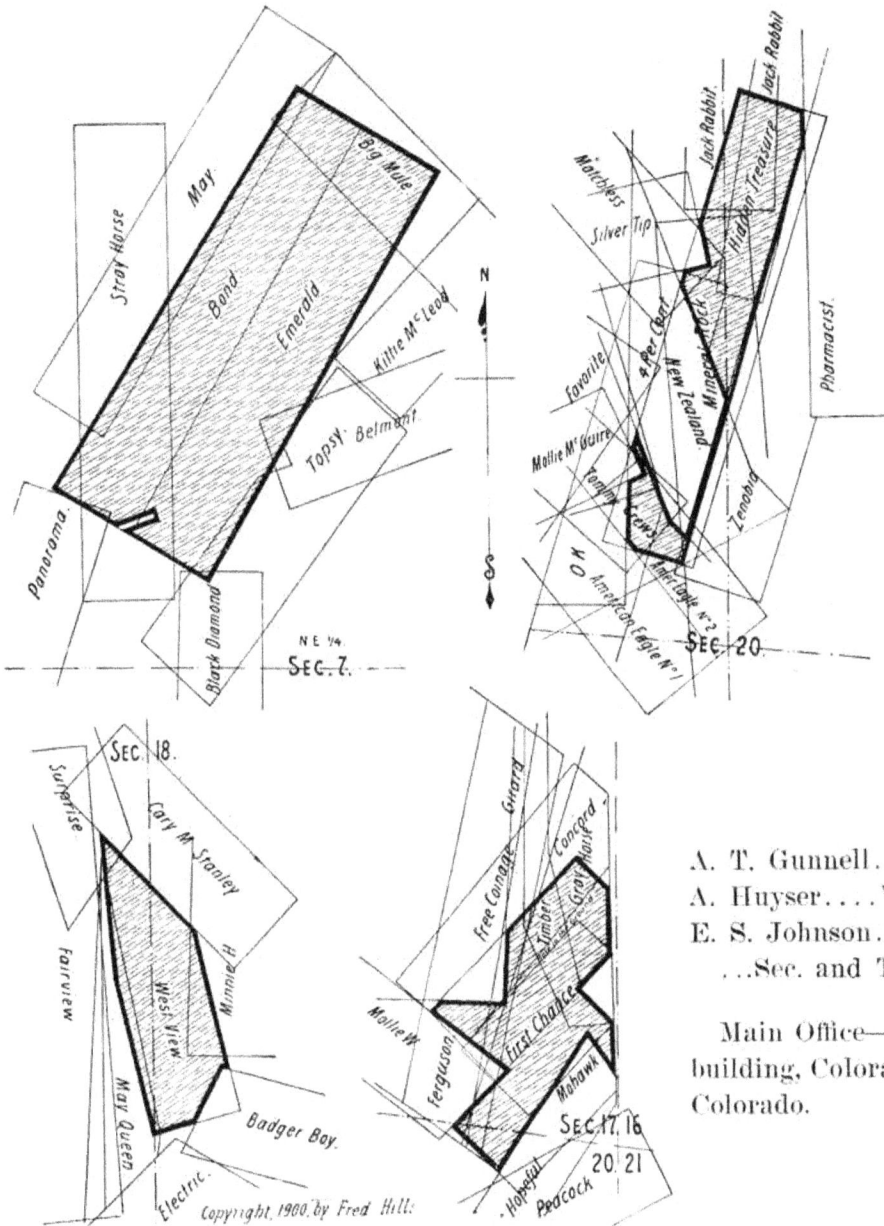

Copyright, 1900, by Fred Hill.

A. T. Gunnell.....Pres Directors
A. Huyser....Vice-Pres
E. S. Johnson.......
...Sec. and Treasurer

Main Office—66 P. O. building, Colorado Spgs. Colorado.

Capitalization

2,000,000 shares. Par value, $1.00.

In treasury January 1, 1900, 30,000 shares.

Property

Owns the Mohawk and Hole in the Ground, in the S. E. 1-4 of section 17, on the north slope of Bull hill; the West View, in the S. E. 1-4 section 18, on the west slope of Globe hill; the Emerald and Bond, in section 7, on Tenderfoot hill; also the Hidden Treasure, in the N. 1-2 section 20, on Bull hill. The total acreage of this company is 42 acres. All patented except the Hidden Treasure, which is in process.

Development

Shaft house and steam plant, sinking shaft 4 1-2x8 in the clear, timbered with square sets 8x8 to a depth of 500 feet, by contract as quickly as three shifts of three men each can perform the work. This work is being done under lease on the south part of Hidden Treasure.

Highest price for stock during 1899, 15 cents; lowest price for stock during 1899, 10 cents.

The Buck Horn Gold Mining Company.

Incorporated July 19, 1892.

Copyright, 1900, by Fred Hills.

Directors

Henry M. Blackmer, President; N. S. Gandy, Vice-President; Asa T. Jones, Secretary and Treasurer; Geo. E. Lindley, F. E. Robinson.

Main Office—Colorado Springs, Colo.

Capitalization

1,250,000 shares. Par value, $1.00. In treasury January 1, 1900, $3,492.00 cash.

Property

Owns the Whippoorwill, comprising 2 1-4 acres in the S. 1-2 of section 20; the Last Chance and Cheyenne, 19 acres, in the S. E. 1-4 of section 21; the Combination No. 2 and the Jeanette and Mule, 25 acres, in the N. 1-2 of section 21, on Bull hill; also the Grass, 9 acres, in the S. W. 1-4 of section 21. The company now owns only the mineral rights of the "Grass," the surface ground being in controversy at the present time. To this they expect very shortly to secure title.

Development

The Whippoorwill, bonded for $20,000.00 and leased, is being worked with machine drills at the present time.

Highest price for stock during 1899, 6 3-8 cents; lowest price for stock during 1899, 2 1-8 cents.

96

The Bull Hill Gold Mining and Tunnel Company.

Incorporated November, 1895.

T. A. Sloane............................President
J. M. Harden.....................Vice-President
F. P. Davis.............................Secretary
John Pederson........................Treasurer
Jacob Bishoff.

Directors

Main Office—Colorado Springs, Colorado.

1,250,000 shares. Par value, $1.00.
In treasury January 1, 1900, 150,000 shares.

Capitalization

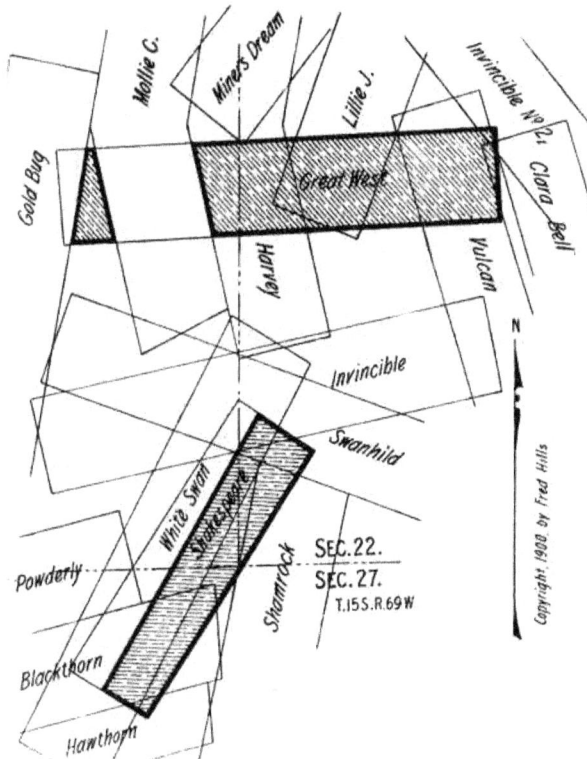

Property

Owns the Great West, 7 1-2 acres, with 2 1-4 acres mineral rights, making a total of 9 3-4 acres, in the south 1-2 section 22; the Shakespeare, 4 1-2 acres, in the S. 1-2 section 22; the M. E., the Coxey and the Saddle Mountain Tunnel Site. These last three, being locations, are not shown on the map. Receiver's receipt held for the Great West. The Shakespeare in process of patenting.

Development

There is a 50-foot shaft on the Shakespeare, a 50-foot shaft on the Great West, a 30-foot shaft on the M. E., a 30-foot shaft on the Coxey, and 30 feet of work has been done on the tunnel on the Saddle mountain site. The Shakespeare and the Great West claims lie to the east of the Isabella mine and have fine showings, assays running as high as $240 to the ton.

Highest price for stock during 1899, 18 cents; lowest price for stock during 1899, 14 3-4 cents.

4

The Cable Consolidated Mining Company.

Incorporated February 7, 1900.

Directors

F. E. Brooks..........................President
W. F. Crosby.....................Vice-President
W. H. Leonard............Secretary and Treasurer
J. W. Wright.....Assistant Secretary and Treasurer
O. B. Willcox.   Guy Phelps Dodge.
Max Straus.

Main Office—No. 63 Hagerman building, Colorado Springs, Colorado.

1,500,000 shares.   Par value, $1.00.

Capitalization

In treasury March 10, 1900, 300,000 shares.

Copyright, 1900 by Fred Hills.

Property

Owns the Posey and Cable lodes, containing 4.62 acres, survey No. 10,923; the St. Peter lode, containing 6.413 acres, survey No. 9,110; the Aztec, containing 3.204 acres, survey No. 10,176; also a part of the Black Hawk, containing 3.964 acres, survey No. 8,157; all the above property being situated in the S. E. 1-4 of section 25, on Beacon hill. Total acreage is 18.201 acres. All patented. The property of this company adjoins the Prince Albert and the Arequa townsite on Beacon hill.

Development

There is a shaft house and horse whim on the Aztec and St. Peter. Machinery will be installed in the main working shaft in the center of the group. The Aztec and the St. Peter have a shaft 185 feet deep and a drift of 60 feet. A vein from three to four feet wide has been disclosed, some ore running as high as $185. In developing these claims about $6,000 have been expended. About $3,000 have also been expended on the Posey and Cable lodes. These have a tunnel of 200 feet, a shaft of 35 feet, open cuts, etc. A network of veins has been exposed, some showing values. In the center of this group of claims a main shaft is being sunk.

History

The present ownership is the result of a consolidation, some of the properties having been long in litigation, which prevented their development. By purchase of all conflicting interests this company has been able to secure perfect title to its large area, and, with the development which the company is prepared to prosecute, it is expected the group will soon enter the list of its neighboring Beacon hill producers.

Highest price paid for this stock to date, 5 cents.

The Cadillac Gold Mining Company.

Incorporated December 24, 1895.

J. A. Sill...............................President

Arthur Cornforth....................Vice-President

H. J. English.............Secretary and Treasurer

F. H. Dunnington...............Assistant Secretary

Main Office—2 North Nevada avenue, Colorado Springs, Colorado.

1,000,000 shares. Par value, $1.00.

In treasury January 1, 1900, 100,000 shares; in treasury January 1, 1900, $400.00 cash.

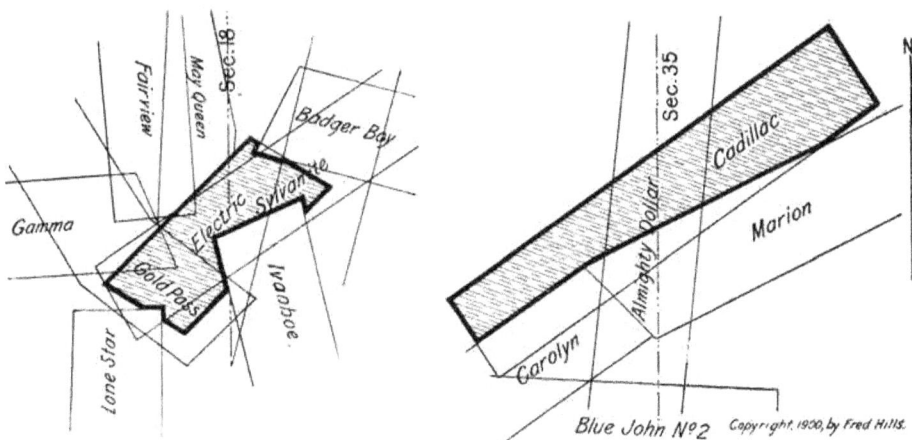

Blue John No 2   Copyright, 1900, by Fred Hills.

Owns the Electric claim, 3.83 acres, in the S. 1-2 section 18, on Gold hill, and the Cadillac claim, 7.853 acres, in the S. W. 1-4 section 34, on the east slope of Cowan Mountain. Patented.

The Electric claim has one shaft 120 feet deep and is still being sunk deeper. Just enough development has been done on the Cadillac to obtain patent. The work is confined to the Electric lode, where very good indications are shown, and active work is now being done by lessees.

Highest price for stock during 1899, 4 cents; lowest price for stock during 1899, 2 1-4 cents.

The Caledonia Gold Mining Company.

Incorporated 1894.

Directors

A. J. Gillis..............................President

Angus Gillis....................Vice-President

A. J. Noonan.........................Secretary

S. P. Dickens..........................Treasurer

Ward Barber.   Con Schott.

Main Office—No. 38 Hagerman building, Colorado Springs, Colorado.

Capitalization

1,000,000 shares. Par value, $1.00.

In treasury January 1, 1900, 150,000 shares.

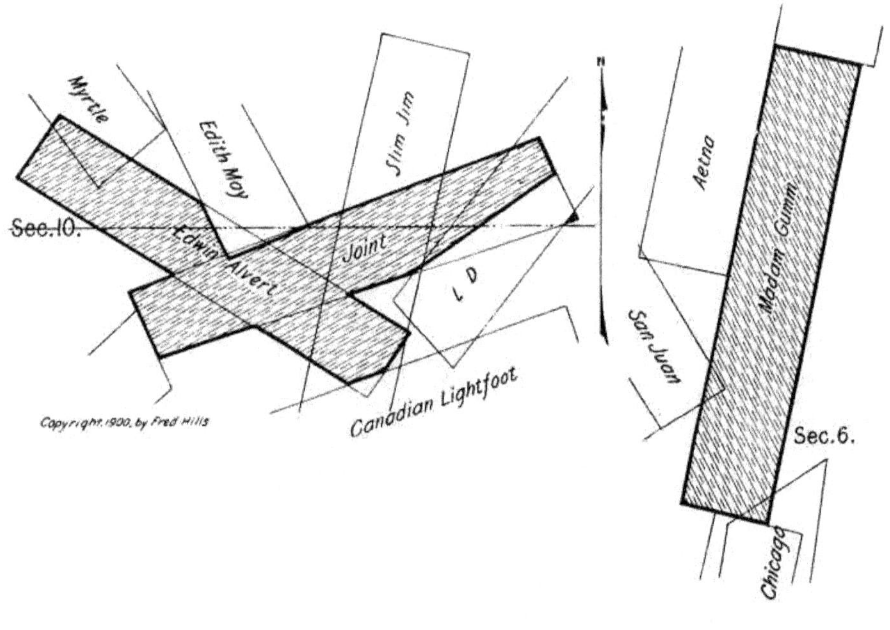

Copyright 1900, by Fred Hills

Property

Owns the Madam Gumm, comprising 10 acres in the S. 1-2 of section 6, township 16 south, range 69 west, on Straub mountain; also the Edwin Alvert and Joint, 14 3-4 acres, in the E. 1-2 of section 10, township 15 south, range 70 west, on Iron mountain. The Madam Gumm is patented. Receiver's receipt held for the other claims.

Development

A 110-foot shaft has been sunk on the Madam Gumm, also 50 feet of drift. The shaft is timbered. On the Edwin Alvert and Joint 100 feet of work has been done.

Highest price for stock during 1899, 1 3-4 cents.

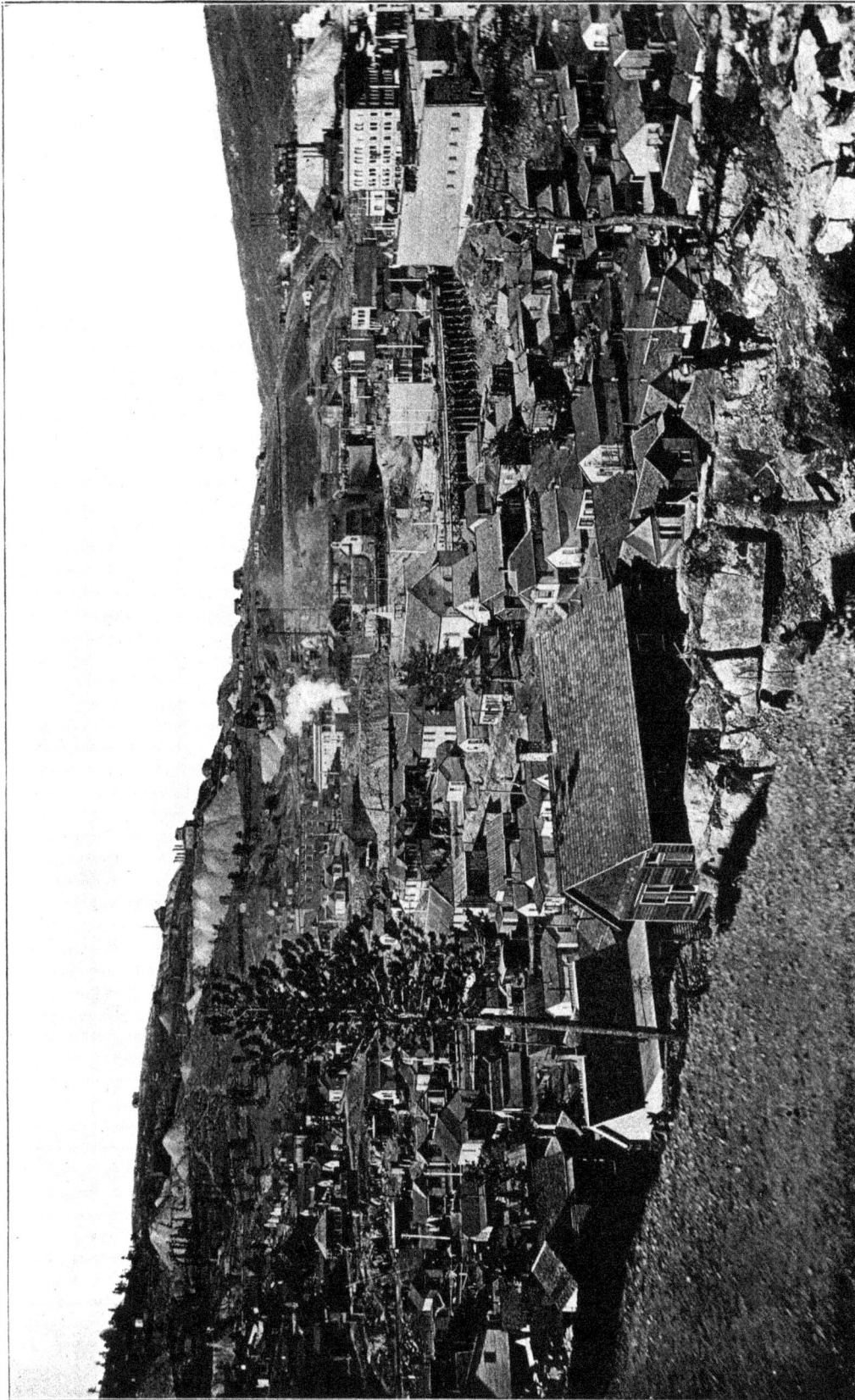

TOWN OF VICTOR, 1900—AS REBUILT AFTER THE FIRE OF 1899.

SHOWING THE PORTLAND, GOLD COIN, STRATTON'S INDEPENDENCE AND STRONG MINES.

The Cameron Mines, Land and Tunnel Company.

Scale

0 100 200 300 400 500 600 700 800 900 1000 ft.

Copyright 1900 by Fred Hills

The Cameron Mines, Land and Tunnel Company.

Incorporated September 3, 1897.

C. F. Rickey............................President

C. L. Arzeno.......................Vice-President

F. M. Woods...........................Secretary

H. E. Woods...........................Treasurer

Directors

     John McConaghy.   Geo. H. Lee.

            A. T. Voll.

Main Office—Colorado Springs, Colorado.

Branch Offices—Victor and Cameron, Colorado.

2,000,000 shares.   Par value, $1.00.

In treasury January 1, 1900, 200,000 shares; in treasury January 1, 1900, $5,000.00 cash.

Capitalization

Owns the Cameron townsite, formerly Grassy, and the Victoria No. 1, in all 184 acres, in section 17; also the Pinnacle Park, in section 17; the company also owns extensive holdings in Fremont and Chaffee counties, comprising over 750 acres. These latter, being outside the Cripple Creek district, are not shown on plat.

Property

The Cameron (Grassy) townsite has graded streets and sidewalks, residences and store buildings, constructed and owned by the company. Has six shafts ranging in depth from 60 to 150 feet, about 300 feet of cross-cutting, 200 feet of drifting, and other minor workings. The Whitehorn townsite has some streets, sidewalks and buildings, constructed by the company, and buildings owned by it. The Cameron tunnel is in 490 feet; the Cincinnatus tunnel, 240 feet; the Mesa tunnel, 85 feet; the Fardown tunnel, 130 feet; the Bull Quartz tunnel, 123 feet; the Corinne tunnel, 289 feet; the Little Rose shaft, 80 feet deep; and numerous other shafts ranging in depth from 10 to 75 feet.

Development

Shaft No. 2, Cameron (Grassy), just producing. The Corinne has produced about 200 tons. Little Rose has pay ore, but none shipped as yet.

Production

Highest price for stock during 1899, 60 cents; lowest price for stock during 1899, 20 cents.

The Cannon Ball Gold Mining Company.

Incorporated November 20, 1894.

Directors

L. E. Kimball.........................President

H. A. McIntyre......................Vice-President

D. Le Duc.............................Secretary

C. W. Buck...........................Treasurer

R. H. Buck.   D. C. Beaman.

Col. Ben. A. Block.

Main Office—Denver, Colorado.

Capitalization

1,500,000 shares.   Par value, $1.00.

In treasury January 1, 1900, 235,000 shares.

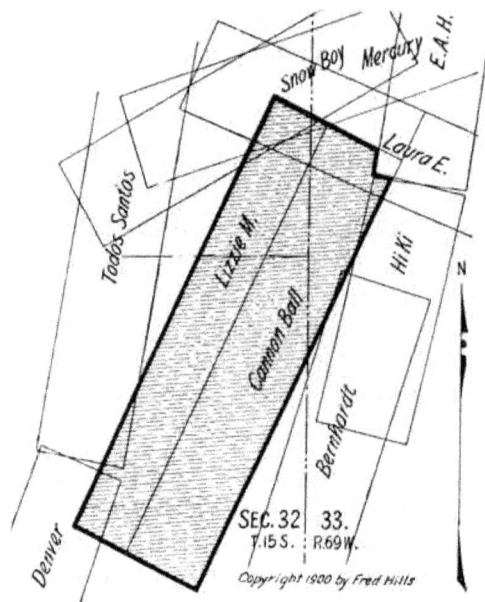

Property

Owns the Lizzie M. and the Cannon Ball, containing 15.8 acres, situated in the E. 1-2 section 32, on Big Bull hill. Both patented. A small fraction of the Hidden Treasure is also held by location.

Development

The company have sunk a shaft about 150 feet deep. From this assays as high as $90 have been received. The average assays on the vein, from 4 to 6 feet wide, are from $6 to $12.

Highest price for stock during 1899, $6 per M; lowest price for stock during 1899, $2 per M.

The Carbonate Hill Gold Mining and Milling Company.

Incorporated December 3, 1895.

W. A. Davis............................President
Wm. Clark........................Vice-President
S. J. Davis...........................Secretary
Mrs. A. N. Frowine....................Treasurer
    J. O. Hardwick.   F. D. Fox.
         Robert Price.

Directors

Main Office—55 Bank building, Colorado Springs, Colorado.

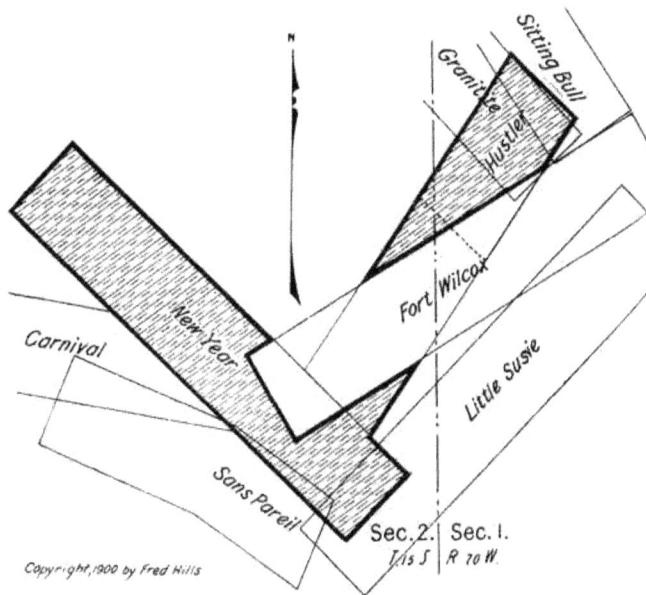

Copyright, 1900 by Fred Hills

1,000,000 shares. Par value, $1.00.   **Capitalization**

In treasury January 1, 1900, 12,000 shares; in treasury January 1, 1900, $25.00 cash.

Owns the New Year and the Hustler lodes, containing 12 acres, in   **Property** the S. E. 1-4 of section 2, on Copper mountain. New Year lode is patented, while the Hustler is in process. By an important decision of the secretary of the interior this company has been awarded 2 acres additional ground on Hustler.

There is a shaft house on the Hustler lode. About 200 feet of tunnel-   **Development** ing and shaft work.

Highest price for stock during 1899, 3 1-2 cents; lowest price for stock during 1899, $5.00 per 1,000.

The Carrie S. Gold Mining Company.

Incorporated.

Directors — J. T. Burkholder, president; Lee Glenn, vice-president; A. C. Wagy, secretary and treasurer. V. A. Chapin. W. P. Wagy.

Main Office—Exchange Bank building, Colorado Springs, Colorado.

Capitalization — 2,000,000 shares. Par value, $1.00.

In treasury January 1, 1900, 52,000 shares; in treasury April 1, 1900, $600.00 cash.

Property — Owns the Evening Star, the Baby P., the Bessie C., containing 6 1-2 acres, in the N. E. 1-4 of section 13, on Mineral hill; the James River, Old Dominion, Old Dominion No. 2, containing 5 1-2 acres, in the N. W. 1-4 of section 18, on Carbonate hill; the Liberty, Pactolus, Pactolus No. 2, 3, 4 and 5, the Midas No. 2, containing about 20 acres, in the N. 1-2 of section 18, in Poverty gulch. Receiver's receipt held for all the above property.

Development — The Pactolus group is being developed with a 100-foot tunnel. The lessee on the Pactolus has a shaft house and steam hoist, with a shaft 70 feet deep. On the other claims sufficient work has been done to obtain a patent. The lessees now sinking a shaft on the Pactolus have found some encouraging assays in cutting through a basalt dyke, dipping at an angle of 35 degrees.

Highest price for stock during 1899, 4 1-8 cents; lowest price for stock during 1899, 3 1-4 cents.

The Catherine H. Gold Mining Company.

Incorporated December, 1895.

Wm. Banning .... President Directors

Franklin E. Brooks...................... Secretary

E. C. Fletcher.   T. H. Devine.

C. S. Thomas.

Main Office—56 Bank building, Colorado Springs, Colorado.

1,250,000 shares.   Par value, $1.00. Capitalization

In treasury January 1, 1900, 132,000 shares.

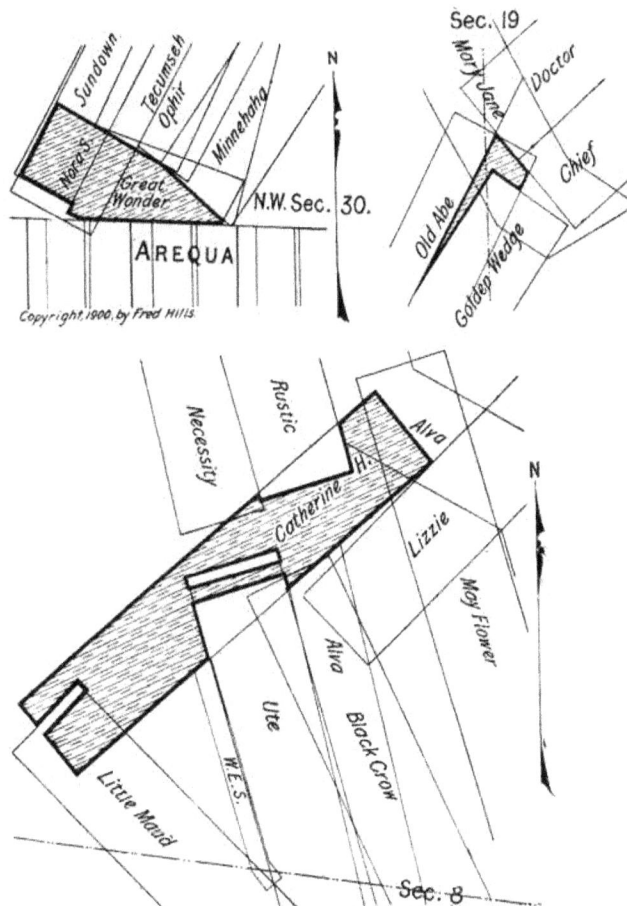

Copyright, 1900, by Fred Hills.

Owns the Great Wonder, 2 acres, in the N. W. 1-4 section 30; Old Abe, Property
0.59 acre, in the S. W. 1-4 section 19, both on Raven hill and patented.
Also Catherine H., 8 1-2 acres, in the N. E. 1-4 section 8, on Galena hill,
patented; and the Dakota Girl, 7 acres, which is held by location. The
latter not shown on plat.

One shaft 50 feet deep on the Great Wonder; one 50 feet deep on Development
Old Abe, and one 60 feet deep, with drifts and levels, on the Catherine H.

The Cedar Hill Gold Mining Company.

Incorporated February, 1900.

Directors   H. Gardner, president; J. D. Flock, vice-president; C. C. Harrison, secretary; Albert R. Gardner, treasurer; A. J. Gillis.

Main Office—54 Hagerman block, Colorado Springs, Colorado.

Capitalization   1,500,000 shares. Par value, $1.00.

In treasury January 1, 1900, 400,000 shares; in treasury January 1, 1900, $4,000.00 cash.

Property   Owns the Home lode, 6 acres, in the S. E. 1-4 section 17, north of and adjoining the Hart group, and adjoining and east of the Jerry Johnson and Damon, on north slope of Ironclad hill; also Cedar Hill lode, located between the Modoc, Rigi and Portland groups, on Battle mountain, about 1 acre surface, and a total of 3 1-4 acres, including vein rights, in the N. E. 1-4 section 29. All patented.

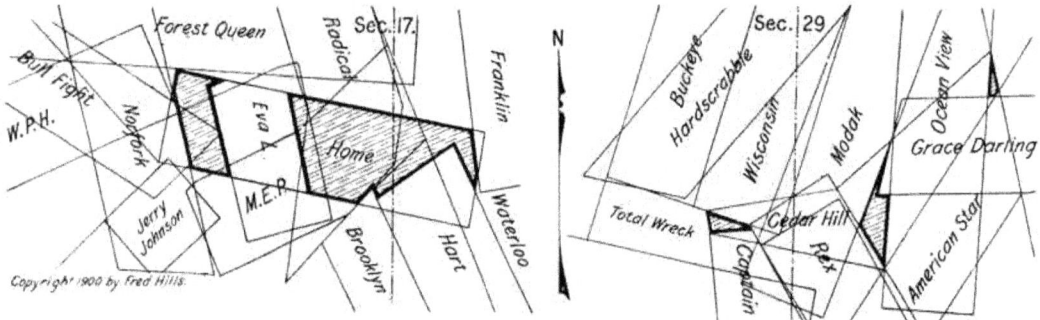

Copyright 1900 by Fred Hills.

Development   There is a small shaft house and good blacksmith shop on the property; 190 feet of sinking; contract out to sink 100-foot double compartment shaft on Home lode; sub-lessees also sinking on Home lode. Lessees are also sinking on Cedar Hill lode. The property has been idle owing to conflicting ideas of the several owners. The entire property has now changed hands and the present management intend to do their best to develop it, as their ownings comprise some of the best prospective property in the Cripple Creek district. Additional property will be purchased whenever any bargains are offered.

As this company was not incorporated no prices can be quoted on stock last year (1899).

The Celestine Gold Mining Company.

Incorporated March 20, 1900.

Directors
W. R. Foley.....................President
C. N. Miller...............Vice-President
S. J. Mattocks....Secretary and Treasurer
W. H. Allen.   A. G. Young.

Main Office—104 Pike's Peak avenue, Colorado Springs, Colorado.

Capitalization   1,250,000 shares. Par value, $1.00.

In treasury January 1, 1900, 350,000 shares.

Property   Owns block No. 15, 5 1-3 acres, First Addition to Fremont, and mineral rights under the balance of the addition, in all about 70 acres, in the N. W. 1-4 of section 24 on Gold hill. All patented.

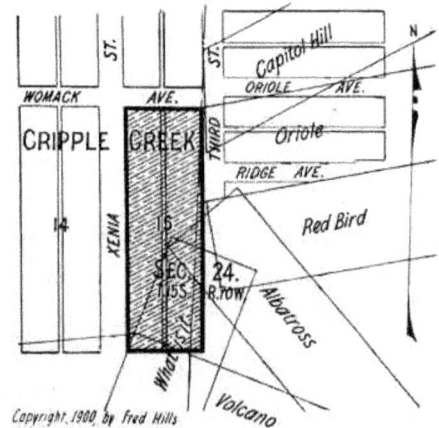

Copyright 1900 by Fred Hills.

Development   Steam hoist; full plant of machinery. Only sufficient work to obtain a patent has been done. The stock was placed at 15 cents per share.

The Central Consolidated Mines Corporation.

Incorporated August 2, 1899.

D. P. Sill.............................President
William Whelan....................Vice-President
H. C. Morse.........................Secretary
Robt. A. Mack......................Treasurer

Samuel Gale.   J. W. Nicholas.

J. C. McKenna.   Matthew Helmer.

Directors

Main Office—Room 3 Mining Exchange building, Colorado Springs, Colorado.

1,500,000 shares. Par value, $1.00.

Capitalization

In treasury January 1, 1900, 300,000 shares; in treasury January 1, 1900, $3,000.00 cash.

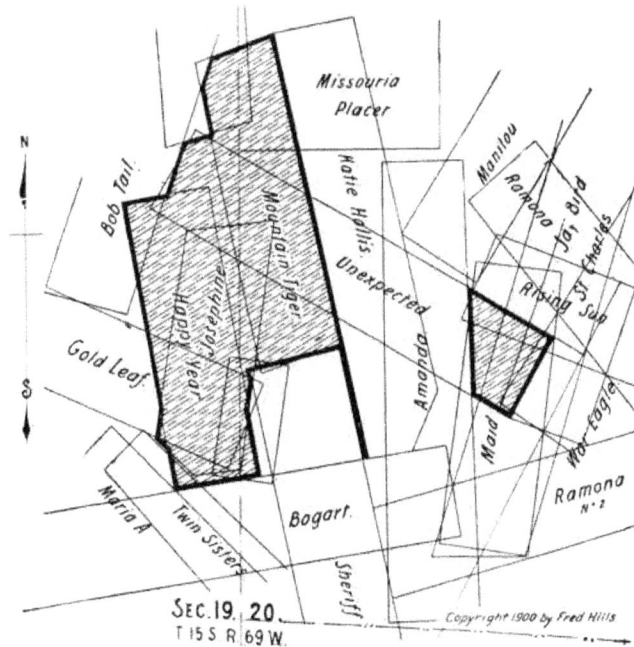

Owns the Happy Year and the Josephine, combined area being 8.67 acres, in the N. E. 1-4 of section 19, on Raven hill; the Unexpected and the Mountain Tiger, combined area 7 acres, in the N. W. 1-4 of section 20, on Bull hill; also the vein rights through the Katie Hollis and the Amanda claims, 3.82 acres. All patented. The north end of Unexpected claim is leased for 18 months from March, 1900, 20 per cent. royalty.

Property

There are several shafts about 50 feet deep and a tunnel on the Happy Year and Josephine claims. The tunnel has a length of about 800 feet and a depth of 250 feet. The greater part of the development work has been done on the Happy Year and the Josephine claims.

Development

Highest price for stock during 1899, 9 cents; lowest price for stock during 1899, 4 1-4 cents.

109

The Central Gold Mining Company.

Incorporated May 13, 1896.

Directors

J. M. Hawkins........................President

J. H. Ryan........................Vice-President

H. M. Mason..............Secretary and Treasurer

L. G. Campbell.   Geo. Saunders.

Main Office—De Graff building, Colorado Springs, Colorado.

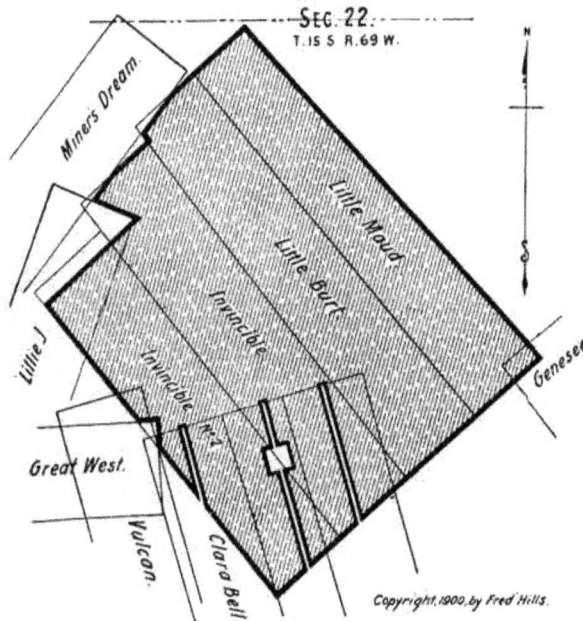

Copyright, 1900, by Fred Hills.

Capitalization

1,250,000 shares. Par value, $1.00.

In treasury January 1, 1900, 40,000 shares.

Property

Owns the Little Maud, the Little Burt, the Invincible and the Invincible No. 2, containing about 37 acres in all, situated in S. E. 1-4 of section 22, on Big Bull hill. All the property is patented.

There is a shaft on each of these claims. The stock is not on the market.

The Champion Consolidated Mining Company.

Incorporated September 8, 1896.

Frank G. Peck........................President

Verner Z. Reed....................Vice-President

J. W. D. Stovell...........Secretary and Treasurer

John T. Hawkins.   Wm. P. Sargeant.

Directors

Main Office—Rooms 21 and 22, P. O. block, Colorado Springs, Colorado.

1,250,000 shares.   Par value, $1.00.

Capitalization

In treasury January 1, 1900, 168,000 shares.

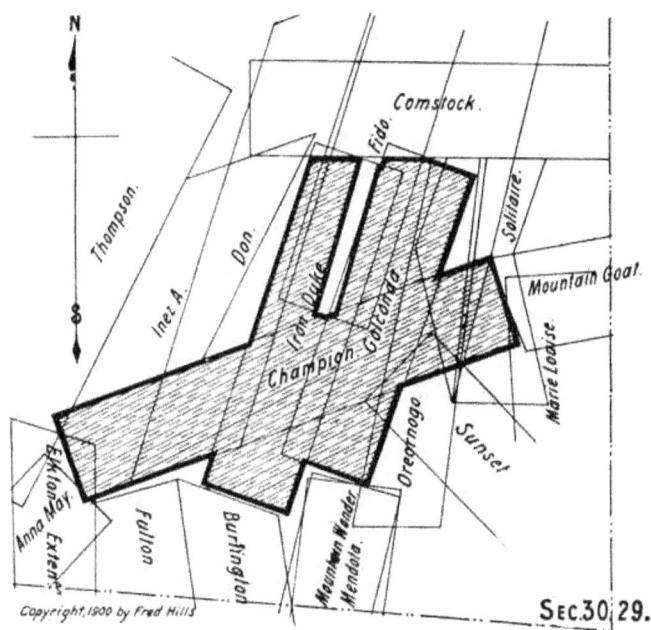

Copyright, 1900 by Fred Hills

SEC. 30 | 29.

Property

Owns Champion, Golconda and Iron Duke, about 17 3-4 acres in all, and patented. Situated in the N. E. 1-4 section 30, on Raven hill.

The south end of the Champion claim has been leased for two years to Becker & Co., of Denver, who are under contract to sink a 200-foot 3-compartment shaft.

Very little development has been done on the property, but lately steps have been taken to prosecute the work vigorously on account of good ore having been discovered on every side of the group.

Highest price for stock during 1899, 10 1-4 cents.

The Chicolo Gold Mining Company.

Incorporated December 7, 1895.

Directors

D. N. Heizer, president; W. G. Purdy, vice-president; C. E. Heizer, secretary; E. A. Colburn, treasurer. C. H. Dudley. W. H. Truesdale.

Main Office—16 N. Nevada avenue, Colorado Springs, Colorado.

Capitalization

1,500,000 shares. Par value, $1.00.

In treasury January 1, 1900, 16,663 shares.

Property

Owns the Forlorn Hope, 10 acres, in the S. E. 1-4 of section 7, on Tenderfoot hill. Patented. Allegany, 7.156 acres, in the S. W. 1-4 of section 32, on Straub mountain, just south of Lawrence. Patented. Gold Bug,

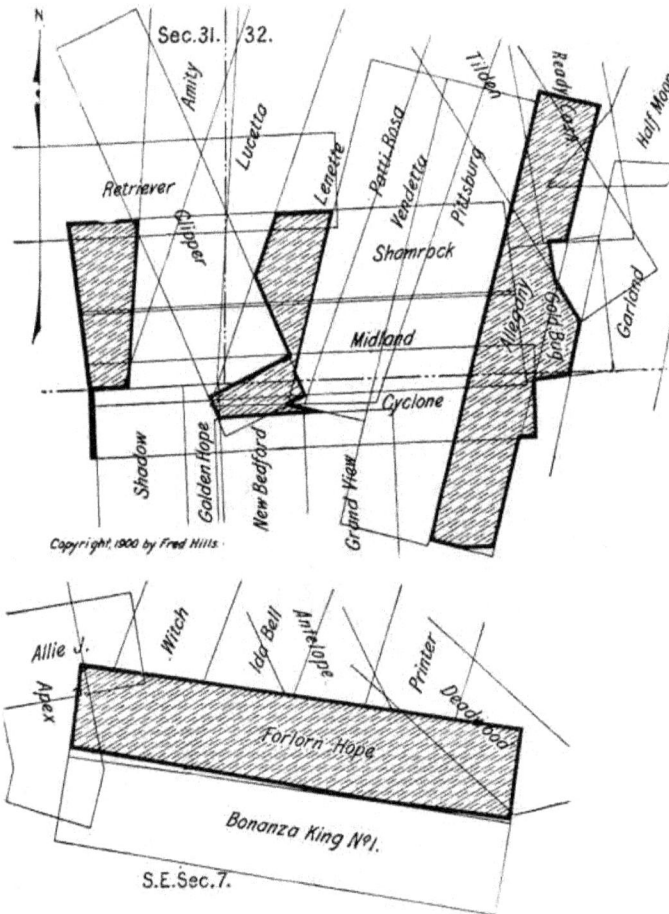

Shamrock, Cyclone and Midland, 15 acres, in the S. W. 1-4 of section 32, on Straub mountain. Receiver's receipt. The company also owns the Blue Bonnet, Emma B., Lone Boy and Mascot, 37.749 acres, in the E. 1-2 of section 30, on the northeast slope of Rhyolite mountain, which can not be shown on plat. Patented.

Development

The Forlorn Hope has a shaft 100 feet deep, with some drifting, while on the other property patent work only has been done. There is a lease on the Forlorn Hope running two years at a royalty of 20 per cent.

Highest price for stock during 1899, 4 3-4 cents; lowest price for stock during 1899, 1 cent.

The C. K. & N. Mining Company.

Incorporated March, 1894.

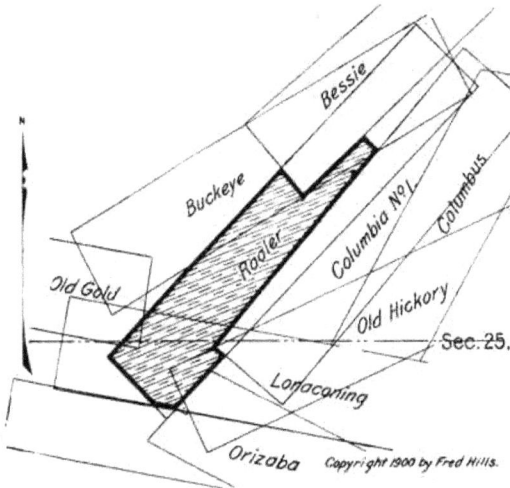

H. Hutchinson..........President   Directors
J. E. Hunter........Vice-President
Jno. L. Semmes..Sec. and Treasurer
K. Macdermid   D. H. Imler.

Main Office—24 and 25 Exchange Bank building, Colorado Springs, Colorado.

1,250,000 shares. Par value, $1.00.   Capitalization

Owns the Raaler, consisting of 5.2   Property
acres, in the N. E. 1-4 of section 25, on the west slope of Beacon hill.

The property is being developed by lessees and is under lease for a   Development
term of three years from November 30th, 1899, to Horace Granfield, who has the means to prosecute development work. As it adjoins the Orizaba of the El Paso G. M. Co., the prospects are good to make a producer.

Highest price for stock during 1899, 2 3-4 cents; lowest price for stock during 1899, 2 cents.

The Clyde Gold Mining Company.

Incorporated March 6, 1900.

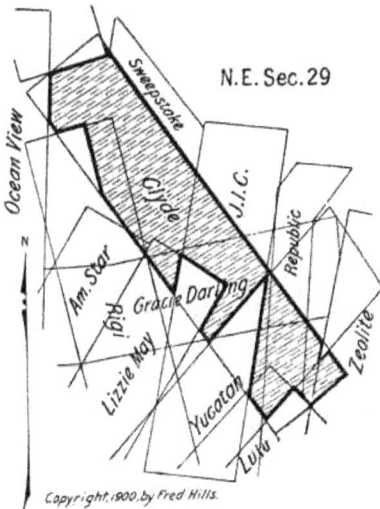

Frank C. Andrews..President and Treasurer   Directors
Frank H. Pettingell.................
....Vice-President and General Manager
Fred S. Osborn..................Secretary
M. F. Brabb.   Chas. A. Brace.

Main Office—11 Bank block, Colorado Springs, Colorado.

1,000,000 shares. Par value, $1.00.   Capitalization

In treasury January 1, 1900, 200,000 shares; in treasury January 1, 1900, $7,000 cash.

Owns the Clyde, about 7 acres, in the N.   Property
E. 1-4 of section 29, Survey No. 8985, situated on Battle mountain. Property is patented.

One shaft, 4x8 in the clear, 600 feet deep, with about 300   Development
feet of drifting; one drift north, one east and one west; also one 200-foot shaft 3 1-2x7, with 435 feet drifting from bottom. About 20 other shafts have been sunk, varying from 10 to 78 feet. Several open cuts and trenches. The greater part of the work has been done near the middle of the west side line; 12x14 steam hoist; 100-horse power boiler; large shaft house of about 75x35 feet; blacksmith shop; powder magazine; office and store house. One 150-gallon pump at 500-foot station, one sinker pump below the 500-foot station, and five machine drills. As this is a private corporation there has been no stock offered for sale.

The C. O. D. Gold Mining Company.

Incorporated 1892.

Directors

P. E. C. Burke...........................President
Chas. L. Tutt.............. .... Vice-President
Chas. E. Howard...........Secretary and Treasurer
H. A. Watson.   C. J. Cover.

Main Office—No. 269 Bennett avenue, Cripple Creek, Colorado.

Capitalization

1,000,000 shares.   Par value, $1.00.

In treasury January 1, 1900, 100,000 shares; in treasury January 1, 1900, $1,000 cash.

Copyright 1900 by Fred Hills.

Property

Owns a part of Vindicator and C. O. D. No. 2, situated in S. W. 1-4 of section 21, on the south slope of Bull hill; also owns 9.8 acres of the Rosario, Mollie Noble and the Lulu M., in the same location.   The total acreage owned by this company is about 15.3 acres.   All patented.   The ownership of the tracts A. and B. is in controversy between this company and the Keystone Mining and Milling Company, and this ground is shown in dotted lines.

Development

There are several shafts ranging from 30 feet to 125 feet deep. The greater part of the development work is being done on the Lulu M. and the C. O. D.   At present the property is being worked by lessees. No stock is being offered for sale in public.

114

The Colfax Mining Company.

Incorporated March 5, 1892.

F. H. Pettingell......................President
A. P. Mackey.....................Vice-President
L. A. Civill.........................Secretary
J. R. McKinnie.....................Treasurer

A. B. Noxon.

Directors

Main Office—No. 11, Bank block, Colorado Springs, Colorado.

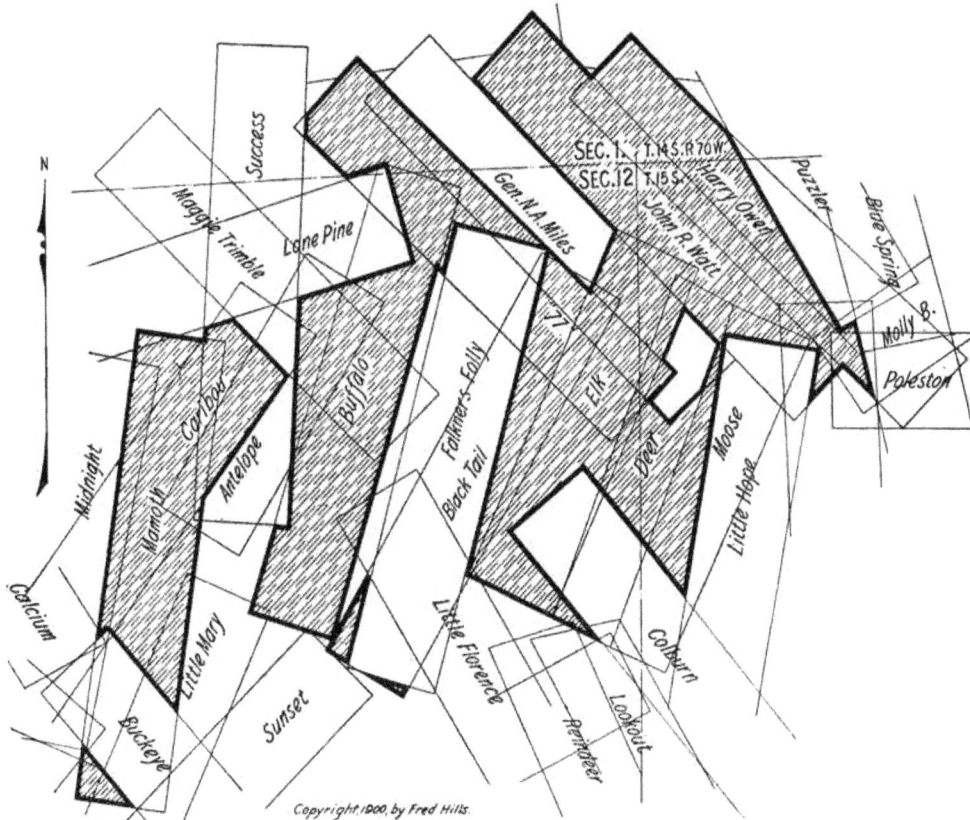

Copyright, 1900, by Fred Hills.

1,000,000 shares. Par value, $1.00.

In treasury January 1, 1900, 30,784 shares.

Capitalization

Owns the Mammoth, survey No. 8,951, containing 7,690 acres, on the north slope of Mineral hill; also the Antelope, Moose, Caribou, Buffalo, Black Tail, 77, Deer, John R. Watt, Harry Owen and the Elk, survey No. 9,572, containing in all 45.590 acres, all in the N. W. 1-4 section 12. The Mammoth is patented. Receiver's receipt held for the remainder.

Property

The Colomokas Gold Mining Company.

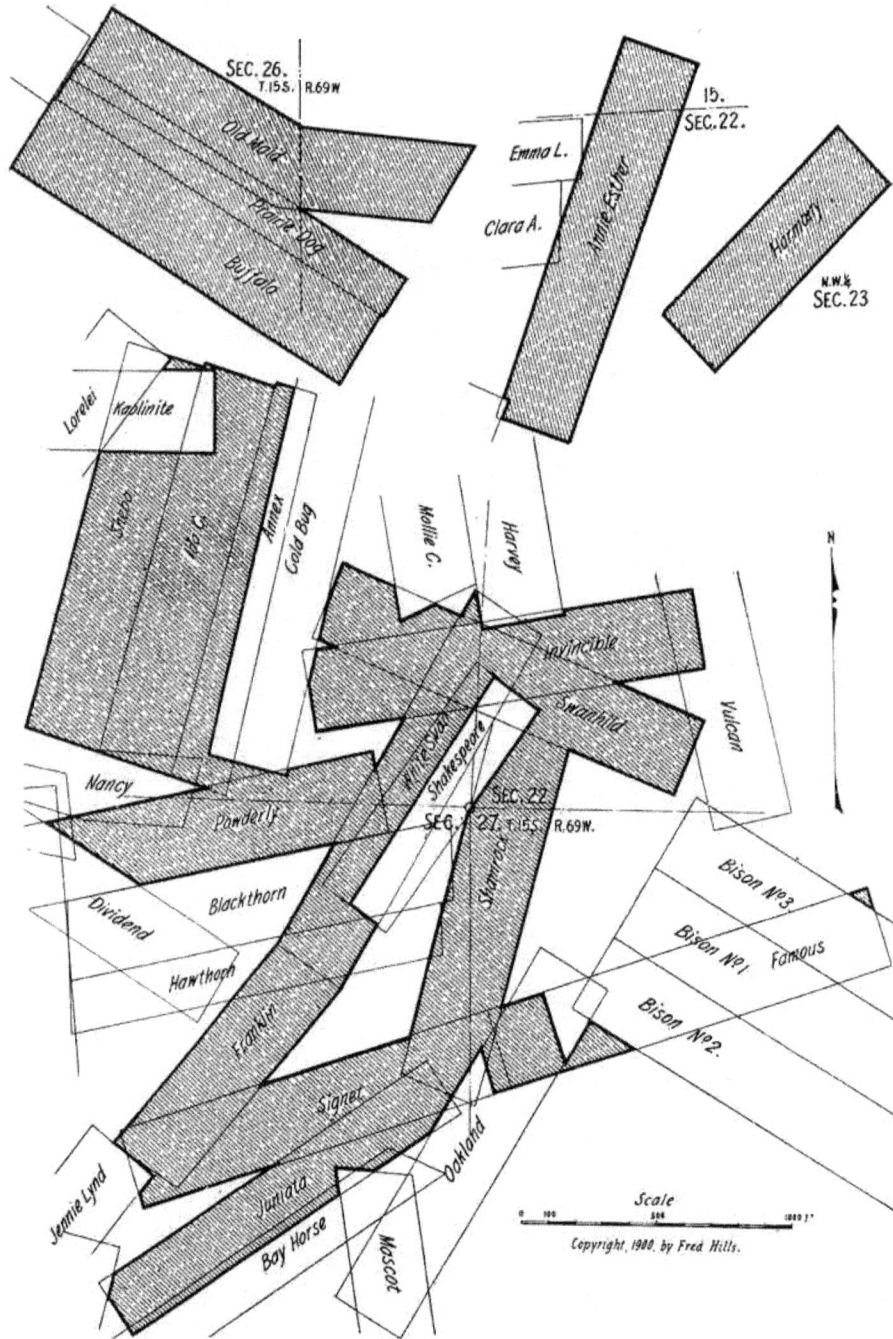

SEC. 26.
T.15 S. R.69 W

Old World

Prairie Dog

Buffalo

15.
SEC. 22.

Emma L.

Clara A.

Annie Esther

Harmony

N.W. ¼
SEC. 23

Lorelei

Kaolinite

Shelo

Ida C.

Annex

Gold Bug

Mollie C.

Harvey

Invincible

Swanhild

Vulcan

N

Shakespeare

Nancy

Powderly

SEC. 22.

SEC. 27. T.15 S. R.69 W.

Bison Nº 3

Dividend

Blackthorn

Shamrock

Bison Nº 1   Famous

Hawthorn

Bison Nº 2.

Franklin

Signet

Oakland

Jeannie Lynd

Juniata

Bay Horse

Mascot

Scale

0   100   500   1000 ¹'

Copyright, 1900, by Fred Hills.

116

The Colomokas Gold Mining Company.

Incorporated January 9, 1896.

E. S. Walters, president; J. W. Sheafor, vice-president; F. W. Kors- meyer, secretary; A. M. Korsmeyer, treasurer; J. W. Ady, E. L. Sumner, H. F. Gourley.

Main Office—Room No. 12, Exchange Bank building, Colorado Springs, Colorado.

5,000,000 shares. Par value, $1.00.
In treasury January 1, 1900, 2,400,000 shares; in treasury January 1, 1900, $100.00 cash.

Owns the Annie Esther, in the N. E. 1-4 section 22, on Cow moun- tain; the Harmony, in the N. W. 1-4 section 23, on Cow mountain; the Buffalo, Prairie Dog and Old Maid, in the N. W. 1-4 section 26; the Signet, Famous, Juiniata, Shamrock, Powderly, Annex, Ida C, Sheba, White Swan, Swanhilde, Franklin and the Invincible, all in the S. 1-2 section 22, and in the N. 1-2 section 27, all on Invincible hill, a half mile east of the Victor mine. Total acreage of this company is 150 acres. The Annie Esther, Harmony, Signet, Famous, Shamrock, Juiniata, Old Maid, Prairie Dog and Buffalo are patented. The remainder in process of patenting.

Horse whim, blacksmith shop, stable. Underground machinery, 100 feet track and two cars. There are 18 shafts, aggregating 1,300 feet of work; also 700 feet of tunnels and levels. The greater part of the development work is being done on the Invincible, Powderly and the Famous.

Highest price for stock during 1899, 5 cents; lowest price for stock during 1899, 2 1-2 cents.

The Colomont Gold Mining Company.

Incorporated August, 1899.

Main Office—24 Bank block, Colorado Springs, Colorado.

Capitalization

2,000,000 shares. Par value, $1.00.

In treasury January 1, 1900, 535,000 shares.

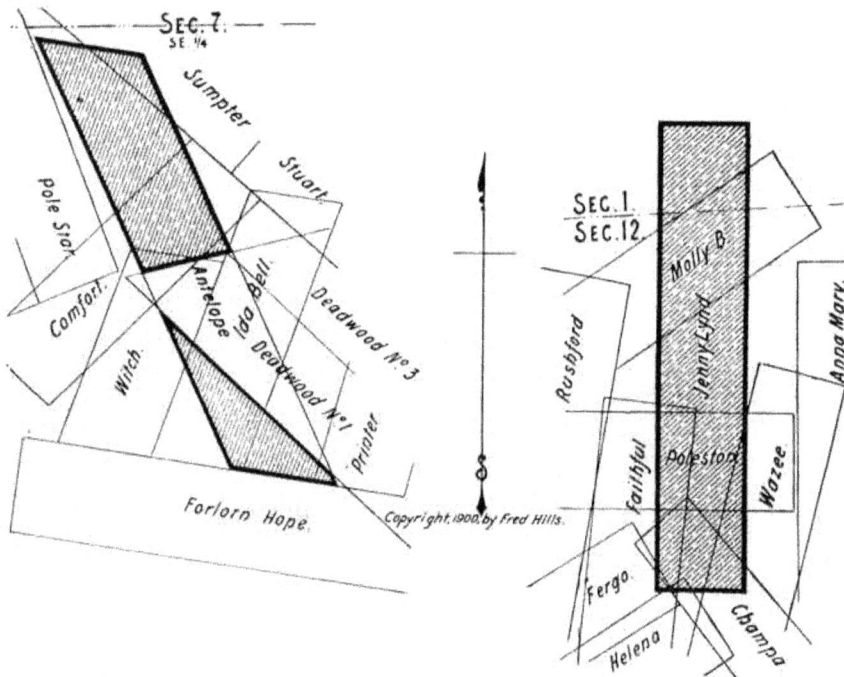

Property

Owns the Antelope, in the S. E. 1-4 of section 7, containing 7 acres, and the Jenny Lynd, in the N. E. 1-4 of section 12, containing 10 acres. Both patented.

Development

On the Antelope is a shaft house and horse whim. Three shafts, ranging in depth from 25 to 50 feet, have been sunk on the Antelope and contract let for 100 feet. A 4-foot vein disclosed. Will run a general average of $16.80 at ten feet deep. The Jenny Lynd has a 50-foot shaft. Mill returns on a 10-ton shipment, caught on plates, Beaver Park mill, 1895, gave $10.80 per ton. Greater part of the development work has been done on the Antelope. On April 24, at a depth of 75 feet, a valuable vein of ore two feet wide, averaging $30 per ton, was opened up.

Highest price for stock during 1899, 2 1-2 cents; lowest price for stock during 1899, 2 1-2 cents.

The Colonial Dames Gold Mining Company.

Incorporated August, 1896.

C. H. Emerson,
        President.
Walter A. Stebbins,
        Vice-President.
Wm. G. Liese,
        Secretary.
Ben A. Metcalf,
        Treasurer.
Carl Johnson.
Geo. Hummer.

Main Office—Colorado Springs, Colorado.

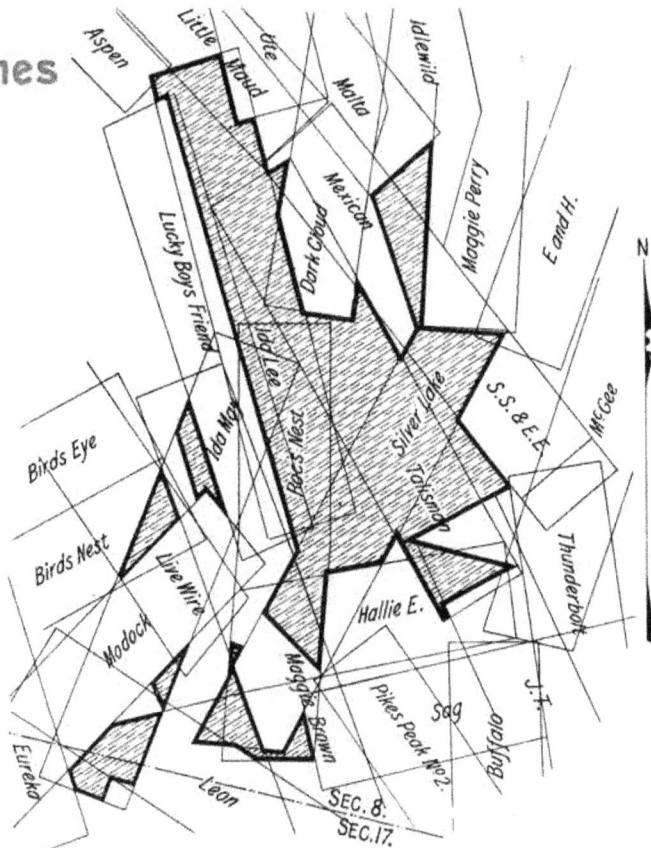

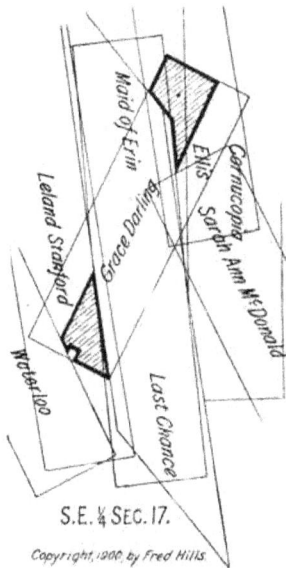

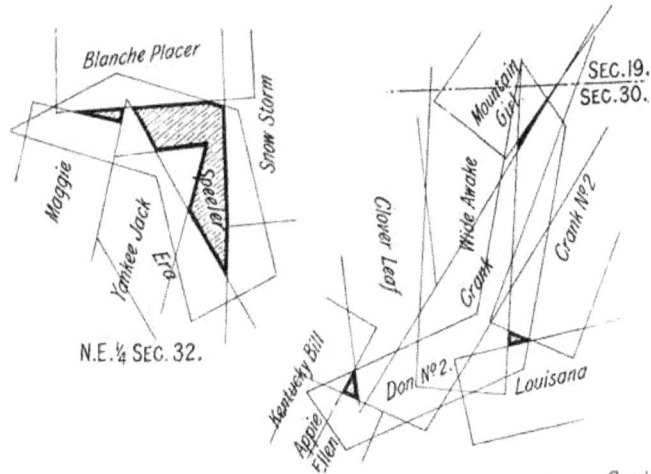

Copyright, 1900, by Fred Hills.

Capitalization   2,000,000 shares. Par value, $1.00. In treasury January 1, 1900, 100,000 shares. In treasury January 1, 1900, $300.00 cash.

Property   Owns the Grace Darling, containing nearly 2 acres, in section 17, on the north slope of Bull hill; the Dames group, containing the Ida Lee, Talisman, Silver Lake, Roc's Nest, Ida May and Talesman, 27 acres in all, situated partly in section 17 and partly in section 8; the Don No. 2, containing nearly one acre, in section 30, on Raven hill; and the Speeler, containing nearly 3 acres, in section 32, on the south slope of Battle mountain. The Grace Darling, Ida Lee, Talisman, Silver Lake, Roc's Nest and Ida May are patented. The others in process.

Development   The Grace Darling has two shafts of 60 and 80 feet respectively. The Ida Lee has a shaft 250 feet deep. The Silver Lake has a shaft of 35 feet; the Roc's Nest one of 50 feet. The greater part of the development work has been done on the Grace Darling.

Highest price for stock during 1899, 3 cents; lowest price for stock during 1899, 2 1-2 cents.

119

The Colorado City and Manitou Prospecting and Mining Company.

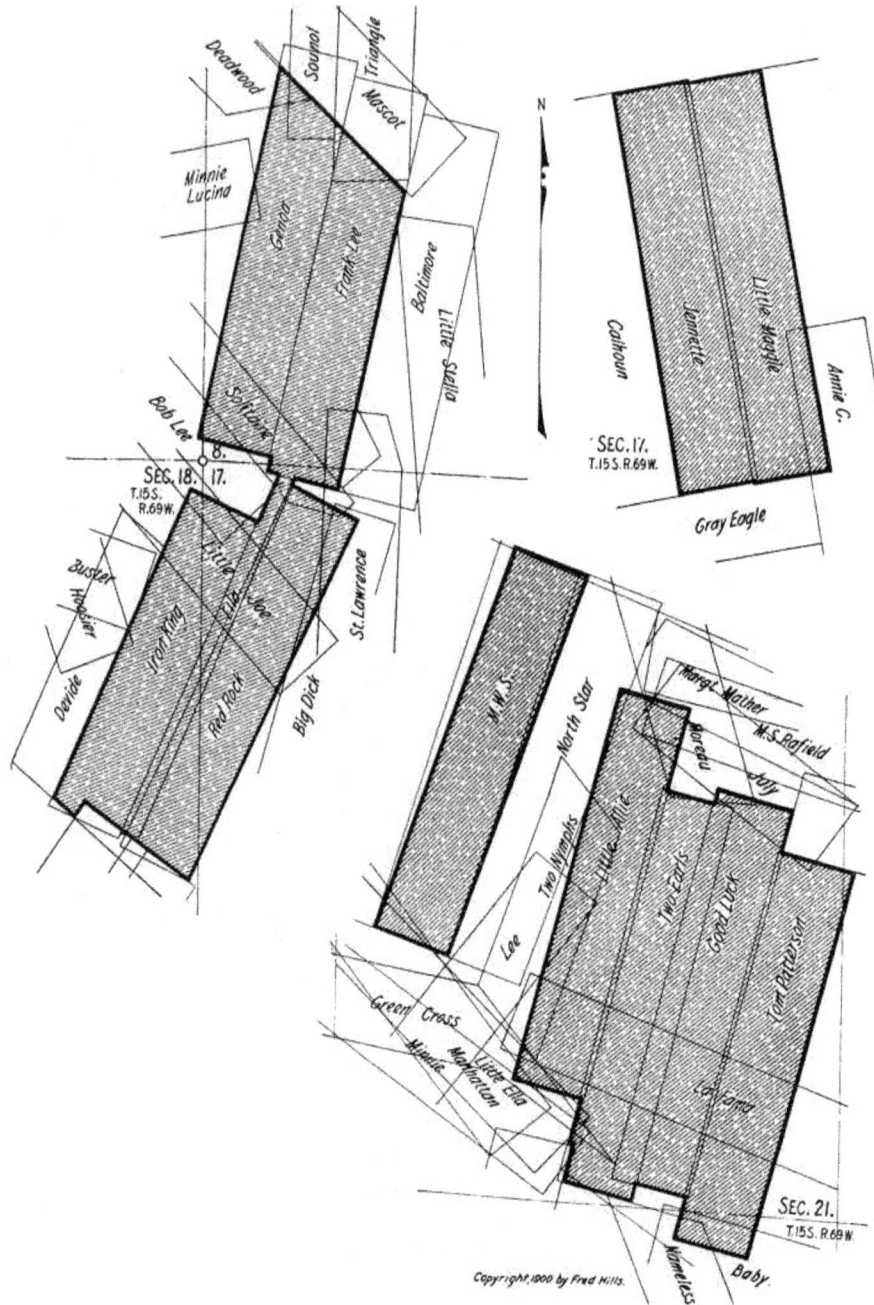

Deadwood · Soapoi · Triangle · Mascot · Minnie Lucina · Genoa · Franklin · Baltimore · Little State · Bob Lee · Solitaire · SEC. 18. 17. · T.15 S. R.69 W. · Buster · Hoosier · Little Joe · Iron King · St. Lawrence · Divide · Red Rock · Big Dick

Calhoun · Jennette · Little Maggie · Annie C. · SEC. 17. T.15 S. R.69 W. · Gray Eagle

M. M. S. · North Star · Mary L. Mather · M. S. Rofield · Moravia · Betty · Two Nymphs · Little Allie · Little Ella · Two Earls · Good Luck · Tom Patterson · Lee · Green Cross · Moravia · Manhattan · San Juan · SEC. 21. T.15 S. R.69 W. · Nameless · Baby.

Copyright, 1900 by Fred Hills.

The Colorado City and Manitou Prospecting and Mining Company.

Incorporated 1895.

A. Z. Sheldon.........................President
T. J. Dalzell...........Vice-President and Manager
J. M. Jackson.........................Secretary
F. E. Robinson.........................Treasurer
Albert Wagner.

Directors

Transfer Office—No. 29 Midland block, Colorado Springs, Colorado.

2,000,000 shares. Par value, $1.00.

Capitalization

In treasury January 1, 1900, 26,197 shares; in treasury April 1, 1900, $200.00 cash; present indebtedness of the company, $1,000.00.

Owns the Iron King, Ella, Red Rock, Genoa and Frank Lee, between 35 and 40 acres, in the N. E. 1-4 section 18, the N. W. 1-4 section 17, and the S. W. 1-4 section 8, on Tenderfoot hill; the Jennette and Little Maggie, 19.69 acres, in the N. W. 1-4 section 17, on Galena hill; the Little Allie, Two Earls, Good Luck and the Tom Patterson, 36 acres, in the N. W. 1-4 section 21, on Bull hill; the M. W. S., 10.329 acres, in the N. W. 1-4 section 21, on Bull hill; in all, over 100 acres. All patented.

Property

The Elks M., M. & L. Co. is working on the south end of the Iron King, Red Rock and Ella claims, on Tenderfoot hill. They have sunk a shaft 135 feet, and there are about 500 feet of cross-cuts. Work is steadily progressing and the indications for pay ore seem favorable. The north end of these claims is leased to lessees, who are sinking 20 feet per month. Lessees are working with a steam plant on the Genoa and Frank Lee, and also on the Jennette and the Little Maggie. On the latter property the shaft is down 280 feet and a large amount of cross-cutting has been done. The Progress G. M. Co. has a bond and lease on the M. W. S. claim, the bond being for $35,000, expiring August 1, 1900. This bond will probably be taken up by the Progress Company, and the money so obtained will be used by the company to develop some of their other properties. The policy of the company is to get all their properties leased to good parties, who will thoroughly develop them.

Development

Highest price for stock during 1899, 21 cents; lowest price for stock during 1899, 2 7-8 cents.

The Colorado Springs Gold Mining Company.

Incorporated 1899.

Directors

L. W. Cunningham, president; J. Bishof, vice-president; J. K. Miller, secretary; K. Macdermid, treasurer; Arthur Knecht.

Main Office—First National Bank building, Colorado Springs, Colorado.

Capitalization

1,500,000 shares. Par value, $1.00. In treasury January 1, 1900, 400,000 shares; in treasury January 1, 1900, $1,200.00 cash.

Property

Owns Oro lode, containing 6.7 acres, in section 17, on Ironclad hill. Patented.

Development

There are four shafts of an average depth of 60 feet on this property.

Since this is a new corporation, very little development work has been done, but the company is preparing to push operations.

Highest price for stock during 1899, 3 cents; lowest price for stock during 1899, 2 1-2 cents.

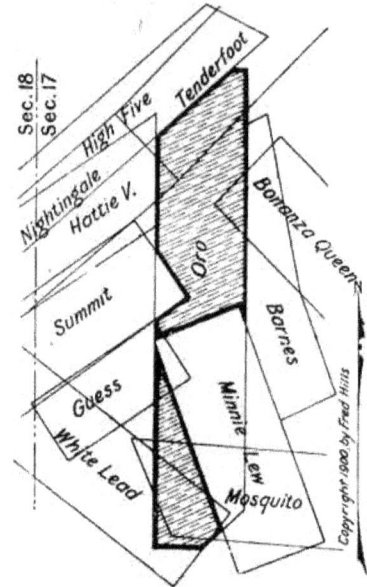

The Columbia and Cripple Creek Gold Mining Company.

Incorporated January 31, 1896.

Directors

L. B. Grafton, president; J. F. Humphrey, vice-president; J. F. Lilly, secretary and treasurer; J. C. Plumb, M. A. Leddy, E. E. Nichols, Jr., F. W. Howbert.

Main Office—Rooms 17 and 18, Exchange Bank block, Colorado Springs, Colorado.

Capitalization

2,500,000 shares. Par value, $1.00.

In treasury January 1, 1900, 700,000 shares; in treasury April 1, 1900, $700.00 cash.

Property

Owns the El Paso placer claims Nos. 1, 2 and 4, in the W. 1-2 section 9; the S. W. 1-4 section 10; the N. W. 1-4 section 15, all in township 15 south, range 69 west, adjoining school section 16, and three-fourths mile northeast of Grassy townsite, a total of 341.003 acres, on Cow, Calf and Trachyte mountains and Galena hill.

Development

Sufficient work to secure patent has been done.

History

Until the past year the company has done all work necessary to perfect title and acquire patent, and the widening of the gold belt in the Cripple Creek district gives to this company splendid prospects for the future. The opening up of good ore in school section 16, and the fact that the veins trend in the direction of the property owned by this company, together with the fact that very encouraging assays have been found upon the property, gives to the company great encouragement, and the new management expects to bring the property to the front. The recent strike upon the property of the Mayflower Gold Mining Company has added much to the outlook for this part of the district.

Highest price for stock during 1899, 5 cents; lowest price for stock during 1899, 1 3-4 cents.

The Columbia and Cripple Creek Gold Mining Company.

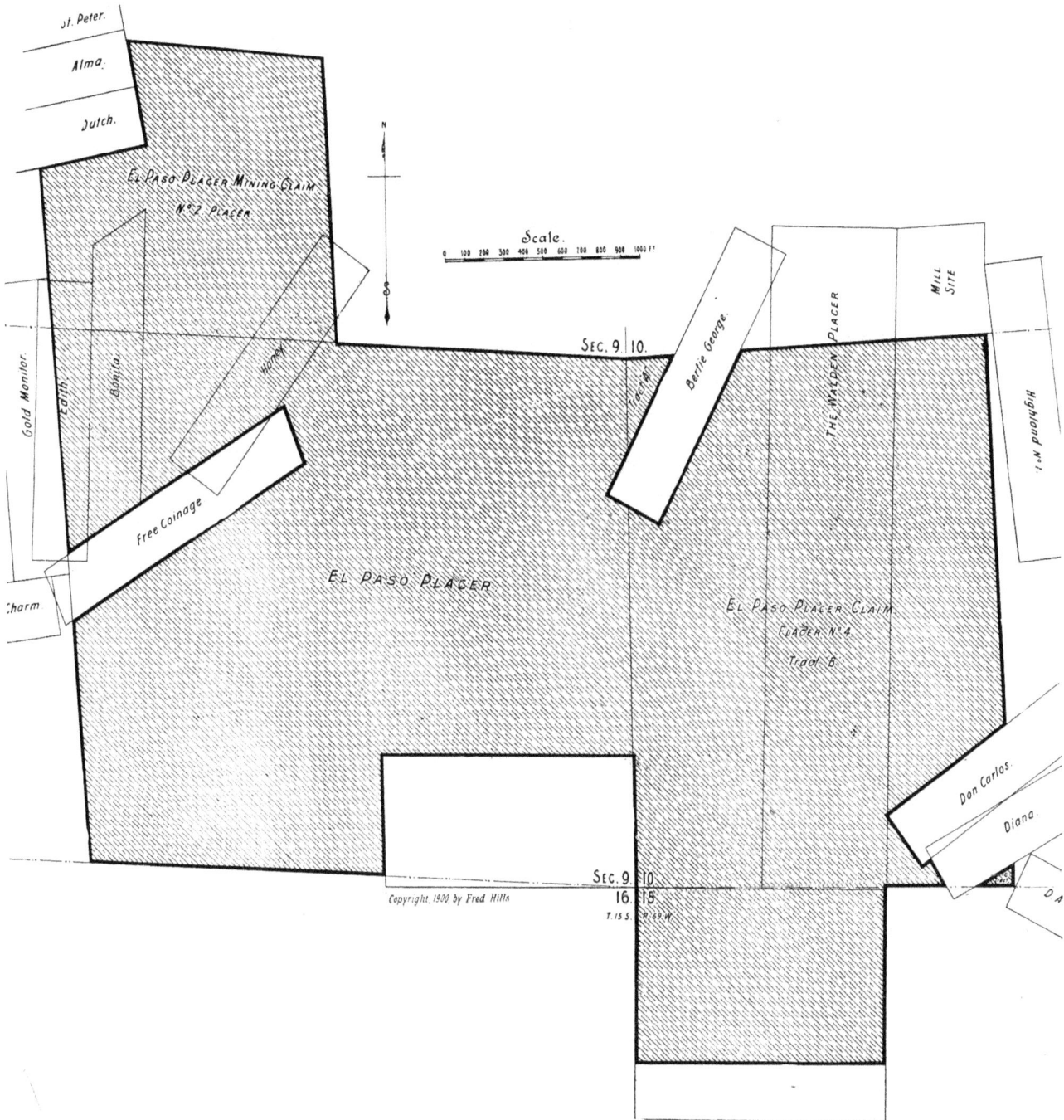

St. Peter.

Alma.

Dutch.

El Paso Placer Mining Claim
N.º 2 Placer

Scale.
0 100 200 300 400 500 600 700 800 900 1000 FT.

N

Gold Monitor

Edith.

Bonita.

Hopper

Free Coinage

Charm

SEC. 9. 10.

Tract #1

Bertie George.

The Walden Placer

Mill Site

Highland N.º 1

EL PASO PLACER

El Paso Placer Claim
Placer N.º 4
Tract B

Don Carlos

Diana

D A

Copyright, 1900 by Fred Hills

SEC. 9. 10.
16. 15.
T. 15 S.   R. 69 W.

The Columbine-Victor Deep Mining and Tunnel Company.

The Columbine-Victor Deep Mining and Tunnel Company.

Incorporated March 1, 1894.

Warren Woods...................President	Directors
H. E. Woods.................Vice-President	
F. M. Woods............Secretary and Treasurer	
C. L. Arzeno. J. M. Allen.	

Main Office—Victor, Colorado.

Branch Office—Colorado Springs, Colorado.

2,000,000 shares. Par value, $1.00.   Capitalization

Owns the Newmarket, 3 acres; the Windy, 2.7 acres; the Rose Bud, Property
3.8 acres; the Brunette, 2.13 acres; the Dorothy, .05 acre; the Zitella,
1.3 acres; the K. P. O. A., 4.2 acres; two-thirds of the Little Maggie;
the May B. and the M. K. & T., 18.6 acres; the Deadwood, 4 acres; one-
eighth of the Canuck, 8 acres; one-eighth of the Lottie May, 3-4 acre;
the Ethel, 1 3-4 acres; the Clemma E., 2 acres; one-half of the Santa
Christo, 3-4 acre; one-sixth of the Mary Schroeder, 1-2 acre; the Bonanza,
4 acres; all situated in sections 29 and 30. This company also controls
the property of the Columbine Gold Mining Company.

The Columbine-Victor tunnel extends 3,740 feet through Squaw Development
mountain. In connection with this tunnel there has also been done
1,700 feet of drifting. On the M. K. & T. there is a shaft 200 feet deep,
and 500 feet of drifting. The K. P. O. A. has a tunnel of 260 feet. In
addition to the above development work there are twenty other shafts,
ranging in depth from 50 to 150 feet.

Highest price for stock during 1899, 24 cents; lowest price for stock
during 1899, 12 cents.

The Comanche Plume Mining Company.

Incorporated March, 1900.

Directors

W. V. Pettit................President
C. H. Bryan................Secretary

Main Office—10 Hagerman building, Colorado Springs, Colorado.

Capitalization

1,500,000 shares. Par value, 1 cent.
In treasury March 1, 1900, 250,000 shares; in treasury March 1, 1900, $10,000 cash.

Property

Owns the Mountain Goat, Magdalene, Little Edith and Stella Girl, 7 1-2 acres in all, in the N. W. 1-4 of section 29, on Battle mountain; also owns 2 1-2 acres additional vein rights. All the property is patented.

Development

One tunnel of 150 feet.

Stock selling at 7 cents, May, 1900.

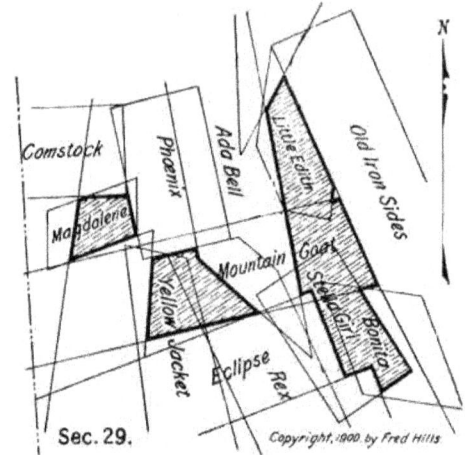

The Combination Gold Mining Company.

Incorporated February 10, 1900.

Directors

H. L. Shepherd...............................President
R. W. Griswold.........................Vice-President
J. E. Jones.....................Secretary and Treasurer

Main Office—267 Bennett avenue, Cripple Creek, Colorado.

Capitalization

100,000 shares. Par value, $1.00.

Property

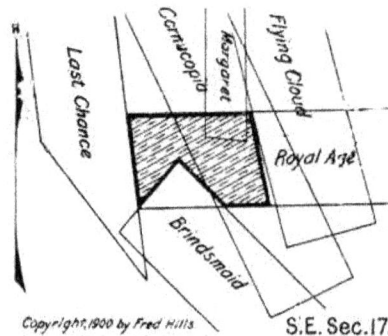

Owns the Royal Age lode claim, consisting of 2 acres, in the S. E. 1-4 section 17, on Bull hill, adjoining the Pinnacle's property on the northwest.

Development

There is a 90-foot surface tunnel and a 50-foot shaft on the property.

No stock having changed hands, there were no prices quoted.

The Commercial Men's Gold Mining and Milling Company.

Incorporated February 12, 1896.

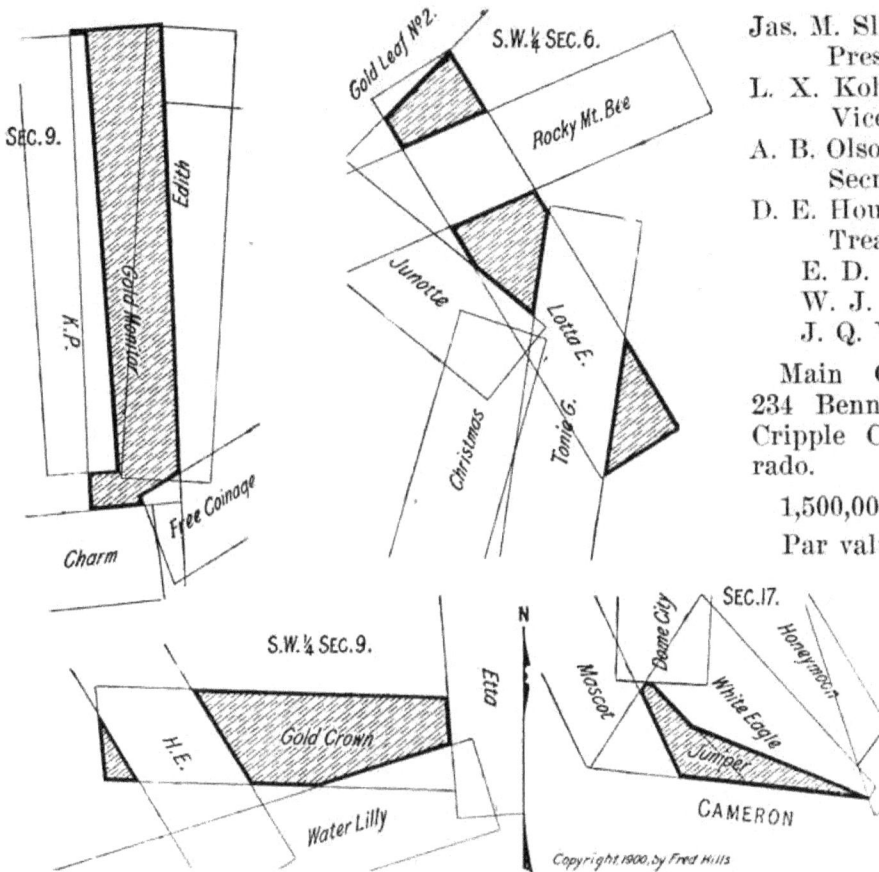

Copyright 1900, by Fred Hills

Jas. M. Slusher,
  President;
L. X. Kohlman,
  Vice-President;
A. B. Olson,
  Secretary;
D. E. Houck,
  Treasurer;
  E. D. Hoopes,
  W. J. Dunigan,
  J. Q. Van Orsdol.

Directors

Main Office — No. 234 Bennett avenue, Cripple Creek, Colorado.

1,500,000 shares. Par value, $1.00.

Capitalization

Owns the Lottie E., containing 2 acres, in the S. W. 1-4 section 6, township 15 south, range 69 west; the Gold Crown, containing 4.608 acres, in the S. W. 1-4 section 9, township 15 south, range 69 west; the Gold Monitor, containing 8.013 acres, in the S. W. 1-4 section 9, township 15 south, range 69 west; and the Jupiter, containing 1.726 acres, in the N. E. 1-4 section 17, township 15 south, range 69 west. Total acreage is 16 acres. Receiver's receipts held for the above property.

Property

The Gold Monitor has a shaft 75 feet deep; the Lottie E. has a shaft 50 feet deep and drift of 25 feet.

Development

Highest price for stock during 1899, 2 1-2 cents; lowest price for stock during 1899, 1-2 cent.

The Common-Wealth Mining and Milling Company.

Incorporated December 20, 1899.

Spencer Penrose, president; C. M. MacNeill, vice-president; J. McK. Ferriday, secretary; Geo. F. Fry, treasurer; R. W. Griswold.

Directors

Main Office—Room 11, El Paso Bank block, Colorado Springs, Colorado.

1,500,000 shares. Par value, $1.00.

In treasury January 1, 1900, 460,000 shares; in treasury January 1, 1900, $16,000.00 cash.

Capitalization

(CONTINUED ON PAGE 128.)

Property

Owns the Deadwood claim, 4.056 acres, in the N. E. 1-4 section 17, on Squaw mountain; the San Francisco, 6.639 acres, and the J. T., 6.685

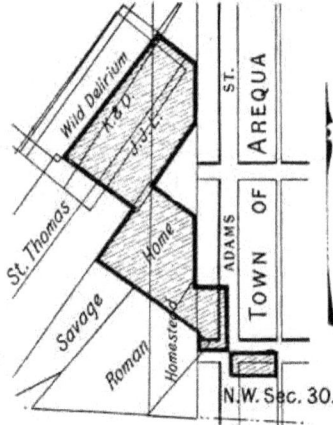

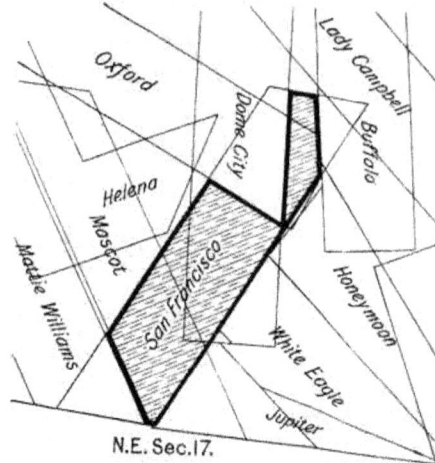

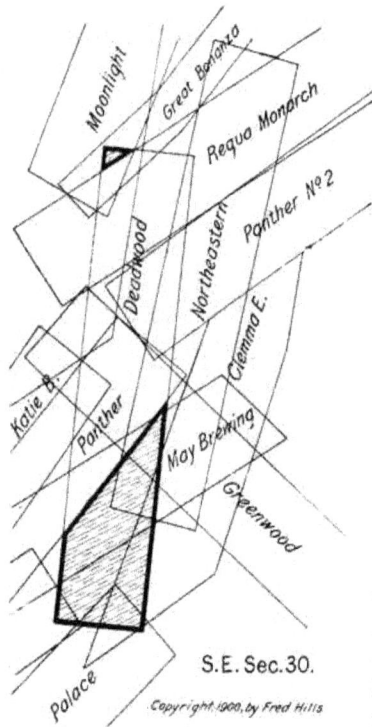

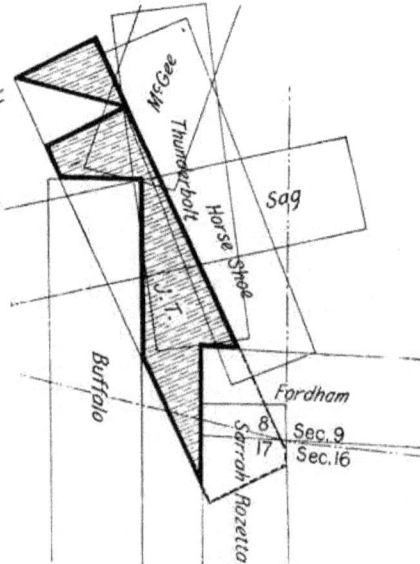

N.W. Sec. 30.

N.E. Sec. 17.

S.E. Sec. 30.

Copyright 1900, by Fred Hills

acres, in the N. E. 1-4 section 17, on Galena hill; also Home, J. J. L., lots 14 to 20 inclusive, block 6, and lots 38, 39 and 40, block 14, town of Arequa; 6.170 acres, E. 1-2 section 30, on Beacon hill. All patented, with the exception of that portion surrounded by a dotted line on the J. T., the company having the mineral rights, however, for the whole length of claim.

Development

There is an electric hoist and shaft house on the Home claim. The total present development consists of 525 feet of tunnels, 405 feet of shafts and 175 feet of drifts. The policy of the company will be to use its money judiciously to open up the J. J. L. and Deadwood. The J. J. L. shaft will be sunk to 460 feet, at an approximate cost of $9,000.00, and a shaft will be sunk on the Deadwood with the idea of opening up the Squaw mountain tunnel No. 4 vein, at an approximate cost of $2,500.00, which will leave a balance in the treasury of $4,500.00 for future use and for such contingent expenses as may occur. No salaries will be paid officers of the company at the present time.

The Consolidated Gold Mines Company.

Incorporated September 23, 1898.

Warren Woods.....................President
H. E. Woods...................Vice-President
F. M. Woods.........Secretary and Treasurer

Main Office—Victor, Colorado.

Branch Office—Colorado Springs, Colorado.

Directors

1,250,000 shares. Par value, $1.00. *Capitalization*

In treasury January 1, 1900, 5,000 shares; in treasury January 1, 1900, $25,000.00 cash.

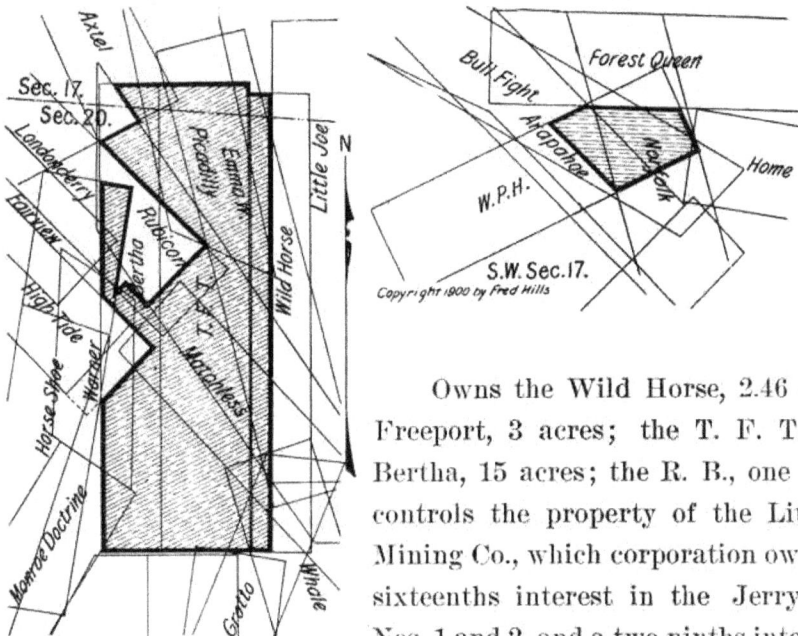

Owns the Wild Horse, 2.46 acres; the *Property* Freeport, 3 acres; the T. F. T. and the Bertha, 15 acres; the R. B., one acre; also controls the property of the Little Pedro Mining Co., which corporation owns a three-sixteenths interest in the Jerry Johnson Nos. 1 and 2, and a two-ninths interest in the Arapahoe and the M. E. P. The company also controls the property of the Bonanza Queen Mining Company, the New Zealand Mining Company, and has various other holdings.

No. 1 shaft on the Wild Horse is 348 feet deep, with 2,500 feet of *Development* underground workings, aside from stopes. The Gleason shaft, a three-compartment shaft on the Wild Horse, is 300 feet deep. On the Wild Horse, the T. F. T. and the Bertha there are 10 other shafts, ranging in depth from 30 to 100 feet. Good buildings, complete plant and equipment on both the Wild Horse shafts.

Gross production to January 1, 1900, $500,000.00; dividends to *Production* January 1, 1900, $60,000.00; last dividend April 25, $10,000.00. *and Dividend*

Highest price for stock during 1899, $1.25 per share; lowest price for stock during 1899, 60 cents per share.

5

The Consolidated Night Hawk and Nightingale Company.

Main Office—Colorado Springs, Colorado.

Capitalization

1,250,000 shares. Par value, $1.00.

Property

Owns the Night-Hawk and the Nightingale, containing about 18 acres, in section 20, on Bull Hill. This property has produced about $20,000.00, and in March, 1900, Mr. W. S. Stratton bought the whole of the stock for 20 cents per share. Unable to state Mr. Stratton's future policy.

The Constantine Consolidated Mining Company.

Incorporated 1899.

Directors

L. R. Ehrich..............President
J. R. McKinnie........Vice-President
N. S. Gandy..............Secretary
W. C. Stark..............Treasurer
Geo. E. Hasey.

Main Office—Hagerman building, Colorado Springs, Colorado.

Capitalization

1,500,000 shares. Par value, $1.00.

In treasury January 1, 1900, about 400,000 shares; in treasury January 1, 1900, about $600.00 cash.

Property

Owns the Constantine and Julia E., containing 10 acres, in the S. E. 1-4 section 19, on Raven hill. Patented.

Development

The company will develop their property through lessees, and the ground is all leased at the present time. There is a shaft and drifts amounting to probably 300 feet on each claim. The Julia E. adjoins the Jack Pot and Doctor, and was located and patented with the Doctor, which adjoins the Jennie Sample of the Gould Company and the Elizabeth Cooper of the Nugget Company. The Constantine lies between the Raven Company's property and the Joe Dandy mine.

Highest price paid during 1899, about 10 cents; lowest price paid during 1899, about 10 cents.

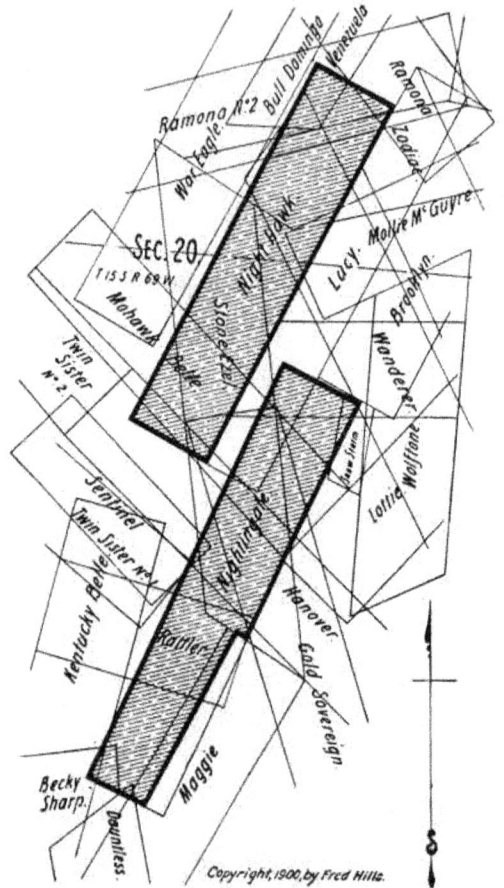

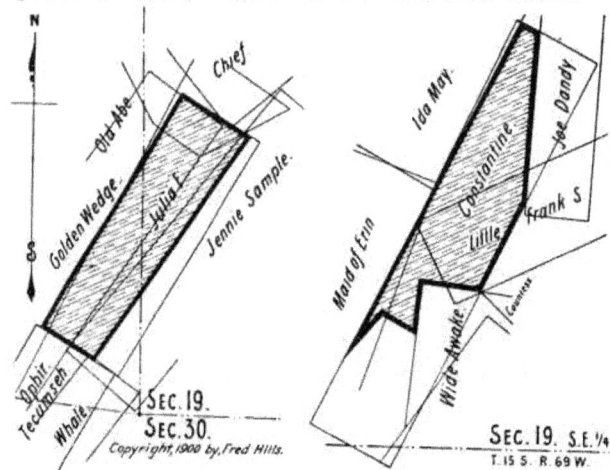

The Constitution Gold Mining Company.

Incorporated February, 1896.

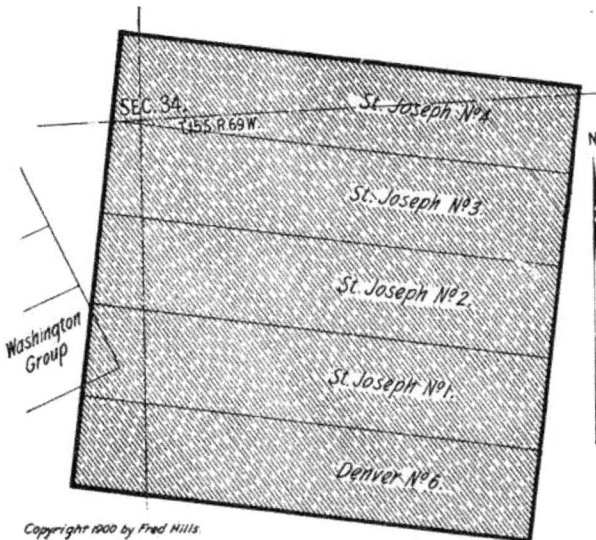

R. A. Handy........President Directors
J. W. Nicholas. .Vice-President
C. E. Tyler. .Sec. and Treasurer
Wm. High.

Main Office—17 N. Tejon street, Colorado Springs, Colorado.

1,250,000 shares. Par value, Capitalization $1.00.

In treasury January 1, 1900, about 150,000 shares.

Owns the Denver No. 6 and Property the St. Joseph Nos. 1, 2, 3 and 4; all full claims, situated in section 34, on Big Bull mountain. Property is held by location.

Four years' assessment work has been done on each claim. No stock sold on market.

The Contact Gold Mining and Tunnel Company.

Incorporated.

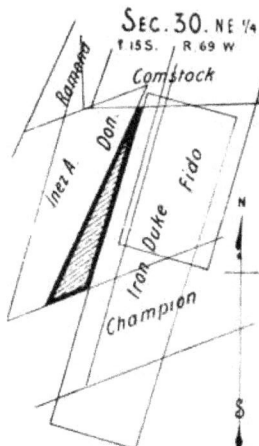

Frank Timmis.............President and Manager Directors
L. F. Stephens........................Secretary

Main Office—Room 2 Safety Deposit building, Cripple Creek, Colorado.

1,500,000 shares. Par value, $1.00. Capitalization

In treasury January 1, 1900, 255,000 shares; in treasury January 1, 1900, $4,500.00 cash.

Owns the Don lode, in the N. E. 1-4 of section 20, Property containing about 3-4 of an acre, patented; the Chicago Nos. 2, 3 and 4, containing 27 acres, on the southwest slope of Grouse mountain, in process of patenting. This property does not show on the map. The company also owns the Bull Cliff Trans. Drainage Tunnel and Tunnel Site, located on the northeast slope of Bull hill.

The Don lode has a shaft 225 feet deep and contract is let to sink it Development an additional 200 feet. The Bull Cliff tunnel is in 210 feet, partly double track.

Gross production to date, 100 tons low-grade ore from the Don lode. Production

Highest price for stock during 1899, 10 cents; lowest price for stock during 1899, 5 cents.

The Copper Mountain Gold Mining Company.

Incorporated.

Directors

M. Kinney..............................President
D. J. Duncan.......................Vice-President
C. M. Miller...........................Secretary
Frank G. Peck.........................Treasurer

Main Office—Exchange Bank block, Colorado Springs, Colorado.

Capitalization

1,000,000 shares. Par value, $1.00.

In treasury January 1, 1900, $100.00 cash; present indebtedness of the company about $1,500.00.

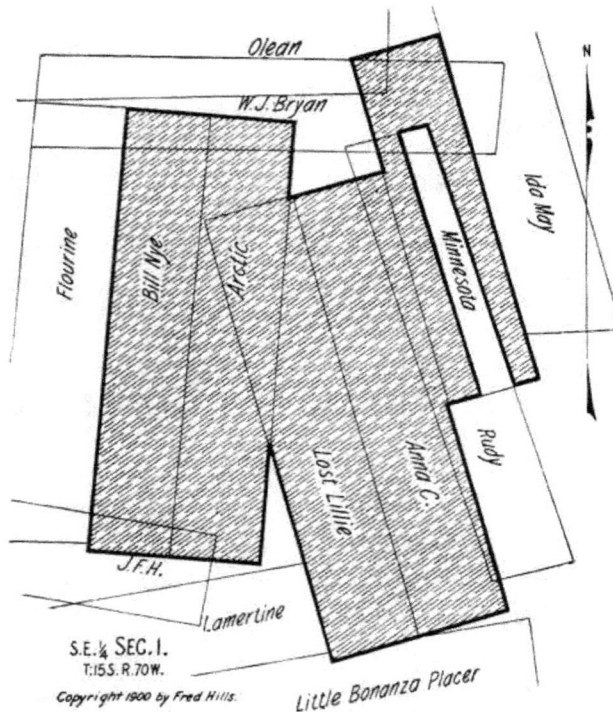

S.E. ¼ SEC. 1.
T:15 S. R 70 W.
Copyright 1900 by Fred Hills.

Property

Owns the Arctic, Bill Nye, Anna C., Lost Lillie and the Minnesota, containing, in all, 42.467 acres, situated in the S. W. 1-4 of section 1, on Copper mountain. All the above claims are patented.

Development

Two small shaft houses and one tool house. The Anna C. has an incline of 200 feet and a perpendicular shaft of 80 feet depth. The Lost Lillie has a shaft 200 feet deep and about 500 feet of drifting. On the Bill Nye there is a shaft 100 feet deep.

Production

Gross production to January 1, 1900, about 40 tons of about one-ounce ore.

Highest price for stock during 1899, 14 cents; lowest price for stock during 1899, 2 cents.

The Copper-Signal Gold Mining Company.

Incorporated September, 1899.

James F. Hadley, president; H. L. Shepherd, vice-president; Wm. M. ^{Directors} Broyles, secretary and treasurer. P. J. Lynch. P. E. C. Burke.

Main Office—Safe Deposit building, Cripple Creek, Colorado.

1,250,000 shares. Par value, $1.00. In treasury January 1, 1900, ^{Capitalization} 200,000 shares; in treasury January 1, 1900, $500.00 cash.

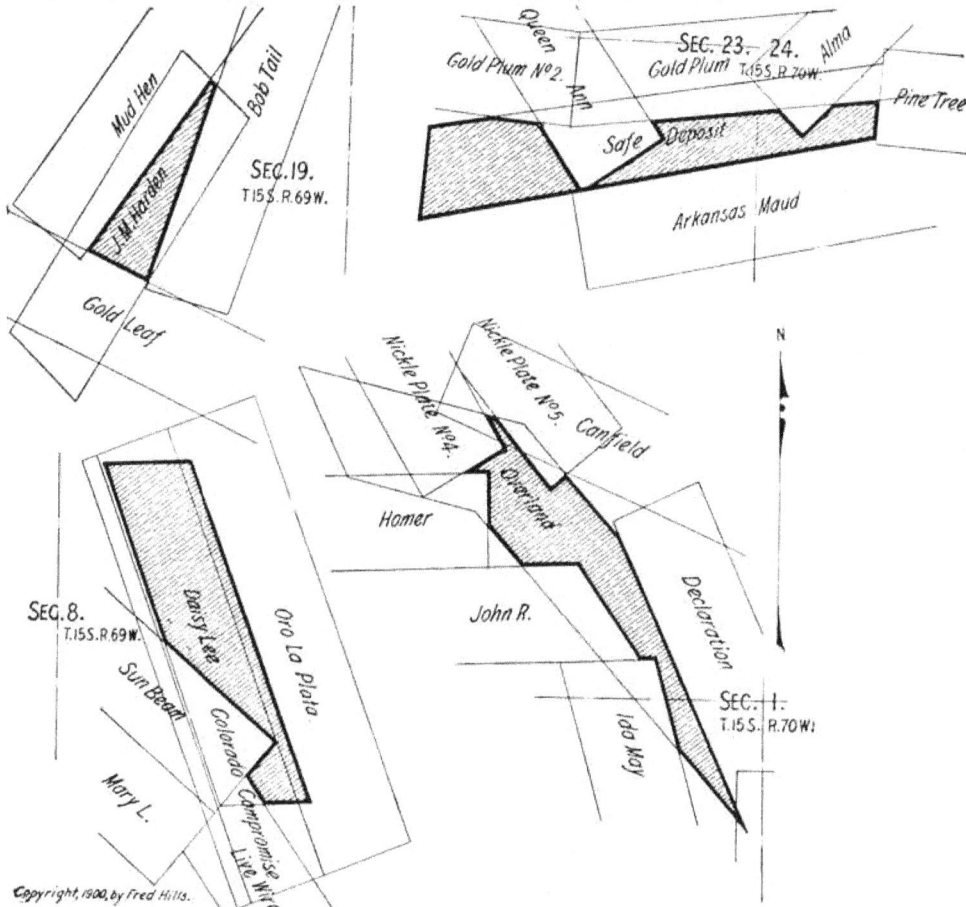

Owns the J. M. Harden, 2.16 acres, in the N. E. 1-4 of section 19, on ^{Property} Raven hill; the Daisy Lee, 5.821 acres, in the S. E. 1-4 of section 8, on Galena hill; the Overland, 5.998 acres, in the N. W. 1-4 of section 1, on Copper mountain; the Safe Deposit, 5.462 acres, in the E. 1-2 of section 23, on Signal hill. A total of 19.441 acres. The J. M. Harden and the Overland are patented. Receiver's receipt held for the Daisy Lee and the Safe Deposit.

Only patent work has been done on the Safe Deposit and the Daisy ^{Development} Lee. The Overland, which is leased, has a shaft 75 feet deep. The greater part of the development work is being done on the Harden. On this claim there is quite an amount of shallow work, but systematic work has only just begun. The lessees are just down to the vein at a depth of 25 feet. They will immediately put in machinery and probably erect an ore house. This claim lies about 650 feet northeast of the Morning Star, of the Enterprise Company, and has opened their vein. The Overland is being developed to catch the Nickle Plate vein, but has a good vein in the west end, carrying $28 per ton; not down below 20 feet. The Daisy Lee, which has made a good showing, will soon be leased.

Highest price for stock during 1899, 3 3-4 or 4 cents; lowest price for stock during 1899, promoted at 2 cents; latterly at 3 cents.

The Coriolanus Gold Mining Company.

Incorporated January 28, 1898.

Directors

> Jas. P. Pomeroy........................President
> W. H. Leonard.......................Vice-President
> Clarence Edsall.............Secretary and Treasurer
> T. B. Burbridge.   J. Arthur Connell.

Main Office—Hagerman building, Colorado Springs, Colorado.

Capitalization

1,500,000 shares. Par value, $1.00.

In treasury January 1, 1900, 150,000 shares; in treasury January 1, 1900, $50,000.00 cash.

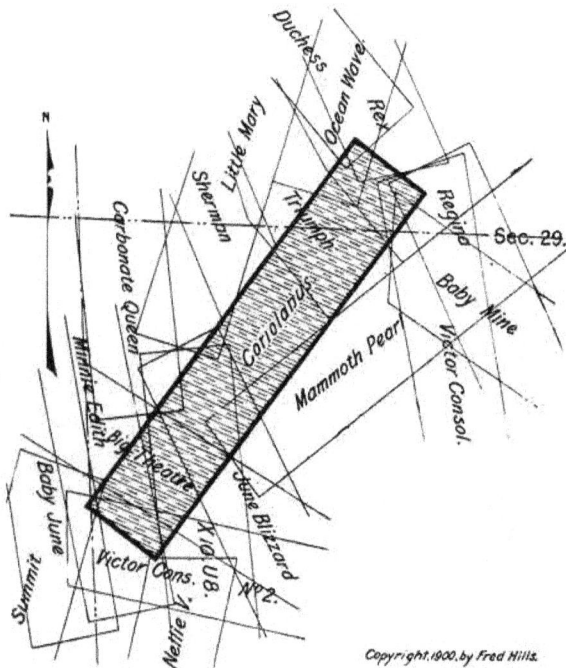

Copyright 1900, by Fred Hills.

Property

Owns the Coriolanus, 10 1-3 acres, in the S. W. 1-4 of section 29, on Battle mountain. Patented.

Development

Two steam hoisting plants on the property and ore bins.

History

In March of this year this company was re-organized and has adopted a policy of development that will likely make the Coriolanus a big paying mine. A new double-compartment shaft has been contracted for and will be sunk 500 feet immediately. It is expected that this shaft will intersect the Ajax vein. The mine has been worked for several years in a desultory sort of way, and yet about $75,000.00 of ore has been taken out of the property. Last year over $12,000 of ore was extracted, which placed the property on the mines list instead of the prospects. The present price of stock is from 22 to 25 cents, and, as the company, until the new organization, was a close corporation, no sales had been made; hence it is impossible to quote prices of stock prior to this time.

The Cortland Gold Mining and Tunnel Company.

Incorporated April. 1896.

E. E. McMahan........................President
Guy Bartlett.....................Vice-President
Francis Capell........................Secretary
Alex Koehler.........................Treasurer
R. F. Ferguson.

Directors

Main Office—604 E. Platte avenue, Colorado Springs, Colorado.

1,250,000 shares. Par value, $1.00.
In treasury January 1, 1900, 118,000 shares.

Capitalization

Owns the Midget, situated in the N. W. 1-4 of section 26, township 15, south of range 69 west, containing 8.50 acres, in process of patenting; also the F. H. & R. claim, held by location. This latter claim is not shown on plat.

Property

The Midget lode has a tunnel. There is a timbered shaft 50 or 60 feet deep on the F. H. & R. claim. These claims have shown good leads of low grade ore.

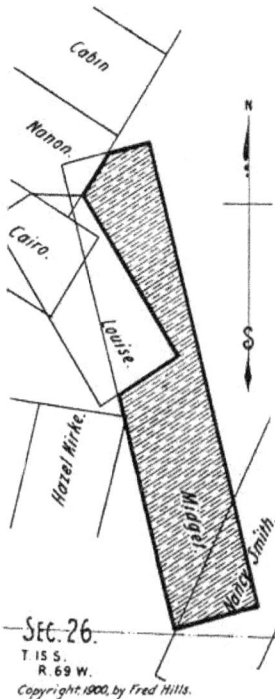

Development

The Cosmos Gold Mining Company.

Incorporated January, 1896.

H. J. Hagerman....................President
R. C. Thayer..............Secretary and Treasurer
K. R. Babbitt.   Sam Strong.

Directors

Main Office—5 Giddings building; transfer, International Trust Company, Colorado Springs, Colorado.

1,000,000 shares. Par value, $1.00.
In treasury January 1, 1900, 75,000 shares; in treasury January 1, 1900, $300.00 cash.

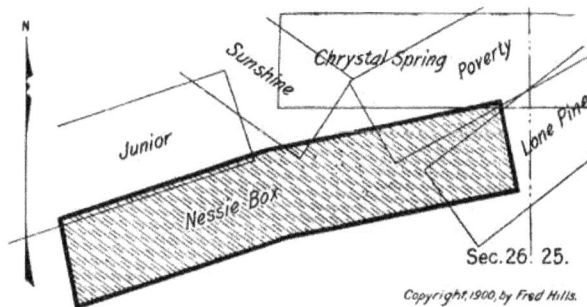

Capitalization

Owns the Nessie Box, 10 1-3 acres, in the N. E. 1-4 section 26, township 15 south, range 70 west, patented, just west of Mound City; also the Christopher Columbus Nos. 1, 2, 3 and 4, containing about 40 acres, patented, north of Rhyolite mountain. As these last-mentioned claims are outside of the Manual territory they can not be shown thereon.

Property

There is about 350 feet of incline on the Nessie Box, and 500 feet of tunneling and shafts on the Christopher Columbus group.

Highest price of stock during 1899, $9.50 per M.

135

The Creede and Cripple Creek Gold Mining Company.

Incorporated.

Directors

James F. Burns.........................President
Frank G. Peck.............Secretary and Treasurer
   John Harnan.   L. F. Curtis.
                 M. Kinney.

Main Office—No. 68 First National Bank building, Colorado Springs, Colorado.

Capitalization

800,000 shares.   Par value, $1.00.

In treasury January 1, 1900, $223.48 cash; in treasury January 1, 1900, 500,000 shares of the Union Bell Gold Mining Company.

Property

Owns the Hillside, survey No. 9,348, containing 3 acres, in the N. W. 1-4 section 19, on Gold hill; the Little Mary and the Ocean Wave, survey No. 8,192, both situated in the N. W. 1-4 section 29, on Battle mountain, containing in all 9 acres.   All patented.

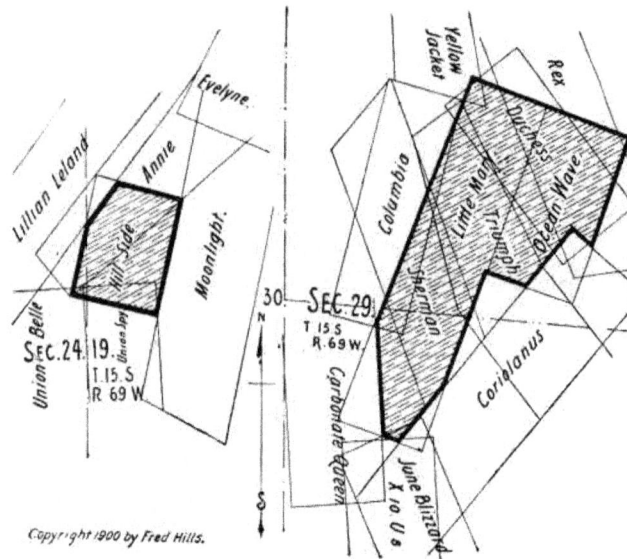

Copyright 1900 by Fred Hills.

Development

Shaft house on the Hillside.   The Hillside has a shaft 210 feet deep and 1,000 feet of cross-cuts and drifts.   On the Little Mary there are several shafts less than 100 feet deep, and about 800 feet of cross-cuts and drifts.   These two latter claims are traversed by the Uinta tunnel. All the claims are leased.

Production

Gross production to January 1, 1900, about $15,000.   Net value of royalties for 1899, $135.72.

Highest price for stock during 1899, 13 1-4 cents; lowest price for stock during 1899, 5c.

The Cresson Consolidated Gold Mining and Milling Company.

Incorporated 1895.

F. H. Whitney, president; K. R. Babbitt, vice-president; K. A. Rea, _{Directors} secretary; J. R. Harbeck, treasurer. D. Ostrander. Eugene Harbeck. J. H. Wattles.

Main Office—205 La Salle street, Chicago, Illinois.

1,000,000 shares. Par value, $1.00. _{Capitalization}

Owns the Mary L., Sadie Bell, Draper, Robin Hood and Friar Tuck, _{Property} 29.68 acres in all, in the S. W. 1-4 of section 20, and in the N. W. 1-4 of section 29. All patented. These claims are all in one group on Raven and Bull hills.

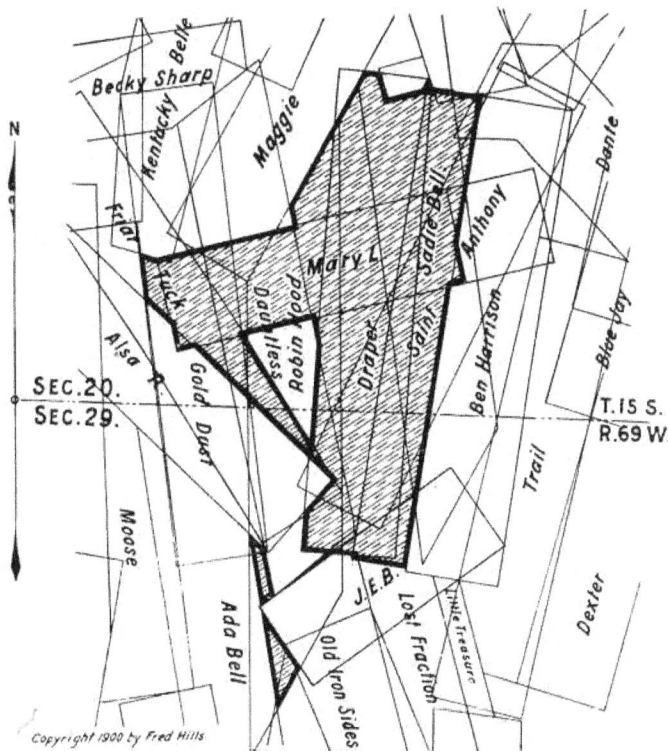

Copyright 1900 by Fred Hills

There are four plants of machinery. Seven leases are being operated. _{Development} Property is not bonded. About 2,000 feet of development work has been done, with no particular regard to claim boundaries. The Draper claim is being worked the most actively. The present lease on the central block of the Mary L. and Draper claims calls for a 600-foot shaft.

The property was operated by the company during 1895-1896. In _{History} 1896 it was leased for a short time. From January, 1897, to September, 1898, the property laid idle. One lease began operations September, 1898. The present leases, seven in number, were given October and November, 1899. Stock is not on the market, being closely held.

Sec.12. Sec.7.

Stray Horse

Ninety-one

93

Jupiter

Monmouth

Panorama

Kansas

Toll Road Placer

McCoy

White Rose

Bon Ton

Venus

Geneva

Hub

Bell Key

Mollie Belle

S.E. Sec. 31.

Etta Chot

J. C.

Tenderfoot

Bertie

Copyright, 1900 by Fred Hills.

7.

13. Sec. 18.

Little Percy

Florence Nightingale

N

Blue Bird

No 2.

Scotia

Mary A.

Old Dominion

Tomes River

Old Dominion

Roanoke

Ella W

Mortality

Kansas City

Wolverine

Eldorado

E.C. & F.

Shoo Fly

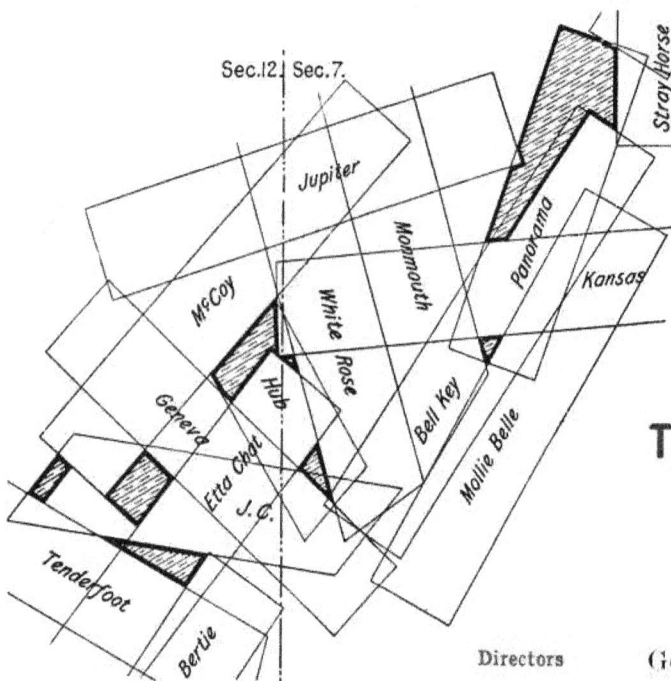

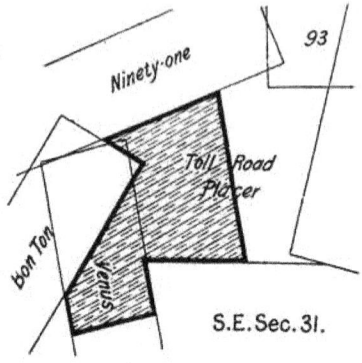

The Cripple Creek Bullion Gold Mining Company.

Incorporated 1895.

Directors   George W. Baxter....President
J. C. Mitchell...Vice-President
J. D. Niederlitz..Sec and Treas

Main Office—Equitable bldg,
Denver, Colorado.

Capitalization   1,500,000 shares. Par value, $1.00. In treasury January 1, 1900, 250,000 shares; in treasury January 1, 1900, about $1,000.00 cash.

Property   Owns the northern portion of the Venus and the western portion of the Toll Road placer, over 5 acres patented and about 2 acres not patented, on Squaw mountain, in the S. E. 1-4 of section 31. The Kansas City, 1 1-4 acres, patented, on Womack hill, in the N. W. 1-4 of section 18, near the Gold King Company. The Hub, J. C. and Panorama, about 5 acres, receiver's receipt, in the S. W. 1-4 of section 7, located on the west slope of Carbonate hill. The S. M. P., about 1 acre, adjoining the Hub and J. C. The Kansas City No. 2, about 2 acres, in the N. W. 1-4 of section 18, adjoining the Kansas City; in process of patenting; 41 acres, patented, on Nipple mountain, and about 8 acres in process. A part of the property is leased and more will be leased later.

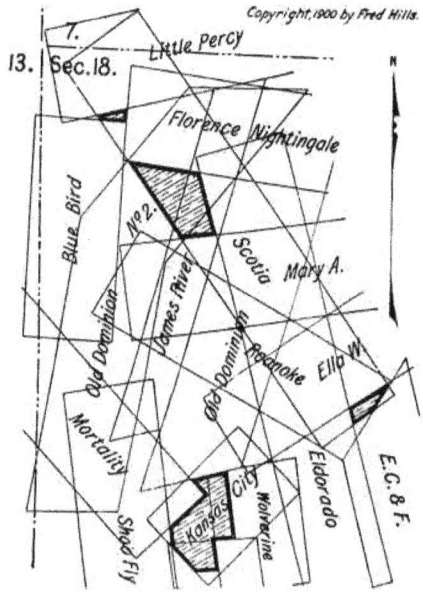

Highest price for stock during 1899, 3 7-8 cents; lowest price for stock during 1899, 2 1-2 cents.

The Cripple Creek-Columbia Mining Company.

Incorporated August 29, 1899.

C. C. Hamlin.........................President   Directors

C. H. Dudley.....................Vice-President

O. H. Shoup..............Secretary and Treasurer

    H. H. Barbee.   E. J. Eaton.

Main Office—First National Bank block, with Reed & Hamlin Investment Company, Colorado Springs, Colorado.

1,250,000 shares. Par value, $1.00.   Capitalization

In treasury January 1, 1900, 200,000 shares; in treasury January 1, 1900, $2,100.00 cash.

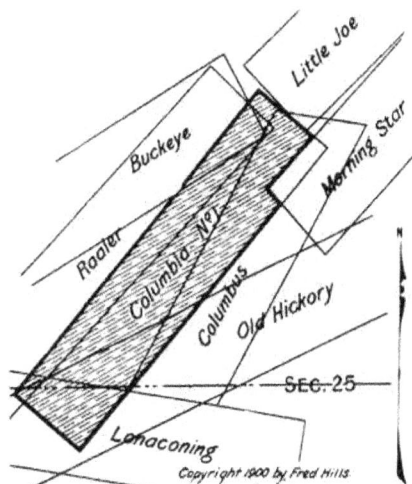

Copyright 1900 by Fred Hills.

Owns the Columbia No. 1, containing about 10 acres, situated in section 25, on Beacon hill. Patented.   Property

This property was incorporated late last year and has been systematically developed under the leasing system. It has a shaft about 285 feet deep, and has a good shaft house and steam hoist on the property. Beacon hill has very few deep shafts, but from the development that has been done the indications are most excellent. It will be the policy of the company to continue active development work under the leasing system.   Development

Highest price for stock during 1899, 13 1-2 cents; lowest price for stock during 1899, 6 1-8 cents.

The Cripple Creek Consolidated Mining Company.

Sec. 18. Monday
Xenia
Delphna
Padtolus
Dead Horse
Eagle
Tam O'Shanter
Queen Bess
Jim Blair

Bonanza
W.P.H.
Queen No 2.
Tutonic
Minnie Law
American Eagle
Mosquito
Axtel
Climax
Fair Sex
Wichita
Sec. 17.
Sec. 20.
Londonderry
High Tide
Henry Clay

Jeff Davis
Hub
La Paloma
Colorado Boss No 2.
Colo Boss
Overlooked
Free Milling
Iron Master
Colo Boss No 3.
Callie
Sec. 19.
Mary McKinney
Morning Glory No 4.

N.E. Sec. 29.
Sitting Bull
Gold Bug
Florence
Trilby
Clyde
Keystone
J.I.C.

Sec. 31.
Viola
Lost Chance
Alpine
Silver Spring
Edith May
Andrews
Adolph

Copyright, 1900, by Fred Hills

Scale

140

The Cripple Creek Consolidated Mining Company.

Incorporated December 15, 1891.

Jos. F. Humphrey........................President
John G. Shields...................Vice-President
Geo. M. Irwin...........................Secretary
D. I. Christopher.......................Treasurer
Irving W. Howbert.   J. A. Hayes.

Directors

Main Office—30 Hagerman building, Colorado Springs, Colorado.

2,000,000 shares. Par value, $1.00.

Capitalization

In treasury January 1, 1900, 17,928 shares; in treasury January 1, 1900, $40,000.00, also 1,000 shares of the Wilson Creek Consolidated M. & M. Co. stock.

Owns the Florence claim, situated in the N. E. 1-4 of section 29, Survey No. 9224, containing 6.874 acres, on Battle mountain; the Colorado Boss Nos. 1, 2 and 3, in the N. 1-2 of section 19, Survey No. 7703, containing 17.210 acres, on Gold hill; the Dead Horse claim, in the N. 1-2 of section 18, Survey No. 7501, containing 10 acres, on Tenderfoot hill, that portion in conflict with the Jim Blaine and the Tam O'Shanter, designated as Tract "A," is in controversy with the Poverty Gulch G. M. Co.; also owns the American Girl, in the S. W. 1-4 of section 17, Survey No. 7513, containing 9.822 acres, on Ironclad hill; the Aeneid, in the S. W. 1-4 of section 17, Survey No. 7681, containing .864 acre, of which this company owns a 1-2 interest; the Andrews claim, in the S. W. 1-4 of section 31, Survey No. 7941, containing 6.67 acres, on Grouse mountain; and the Alpine and the Viola claims, in the S. W. 1-4 of section 31, Survey No. 7937, containing 20.61 acres, on Squaw mountain. The company owns, in all, 72.050 acres. All patented. This company formerly owned the May Queen and Geneva, in the S. W. 1-4 of section 18, in Poverty gulch. Early in 1900 these claims were sold to Mr. W. S. Stratton. (See Stratton's group.)

Property

The Colorado Boss Nos. 1 and 2 and a part of No. 3 is leased. A part of the Dead Horse is also leased. The company is now working on the Colorado Boss No. 3. On the Florence lode the company will sink a shaft 500 feet deep.

Development

On March 24, 1900, a dividend of 8 cents per share ($160,000) was paid the stockholders from the sale of the May Queen and Geneva claims.

Dividend

Highest price for stock during 1899, 19 3-8 cents; lowest price for stock during 1899, 8 cents.

The Cripple Creek Free Gold Mining and Milling Company.

Incorporated November, 1894.

Directors

W. H. Anderson..............President
F. W. Ford.Vice-President and Treasurer
M. E. Anderson..............Secretary
A. A. Ford.   A. Ford.
Sarah W. Ford. Bertha E. Ford.

Main Office—Nos. 1 and 2, Safe Deposit building, Cripple Creek, Colorado.

Capitalization

1,000,000 shares. Par value, $1.00.

In treasury January 1, 1900, 57,004 shares; in treasury January 1, 1900, $500.00 cash.

Property

Owns the Australia lode claim, containing 10 1-3 acres, situated in the E. 1-2 section 25, between the Fanny B., Orizabas and the Kimberly Company's property on the one side, and the Black Bell on the other. The patent has been delayed owing to an adverse decision against the Little May, included in the same group.

Development

The property is divided into five blocks, and there are shafts of from 50 to 250 feet on each. The main workings at the north end are 250 feet deep. There are several drifts and cross-cuts. A tunnel from outside ground cuts the vein near the center of the claim. Some drifting and stoping has been done. There are three other shafts of from 50 to 100 feet deep, which have cut the vein.

Production

Gross production to January 1, 1900, about $6,000.00. Net profit on ore mined during 1899, about $100.00.

History

This company is at present involved in litigation with Burris et al., in which they seek to purchase it at $7,500.00. The company has enforced forfeiture of contract, and a final decision in its favor is expected within a very short time.

The stock is not listed, all being held by a few individuals.

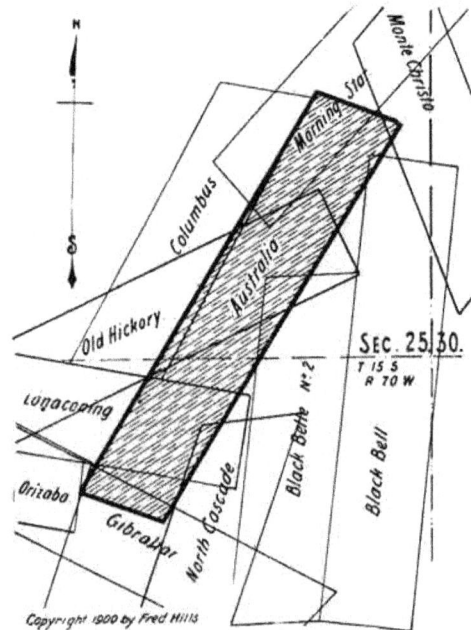

The Cripple Creek and Georgetown Gold Mining Company.

Organized 1899.

Richard J. Bolles, president; Geo. C. Hewett, vice-president; Chas. S. Wilson, secretary; James R. Lane, treasurer. Directors

Main Office—No. 9 De Graff block, Colorado Springs, Colorado.

1,250,000 shares.   Par value, $1.00. Capitalization

Owns the Lucky No. 2 and the Oasis No. 2, on Tenderfoot hill; the Property Algiers Nos. 1 and 2, and the Hosmer, on Lincoln hill and Rhyolite mountain, containing in all about 22 acres, in process of patenting. The company also owns a group, containing about 8 1-2 acres, in the

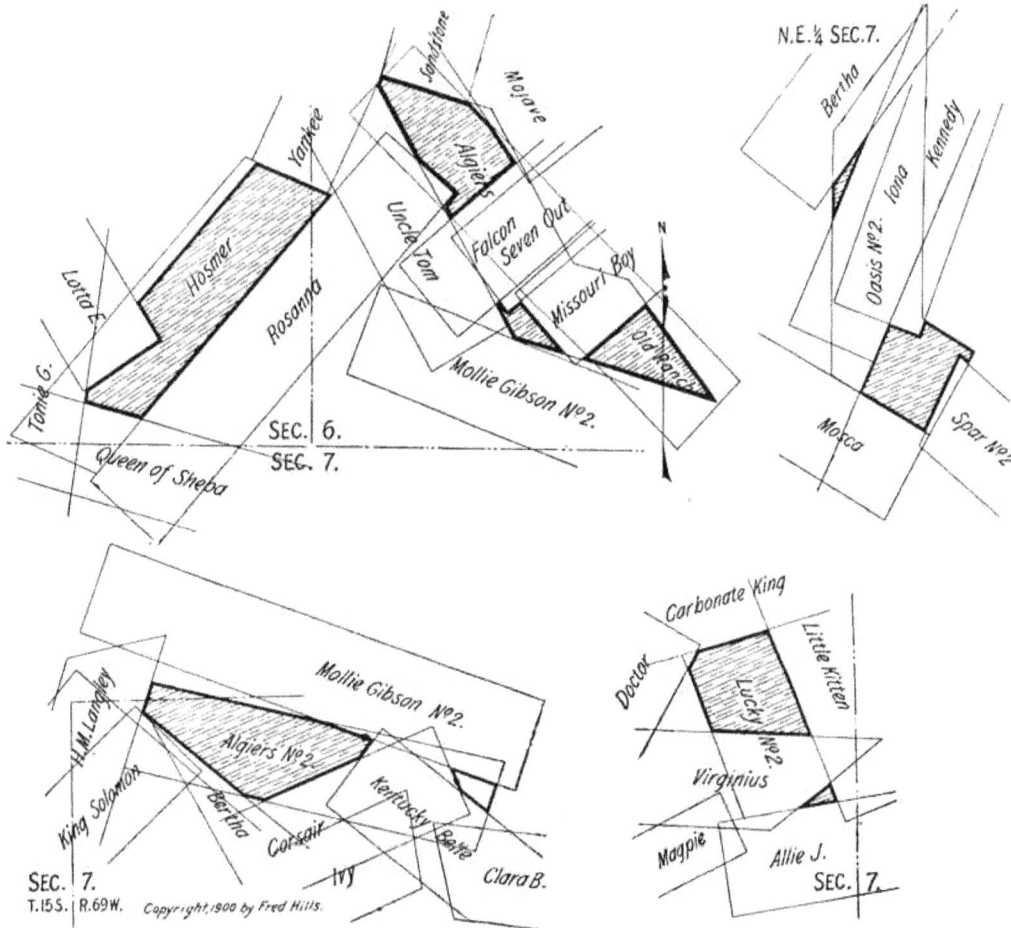

Griffith mining district, Georgetown, Colorado, which has produced about $20,000 from surface workings, having the same vein and adjoining the Centennial mine, which has shipped $300,000 and is still producing.

The Lucky No. 2 has a shaft 30 feet deep, and 40 feet of drifting Development from bottom. The Oasis No. 2 has a 30-foot shaft. On the Hosmer 25 feet of trenching and other work has been done. The Algiers Nos. 1 and 2 have 20 feet of work and trenching. These properties are on phonolite and quartz lodes and are well located, being between the Hoosier property on Tenderfoot hill and the Montreal and Lost Lillie on Copper mountain, and the Mayflower, near Galena mountain. Good assays have been obtained from these lodes, and the outlook for them is as good as that of most paying lodes in the camp before development.

143

The Cripple Creek Gold Exploration Company.

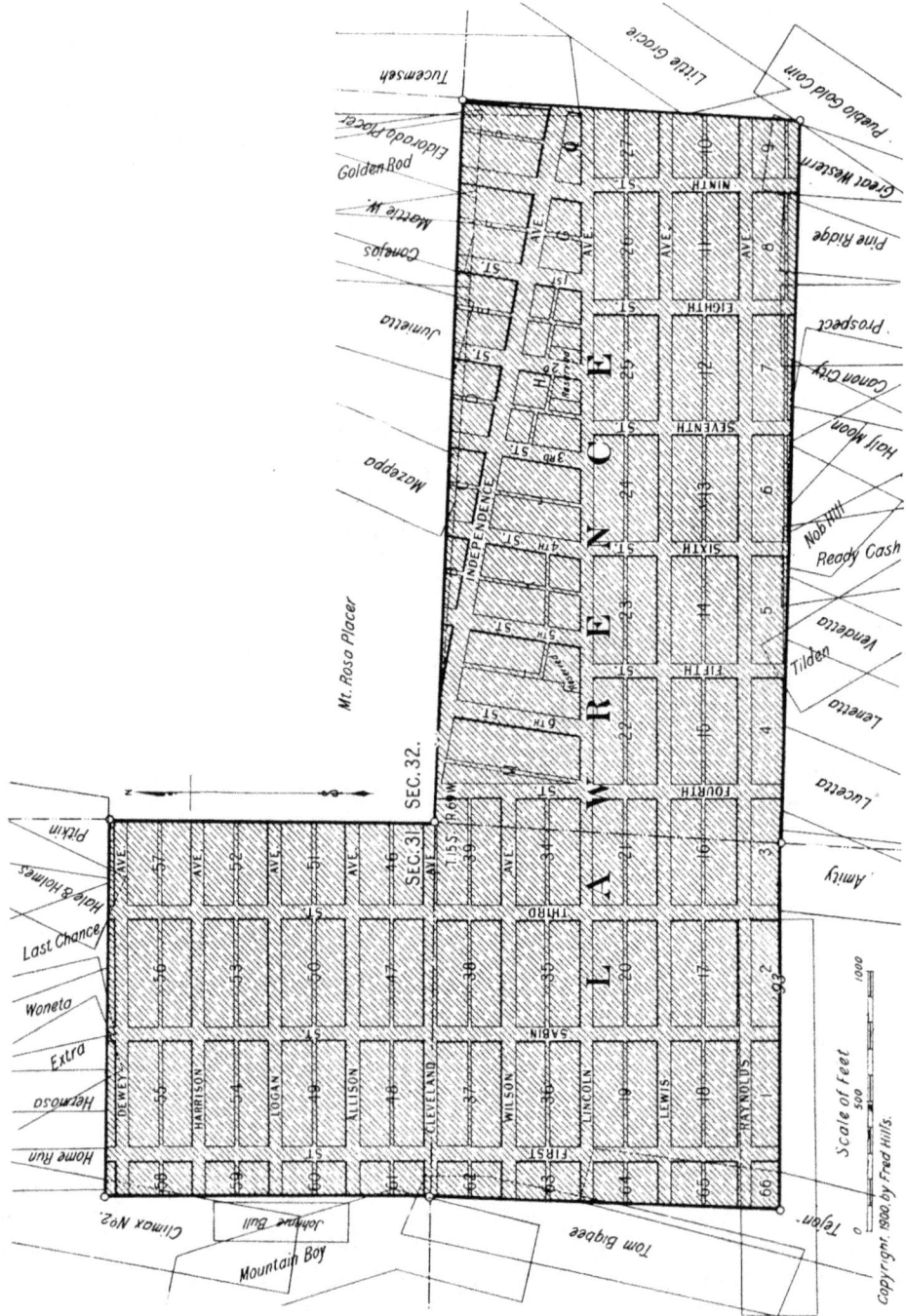

The Cripple Creek Gold Exploration Company.

Incorporated September 18, 1895.

Warren Woods........................President   Directors
W. F. Crosby.......................Vice-President
L. R. Ehrich...........................Secretary
H. E. Woods..........................Treasurer
F. M. Woods....................General Manager

Main Office—Council Bluffs, Iowa.
Branch Office—Colorado Springs, Colorado.

1,800,000 shares. Par value, $1.00.   Capitalization
In treasury January 1, 1900, $17,500.00 cash.

Owns the Lawrence townsite, containing 160 acres, situated in sec-   Property
tion 32. Patented.

On the May Bell, 5 acres, which is leased and bonded, are four   Development
shafts ranging in depth from 50 to 350 feet. Very rich assays have
been taken from this territory, on which active development work is now
being prosecuted. The surface of this property, which is adjacent to the
town of Victor, will bring in a very large treasury surplus from the sale
of the lots.

Highest price for stock during 1899, 17 3-4 cents; lowest price for
stock during 1899, 9 3-8 cents.

The Cripple Creek Imperial Mining Company.

Incorporated November 21, 1895.

Directors

L. C. De Morse, president; J. M. Calkins, vice-president; D. Le Duc, secretary and treasurer; D. C. Beaman, Frank Earle.

Main Office—Mining Exchange, Denver, Colorado.

Capitalization

1,000,000 shares. Par value, $1.00 per share. In treasury January 1, 1900, 50,000 shares.

Property

Owns the Maggie V., the Alice M., the Julia A. and the Virginia B., containing about 27 acres, situated in the N. E. 1-4 section 9, and in the S. W. 1-4 section 3, on Trachyte mountain. All patented.

The property was developed by tunnel.

Highest price for stock during 1899, $7.00 per M; lowest price for stock during 1899, $3.00 per M.

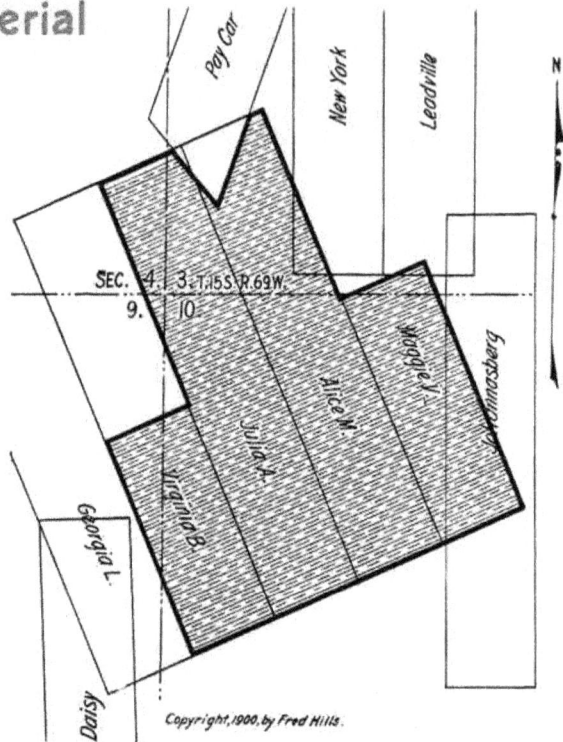

Copyright, 1900, by Fred Hills.

The Cripple Creek Madonna Mining Company.

Incorporated November, 1895.

Directors

Ramsay C. Bogy, president; D. D. Muir, vice-president; J. D. Niederlitz, secretary and treasurer.

Main Office—Equitable building, Denver, Colorado.

Capitalization

1,000,000 shares. Par value, $1.00. In treasury January 1, 1900, 375,000 shares; in treasury January 1, 1900, some cash.

Property

Owns the Madonna and the Madonna Extension, 10 acres each, located in the N. 1-2 of section 1, township 16 south, range 70 west, on south slope of Grouse mountain. All patented.

Development

About 300 feet of development work has been done on the Madonna and about 150 feet on the Madonna Extension. About two-thirds of the Madonna and one-third of the Madonna Extension is leased. Work is not being prosecuted at present.

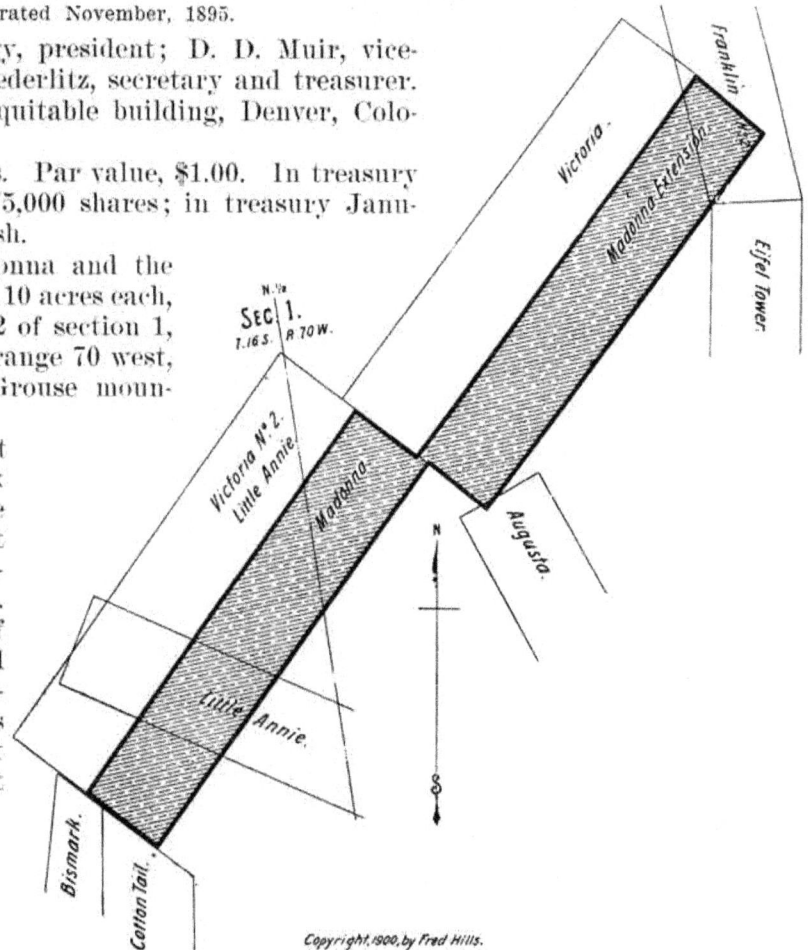

Copyright, 1900, by Fred Hills.

The Croesus Gold Milling and Tunnel Company.

Incorporated 1895.

J. Arthur Connell, president and treasurer; Geo. Rex Buckman, Directors
vice-president; D. D. Lord, secretary; Wm. P. Bonbright.

Main Office—67-68 P. O. building, Colorado Springs, Colorado.
Transfer Office—International Trust Company, Colorado Springs.

2,000,000 shares.   Par value, $1.00. Capitalization
In treasury January 1, 1900, about 700,000 shares; in treasury January 1, 1900, about $800.

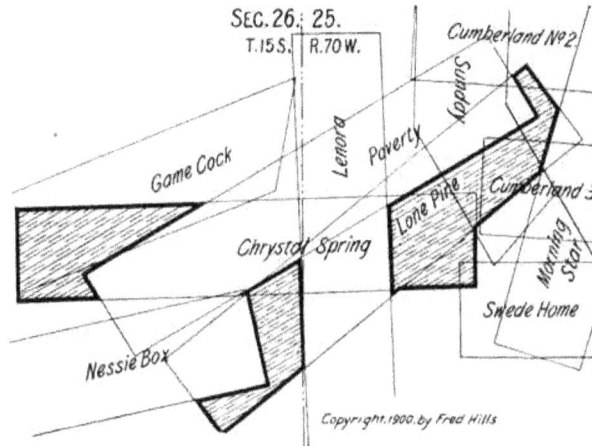

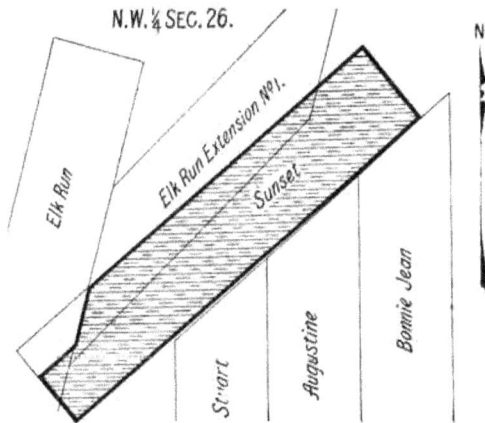

Copyright.1900.by Fred Hills

Owns the Sunset, Little Chief, Chrystal Spring and Lone Pine, in Property
all about 17 acres, in the N. W. 1-4 of section 25, and in the N. E. 1-4 of
section 26, west of Mound City; also the Esmeralda Nos. 1, 2, 3 and 4.
in all about 40 acres, in section 30, township 14 south, range 69 west, on
Rhyolite mountain.   All the property is patented.

But little more work has been done on the property than was re- Development
quired for a patent. No work at present is being prosecuted.

Highest price for stock during 1899, 1 1-8 cents; lowest price for
stock during 1899, $4.00 per thousand.

The Currency Mining Company.

Scale:
0   100   400   800 FT.

Copyright, 1900, by Fred Hills.

The Currency Mining Company.

Incorporated May 4, 1892.

Wm. A. Otis............................President
Philip B. Stewart....................Vice-President
C. E. Titus...........................Secretary
Wm. A. Otis & Co......................Treasurer
J. P. Pomeroy.   Leonard E. Curtis.

Directors

Main Office—Wm. A. Otis & Co., Giddings block, Colorado Springs, Colorado.

1,250,000 shares. Par value, $1.00.

Capitalization

In treasury January 1, 1900, 6,072 shares.

Owns the Fairfax, 3 acres, in the N. E. 1-4 of section 19, on Raven hill; the Engineer, 6 acres, in the N. W. 1-4 of section 30; the Amy, 8 acres, in the S. W. 1-4 of section 19, on Guyot hill; a one-half interest in the Index, 4 acres, in the S. W. 1-4 of section 18, on Gold hill; a one-half interest in the Hale and Holmes, 5 acres, in the N. E. 1-4 of section 31, on Squaw mountain; the Four Brothers, 3 acres, in the N. W. 1-4 of section 30, on Beacon hill; the Modoc, 1 acre, in the N. W. 1-4 of section 30; the Benton and a part of the Catalpa, 2 acres, in the S. E. 1-4 of section 18, on Globe hill; the Printer, in the E. 1-2 of section 7, and the Amity, 9 acres, in the S. W. 1-4 of section 32. All the above property is patented.

Property

The greater part of the development work has been done on the Amy claim. This property has three shafts, 500 feet deep, and about 800 feet of drifting. The Amity also has some shafts and drifts. The policy of the company is to develop its properties by leasing, which is now being done, three leases having recently been granted.

Development

Highest price for stock during 1899, 8 3-8 cents; lowest price for stock during 1899, 4 cents.

The Damon Gold Mining Company.

Incorporated 1894.

Directors

Warren Woods............... ....President
H. E. Woods..........Vice President and Treasurer
F. M. Woods........Secretary and General Manager
  J. M. Hawkins.   J. M. Allen.

Main Office—Victor, Colorado.

Capitalization

2,000,000 shares. Par value, $1.00.

In treasury January 1, 1900, 141,139 shares; in treasury January 1, 1900, $7,188.02 cash.

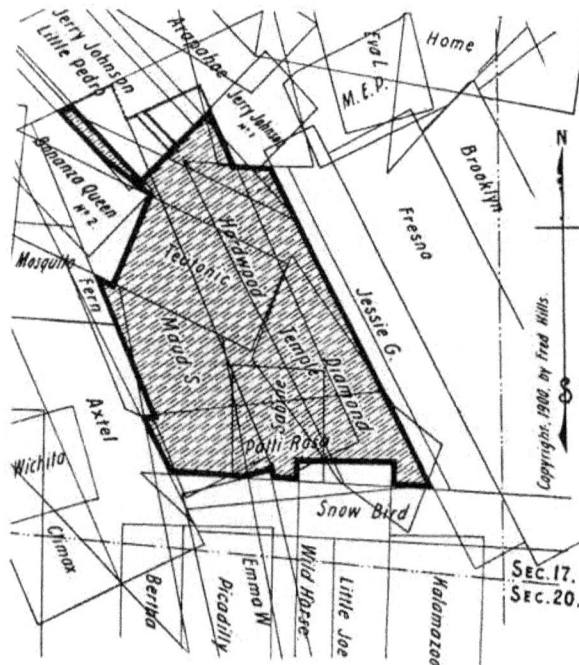

Property

Owns the Diamond, patented; the Patti Rosa, the Maud S. and the Hardwood, the last three of which are held by receiver's receipts; in all about 20 1-4 acres, situated on Bull hill, in the S. W. 1-4 section 17. The northern portion shown in dotted lines in plat has mineral rights thereunder belonging to this company, but not the surface.

Development

The northeast portion of this property, containing 2 3-4 acres, was leased to the Alert Gold Mining Company in December, 1898, and the lease will continue until May, 1901. These lessees have opened up a fine body of ore and have shipped about $145,000 worth. They are now operating in a vertical shaft 4 1-2x10 feet in the clear, at a depth of 275 feet. They have also opened up about 1,500 feet of drifts and cross-cuts. There are several other leases being operated on the property, but none have found ore in paying quantities. On another page will be found description of the Alert Gold Mining Company.

Highest price for stock during 1899, 29 3-8 cents; lowest price for stock during 1899, 14 7-8 cents.

The Dante Gold Mining Company.

Incorporated September 20, 1895.

L. L. Aitken, president; W. Arthur Perkins, vice-president; O. H. Shoup, secretary and treasurer. C. C. Hamlin. S. D. Johnson. Directors

Main Office—First National Bank block, with Reed & Hamlin Investment Company, Colorado Springs, Colorado.

1,250,000 shares. Par value, $1.00. Capitalization

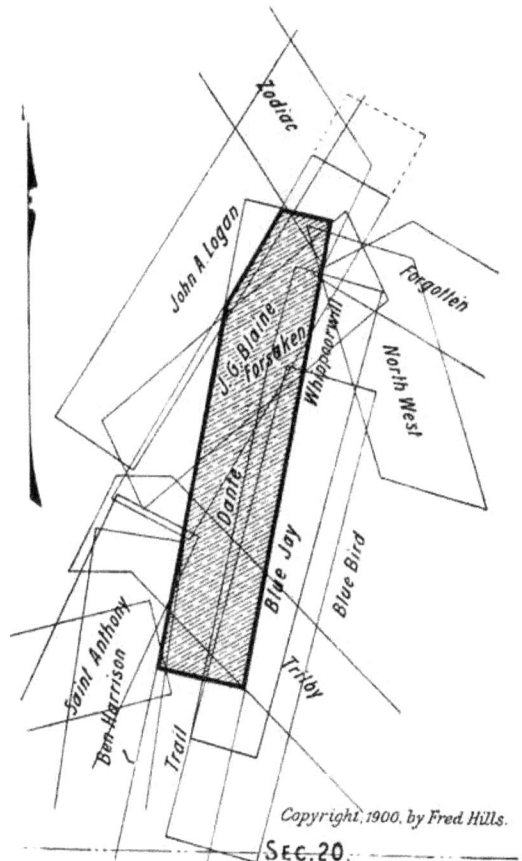

Copyright, 1900, by Fred Hills.

Owns the Dante claim, containing nearly 10 acres, located in section 20, on Bull hill. Patented. Property

A good shaft house, ore bins and plant of machinery is on the property, which belongs to the company. The main shaft is 450 feet deep, with several other shafts from 150 to 250 feet deep. The entire claim is under lease, there being eight sets at work, with royalties ranging from 10 per cent. upward. No dividends have ever been declared by the company. The principal work is being done on the fourth and sixth levels. On the fourth level there is a very good stope of ore about 120 feet in length. The Dante is a well-known property and the big dyke which runs through Bull hill from south to north and which appears to be a feeder of the great mines, runs through the property. Development

Production to January 1, 1900, $90,000.00. Production

Highest price for stock during 1899, 23 1-2 cents; lowest price for stock during 1899, 6 1-2 cents.

The Dead Shot Gold Mining Company.

Incorporated November, 1899.

Directors

William Shemwell......................President
E. J. Eaton.......................Vice-President
A. F. Woodward.......................Secretary
John M. Pring.......................Treasurer
R. J. Gwillim.

Main Office—El Paso Abstract Office, 113 East Kiowa street, Colorado Springs, Colorado.

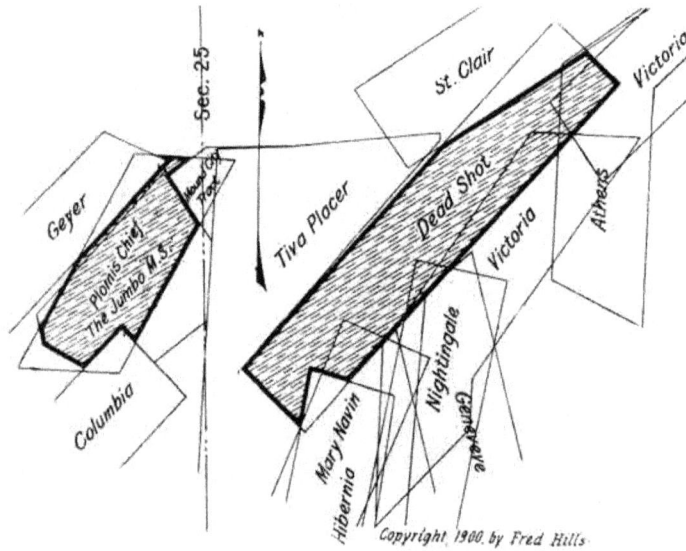

Copyright, 1900, by Fred Hills.

Capitalization

1,250,000 shares.   Par value, $1.00.

In treasury January 1, 1900, 146,000 shares; in treasury January 1, 1900, $800.00 cash.

Property

Owns the Dead Shot, containing 9.40 acres, and the Plomis Chief, containing 4 acres, both situated in the N. E. 1-4 of section 25.   Patented.

Development

There is one 400-foot shaft and drift and one whim on property. The greater part of the development work has been done on the Dead Shot.   The company is at present sinking and cross-cutting for a vein to the depth of 100 feet.

Highest price for stock during 1899, 5 cents; lowest price for stock during 1899, 5 cents.

The De Beers Gold Mining Company.

Incorporated December 27, 1895.

C. L. Arzeno...........................President   Directors

H. E. Woods.........Vice-President and Treasurer

F. M. Woods..........................Secretary

J. M. Allen.................Assistant Secretary

Main Office—Giddings building, Colorado Springs, Colorado.

1,250,000 shares.   Par value, $1.00.   Capitalization

In treasury January 1, 1900, 40,625 shares.

Owns the Blue Rooster, the Happy Tom and the Morning Star, Property situated in the N. 1-2 of section 28, on Big Bull mountain, and containing about 13 acres.

This property being situated near the Gold Knob territory makes Development it valuable, and with development good results may be expected.

The Defender Gold Mining Company.
Incorporated 1895.

Directors
Edwin Arkell, president; R. Harding Loper, vice-president; Jos. P. Walsh, secretary and treasurer; E. C. Stoddard, W. B. McArthur.

Main Office—37 Postoffice building, Colorado Springs, Colorado.

Capitalization
2,000,000 shares. Par value, $1.00. In treasury March 1, 1900, 250,000 shares; in treasury March 1, 1900, $200.00 cash.

Property
Owns the Broadmore, Mandolin and part of the Defender, 16 3-4 acres in all, patented, in the S. W. 1-4 section 9, on Galena hill. The company also own the John Hancock, Casco, Puritan and Ashmont lodes, 31 acres, in process, on the north side of Copper mountain, which are not shown on map.

Development
300 feet of shafting, tunnels, etc., on the Broadmore group. Work enough done to patent the John Hancock group. The Broadmore claim is being developed by lessees. The Defender group was purchased from Homer Hart of Cripple Creek. The company made application for patent and four adverses were filed and serious litigation was encountered; debts were contracted; finally an agreement was reached where the Defender Company gave up one-half of its territory in order to get patent. This being accomplished, the company sent out notice to stockholders for an increase of the capital to pay up debts and buy additional property and place sufficient stock in the treasury to further develop the property; this has been done and the company is once more on good footing. Highest price for stock during 1899, 1 5-8 cents; lowest price for stock during 1899, $6.00 per 1,000.

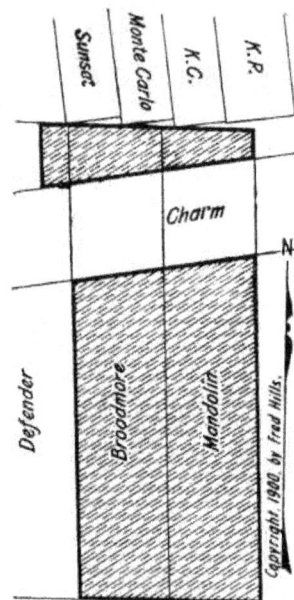

S.W. Sec. 9.

The Del Coronado Gold Mining Company.
Incorporated November 11, 1895.

Directors
G. B. Bish............President
Geo. F. Harbaugh.....Vice-Pres
Taylor J. Downer..Sec and Treas
Alexander Koehler. Geo. Peyser.

Main Office—104 1-2 E. Pike's Peak avenue, Colorado Springs, Colorado.

Capitalization
900,000 shares. Par value, $1.00. In treasury January 1, 1900, 5,000 shares.

Property
Owns the Rock Island and Lulu lodes, containing 17 acres, in the S. W. 1-4 of section 27, on Big Bull mountain. Patented.

Development
One shaft on the property 65 feet deep. A good vein has been opened up, assays on which ran from $13.00 to $20.00 per ton, and the prospects for the future are encouraging.

Highest price for stock during 1899, $15.00 per 1,000; lowest price for stock during 1899, $7.50 per 1,000.

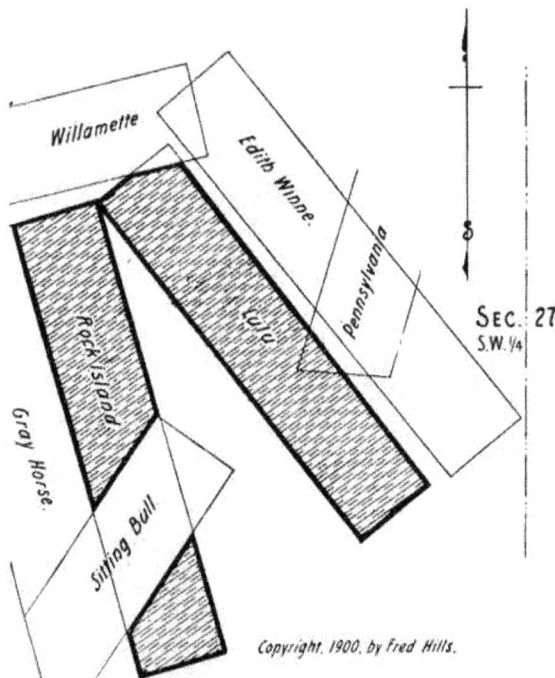

SEC. 27.
S.W. 1/4

Copyright, 1900, by Fred Hills.

154

The Delmont Gold Mining Company.

Incorporated February, 1896.

Main Office—16 North Nevada avenue, Colorado Springs, Colorado.

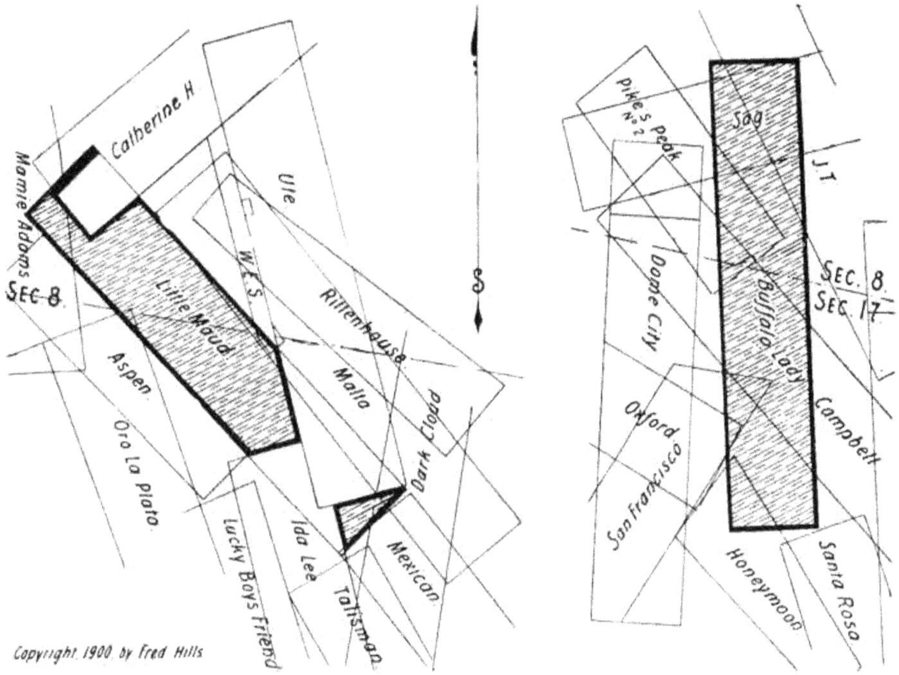

Copyright, 1900, by Fred Hills

Capitalization

1,250,000 shares.   Par value, $1.00.

In treasury January 1, 1900, 386,500 shares; in treasury January 1, 1900, $600.00 cash.

Property

Owns the Little Maud, 8 acres, in the E. 1-2 of section 8, township 15, south of range 69 west. The Buffalo, 10 acres, in the N. E. 1-4 of section 17, and the S. E. 1-4 of section 8, township 15, south of range 69 west, all north of Grassy, on the south slope of Galena hill. Patented.

Development

Shaft about 65 feet deep, with some drifting. The company has not done much work as yet in the way of development; immense and perfect veins are found, and enough has been seen by way of assays and development to demonstrate that the prospects are excellent. The property will be developed through lessees.

The Des Moines Gold Mining Company.

Incorporated.

Main Office—6 P. O. building, Colorado Springs, Colorado.

Capitalization

750,000 shares. Par value, $1.00.

In treasury January 1, 1900, 151,335 shares.

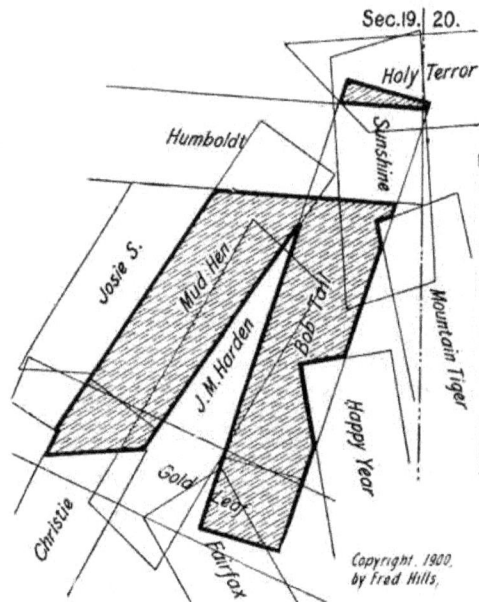

Property

Owns the Mud Hen and the Bob Tail lode claims, containing 18 acres, situated in the N. E. 1-4 of section 19, on the north slope of Raven hill. All patented.

Development

One shaft on Bob Tail 162 feet deep, from which assays as high as $500.00 per ton have been obtained, but these values have been from "pockets" in the vein. One shaft 100 feet deep on the Mud Hen, on a vein showing values, but no pay ore encountered in quantities sufficient to ship. One tunnel 100 feet long. The company has vein rights which are not shown on map.

Highest price for stock during 1899, 7 3-4 cents; lowest price for stock during 1899, 2 1-2 cents.

The Dexter Gold Mining Company.

Incorporated March, 1896.

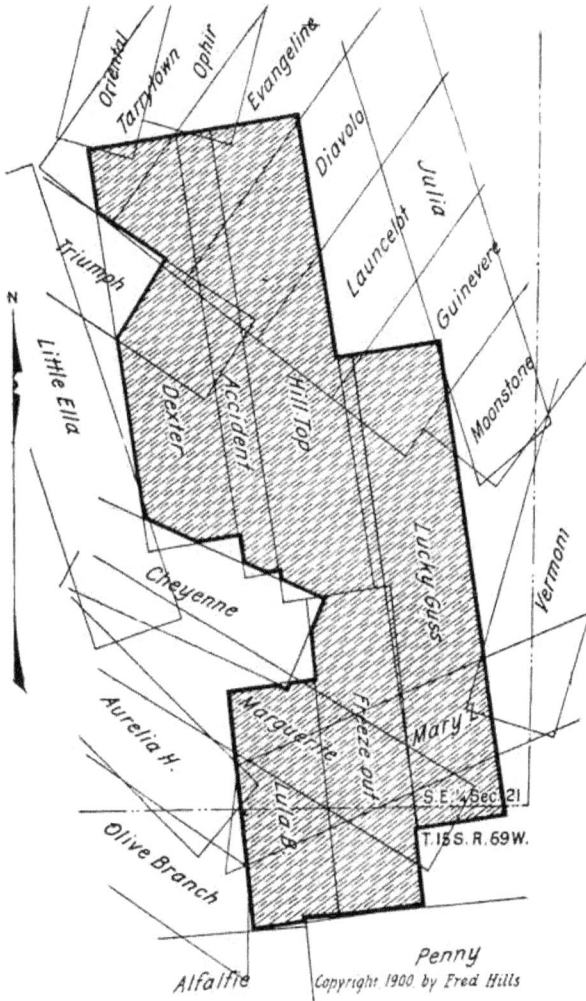

Map labels:
Oriental, Tarrytown, Ophir, Evangeline, Diavolo, Julia, Launcelot, Guinevere, Triumph, Little Ella, Dexter, Accident, Hill Top, Moonstone, Lucky Guss, Vermont, Cheyenne, Aurelia H., Marguerite, Freeze out, Mary L., Olive Branch, Lula B., S.E. ¼ Sec. 21, T. 15 S. R. 69 W., Penny, Alfalfa, N

Copyright 1900 by Fred Hills

Directors

D. Miniun,
  President.
J. H. Morse,
  Vice-President.
E. C. Fletcher,
  Secretary and Treasurer.
E. S. Bach.
Fred Gibbon.
  Main Office—No. 54 P. O. building, Colorado Springs, Colorado.

Capitalization

1,500,000 shares.
Par value, $1.00.

In treasury January 1, 1900, 140,000 shares.

In treasury January 1, 1900, $1,500.00 cash.

Property

Owns the Dexter, the Accident, the Hill Top, the Freeze Out, the Lula B. and the Lucky Guss, containing in all about 47 acres, situated in section 21, on the east slope of Bull hill. All patented.

Development

The Lucky Guss has a shaft 200 feet deep; the Dexter, one of 90 feet; the Hill Top, one of 75 feet; the Lula B., one of 50 feet, with about 50 feet of drifts. The Lucky Guss is receiving the greater part of the development work.

History

At a depth of 150 feet the main shaft on the Lucky Guss encountered a narrow seam from which values averaging $1,700 were obtained. Continuing down 50 feet, the same streak was again encountered in a cross-cut from the bottom of the shaft, the values ranging the same as at the 150-foot point. Owing to a flow of water which the facilities of the company were unable to handle underground work was abandoned until adequate machinery could be placed over the shaft.

Formation

Contacts between breccia, granite, phonolite and andesite, being in this respect identical with that which is produced from the best mines in Cripple Creek.

Highest price for stock during 1899, 5 cents; lowest price for stock during 1899, 2 cents.

The Dictator Gold Mining Company.

Incorporated December 21, 1895.

Directors

M. J. McNamara......President
H. E. Insley......Vice-President
C. C. Lunt..........Secretary
H. F. Brooks.........Treasurer
Jos. H. Ryan..........Manager
   Thos. Milroy. L. C. De Morse.
   J. S. Stahl. Geo. E. Hannan.

Main Office—206 Mining Exchange building, Denver, Colorado.

Capitalization

1,500,000 shares. Par value, $1.00.

In treasury January 1, 1900, 59,000 shares.

Property

Owns Stella A., Mountain Peak and Grey Eagle, containing about 15 1-2 acres, located in N. W. 1-4 of section 17, on Galena hill, patented; and the Celestine R. claim, 8 1-2 acres, in section 24, west of Mound City, in process of patenting. This latter claim is not shown on map.

Development

The company is very hopeful as to what future development will prove, on account of recent rich strikes made on the property of the Cameron townsite. The 350 feet of the east end of the claims has been leased for a period of three years from January 1, 1900. There is a tunnel 130 feet in, and a winze 80 feet, with about 100 feet of cross-cuts, levels, etc.

Highest price for stock during 1899, $60.00 per 1,000; lowest price for stock during 1899, 75 cents per 1,000.

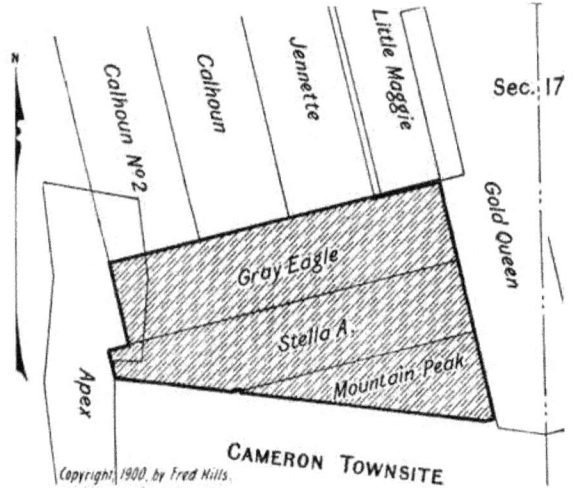

The Dillon Gold Mining Company.

Incorporated September 29, 1899.

Directors

M. Finnerty.............President and Treasurer
Wm. H. Day, Jr...................Vice-President
Edward Kent.......................Secretary
   John E. Phillips.   Dr. J. W. Graham.
   Sherwood Aldrich.   Chas. Donnelly.

Main Office—Denver, Colorado.

Capitalization

1,250,000 shares. Par value, $1.00.

In treasury January 1, 1900, 180,000 shares; in treasury January 1, 1900, about $10,000 cash.

Property

Owns the Dillon mining claim, survey No. 8,786, and the Amazette mining claim, survey No. 8,703, total acreage being about 7 1-2 acres, in the S. W. 1-4 section 29, adjoining the Portland and the Strong mines on the east, and the Gold Coin and Granite mines on the west. Both claims are patented.

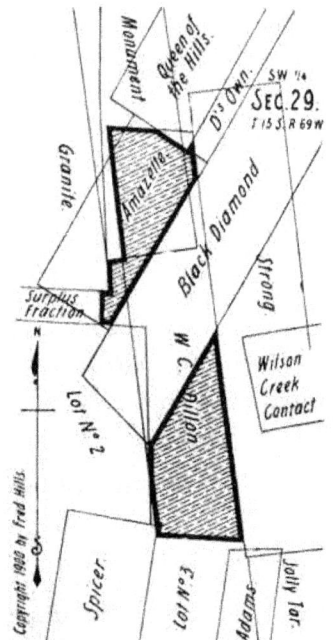

(CONTINUED ON PAGE 159.)

The Dillon Gold Mining Company—Continued.

The north end of the Dillon has a two-compartment shaft 700 feet deep. Four separate veins have been cut, from which ore has been shipped that has run as high as 7 1-2 ounces in gold. The surface improvements consist of a large shaft house, a hoisting plant capable of sinking 1,000 feet, an air-compressor, and everything that goes to make up a first-class mining plant. Underground machinery, three air-drills. The development of this property did not commence until in 1899. The greater part of the development work is now being done on the Dillon. *Development*

Gross production to January 1, 1900, $20,000.00. *Production*

Highest price for stock during 1899, 40 cents; lowest price for stock during 1899, 40 cents.

The Dold Mining and Milling Company.
Incorporated 1894.

Copyright 1900 by Fred Hills

J. L. Middagh.........President *Directors*
E. D. Marr.......Vice President
C. B. Gunn..........Secretary
P. W. Middagh. J. J. O'Connor.

Main Office—Room 35, P. O. building, Colorado Springs, Colo.

1,000,000 shares. Par value, $1.00. In treasury January 1, 1900, 50,000 shares; in treasury January 1, 1900, $60.00 cash. *Capitalization*

Owns the Great Eastern, Mount Forest and Spring Gulch, situated in the N. E. 1-4 of section 12, containing 28 acres, on Carbonate hill. All patented. *Property*

The Mount Forest has a shaft 60 feet deep, 50 feet of cross-cutting, and 40 feet of drifting. The Great Eastern has a shaft 50 feet deep and 40 feet of drifting. The Spring Gulch has two shafts, one 40 feet deep, the other 50 feet deep. *Development*

Highest price for stock during 1899, 2 cents; lowest price for stock during 1899, $4.00 per M.

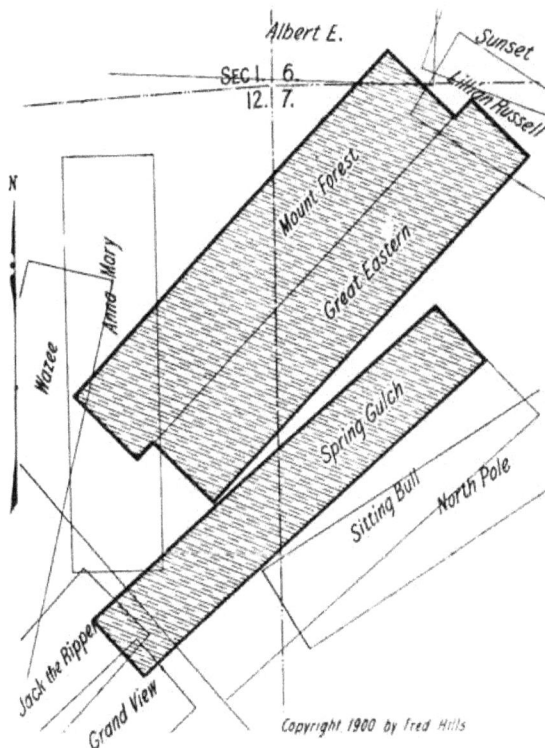

The Door Key Gold Mining Company.

Incorporated April 12, 1900.

Directors

A. F. Woodward.........................President
J. W. Finkbiner....................Vice-President
W. W. Williamson.....................Secretary
O. L. Godfrey...........................Treasurer

E. C. Bale.

Main Office—25 1-2 N. Tejon street, Colorado Springs, Colorado.

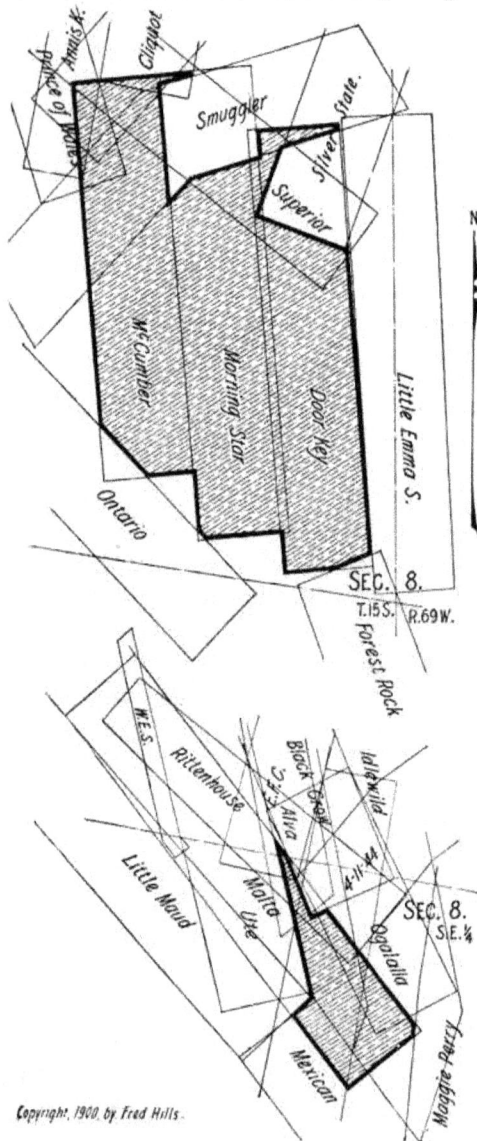

Copyright, 1900, by Fred Hills.

Capitalization

1,500,000 shares. Par value, $1.00. In treasury May 1, 1900, 365,000 shares.

Property

Owns the McCumber, Morning Star and Door Key, Survey No. 11041, containing 22.929 acres, patented, in the N. 1-2 of section 8, on Tenderfoot hill; the Malta, Survey No. 10942, containing 4.558 acres, on Galena hill, for which receiver's receipt is held. A part of the Door Key, which is still in litigation, is omitted from the above. In case the company wins, they will have more ground than the above shows. These are simply prospects, only sufficient work having been done for patents.

The Duchess Mining Company.

Incorporated February, 1896.

Henry P. Steele, president; Wm. C. Hanks, vice-president; Geo. E. Green, secretary and treasurer. Fred E. Gates. John H. Martin. Bruce F. Johnson. — Directors

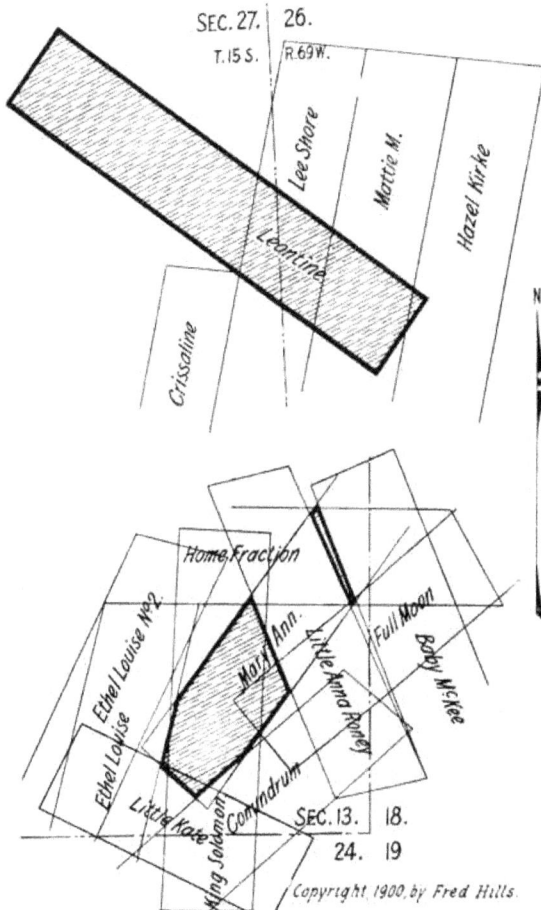

SEC. 27. | 26.
T.15 S. | R.69 W.

Copyright 1900, by Fred Hills.

Main Office—1531 Market street, Denver, Colorado.

Capitalization

1,500,000 s h a r e s. Par value, $1.00. In treasury January 1, 1900, 220,000 shares; in treasury January 1, 1900, $200.00 cash.

Property

Owns the Mary Ann, Survey No. 9102, containing 3.44 acres, situated in the S. E. 1-4 of section 13, on Gold hill; also the right to follow and work all veins discovered on the Mary Ann, in that part of the Little Anna Roney which crosses the Mary Ann, said ground consisting of 2.15 acres. The company also owns the Leontine, containing 10.331 acres, situated in the N. E. 1-4 of section 27, on Big Bull hill. All the above property is patented.

Development

On the Leontine merely enough work to secure patent has been done. The principal work on the Mary Ann consists of a well-timbered shaft, 185 feet deep, and drifts of about 400 feet. The company has been operating with a whim on the Mary Ann. Hereafter it will be the policy of the company to develop its property by leasing. Prominent Eastern capitalists have already secured a lease, without bond, on the Mary Ann, and it is their intention to install a plant of machinery and begin active operations at once. The debt that formerly embarrassed the company has now been liquidated. All properties owned by the company are free and clear of any and all incumbrances.

History

The majority of this stock is held by few people, very little of it being scattered about. Sales have been by private subscription. The Mary Ann adjoins both the Moon-Anchor and the Temomj properties and seems to give promise of making a good mine.

6

The Easter Bell Gold Mining and Milling Company.

Incorporated 1900.

Directors

J. P. Sweeney.........................President
J. R. McKinnie....................Vice-President
R. P. Davie...............Secretary and Treasurer

W. S. Montgomery.   John Harnan.   S. R. Bartlett.

Main Office—Colorado Springs, Colorado.

Capitalization   3,000,000 shares. Par value, $1.00. In treasury May 1, 750,000 shares; in treasury May 1, $30,000 cash.

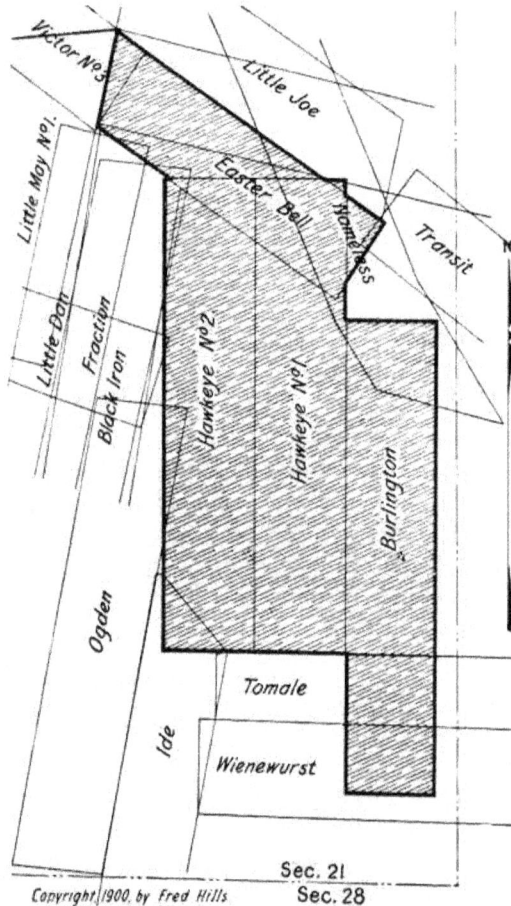

Copyright 1900, by Fred Hills

Property   Owns the Easter Bell, Hawkeye No. 1 and No. 2 and the Burlington, about 35 acres, patented; also owns the Yellow Jacket, a fraction, in process of patenting. All the above property situated on the southeast slope of Bull hill, in the S. 1-2 of section 21.

History   This company will develop its property by the leasing system. The greater portion of the ground will be leased, but a certain part of the ground will be reserved for the company's own work in development, for which enough cash to meet all requirements is in the treasury. There is every indication that the stock will be listed with the Colorado Springs Mining Stock Association. The development work heretofore has not been of such a nature as to need any special comment.

162

The Eclipse Consolidated Gold Mines Company.

Incorporated September 29, 1899.

Fred L. Ballard.........................President
J. S. Tucker......................Vice-President
A. L. Houck..............Secretary and Treasurer
Irving Howbert.   H. L. Shepherd.

Directors

Main Office—No. 109 East Pike's Peak avenue, Colorado Springs, Colorado.

1,250,000 shares. Par value, $1.00. In treasury January 1, 1900, 125,000 shares; in treasury January 1, 1900, $6,000.00 cash. Capitalization

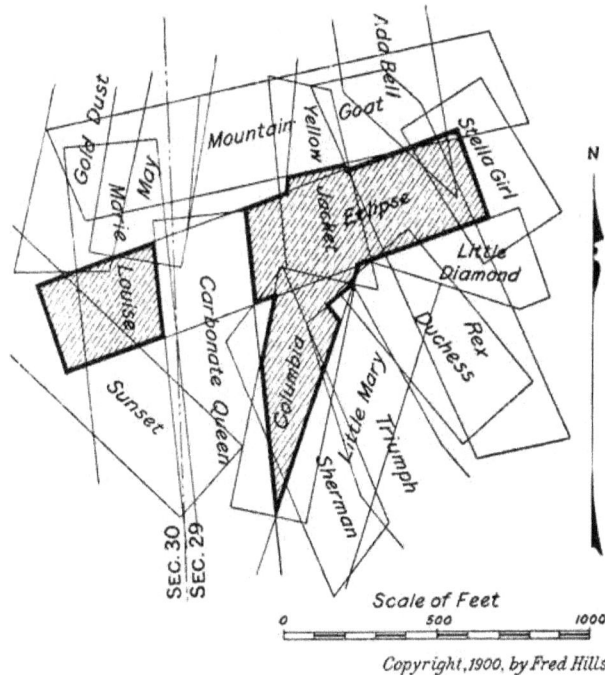

Copyright, 1900, by Fred Hills.

Owns the Eclipse, Yellow Jacket and Columbia, containing 10.5 acres, located in the N. W. 1-4 section 29, on Battle mountain. All patented. Property

There is $5,000 worth of machinery, shaft houses and other improvements on the property, with 900 feet of shafts, drifts, tunnels, etc. The shaft is now 200 feet deep, and the depth is being increased. There are three sets of lessees working the property, and the production has been about $50,000.00 to date. One lessee is shipping and the prospects for the future are flattering. This property was one of the first producers in the Cripple Creek district, and shipped at that time very rich ore. Little work has been done for some years past, but with this newly-organized company active operations are being pushed, and prospects for an early renewal of shipments are good. Development and Production

Highest price for stock during 1899, 11 1-2 cents; lowest price for stock during 1899, 8 cents.

The Eclipse Mining and Milling Company.

Incorporated February 7, 1893.

E. N. Bement..........................President
Samuel P. Beall....................Vice-President
C. C. Lunt.................Secretary and Treasurer
L. C. De Morse.   M. G. Hawley.

Main Office—206 Mining Exchange building, Denver, Colorado.

Capitalization

1,000,000 shares.   Par value, $1.00.

In treasury January 1, 1900, 17,544 shares.

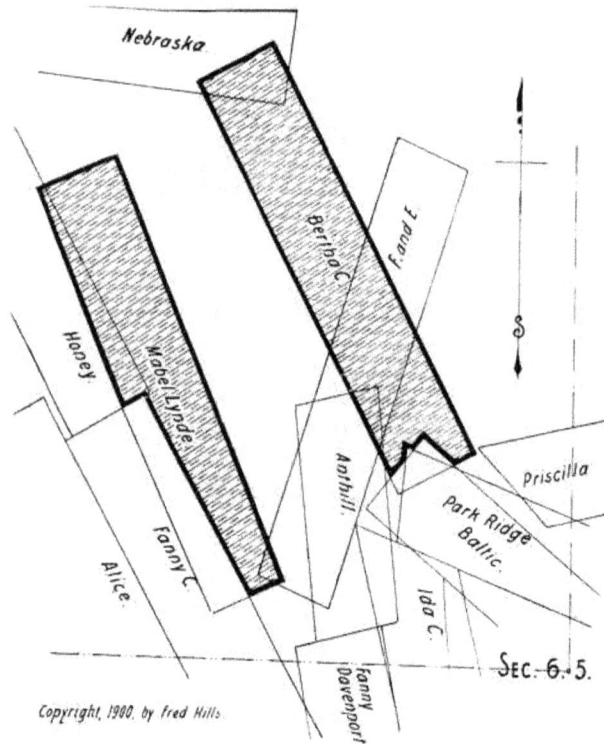

Copyright, 1900, by Fred Hills.

Property

Owns the Mabel Lynde, 7 1-2 acres, and the Bertha C., 9 1-2 acres, in the N. E. 1-4 section 6, on the west slope of Lincoln hill.   Receiver's receipt.

Development

About $2,500 expended on the property in the way of development work.   The company owes about $400.00 on account of patenting; this is in the form of a note given by the officers of the company.

Highest price for stock during 1899, $5.00 per 1,000; lowest price for stock during 1899, $1.00 per 1,000.

The E. F. C. Mining and Milling Company.

Incorporated 1895.

F. M. Woods, president; F. O. Ganson, vice-president; C. L. Arzeno, Directors secretary; H. E. Woods, treasurer; S. H. Ingersoll.

Main Office—Cameron, Touraine P. O., Colorado.

1,000,000 shares.   Par value, $1.00.   Capitalization

In treasury January 1, 1900, 80,000 shares.

Owns the E. F. C., Alva, Rittenhouse, W. E. S. and Dark Cloud, Property survey No. 10,456; the Mexican, 4-11-44 and the Idlewild, survey No. 13,197, in section 8, township 15 south, range 69 west, containing 14.629 acres; and the Annie Danks, survey No. 13,827.   Receiver's receipts are

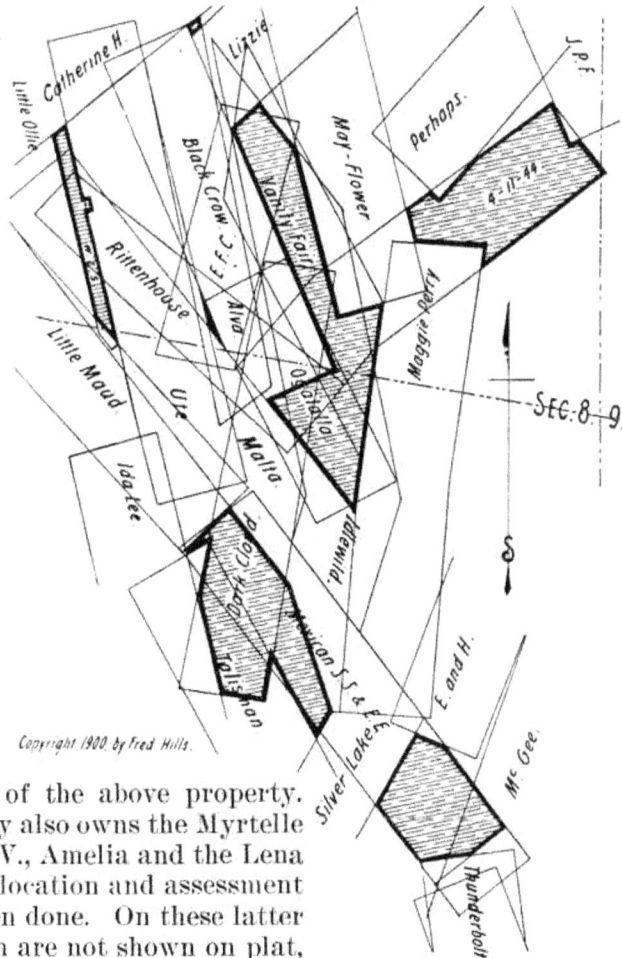

Copyright 1900 by Fred Hills.

held for all of the above property. This company also owns the Myrtelle G., Glennie W., Amelia and the Lena B., on which location and assessment work has been done.   On these latter claims, which are not shown on plat, patent work is now in progress.

The E. F. C. has three shafts, one 110 feet deep, one 85 feet deep, Development and one 25 feet deep.   The Rittenhouse has a shaft 125 feet deep.   On the W. E. S. there is a shaft 25 feet in depth and some trenching.   The Alva has a shaft 25 feet deep; also some trenching.   The Dark Cloud is developed by a tunnel 65 feet in length and 20 feet of cross-cuts.   The Idlewild has two shafts of 75 and 60 feet depth respectively.   The 4-11-44 has five 12-foot holes.   The Mexican has one shaft 75 feet deep, with 20 feet of drifting; also three other shafts ranging in depth from 12 to 18 feet.   On the Annie Danks is a tunnel 35 feet in length and a shaft 50 feet deep.

The El Dorado Gold Mining and Milling Company.

Incorporated June, 1892.

Directors

C. B. Seldomridge.....................President

C. H. Dudley......................Vice-President

W. H. McIntyre...........Secretary and Treasurer

E. M. De La Vergne.   J. F. Maybery.

Main Office—44 Hagerman building, Colorado Springs, Colorado.

Capitalization

1,000,000 shares.   Par value, $1.00.

In treasury January 1, 1900, 149,000 shares.

Property

Owns the El Dorado claim, 10.331 acres, in the N. W. 1-4 section 18, north and west of the El Paso, on Womack hill, patented; also the

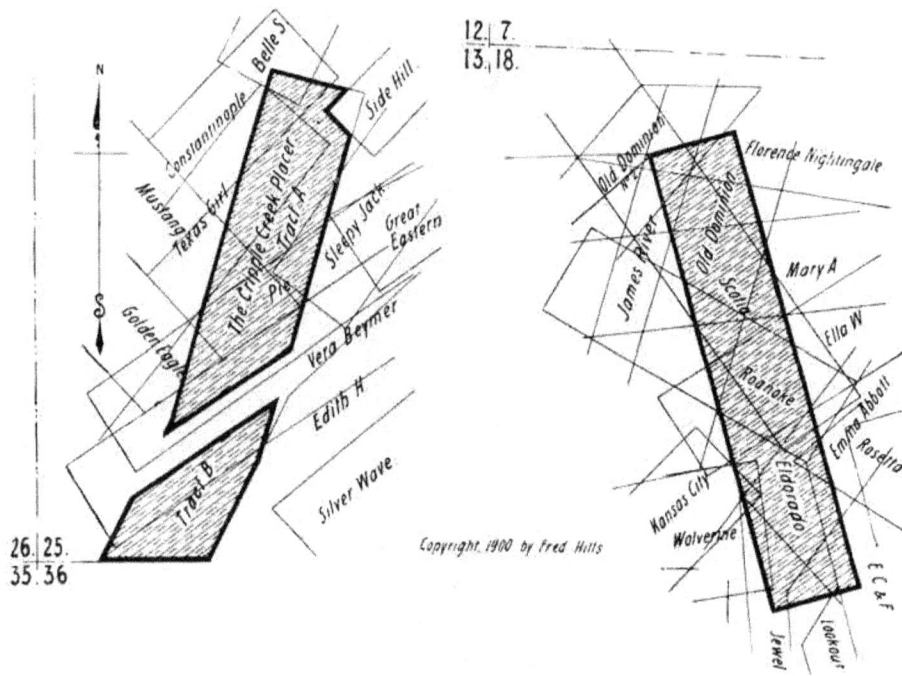

Copyright, 1900 by Fred Hills

Cripple Creek placer, below junction of Cripple Creek and Arequa gulch, about 9 acres, in the S. W. 1-4 section 25, for which receiver's receipt has been issued.

Development

There is a shaft about 60 feet deep and surface prospecting on the El Dorado claim, while trenching and ditching work has been done on the Cripple Creek placer.   The El Dorado claim and the Cripple Creek placer were among the first locations made in the district.   The stock is closely held by the original owners.   It is likely an effort will be made to either have the properties worked under lease or by the company themselves.

Highest price for stock during 1899, 5 cents; lowest price for stock during 1899, 5 cents.

MAIN BUILDINGS OF THE PORTLAND GOLD MINING COMPANY.

The Elkton Consolidated Mining and Milling Company.

Incorporated 1892.

Directors

Geo. Bernard, president; Wm. S. Jackson, vice-president; J. H. Avery, secretary and treasurer; Daniel Thatcher, assistant secretary and assistant treasurer; S. S. Bernard; Richard Clough; Wm. Shemwell; J. W. Graham.

Main office—Rooms 15, 16 and 17 El Paso Bank building, Colorado Springs, Colorado.

Capitalization

1,250,000 shares. Par value, $1.00. Shares in treasury January 1, 1900, 125,000.

Property

Owns the Walter, Katherine, Elkton, Kentucky Bill, Appie Ellen No. 2, and Thompson, located in the N. E. 1-4 section 30, on Raven hill. Property is all patented. In April, 1900, a consolidation of this company with the Raven G. M. Co. and the Tornado G. M. Co. was practically effected, under the name of the Elkton Consolidated Gold Mining Company, with a capitalization of 3,000,000 shares of a par value of $1.00 per share. All that is needed to perfect this consolidation is the ratification of the stockholders. By the terms of the deal the property of the Raven company is sold to the Elkton company for 625,000 shares of the consolidated stock, and the Tornado for 500,000 shares of the new stock. The old Elkton company gets 1,375,000 shares of the new stock, which will be issued to the stockholders in the form of a stock dividend amounting to about 22 per cent.

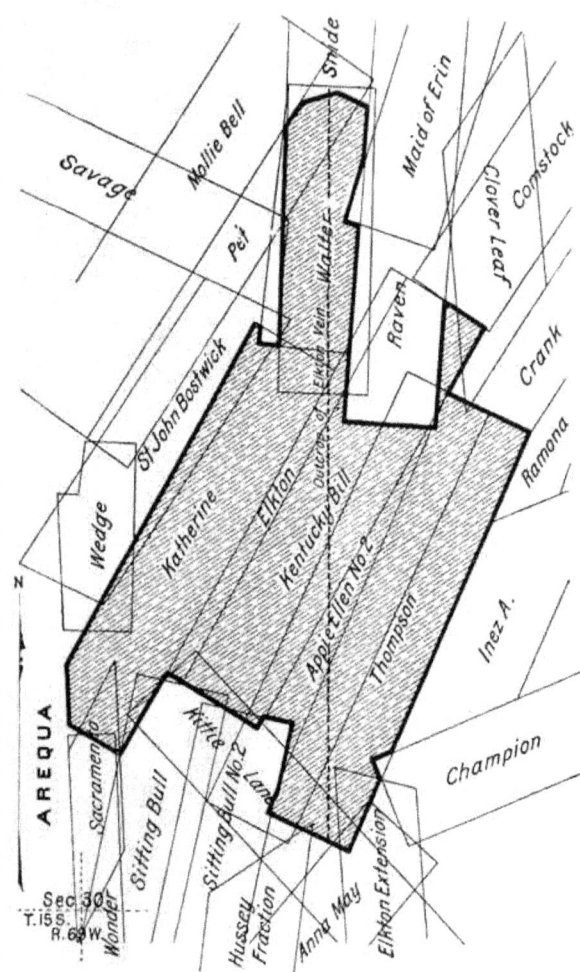

Copyright,1900, by Fred Hills

Development

Improvements consist of two shaft houses, boiler rooms, compressor room, ore house, magazine, office and assay office buildings; six boilers of 625-horse power, one 80-horse power hoist, two six-drill Norwalk compressors, one 80-light dynamo plant, one Gates crusher and engine, assay furnaces complete, etc.; two 600-gallon Knowles triple expansion station pumps; one 400-gallon Snow duplex station pump; two No. 9 Cameron sinkers; one 15-horse power hoist; one 1,000-gallon triple expansion Prescott station pump. Development on the Elkton, Kentucky Bill and Walter consists of a 735-foot shaft, 11,500 feet of drifting, 1,000 feet of cross-cutting. On the Thompson there is a 485-foot shaft, 700 feet of drifting and 400 feet of cross-cutting. On the Appie Ellen No. 2 is a 300-foot shaft and 600 feet of drifting. The Katherine has a 500-foot shaft, with 300 feet of cross-cutting. Originally the Elkton company

(CONTINUED ON PAGE 169.)

claimed ownership by location to the Elkton, Walter and Kentucky Bill claims, and was capitalized for 500,000 shares, par value $1.00. The Walter claim was in conflict with the Snide and Tornado claims, owned by J. W. Graham and Job A. Cooper. To prevent litigation the Walter company was organized, capitalized for 500,000 shares, of which the Elkton company received 250,000 shares. Afterwards the Elkton company increased its capital stock to 1,250,000 shares, placing 250,000 shares in the treasury, and purchased the Walter claim. The Elkton was one of the first producers in the district and has been the most regular shipper in the camp, as is shown by the gross output.

Production and Dividend Production to January 1, 1900, $2,418,569.15; dividends to January 1, 1900, $720,710.57. Last dividend was paid March 20, 1900, $33,750.00. Highest price for stock during 1899, $1.31 per share; lowest price for stock during 1899, 70 cents per share.

The El Paso Gold Mining and Milling Company.
Incorporated 1894.

Directors Geo. Bernard, president; J. W. Mahoney, vice-president; S. S. Bernard, secretary and treasurer; J. M. Jordan, assistant secretary and treasurer. J. W. Graham.

Main Office—115 East Pike's Peak avenue, Colorado Springs, Colorado.

Capitalization 900,000 shares. Par value, $1.00. In treasury January 1, 1900, 89,750 shares; in treasury February 12, 1900, $42,615.83 cash.

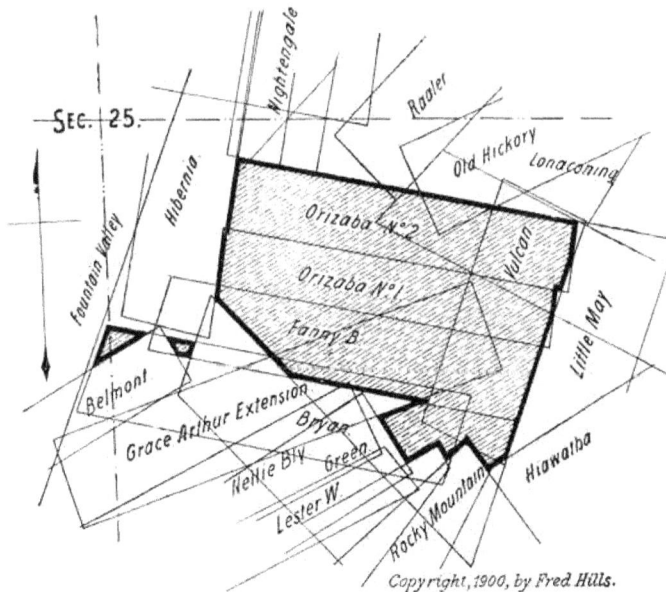

Copyright, 1900, by Fred Hills.

Property Owns the Orizaba Nos. 1 and 2, the Fannie B., the Vulcan and the Bryan Fraction, containing, in all, about 25 acres, all patented, situated in section 25, on Beacon hill.

Development There is a steam plant on the property capable of mining 500 feet deep; also ore house, shaft house, blacksmith shop, etc. The main shaft on the Orizaba No. 2 is 330 feet deep, with 3,000 feet of drifting. Prior to November, 1898, the property was worked exclusively by lessees; since which time the company has been actively engaged in opening up the mine, with gratifying success.

Production and Dividend Gross production to January 1, 1900, $375,000.00; dividends up to January 1, 1900, $10,795.00; last dividend, for $5,000, was paid December 10, 1897.

Highest price for stock during 1899, 50 cents; lowest price for stock during 1899, 9 1-2 cents.

The Emma-Aimee Gold Mining Company.

Incorporated.

Directors   R. H. Magee, vice-president; Will J. Mathews, secretary; Gordon Bish, treasurer; E. W. Case.

Main Office—Rooms Nos. 10 and 11, Exchange National Bank building, Colorado Springs, Colorado.

Capitalization   1,250,000 shares. Par value, $1.00.

In treasury January 1, 1900, 309,000 shares; in treasury January 1, 1900, $121.39 cash.

Property   Owns the Emma-Aimee claim, in the N. E. 1-4 of section 19, containing 7 3-4 acres, on Iron-clad hill. Patented.

Development   There are two tunnels, one of 165 feet, and one of 55 feet. Three shafts have been sunk, one of 40 feet, with drift of 35 feet, one of 60 feet, in the main tunnel, with a drift of 60 feet, and one of 37 feet.

Production   Gross production to January 1, 1900, 20 tons. No ore mined during 1899.

Highest price for stock during 1899, 3 3-8 cents; lowest price for stock during 1899, $4.50 per M.

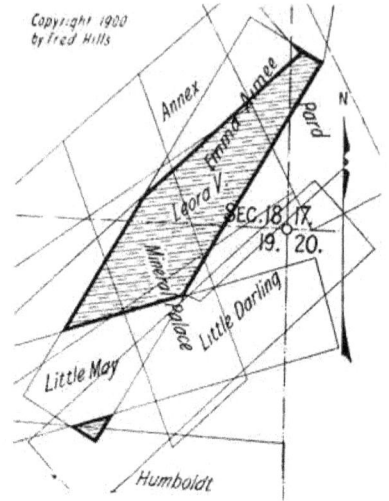

The Empire State Gold Mining Company.

Incorporated December 31, 1895.

Directors   J. E. Phillips, president; L. A. Civill, vice-president; F. H. Pettingell, secretary and treasurer.

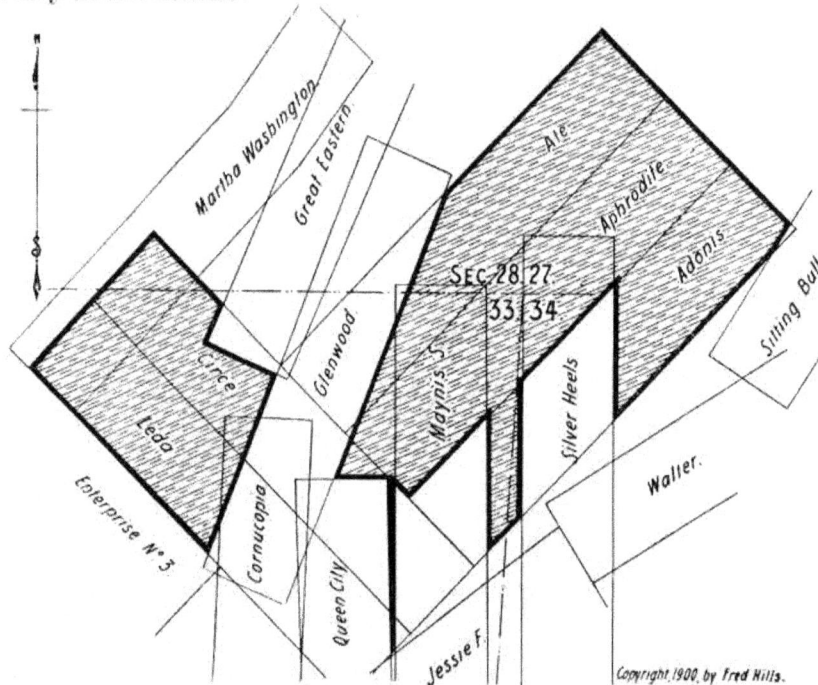

Main Office—11 Bank building, Colorado Springs, Colorado.

Capitalization   1,000,000 shares. Par value, $1.00.

In treasury January 1, 1900, 180,000 shares; in treasury January 1, 1900, $60.00 cash.

Property   Owns the Adonis, Aphrodite, Circe, Ate and Leda, 33.929 acres, in sections 27, 28, 33 and 34, on Big Bull mountain. Patented.

The Enola Mining and Milling Company.

Incorporated.

W. S. Dexter.........................President   Directors

G. L. Pike..........................Secretary

Main Office—Colorado Springs, Colorado.

1,000,000 shares. Par value, $1.00.   Capitalization

Owns the Silver Tip, containing 9 acres, in the N. W. 1-4 section 20.   Property
on Bull hill; also the John Adams and the Bunker Hill, located west of
Cripple Creek.   These latter claims do not appear on the map.   The
Silver Tip is patented.   John Adams and Bunker Hill held by location.

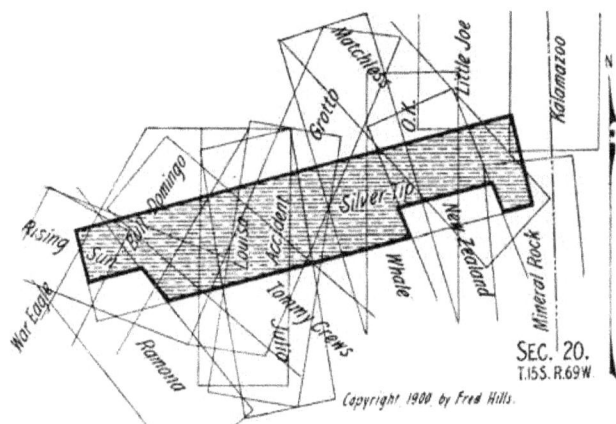

Copyright 1900 by Fred Hills.

Small shaft house. The Silver Tip has three shafts, one 125 feet   Development
deep, with several drifts; one 60 feet deep, and one 75 feet deep, with
drifts and considerable surface prospecting. The greater part of the
development work is being done on the Silver Tip. Two good veins are
shown in the discovery shaft, and one on the east end of the Silver Tip.

Gross production to January 1, 1900, about $5,000.00.   Production

Highest price for stock during 1899, 6 cents; lowest price for stock
during 1899, 3 1-2 cents.

The Enterprise Mining and Land Company.

Incorporated 1892.

Directors

A. M. Ripley..............................President
E. A. Colburn.....................Vice-President
C. H. Dudley.............Secretary and Treasurer
J. A. Wright.

Main Office—14 N. Nevada avenue, Colorado Springs, Colorado.

Capitalization

800,000 shares. Par value, $1.00.

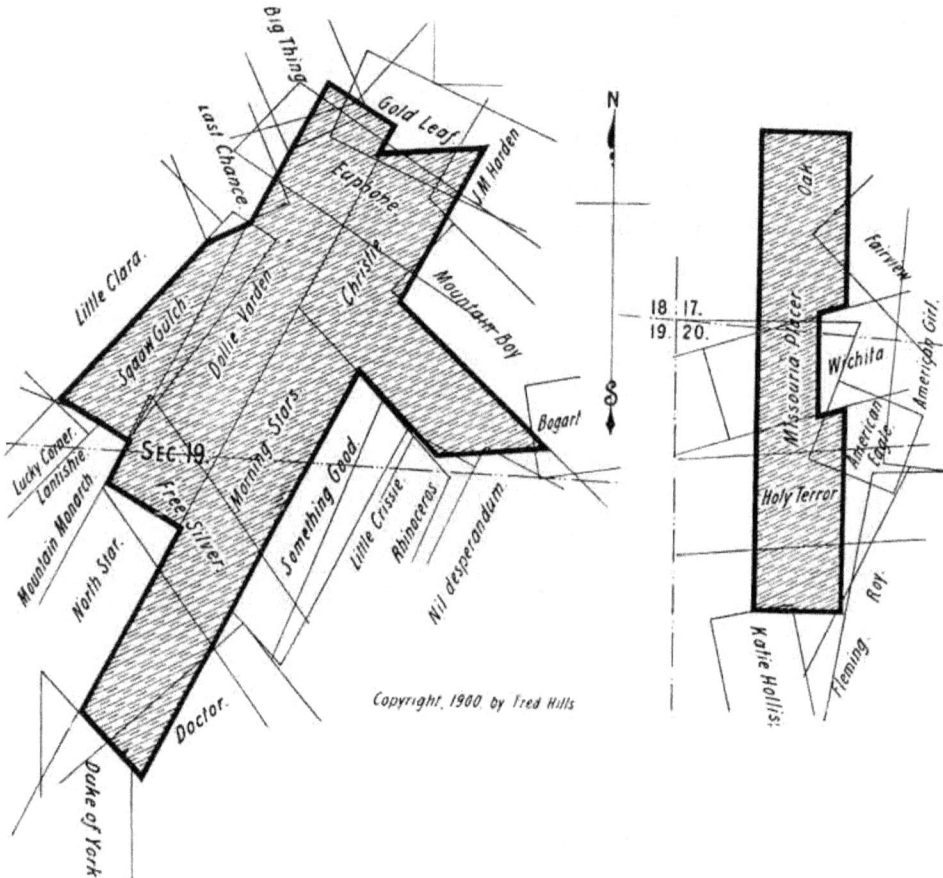

Copyright. 1900 by Fred Hills

Property

Owns the Dollie Varden, Morning Stars, Christie and Squaw Gulch, 30 acres, patented, on Raven hill, in the E. 1-2 of section 19, near the Doctor mine; also the Oak, 10 acres, patented, in the N. W. 1-4 of section 20, on Ironclad hill.

Development

There is an ore house, blacksmith shop and small hoist on the property. Main shaft is about 250 feet deep, with 1,000 feet of drifts and tunnels. About five shafts, 50 to 100 feet deep. A portion of the property is being developed by lessees.

Production to January 1, 1900, $100,000.00.

Highest price for stock during 1899, 21 cents; lowest price for stock during 1899, 5 7-8 cents.

The Erie Gold Mining Company.

Incorporated November 17, 1894.

D. B. Stambaugh........................President   Directors

Myron Wood....................Vice-President

H. C. Cassidy..........................Secretary

Mason Evans..........................Treasurer

H. P. Heedy.   F. H. Schmidt.

Main Office—No. 30 Hagerman building, Colorado Springs, Colorado.

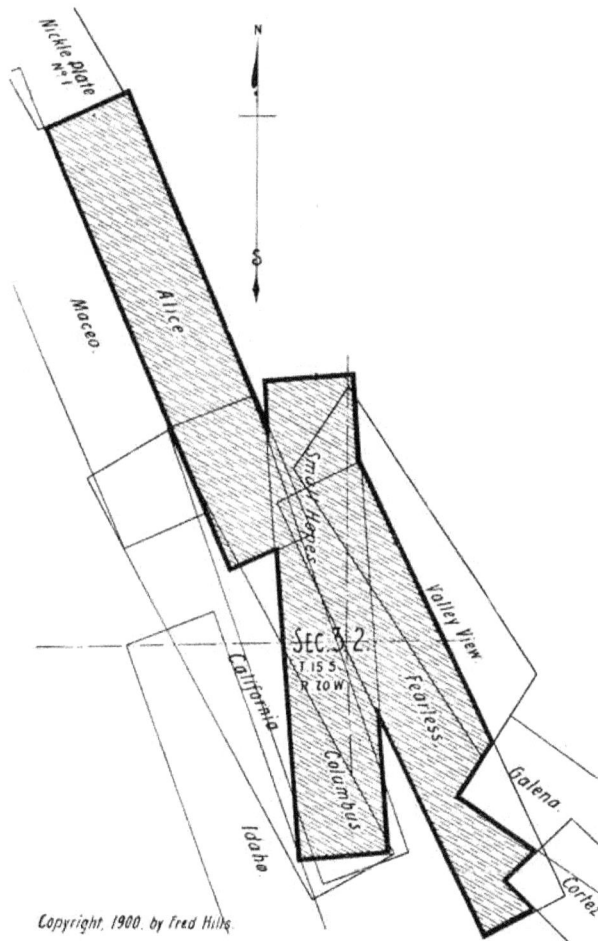

Copyright, 1900, by Fred Hills.

1,200,000 shares. Par value, $1.00.   Capitalization
In treasury January 1, 1900, 220,000 shares.

Owns the Alice, the Fearless and the Small Hopes, containing 26 1-2   Property
acres, situated on Red mountain on line between sections 2 and 3, township 15 south, range 70 west. All patented.

The Alice has a shaft 100 feet deep. On the Small Hopes is a shaft   Development
65 feet deep, and one 135 feet deep on the Fearless. The property was developed as above indicated in 1895, 1896 and 1897. Since December, 1897, no work has been done.

The Ernestine Gold Mining Company.

Incorporated 1895.

Directors

Edwin Arkell, president; R. H. Loper, vice-president; Chas. A. Brooker, secretary and treasurer; J. P. Walsh, assistant secretary.

Main Office—37 and 38 Post-office building, Colorado Springs, Colorado.

Capitalization

1,500,000 shares. Par value, $1.00.

In treasury January 1, 1900, 997 shares.

Property

Owns the Oxford lode, 8 acres, patented, and the Ferrell, 4.5 acres, held by location, in the N. E. 1-4 of section 17, on Galena hill, adjoining the Cameron Townsite and school section 16.

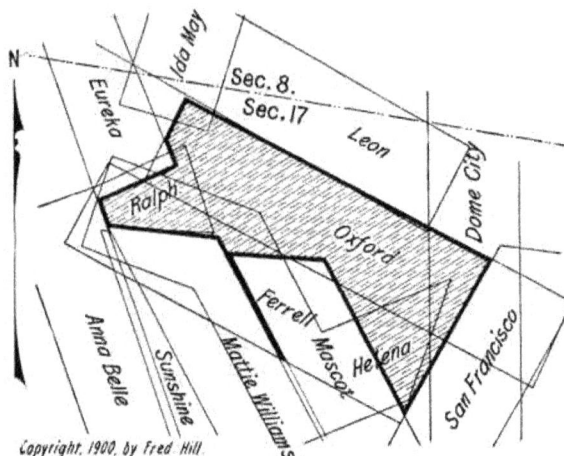

Copyright, 1900, by Fred Hill.

Development

One shaft on the Oxford lode 70 feet deep, with vein full width of shaft showing values from $12.00 to $45.00 per ton; also shaft on the Ferrell 42 feet deep, being sunk to obtain patent. The property was located in 1894. Bought from Homer Hart for $12,500.00. The company made application for patent in 1895; two adverses were filed; the company won out by giving up two acres of ground. The company worked the property as long as the treasury stock held out, but are now leasing. No debts or liabilities.

Highest price for stock during 1899, 1 3-4 cents; lowest price for stock during 1899, $5.00 per 1,000.

The Esperanza Gold Mining Company.

Incorporated December 17, 1895.

Directors

Geo. Macklin, president; A. A. Doyle, secretary and treasurer. E. D. Smith. W. C. Frost. D. I. Christopher. W. L. Douglas.

Main Office—112 E. Pike's Peak avenue, Colorado Springs, Colorado.

Capitalization

1,000,000 shares. Par value, $1.00.

In treasury January 1, 1900, 25,000 shares; no cash.

Property

Owns the Louisa E., in the N. W. 1-4 of section 31, containing about 3 acres, between Squaw and Grouse mountains.

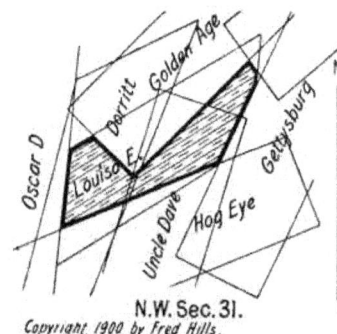

N.W. Sec. 31.
Copyright, 1900 by Fred Hills.

Development

There is a steam plant on the property, 25-horse power upright boiler, 6x8 Fairbanks & Morse hoist and No. 5 Cameron sinking pump. Two shafts on the grounds, one 180 feet deep and one 90 feet deep, with 75 feet of drifting. Vein disclosed in both shafts. Vein in 90-foot shaft 4 to 7 feet; assay, $12.00. Walls granite and porphyry, well defined; can be traced on surface of claim a distance of 700 feet.

Highest price for stock during 1899, 4 cents; lowest price for stock during 1899, 4 cents.

The Estella Gold Mining Company.

Incorporated.

M. J. Monett, president; Henry Bauman, vice-president; John R. Newby, secretary; H. S. Sommers, treasurer; W. H. Picking. *Directors*

Main Office — 53 Bank block, Colorado Springs, Colorado.

1,250,000 shares. *Capitalization* Par value, $1.00. In treasury January 1, 1900, 175,000 shares; in treasury January 1, 1900, $1,200.00 cash.

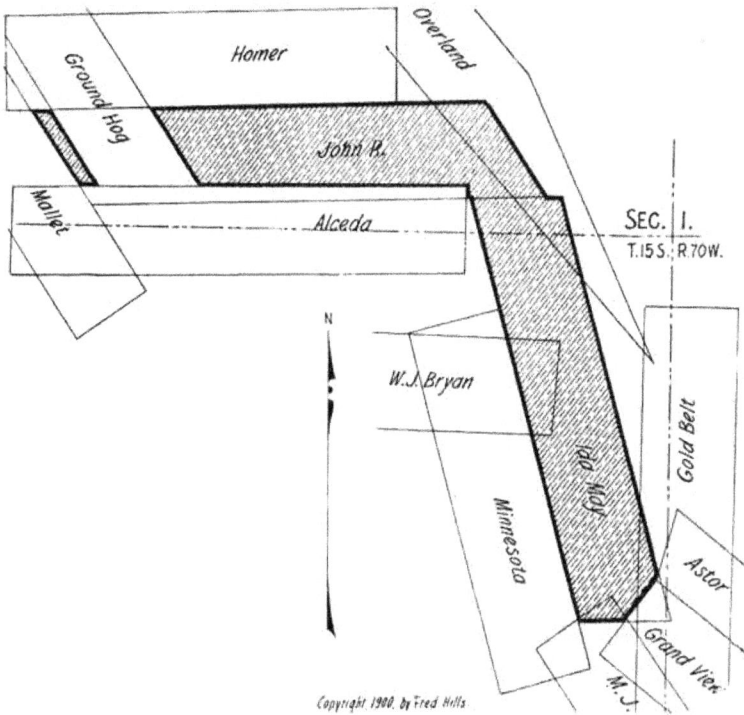

Owns the John R. and the Ida May, containing 16 acres, in the W. *Property* 1-2 of section 1, on Copper mountain. Both patented.

Steam hoist. The John R. has a 225-foot shaft. Some development *Development* work has been done on the Ida May.

The Estelline Gold Mining Company.

Incorporated January 24, 1893.

George W. Cramer, president; D. C. Dodge, *Directors* vice-president; J. B. Andrews, secretary and treasurer.

Main Office—321 Equitable building, Denver, Colorado.

200,000 shares. Par value, $1.00. In treas- *Capitalization* ury January 1, 1900, a nominal amount of cash.

Owns the Estelline claim, containing 8 3-4 *Property* acres, in E. 1-2 of section 32, on Prospect hill, adjoining on the south the Washington and May Raymond claims of Stratton's Independence group.

Two small sized shafts have been sunk on *Development* the claim, one 40 feet deep, the other some 250 feet deep; aside from this but little work has been done on the claim and none is, at present, in progress. The property has been worked in a perfunctory way from time to time by three or four sets of lessees. A few small shipments of ore in less than carload lots have been made. A strong well-defined dyke runs through the property from end to end, and the property is also traversed by two or more dykes of lesser dimensions, from which assays running from $5.20 to as high as $320.00 have been obtained. It seems almost assured that a branch of the Independence or Washington veins, and possibly the main veins themselves, pass through the Estelline claim from end line to end line.

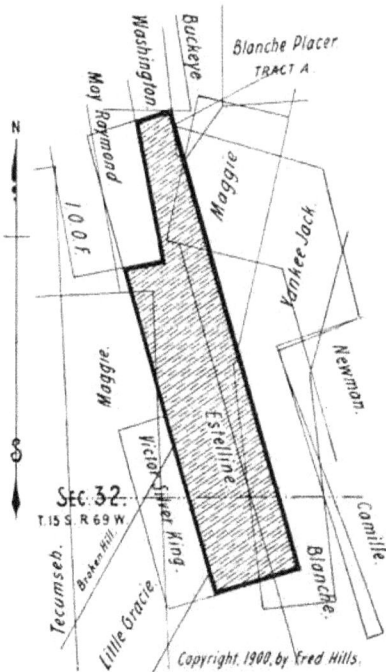

175

The Eudora Mining and Milling Company.

Incorporated December 27, 1895.

Directors

Chas. F. Rickey, president; T. F. McCarthy, vice-president; H. C. Shimp, secretary; E. C. Sheldon, treasurer. H. M. Sturdeyvant.

Main Office—4 Board of Brokers' building, Colorado Springs, Colorado.

Capitalization

1,200,000 shares. Par value, $1.00.

In treasury January 1, 1900, 11,000 shares; in treasury January 1, 1900, $100.00 cash.

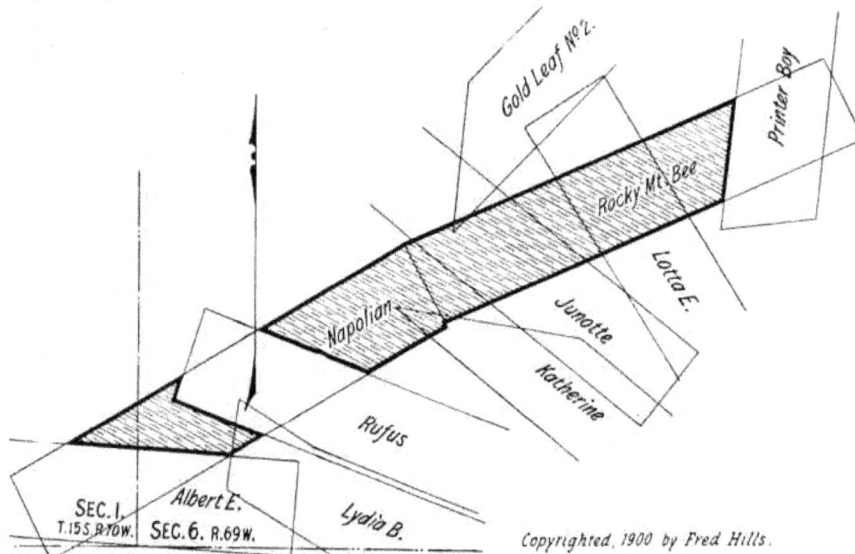

Copyrighted, 1900 by Fred Hills.

Property

Owns the Napolian and the Rocky Mountain Bee, containing 12 acres, situated in the S. E. 1-4 of section 1 and the S. W. 1-4 of section 6, on Copper mountain. All patented.

Highest price for stock during 1899, 1 1-2 cents.

The Eureka Gold Mining Company.

Incorporated March 10, 1894.

Directors

Geo. M. Carter, president; C. C. Smith, vice-president; H. A. Young, secretary and treasurer; P. L. Delany, A. L. Hunter, F. B. Gladding, V. A. Hunter.

Main Office—Room No. 5, Mining Exchange building, Colorado Springs, Colorado.

Capitalization

750,000 shares. Par value, $1.00.

In treasury January 1, 1900, 22,000 shares; in treasury January 1, 1900, $32.30 cash.

Property

Owns the Northwestern claim, 10 acres, in the S. E. 1-4 section 3, township 15 south, range 70 west; also the Best Friend, the Little Giant and the Early Bird, containing in all 6 acres, in the S. E. 1-4 section 12. All patented. The Northwestern not shown on plat.

Only enough development work to obtain a patent has been done.

Highest price for stock during 1899, 2 cents; lowest price for stock during 1899, 1 cent.

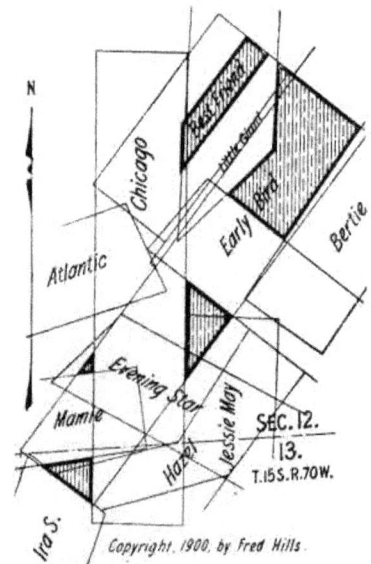

Copyright, 1900, by Fred Hills.

The Fauntleroy Gold Mining Company.

Incorporated October 21, 1899.

Verner Z. Reed......................President
C. C. Hamlin......................Vice-President
O. H. Shoup..............Secretary and Treasurer
F. J. Campbell.   A. T. Gunnell.

Directors

Main Office—Bank block, Colorado Springs, Colorado.

1,250,000 shares.   Par value, $1.00.
In treasury January 1, 1900, 250,000 shares.

Capitalization

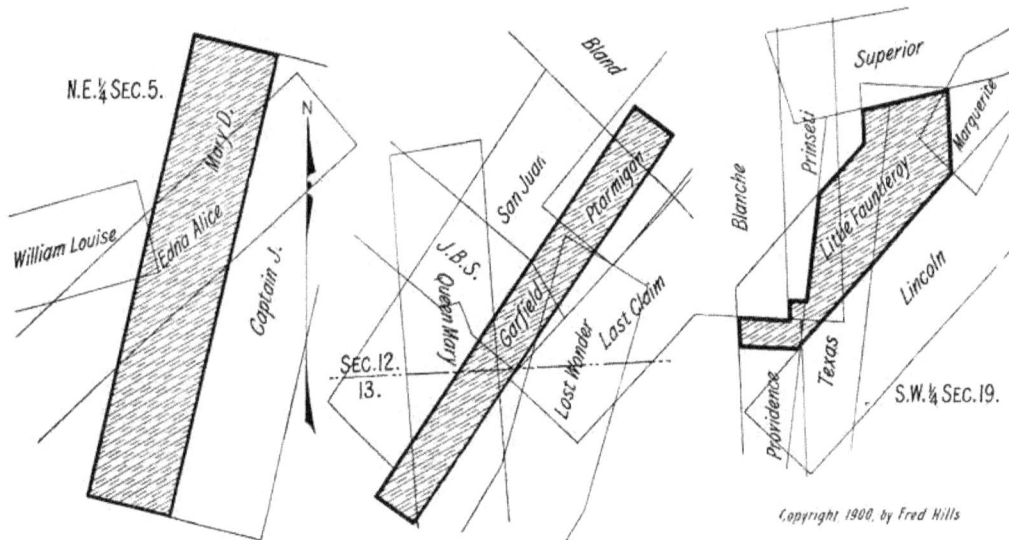

Copyright 1900, by Fred Hills

Owns the Little Fauntleroy, in the S. W. 1-4 of section 19, containing 6 acres on Gold hill; part of the Garfield, in the S. W. 1-4 of section 12, containing 5 1-2 acres, on Mineral hill; the Mary D., in the N. E. 1-4 of section 5, township 16 south, range 69 west, containing 10 acres, south of the town of Victor.   All the above property is patented.

Property

There is a hoist and shaft house on the Little Fauntleroy, and a shaft is now being sunk to encounter ore. The greater part of the development work is being done on the Little Fauntleroy. The shaft on this claim is now about 200 feet deep, and drifting will begin at once. A few shipments have been made.

Development

Highest price for stock during 1899, 9 cents; lowest price for stock during 1899, 8 1-2 cents.

The Favorite Gold Mining Company.

Incorporated 1894.

Directors

H. K. Devereux, president; J. W. Miller, vice-president; J. K. Miller, secretary and treasurer; W. L. Clark, L. M. Barney, W. C. Young, K. R. Babbitt.

Main Office—Colorado Springs, Colorado.

Capitalization

1,200,000 shares. Par value, $1.00. In treasury January 1, 1900, $2,000.00 cash.

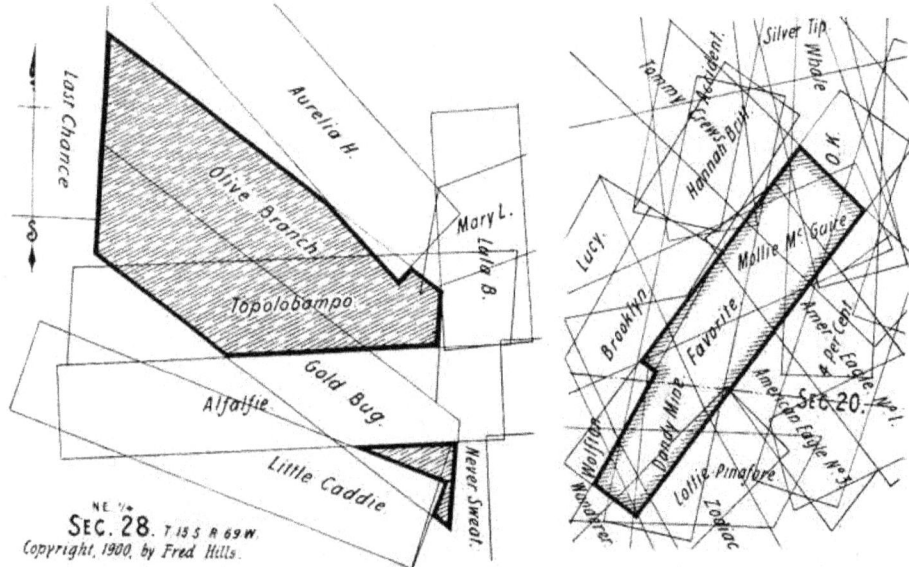

NE. 1/4
SEC. 28. T.15 S. R.69 W.
Copyright, 1900, by Fred Hills.

Property

Owned the Favorite, 8 acres, in the S. E. 1-4 section 20, on Bull hill, which was sold in March, 1900, to Mr. W. S. Stratton, of Colorado Springs, for the sum of $90,000. This property is shown on folding map under "Stratton's Favorite." The company also owns the Gold Leaf, 2 acres, in the S. W. 1-4 section 8, on Galena hill, in process of patenting; the Nugget, 0.723 acre, adjoining the Portland, on Battle mountain, in section 29, patented, for which $8,500.00 cash was paid; also the Olive Branch and Gold Bug, 18 acres, on Bull hill, in sections 21 and 22, for which $18,000.00 cash was paid.

History

While this company owned the Favorite it produced, up to the time of the sale, about $75,000.00 worth of ore. After the sale, as above recited, a dividend of 4 cents per share was declared, payable April 20, 1900, which left a handsome balance in the treasury. The policy of the company will be to develop the property under its own supervision and also to grant liberal leases. Highest price for stock during 1899, 7 cents; lowest price for stock during 1899, 3 1-2 cents.

The Figaro Gold Mining Company.

Incorporated 1896.

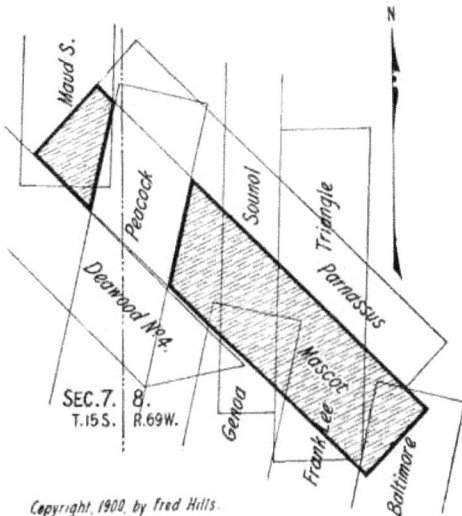

Copyright, 1900, by Fred Hills.

J. W. Miller................President   Directors
W. C. Young...........Vice-President
J. K. Miller....Secretary and Treasurer

Main Office—Colorado Springs, Colorado.

1,250,000 shares. Par value, $1.00.   Capitalization
In treasury January 1, 1900, 150,000 shares.

Owns a two-thirds interest in the Mascot, containing 8 acres, on Tenderfoot hill. The property is patented.   Property

The property is leased and bonded for $6,665; bond expires July 1, 1900.

The Findley Gold Mining Company.

Incorporated December 18, 1895.

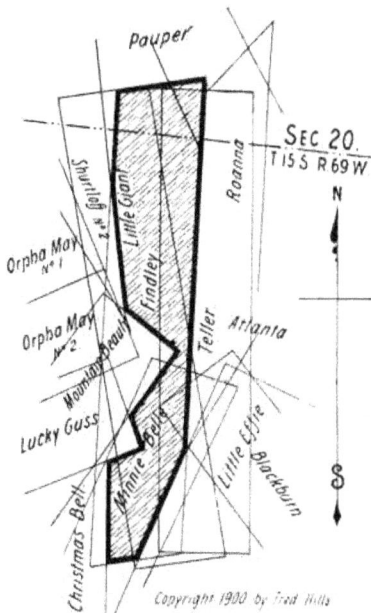

Copyright 1900 by Fred Hills

Geo. E. Lindley..................President   Directors
Geo. R. Buckman.............Vice-President
Asa T. Jones.........Secretary and Treasurer
Henry M. Blackmer.   Wm. A. Otis.
Sherwood Aldrich.   E. A. Richards.

Main Office—First National Bank building, Colorado Springs, Colorado.

1,250,000 shares. Par value, $1.00.   Capitalization
In treasury January 1, 1900, 132,500 shares.

Owns the Findley lode claim, containing 7.5 acres, in the S. E. 1-4 section 20, on Bull hill. All patented.   Property

One steam hoisting plant and machine drills. Two shafts have been sunk to depths of 600 feet and 415 feet respectively, and several drifts have been run on the veins. The greater portion of the property is being developed under one lease.   Development

In March, 1900, a valuable strike was made on the Steelsmith lease on this property, causing a heavy advance in the price of this stock.

Gross production to January 1, 1900, $57,000.00.   Production

Highest price for stock during 1899, 27 cents; lowest price for stock during 1899, 10 cents.

The Flower of the West Gold Mining Company.

Incorporated 1892.

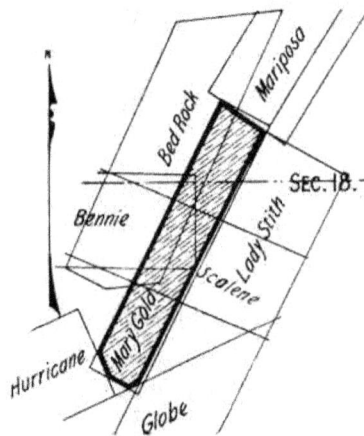

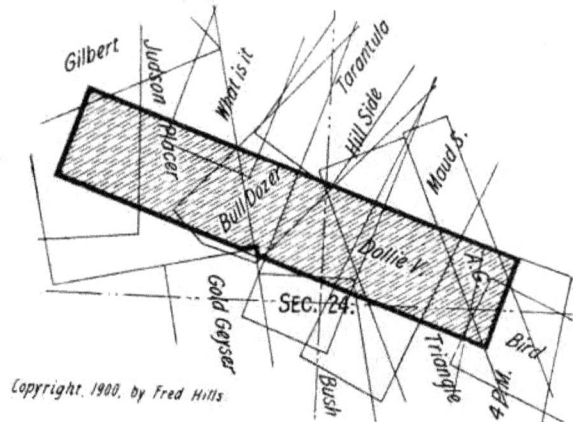

Copyright. 1900. by Fred Hills

Directors

K. C. Schuyler,
        President.

Geo. Rex Buckman,
        Vice-President.

N. S. Gandy,
        Secretary and Treasurer.

Mort Parsons.

P. A. McCurdy.

Main Office — Exchange Bank building, Colorado Springs, Colorado.

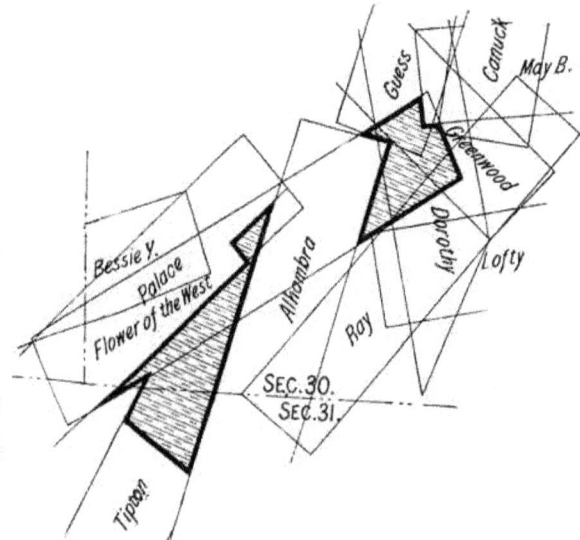

Capitalization

1,500,000 shares.   Par value, $1.00.

In treasury January 1, 1900, 80,000 shares; in treasury January 1, 1900, a nominal sum in cash.

Property

Owns a portion of the Tipton, containing 1 1-2 acres, in the S. W. 1-4 section 30; also the Flower of the West, containing 3 1-2 acres, all on Squaw mountain; also the mineral rights in conflict with the Alhambra, in the S. W. 1-4 section 30; the Dollie V., containing 10 acres, in the center of section 24, on Gold hill; and the Mary Gold, containing 3 acres, in the E. 1-2 section 18, on Globe hill.   All the above property is patented.

Development

Several shafts of from 25 to 150 feet deep have been sunk.   The Tipton claim, Flower of the West and the Dollie V. are leased for 15 months at a 20 per cent. royalty.   Seventy-five shifts are being worked.

Highest price for stock during 1899, 5 1-2 cents; lowest price for stock during 1899, 1 7-8 cents.

180

The Forepaugh Gold Mining Company.

Incorporated February 8, 1900.

C. W. Fairley.........................President   Directors

W. J. Hendrickson..................Vice-President

Phil S. Delany..............Secretary and Treasurer

G. D. B. Bonbright.   W. H. Gandy.

Main Office—35 and 37 Hagerman building, Colorado Springs, Colorado.

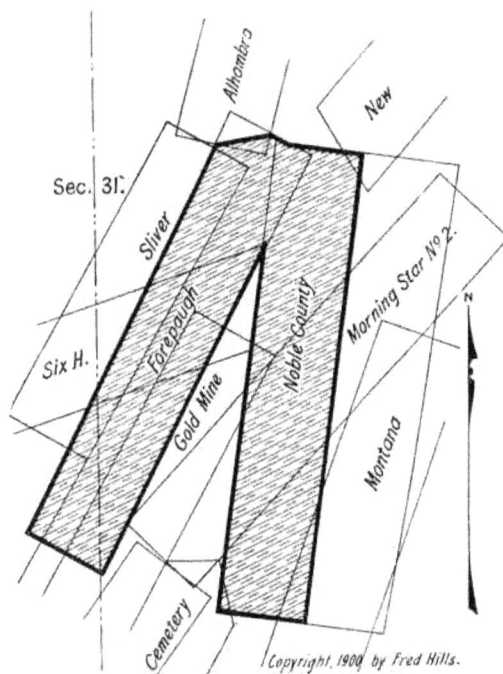

Copyright, 1900 by Fred Hills.

1,250,000 shares.   Par value, $1.00.   Capitalization

In treasury January 1, 1900, 250,000 shares.

Owns the Forepaugh and Noble County, containing 20 acres, in the   Property
N. E. 1-4 section 31, on Squaw mountain.   Patented.

A shaft house is on the property and there has been $2,500.00 spent   Development
on each claim in the way of development.

This company being so recently incorporated no stock has changed
hands, hence no prices quoted.

The Fort Wilcox Gold Mining Company.

Incorporated October 24, 1899.

Directors

W. F. Rock..............................President
M. W. Levy.........................Vice-President
W. A. Delany..............Secretary and Treasurer
D. R. McArthur.   E. J. Amann.

Main Office—No. 365 Bennett avenue, Cripple Creek, Colorado.

Capitalization

1,250,000 shares. Par value, $1.00.

In treasury January 1, 1900, 209,000 shares; in treasury January 1, 1900, about $500.00 cash.

Property

Owns the Fort Wilcox lode mining claim, patent No. 10,180, containing 8 1-2 acres, on Copper mountain, in the S. W. 1-4 section 1, township 15 south, range 70 west. Patented.

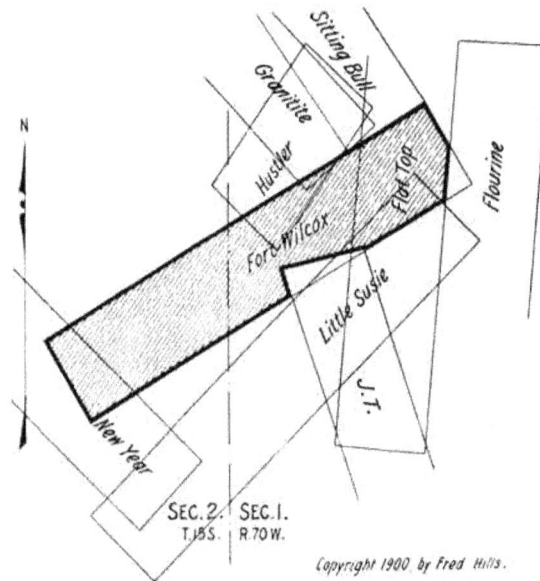

Copyright 1900, by Fred Hills.

Development

Three air-drill compressor with an 80-horse power boiler; shaft house; necessary water tanks and blacksmith shop. The Fort Wilcox claim has a tunnel of 500 feet, and several shafts, the deepest of which is 70 feet. Also a number of cross-cuts and drifts. This property is under lease to a leasing company, backed by the Standard Oil company. The lessee is driving a tunnel and will cut the Fort Wilcox claim at a depth of about 350 feet. The lease calls for 100 feet of work on this tunnel per month.

History

The Fort Wilcox claim was discovered on February 13, 1893, by John Wilson and Fortis Wilcox. United States patent was issued to same parties March 2, 1898. There is a porphyry dyke that runs the full length of the claim, and another that crosses the east end. All the ore the Fluorine claim shipped was taken from a small triangular piece of ground at the N. E. end line of the Fort Wilcox claim, that it lost in going to patent. The Fluorine stoped up to the Wilcox line. The Fort Wilcox claim has 350 feet of the dyke that produced the ore on the Fluorine claim of the Montreal Gold Mining company. The Fort Wilcox claim has shipped ore.

Highest price for stock during 1899, 3 5-8 cents; lowest price for stock during 1899, 2 cents.

The Franklin Gold Mining Company.

Incorporated April 12, 1894.

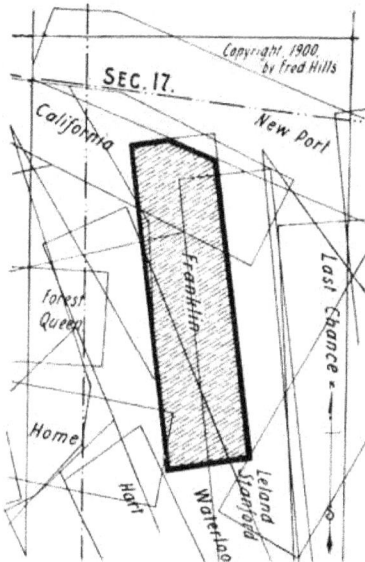

W. L. Cook, president; S. B. Dickens, secretary; E. R. Clark, treasurer; Russell Prentice; W. C. Grafton; P. Pedersen. Directors

Main Office—25 North Tejon street, Colorado Springs, Colorado.

1,000,000 shares. Par value, $1.00. Capitalization

In treasury January 1, 1900, over 50,000 shares; in treasury January 1, 1900, over $100.

Owns the Franklin lode, on Bull hill, 6 62-100 Property acres, in the S. E. 1-4 section 17. Patented.

One shaft has been sunk 115 feet, one 50 feet, Development one 40 feet, one 20 feet, also 175 feet of drifting; has also a whim and shaft-house. The showing has been good.

The property has been worked by the company and lessees. Owing to lack of funds the lessees have just stopped work.

Highest price paid for stock during 1899, 5 cents; lowest price paid for stock during 1899, 2 cents.

The Free Coinage Gold Mining Company.

Incorporated May, 1892.

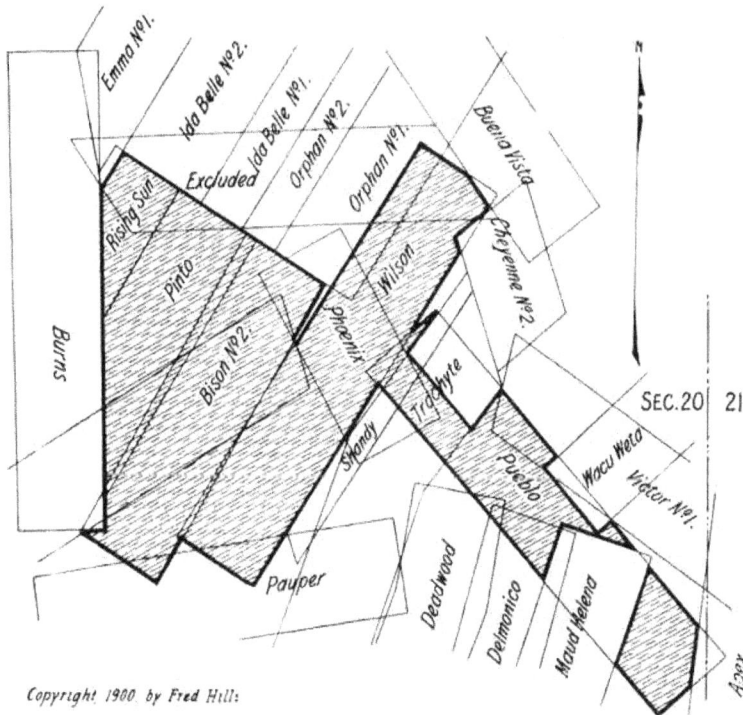

Sam Strong, president, general manager and treasurer; Chas. Cavender, vice-president; J. B. Neville, secretary. Directors

Main Office — Altman, Colorado.

1,000,000 shares. Capitalization Par value, $1.00.

In treasury January 1, 1900, 2,650 shares; also a large working capital.

Owns the Pinto, the Property Rising Sun, the Wilson, the Bison No. 2 and the Pueblo, all situated in the N. E. 1-4 section 20, in a group on Bull hill. All patented.

The main working Development shaft is 550 feet deep, with about 5,000 feet of drifts and cross-cuts. There is a complete plant of machinery, including compressor. All work is being done by the company excepting a few leases on the Wilson.

During 1899 there was about 3,000 tons of good ore shipped, and Production the property is now producing to the full capacity of the machinery. and Dividend This company is not a marketable stock proposition, but it is developing its property in such a way as to make it a dividend payer. It is a close corporation.

The Free Gold Mining Company.

Incorporated 1892.

Directors

Wm. Whalen...........................President

M. S. O'Rourke.....................Vice-President

Phil S. Delany.......................Secretary

T. C. Delany.........................Treasurer

Wm. Bohanna.

Main Office—35 and 37 Hagerman building, Colorado Springs, Colorado.

Capitalization

1,000,000 shares.   Par value, $1.00.

In treasury January 1, 1900, 70,000 shares; in treasury January 1, 1900, $50.00 cash.

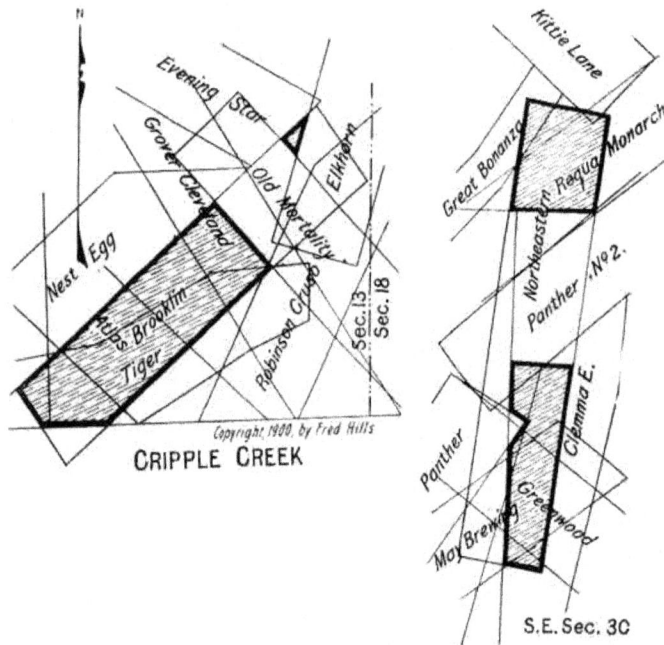

CRIPPLE CREEK

Copyright 1900, by Fred Hills

S.E. Sec. 30

Property

Owns Northeastern, containing 7 3-4 acres, in the S. E. 1-4 section 30, on Squaw mountain, and the Brooklyn, containing 9 acres, in the N. E. 1-4 section 13, on Mineral hill. All patented.

Development

There are several shafts on the property, the deepest being about 100 feet. On the Northeastern the company owns the mineral rights, but not the surface between the northern and southern tracts, which are shaded on plats.

Highest price for stock during 1899, 1 cent; lowest price for stock during 1899, 1 cent.

The Freeport and Cripple Creek Gold Mining Company.

Incorporated 1896.

Daniel A. Freeman.....................President
W. B. Boardman...................Vice-President
J. F. Thompson.......................Secretary
Henry Sachs ....Treasurer
W. E. Nickerson.   C. A. B. Halvorson.

Directors

Main Office—No. 220 Devonshire street, Boston, Mass.

1,250,000 shares. Par value, $1.00.   Capitalization

In treasury January 1, 1900, 123,500 shares; in treasury January 1, 1900, $1,436.62 cash.

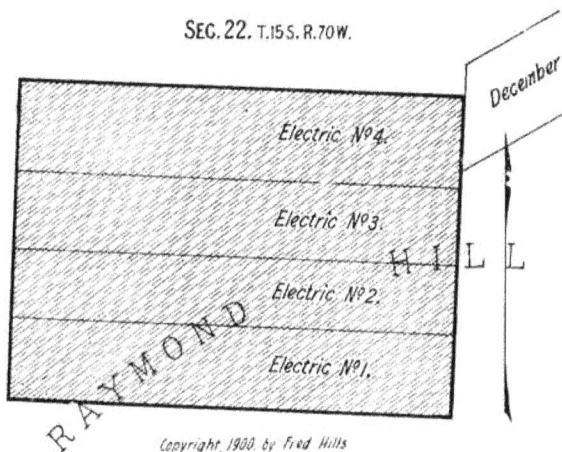

SEC. 22. T.15 S. R.70 W.

Copyright. 1900. by Fred Hills

Owns four claims of 9 acres each, known as Electric Nos. 1, 2, 3 and 4, in the E. 1-2 of section 22, township 15 south, range 70 west, on Raymond hill.   Total acreage owned by the company is 36 acres.   All patented.   The company also holds by lease the Bonanza King claim on Gold hill, and the Fourth Brothers claim on Beacon hill.   Property

There is one steam hoist and one electric hoist, and all necessary buildings.   The Electric No. 2 has a shaft 150 feet deep and a cross-cut of 40 feet.   There are other surface shafts and cuts.   At present the greater part of the development work is being prosecuted on the Four Brothers and the Bonanza King, and from this latter the company commenced shipping ore January 15, 1900.   Development

Highest price for stock during 1899, 10 cents; lowest price for stock during 1899, 10 cents.

The Fulton-Marguerite Mining Company.

Incorporated 1892.

Directors

Walter F. Crosby.........................President
L. Wagner.........................Vice-President
Louis R. Ehrich...........Secretary and Treasurer
J. W. Wright.................Assistant Secretary
John I. McCombs.

Main Office—No. 64 Hagerman building, Colorado Springs, Colorado.

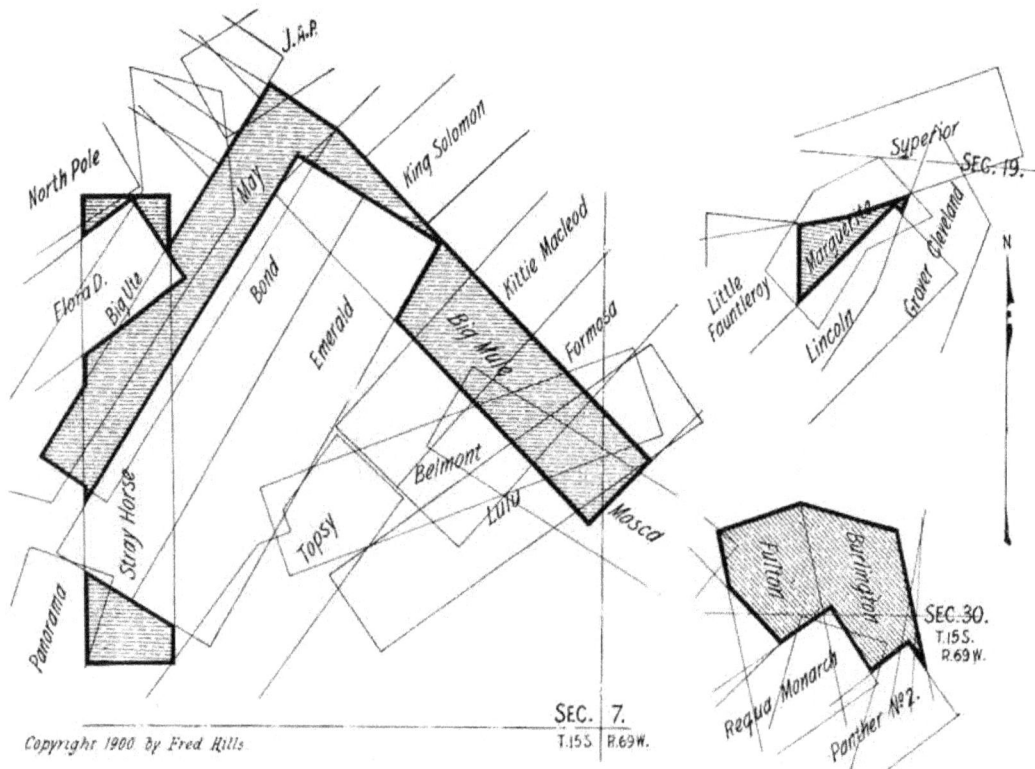

Copyright 1900 by Fred Hills.

SEC. 7.
T.15S R.69W.

Capitalization

1,500,000 shares. Par value, $1.00.

In treasury January 1, 1900, 300,000 shares; in treasury January 1, 1900, $5,000.00 cash.

Property

Owns the Burlington and the Fulton, 2.5 acres, in the E. 1-2 section 30, on Raven hill; the Marguerite, 2 acres, in the W. 1-2 section 19, on Gold hill; the Big Mule, the Stray Horse and the May, in the N. W. 1-4 of section 7, containing 15 acres, on Carbonate hill. All the above property is patented.

The greater part of the development work is being done on the Fulton.

The Galesburg Gold Mining Company.

Incorporated January 20, 1896.

S. A. McEvers, president; J. W. Greiner, vice-president; F. W. Directors Jacques, secretary; Dr. Wm. H. Heisen, treasurer; D. A. Clark, Nathan O. Walker, A. A. Pierce.

Main Office—No. 1332 Tremont street, Denver, Colorado.

1,000,000 shares. Par value, $1.00. In treasury January 1, 1900, Capitalization 200,000 shares; in treasury January 1, 1900, over $100.00 in cash.

Owns the Sherman claim, containing 0.307 acre, situated in the W. Property 1-2 section 29, township 15 south, range 69 west, on Battle mountain, adjoining the Coriolanus mine; the Saddle and the Hershaw, containing

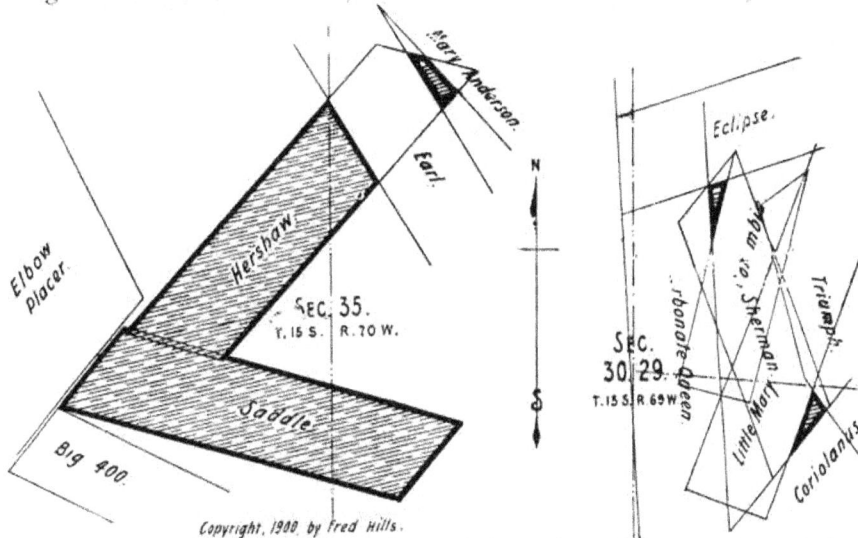

Copyright, 1900, by Fred Hills.

15.997 acres, situated in section 35, township 15 south, range 70 west; also the Brooklyn. This latter is presumably a location only. Receiver's receipts held for the Sherman and the Saddle. The Hershaw in process of patenting.

On the Sherman lode is a shaft 100 feet deep and 15 feet of drifting. Development The Saddle lode has a 60-foot tunnel, which is receiving the greater part of the development work. A progressive leasing policy has been outlined by the present management, which will be vigorously pursued.

Highest price for stock during 1899, 2 cents; lowest price for stock during 1899, 2 cents.

The Garfield Consolidated Mining Company.

Incorporated January 26, 1898.

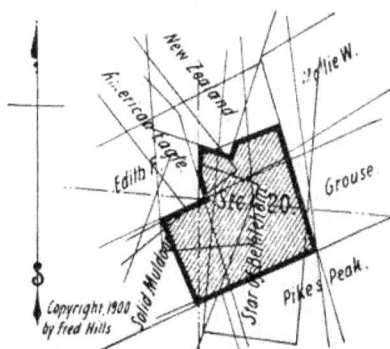

Copyright, 1900 by Fred Hills

Charles J. Hughes, Jr., president; Wm. Trustees H. Chittenden, vice-president; Albert Smith, secretary; Gerald Hughes, treasurer; G. I. Chittenden, H. K. Chittenden.

Main Office—Room 12, Hughes block, Denver, Colorado.

1,250,000 shares. Par value, $1.00. In Capitalization treasury January 1, 1900, $10,867.29 cash; in treasury January 1, 1900, no stock.

Owns the Mineral Rock, the Garfield, and a one-half interest in the Property Lamar, a total acreage in all of about 3 1-2 acres, in the center of section 20. All the above property is in process of patenting.

(CONTINUED ON PAGE 188.)

The Garfield Consolidated Mining Company—Continued.

Development

There is a shaft house, an engine house, an ore house, 6x8 friction hoist, a 25-horse power boiler and feed pump for fire protection. A working shaft has been sunk 600 feet; there are also about 450 feet of drifts and cross-cuts. The greater part of the development work is being done on the Mineral Rock and the Garfield. The upper levels and dump are leased.

History

Up to August 5, 1899, the lessee paid over $20,000 royalty. Since that time the company has worked the property. About $10,000 worth of ore has been found. In developing the property about $20,000 has been spent. A consolidation between this company and the Bankers company is rumored, but no definite information can be obtained.

Dividend

Dividends up to January 1, 1900, $12,000.00.

Highest price for stock during 1899, 23 cents; lowest price for stock during 1899, 12 cents.

The Geneva Gold Mining and Milling Company.

Incorporated November 12, 1895.

Directors

L. C. Hall, president; R. Hillhouse, vice-president; P. Wollesen, secretary and treasurer; J. C. Manchester, F. P. Davis.

Main Office—Colorado Springs, Colorado.

Capitalization

1,500,000 shares. Par value, $1.00. In treasury January 1, 1900, 22,000 shares.

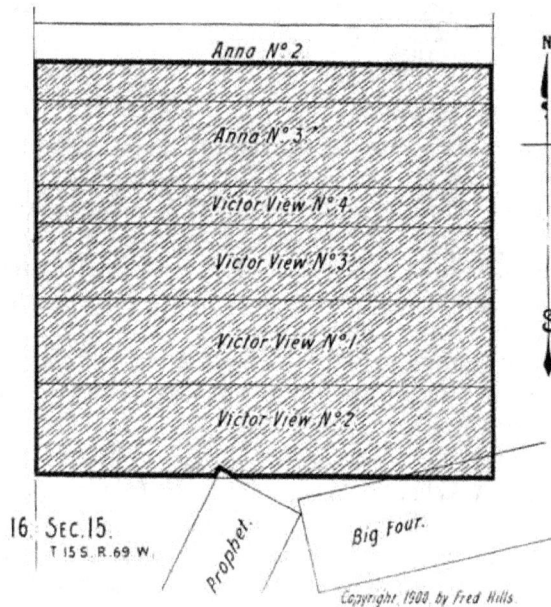

Copyright 1900 by Fred Hills.

Property

Owns the Victor View Nos. 1, 2, 3 and 4, and the Anna No. 3 and a part of Anna No. 2, containing an area of 43 acres, all patented, on the E. line of school section No. 16, on Calf mountain.

Development

There is a shaft 50 feet deep on each claim. The Victor View Nos. 1 and 2 are leased to November 15, 1901, and bonded for $25,000. On these leased claims 40 feet of development work has been done, in addition to the 50-foot shaft. These claims are receiving the greater part of the development work.

History

The company formed owned a total of 68.84 acres, but in settling adverse claims deeded away about 25 acres.

Highest price for stock during 1899, $6.00 per M.; lowest price for stock during 1899, $5.00 per M.

The George Washington Gold Mining Company.

Incorporated January 13, 1896.

Geo. M. Mitchell........ ....President

Wm. Young............Vice-President and Treasurer

Chas. H. Peters.........................Secretary

W. D. Merrill.   Chas. A. Whitescarver.

A. B. McGill.

Directors

Main Office—No. 308 Mining Exchange, Denver, Colorado.

1,250,000 shares.   Par value, $1.00.   In treasury January 1, 1900,   Capitalization
26,500 shares; in treasury January 1, 1900, a nominal sum in cash.

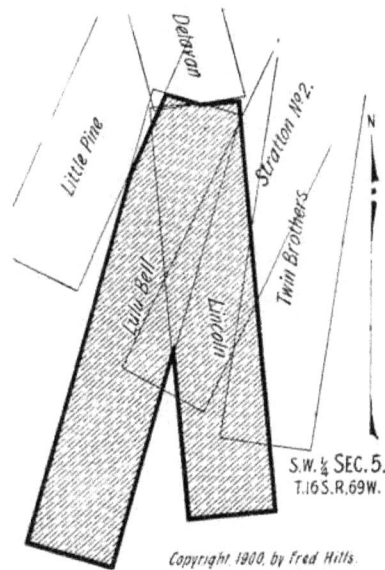

Copyright 1900, by Fred Hills.

Owns the Lincoln, the Lulu Bell and the "V," containing in all 21.56   Property
acres, in the S. W. 1-4 section 5, on the southwest slope of Straub mountain. The Lincoln and the Lulu Bell, containing 16.156 acres, are patented. The "V" is held by location and is not shown on map.

There are several shafts, the deepest being 100 feet; also some   Development
trenching. The greater part of the development work is being done on
the Lincoln.

The German American Gold Mining Company.

Reorganized March, 1900.

Directors

Zeno Felder ....President

J. F. Ensminger...........Secretary and Treasurer

Main Office—No. 211 Bennett avenue, Cripple Creek, Colorado.

Capitalization

1,250,000 shares. Par value, $1.00. In treasury January 1, 1900, 335,000 shares.

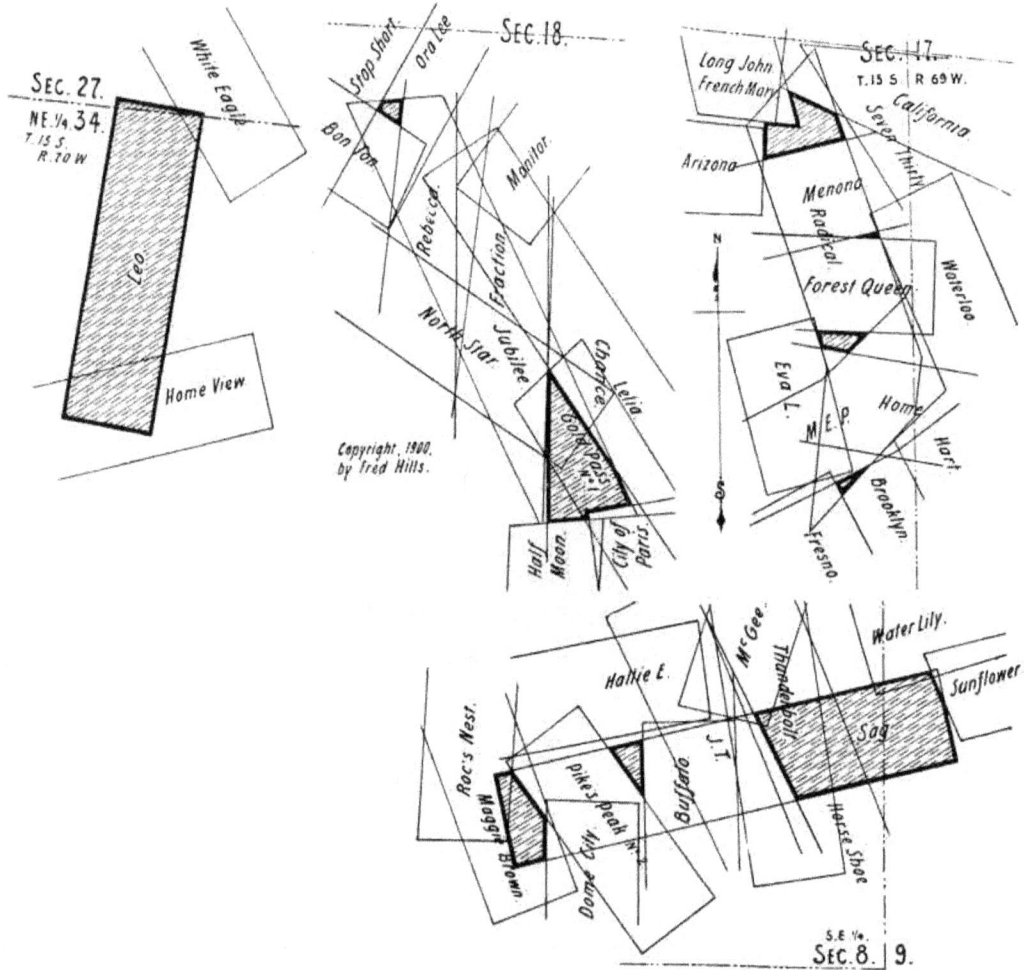

Property

Owns the Jubilee (fraction), containing 1.163 acres, situated in the S. W. 1-4 section 18, township 15 south, range 69 west; the Sag, containing 8.047 acres, situated in the S. E. 1-4 section 9, township 15 south, range 69 west; and the Leo, containing 6.769 acres, in the N. E. 1-4 section 34, township 15 south, range 70 west; also a two-thirds interest in the Radical, located on Bull hill. All patented except the Radical.

Development

The Jubilee has a shaft 200 feet deep; the Sag one of 100 feet. On the Radical there are three shafts ranging in depth from 20 to 50 feet.

The Gilpin and Cripple Creek Gold Mining Company.

Incorporated February, 1898.

John S. Gould, president; F. E. Higgins, vice-president; Geo. W. Directors Merrill, secretary; Erving J. Knight, treasurer; Ramsay C. Bogy.

Main Office—Cripple Creek, Colorado. Branch Office—Providence, R. I.

1,000,000 shares. Par value, $1.00. In treasury January 1, 1900, Capitalization 100,000 shares; in treasury January 1, 1900, $2,000.00 cash.

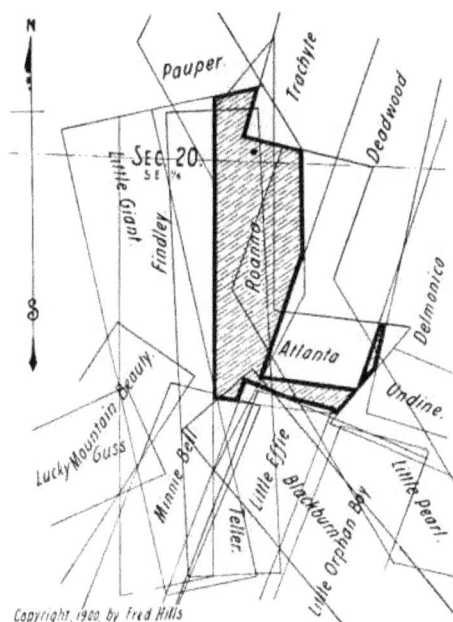

Copyright, 1900, by Fred Hills

Owns the Atlanta and the Roanna, containing 9.250 acres, in the Property S. E. 1-4 section 20, township 15 south, range 69 west, on the south slope of Bull hill, and adjoining the Deadwood, Hull City placer, Findley, etc. The company also owns the Montgomery, in Gilpin county, 10 acres. All patented.

The company is supplied with all the machinery and other facilities Development necessary for developing their properties. The Atlanta has a shaft 360 feet deep and drifts of 700 feet. The Montgomery has a shaft 370 feet deep and drifts of 1,000 feet.

Gross production to January 1, 1900, $6,000.00. Production

Highest price for stock during 1890, 51 cents; lowest price for stock during 1899, 25 cents.

The Gilt Edge Gold Mining, Milling and Prospecting Company.

Incorporated January 1, 1896.

Directors

L. A. Keys..............................President

D. McNeill............Vice-President and Treasurer

E. M. Purdy.............................Secretary

Frank Van Duyne.　　　　　　A. Loberge.

Main Office—Room No. 7, Freeman building, Colorado Springs, Colorado.

Capitalization

1,250,000 shares. Par value, $1.00. In treasury January 1, 1900, 165,000 shares. In treasury January 1, 1900, $125.00 cash.

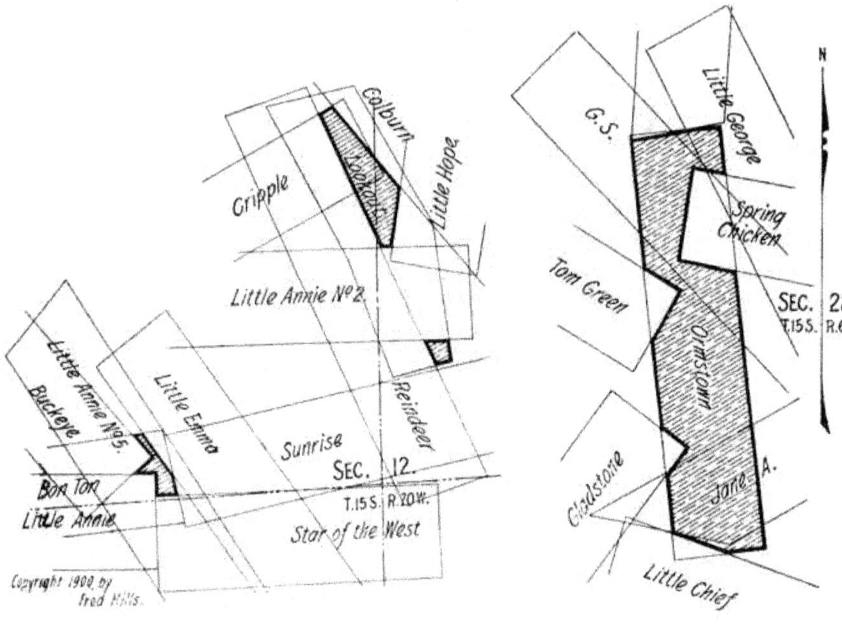

Property

Owns the Ormstown, 8 acres, in the S. E. 1-4 section 28, on Big Bull hill; the Lookout and the Bon Ton, 3 acres, in the center of section 12, on Mineral hill. The Ormstown is patented. The Bon Ton and the Lookout are held by location.

Development

The Ormstown has two shafts, one 80 feet deep and one 65 feet deep, with 50 feet of cross-cutting. The Lookout and the Bon Ton have three shafts, ranging from 40 to 60 feet deep.

The Gladys A. Mining Company.

Incorporated July, 1892.

T. B. Burbridge.........................President   Directors

Chas. N. Miller.............Secretary and Treasurer

J. L. Kinna.   W. W. Kirby.

J. E. Hunter.   J. M. Harden.

Main Office—No. 233 Bennett avenue, Cripple Creek, Colorado.

1,000,000 shares. Par value, $1.00. In treasury January 1, 1900,   Capitalization
51,700 shares; in treasury January 1, 1900, $469.00 cash.

Owns the Montana, 3 3-4 acres, in the N. E. 1-4 section 31, on Squaw   Property
mountain; the None Such, 3 acres, in the S. 1-2 section 24, on Gold hill.
The None Such is patented.

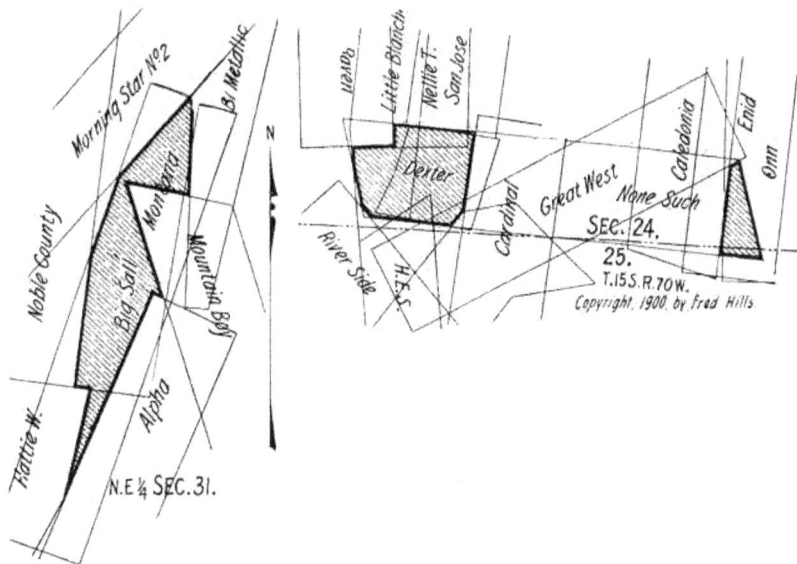

The None Such has a shaft about 65 feet deep. The shaft on the   Development
Montana is about 50 feet deep. The greater part of the development work
is being done on the None Such, which is leased.

In December, 1895, the capital stock was increased from 500,000   History
shares to 1,000,000, some of the new capital stock being utilized for pay-
ment of outstanding debts and for the acquisition of the Montana and
None Such claims. Both the None Such and the Montana have been
leased at certain periods, but no pay ore has as yet been discovered.
The other claims at one time owned by the company have been allowed
to lapse.

Highest price for stock during 1899, 1 1-4 cents; lowest price for
stock during 1899, $4.00 per M.

7

The Glasgow Gold Mining and Milling Company.

Incorporated 1893.

F. E. Robinson, president; J. Arthur Connell, vice-president; D. D. Lord, secretary and treasurer; William P. Bonbright, Chas. L. Zobrist.

Main Office—Room 68, P. O. building, Colorado Springs, Colorado; transfer office, International Trust Company, Colorado Springs, Colorado.

1,000,000 shares. Par value, $1.00. In treasury January 1, 1900, about 90,000 shares; in treasury January 1, 1900, $100.00 cash.

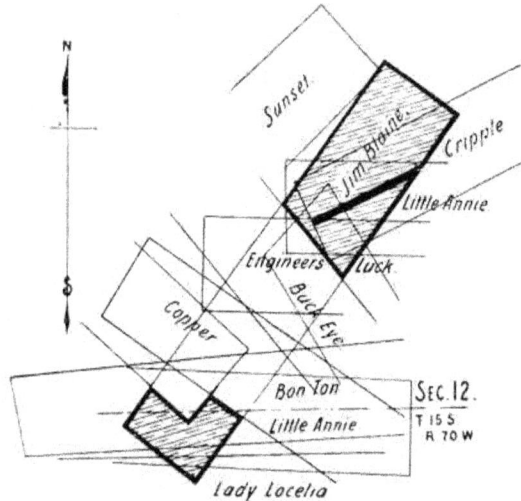

Copyright 1900 by Fred Hills

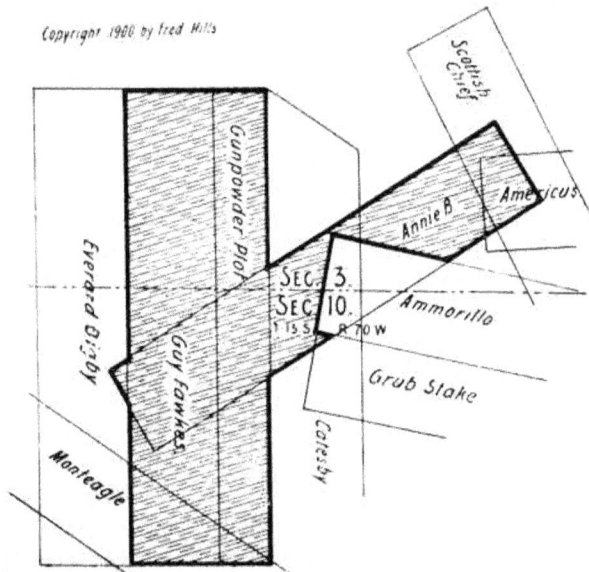

Owns the Guy Fawkes, Gunpowder Plot, and Annie B., comprising about 22 acres in the N. E. 1-4 of section 10, township 15 south, range 70 west, on Iron mountain; and the Jim Blaine, of about 6 acres, in the N. W. 1-4 of section 12, on Mineral hill. All the property is patented. This company now owns all interests formerly held by the Lottie Gibson Mining and Milling Company.

Highest price for stock during 1899, 1 cent; lowest price for stock during 1899, $7.00 per M.

The Globe Hill Gold Mining Company.

Incorporated 1897.

Geo. J. Blakeley, president; J. H. Estabrook, secretary; W. C. Calhoun, treasurer; B. Weinberg, assistant secretary. — Directors

Main Office — No. 517 Mining Exchange Building, Denver, Colorado.

1,000,000 shares. Par value $1.00. In treasury January 1, 1900, 49,000 shares. — Capitalization

Owns the Lottie and the Annie, containing about 15 1-2 acres, in the S. E. 1-4 section 2, township 15 south, range 70 west, on Copper Mountain, survey No. 8,184. Also the Short Topsy, 7 1-2 acres in N. E. 1-4 section 18, on Globe Hill, all patented. Just as this Manual goes to press, the company has sold, so it is reported, the Short Topsy claim to Mr. W. S. Stratton, of this city, for $45,000.00. — Property

About $30,000.00 have been expended in developing the property.

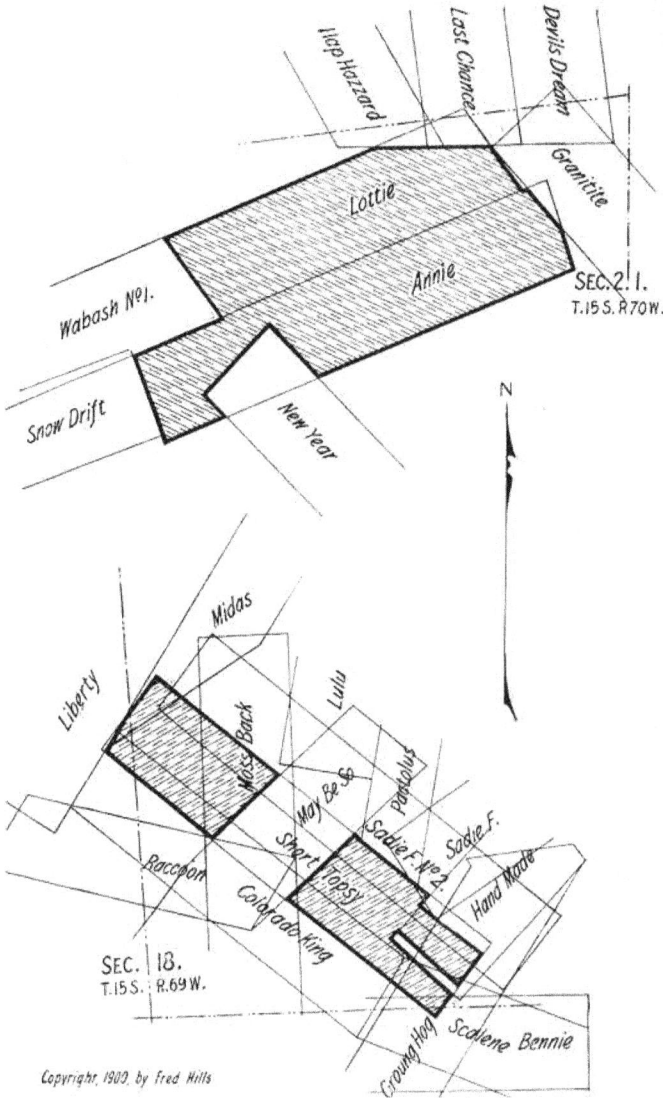

Highest price for stock during 1899, 5 cents; lowest price for stock during 1899, 3 cents.

The Golconda Gold Mining and Tunnel Company.

Incorporated March 18, 1896.

Herbert Johnson, president; John Pederson, vice-president; W. H. Bacon, secretary; D. De Graff, treasurer; W. B. Boardman. — Directors

Main Office—112 East Pike's Peak avenue, Colorado Springs, Colo.

1,250,000 shares. Par value, $1.00. In treasury January 1, 1900, 256,000 shares. — Capitalization

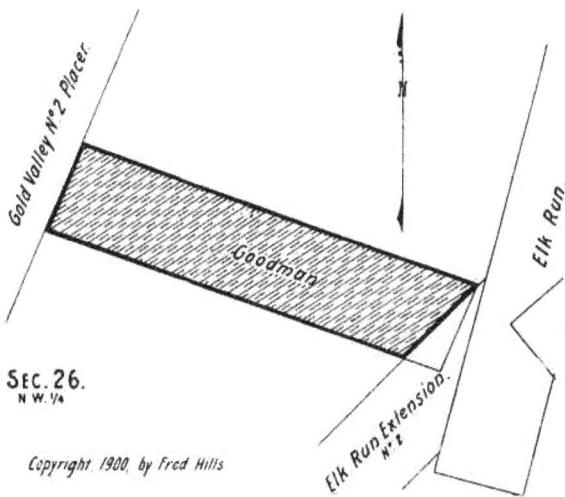

Owns the Goodman claim, 10 acres, in the N. W. 1-4 section 26, township 15 south, range 70 west, one mile west of Mound City; in process of patenting. It has one shaft 82 feet deep on the property. Debts, $125.00. — Property

Highest price for stock during 1899, $7.00 per 1,000; lowest price for stock during 1899, $4.50 per 1,000.

The Gold Band Mining Company.

Incorporated 1895.

Directors

F. L. Sigel................................President
Chas. F. Potter.....................Vice-President
Frank J. Campbell....Secretary and General Manager
Adolph J. Zang........................Treasurer

Main Office—Denver, Colorado.

Capitalization

1,000,000 shares. Par value, $1.00. In treasury April 1, 1900, 384,000 shares; in treasury April 1, 1900, about $4,400 cash.

Property

The company is leasing blocks 11, 21, 22, 23, 27, 28 and 29, in section 16 of state school lands, lying east of the town of Cameron and N. E. of the Pinnacle properties.

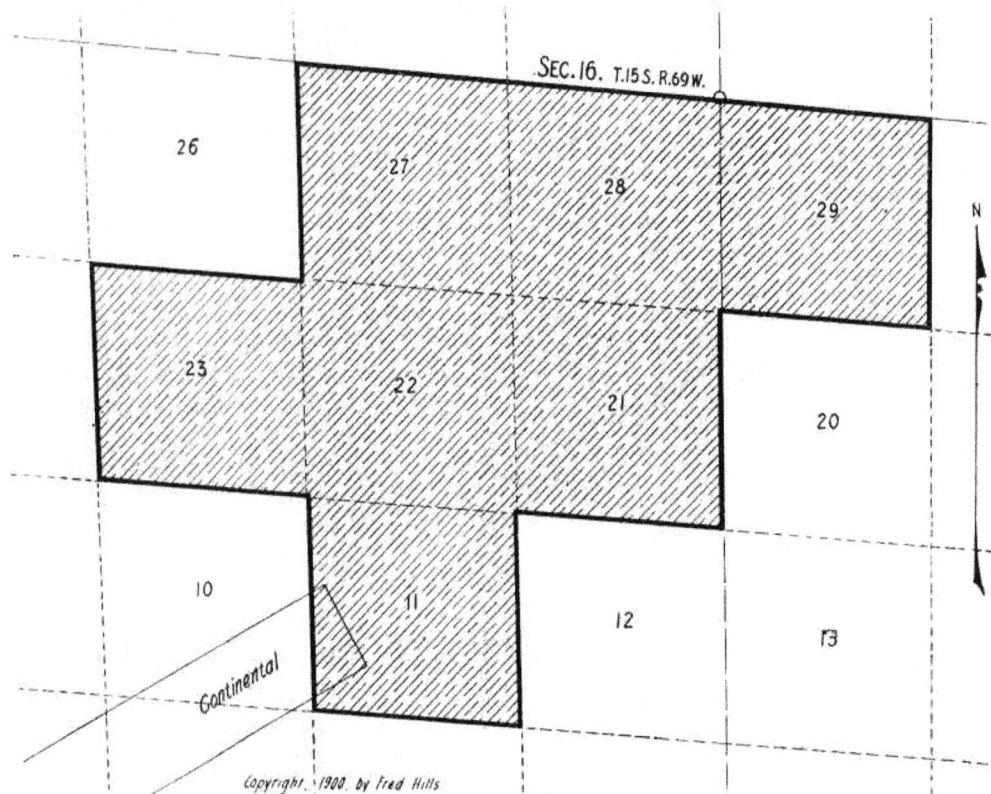

Development

The property is now being actively developed by the company. Several good veins and dykes have been cut, but no pay mineral yet encountered. About $7,000 worth of development work has been done on the property and contracts have just been let for 200 feet of sinking and drifting. As soon as the new working shaft reaches a depth of 100 feet, a plant of machinery will be installed. The properties are held under the most favorable contracts, made by the State Land Board, and leases have from 14 to 16 years yet to run. This corporation has the same officers and directors as the Vindicator Company, and is practically owned and controlled by this latter company.

No stock on the market except treasury stock, which is being sold readily at 10 cents per share for development purposes.

The Gold Belt Consolidated Mining Company.

Incorporated March, 1894.

T. F. Bowman, president; W. P. Stanley, vice-president; Chas. R. Bell, secretary; James T. Stewart, treasurer; James Monyhan. **Directors**

Main Office—No. 602 McPhee block, Denver, Colorado.

1,000,000 shares. Par value, $1.00. In treasury January 1, 1900, 50,653 shares; in treasury January 1, 1900, $500.00 cash. **Capitalization**

Owns the Geneva, the Sumatra, the Hindoo, the Lady Washington, **Property** Athens, Chicago, Georgia and the Blue Ridge, containing a total of 28 acres, patented, situated in the S. E. 1-4 section 12, all on Carbonate hill.

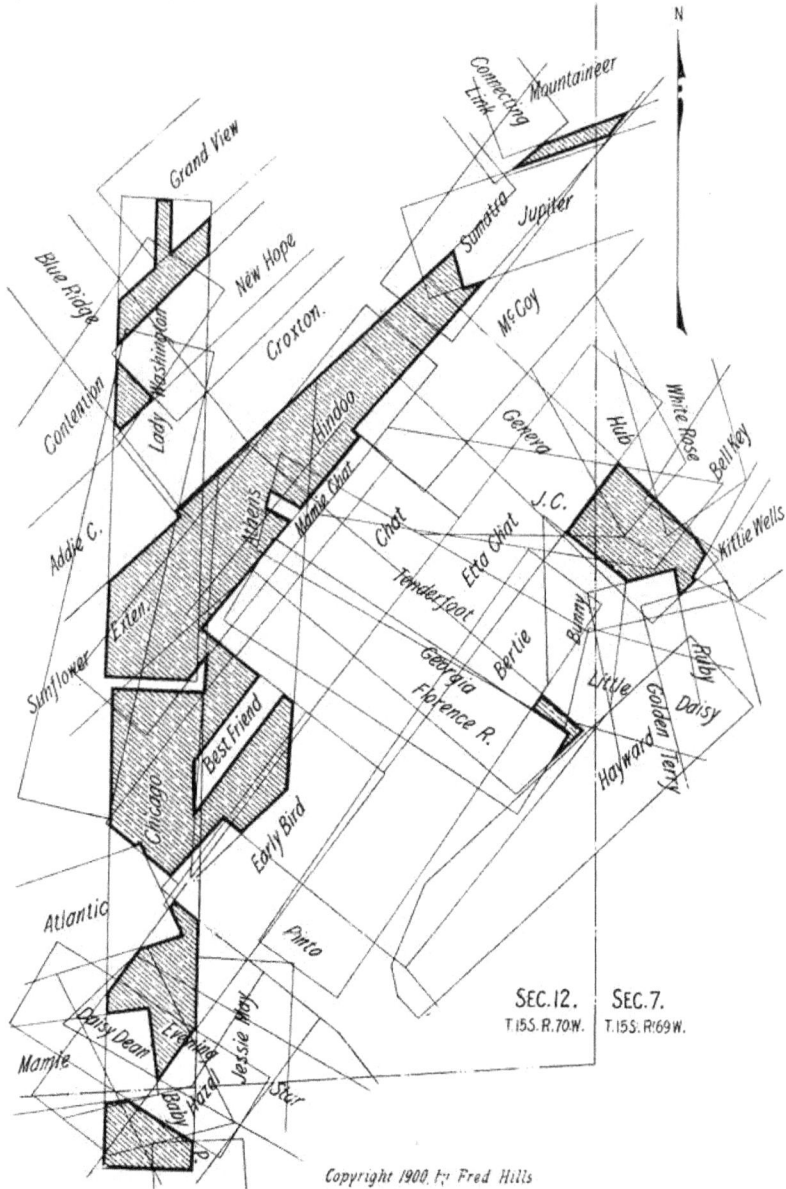

Copyright 1900, by Fred Hills

The Geneva has 225 feet of shafts; the Hindoo, 125 feet of shafts; the **Development** Georgia, 50 feet of shafts; Blue Ridge, 65 feet of shafts; Chicago, 100 feet of shafts and a tunnel of 150 feet; the Sumatra has shaft and tunnel, 75 feet, and 300 feet of drifting. The total improvements of these claims cost more than $10,000.00. The greater part of the development work is being done on the Lady Washington, the Chicago and the Geneva.

Highest price for stock during 1899, $100 per M.; lowest price for stock during 1899, $100 per M.

The Gold Bullion Mining and Milling Company.

Incorporated 1895.

Directors

J. H. Ryan............................President
Frank Sheafer.....................Vice-President
G. E. Kennedy.............Secretary and Treasurer
Jacob Kier.   Wm. Oliphant.

Main Office—Colorado Springs, Colorado.

Capitalization

1,000,000 shares. Par value, $1.00. In treasury January 1, 1900, 39,000 shares.

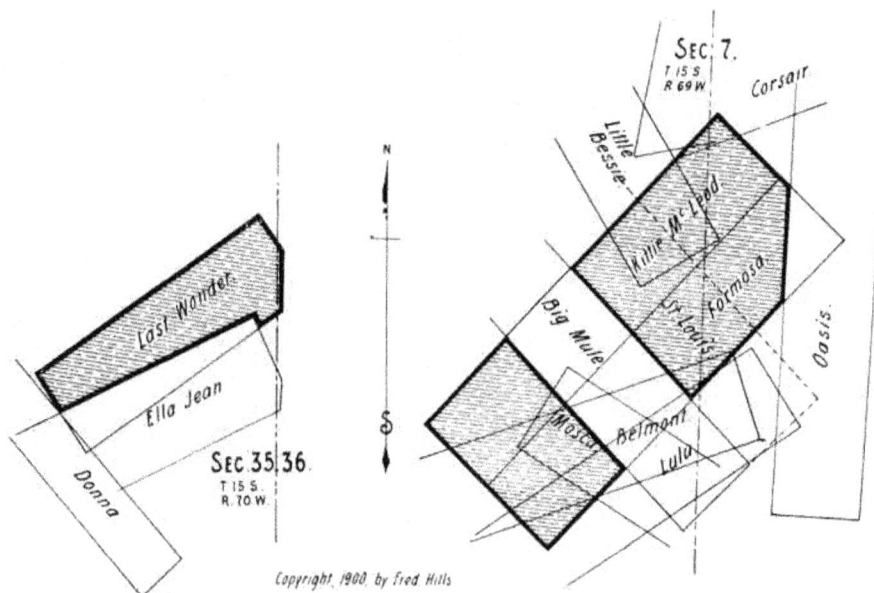

Copyright, 1900, by Fred Hills

Property

Owns the Formosa and the Kitty McLeod, containing 14 acres, in the N. 1-2 section 7, on Carbonate hill; also the Last Wonder, containing 5 3-4 acres, in the S. E. 1-4 section 35, on Grouse mountain. Total acreage owned by this company is 19 3-4 acres; all patented.

Development

The Formosa is leased. It is expected that some development will be made by the Spring Creek Tunnel Company, who have procured a right of way from this company to tunnel through the Kitty McLeod and the Formosa claims.

Highest price for stock during 1899, 1 cent; lowest price for stock during 1899, 1 cent.

The Gold Button Mining and Milling Company.

Incorporated April 15, 1896.

W. P. Carstarphen, Sr.........President   Directors
W. F. Hallam.............Vice-President
G. M. Hossack....Secretary and Treasurer
W. P. Carstarphen, Jr...Assistant Secretary
F. W. Feldwisch.   R. W. Hodden.
F. E. Carstarphen.

Main Office—No. 313 Sixteenth street, Denver, Colorado; secretary's office, No. 204 Times-Herald Building, Chicago, Illinois.

1,500,000 shares. Par value $1.00. In   Capitalization treasury, February 7, 1900, 200,000 shares; in treasury, February 7, 1900, $383.89 cash.

Owns the Flag of Truth, 10 acres; the   Property Mollie McGuire, 10 1-3 acres, and the American Flag, 5 acres, located on the S. W. slope of Sheep Mountain, running down to Oil creek, about three miles from the town of Gillett. Property is crossed by the Midland Terminal Railroad. In process of patenting. Total acreage owned by this company is 25 1-3 acres. Being outside of the territory covered by the colored map, this property is not shown thereon.

One log house, 14x16, and tools. The   Development Flag of Truth has a well timbered tunnel of 175 feet, also drifts, where the greatest development has been done. The Mollie McGuire has 70 feet of shafts and some open cuts. The American Flag has a shaft 35 feet deep, and some open cuts. The property has not yet become a producer. Assays have been obtained from $168 down to a trace of gold.

These properties were located at a time when the locators had the   History pick of the ground in this part of the district. Oil creek, a bold mountain stream, runs within a few feet of the mouth of the tunnel and will furnish ample water power for mill. There is an abundance of timber on the property.

Highest price for stock during 1899, 10 cents; lowest price for stock during 1899, 3 cents.

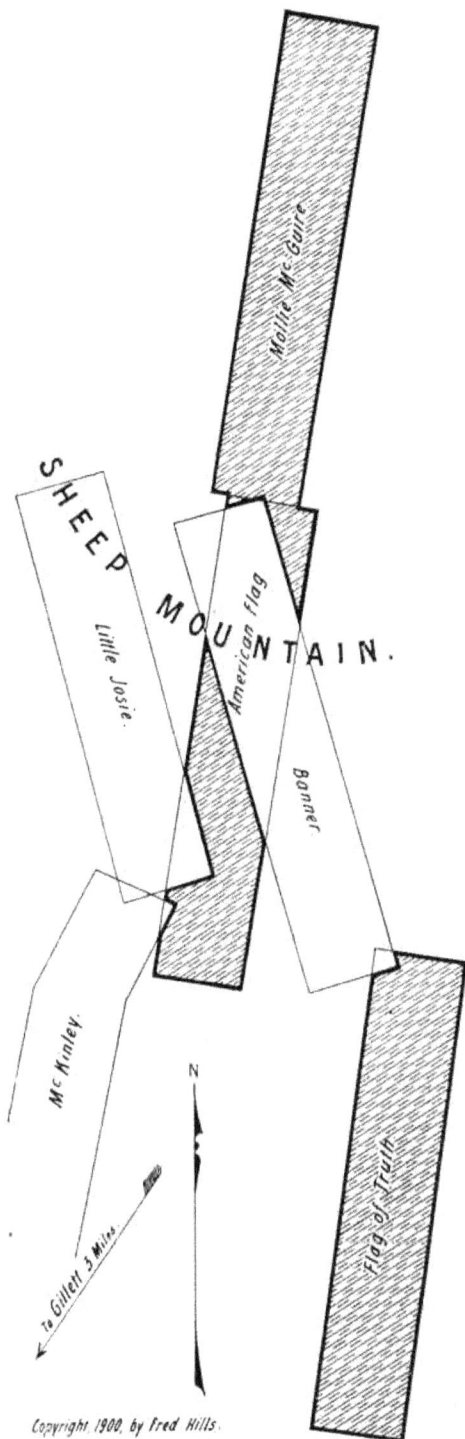

The Gold Coin Mining and Leasing Company.

Incorporated November 19, 1895.

Directors

H. E. Woods........President
Warren Woods..Vice-President
F. M. Woods..............
.....Secretary and Treasurer
J. M. Allen......W. D. Hatton

Main Office—Giddings Building, Colorado Springs, Colorado; branch office, Victor, Colorado.

Capitalization

1,000,000 shares. Par value, $1.00.

Property

Owns the Gold Coin, 6.2 acres; the Little Montana, 5.7 acres; the Golden Discovery, 1.4 acres; the Surprise, .4 acre—all situated on Battle mountain, in sections 29 and 32; also the Last Chance and the Pine Knot, containing about 5 acres, situated on Squaw mountain, in section 31; also the Golden Dawn Nos. 1, 2 and 3, containing 31 acres, located in Fremont county. As this last mentioned property is outside the Cripple Creek district, the plat thereof is not shown.

The company also controls the property of the Gold Deposit Mining Company, and has other interests, including a controlling interest in the Mt. Rosa M. & L. Co.

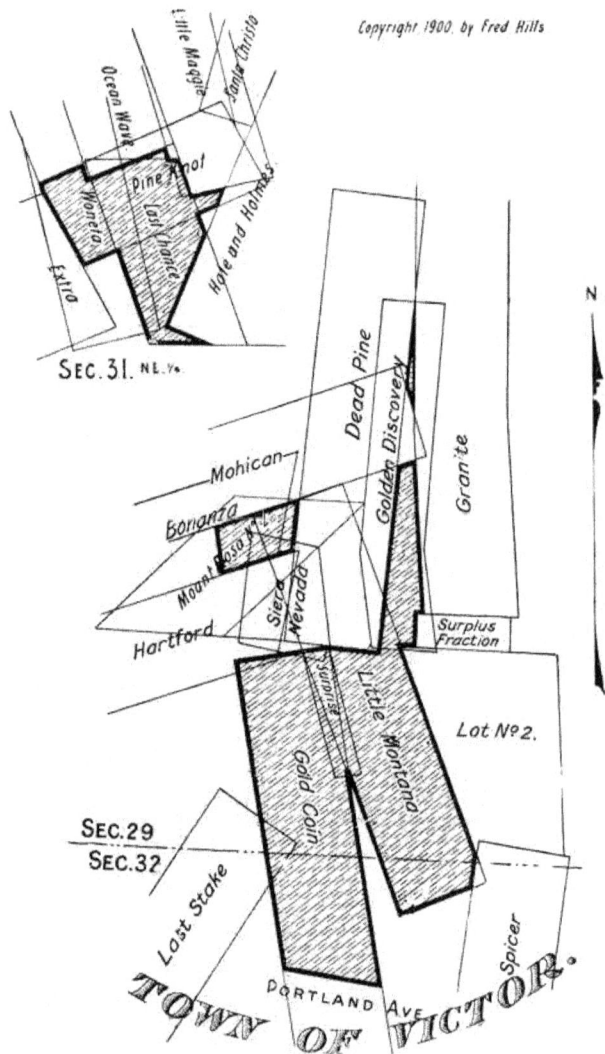

Copyright 1900, by Fred Hills

Development

This company started with a lease and bond on the Gold Coin claim at the foot of Battle mountain, practically in the town of Victor, early in 1895. After spending a large amount of money in development work an ore shoot was finally discovered in 1896, and the bond paid off in September of that year. Since then a heavy production has been made. Dividends were commenced in 1897, and have been continued at the rate of 1 cent per share monthly until January, 1900, when 2 cents per share was started and has continued monthly since that time. The shaft house and ore buildings were destroyed by fire in September, 1899, and new buildings are now being erected of steel and brick, which will be the most commodious in the district, containing the largest compressor and hoist plant in Cripple Creek.

Production and Dividend

Total dividends to May 1, 1900, $440,000.00. Highest price for stock during 1899, $2.60; lowest price for stock during 1899, $1.76; on May 1, 1900, price was $4.25.

The Gold Crater Mining Company.

Incorporated.

This company is owned and controlled by Mr. W. S. Stratton, of History Colorado Springs.

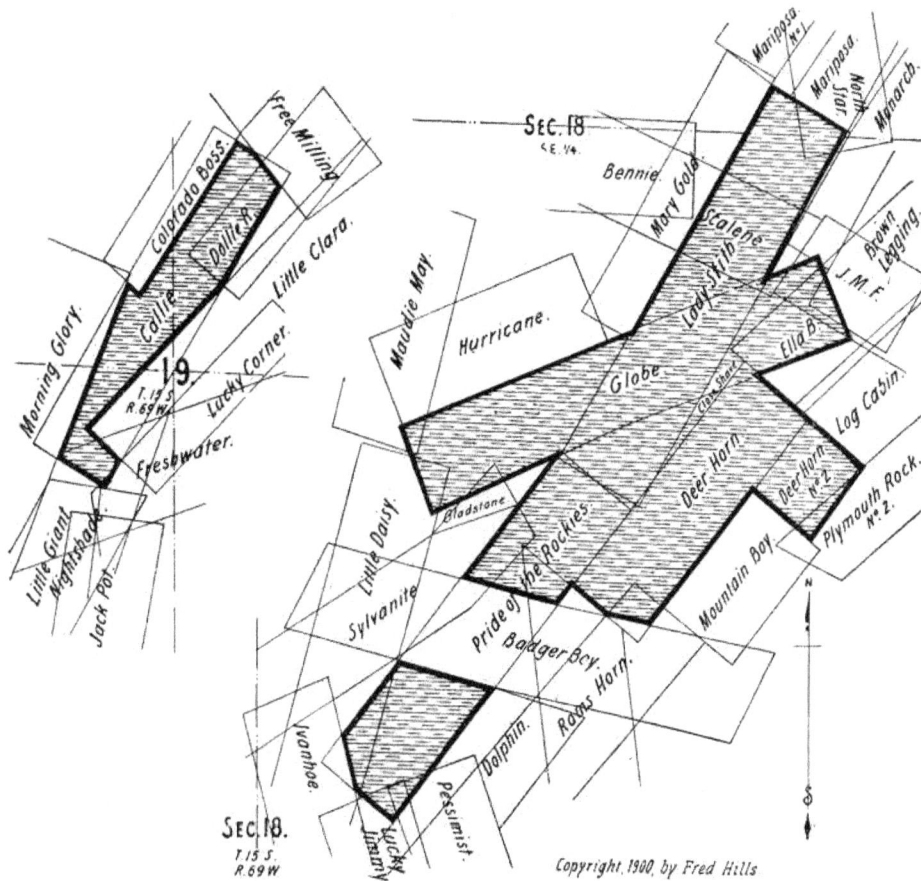

Copyright, 1900, by Fred Hills

The property thus owned consists of the Globe, Close Shave, Lady Property Stith, Deerhorn, Deerhorn No. 2 and Pride of the Rockies, in all about 35 acres, all patented, in section 18, on Globe hill; also owns the Callie, about 5 acres, in section 19, on Gold hill, adjoining the Anaconda.

The Gold Dust Mining Company.

Incorporated 1899.

Directors

Alexander P. Moore..........President and Treasurer
David H. Franks...............Vice-President
Edward McDaniel.....................Secretary
U. Grant Danford.   John W. Wray.
C. A. Richardson.   E. B. Harrold.

Main Office—Cripple Creek, Colorado.

Capitalization

2,500,000 shares. Par value, $1.00. In treasury January 1, 1900, 900,000 shares.

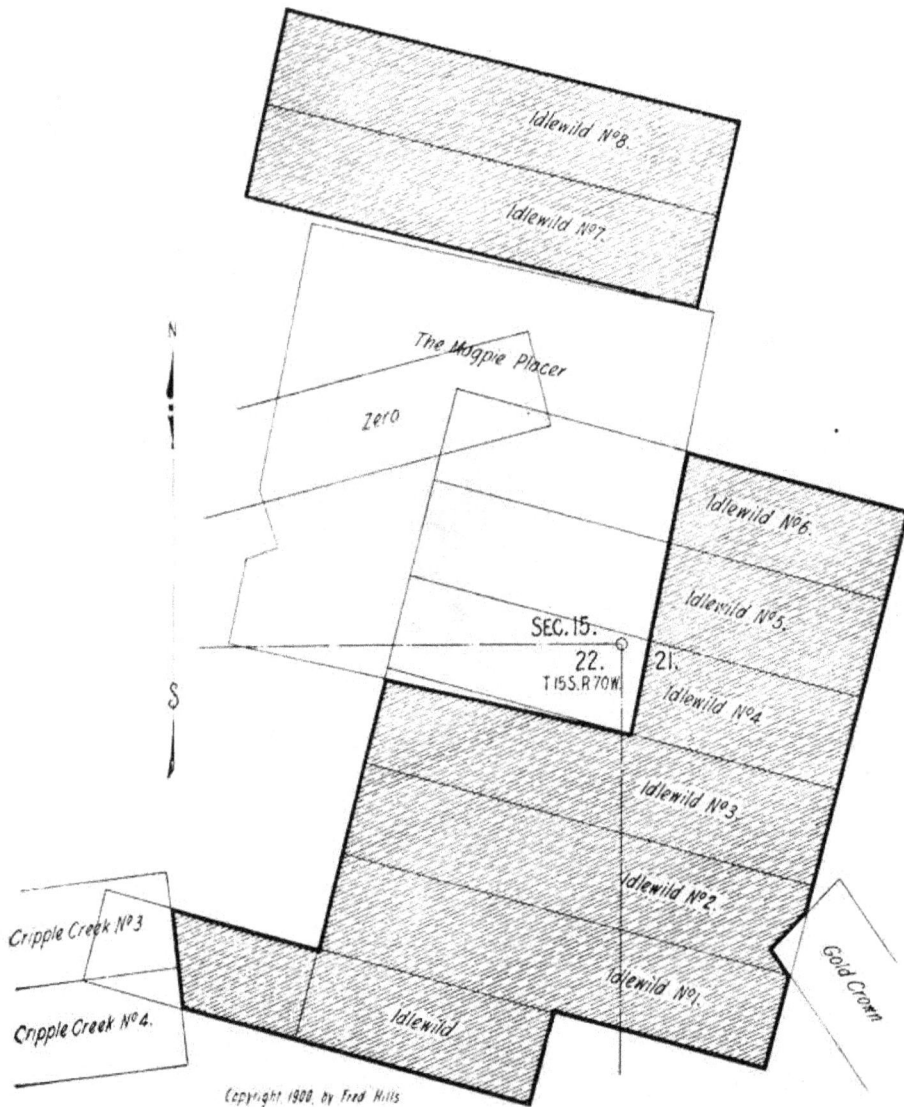

Copyright 1900, by Fred Hills

Property

Owns the Idlewild, the Idlewild Nos. 1, 2, 3, 4, 5, 6, 7 and 8, containing in all about 70 acres, situated in the N. W. 1-4 section 23, the N. E. 1-4 section 22, and the S. E. 1-4 section 15, on Raymond and Gold Quartz hills. Above property held by location.

About $4,500 has been expended in development work on this group of claims, and they are all in process of patenting.

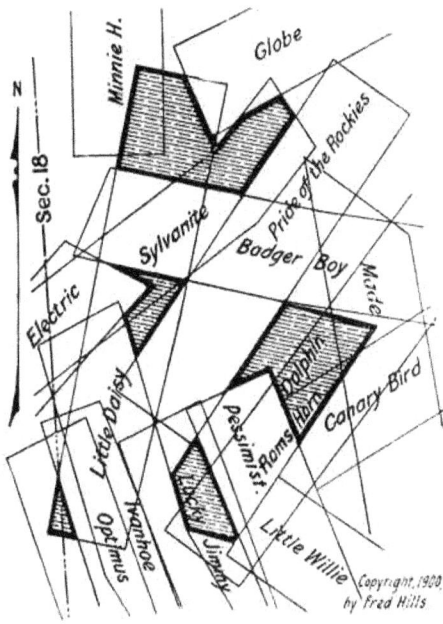

The Golden Age Gold Mining Company.

Incorporated.

H. E. Insley, president; E. C. Shelden, vice-president; H. H. Dorsey, secretary; J. A. Sill, treasurer; C. P. Bently. **Directors**

Main Office—First National Bank Building, room No. 53, Colorado Springs, Colorado.

1,000,000 shares. Par value, $1.00. In treasury January 1, 1900, 51,000 shares; in treasury January 1, 1900, about $100.00 cash. **Capitalization**

Owns the Sylvanite, the Little Daisy, and the Dolphin, containing in all about 9 acres, in a group on Ironclad hill. All patented. On these claims there are about 250 feet of drifts and shafts. The Sylvanite claim is leased. **Property**

Highest price for stock during 1899, 4 7-8 cents; lowest price for stock during 1899, 1 cent.

The Golden Clipper Mining Company.

Incorporated 1895.

A. F. Woodward......President and Treasurer
E. C. Bale.....................Vice-President
M. F. Clark........................Secretary **Directors**

Main Office—The El Paso Abstract Co., 113 E. Kiowa street, Colorado Springs, Colorado.

1,000,000 shares. Par value, $1.00. In treasury January 1, 1900, 28,000 shares. **Capitalization**

Owns the Charley Ross, Phelps, Peterson, and the J. H. Morris, a total of 25 acres, situated in section 3, township 16 south, range 69 west, on the south slope of Little Bull mountain. All patented. **Property**

Simply patent work has been done in development.

Highest price for stock during 1899, $10.00 per 1,000; lowest price for stock during 1899, $3.00 per 1,000.

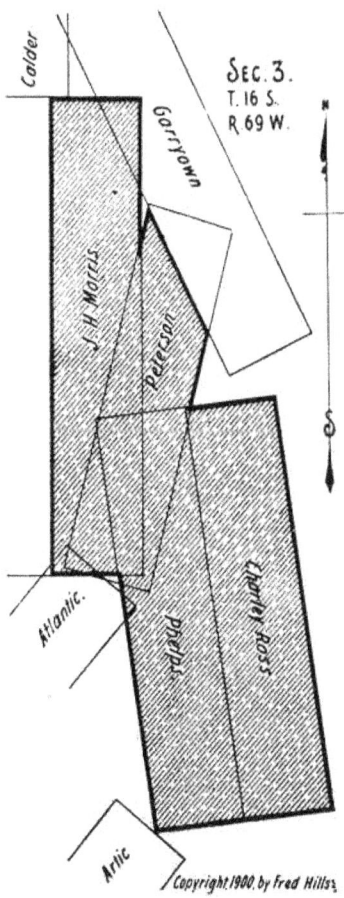

The Golden
Crescent Water
and Light
Company.

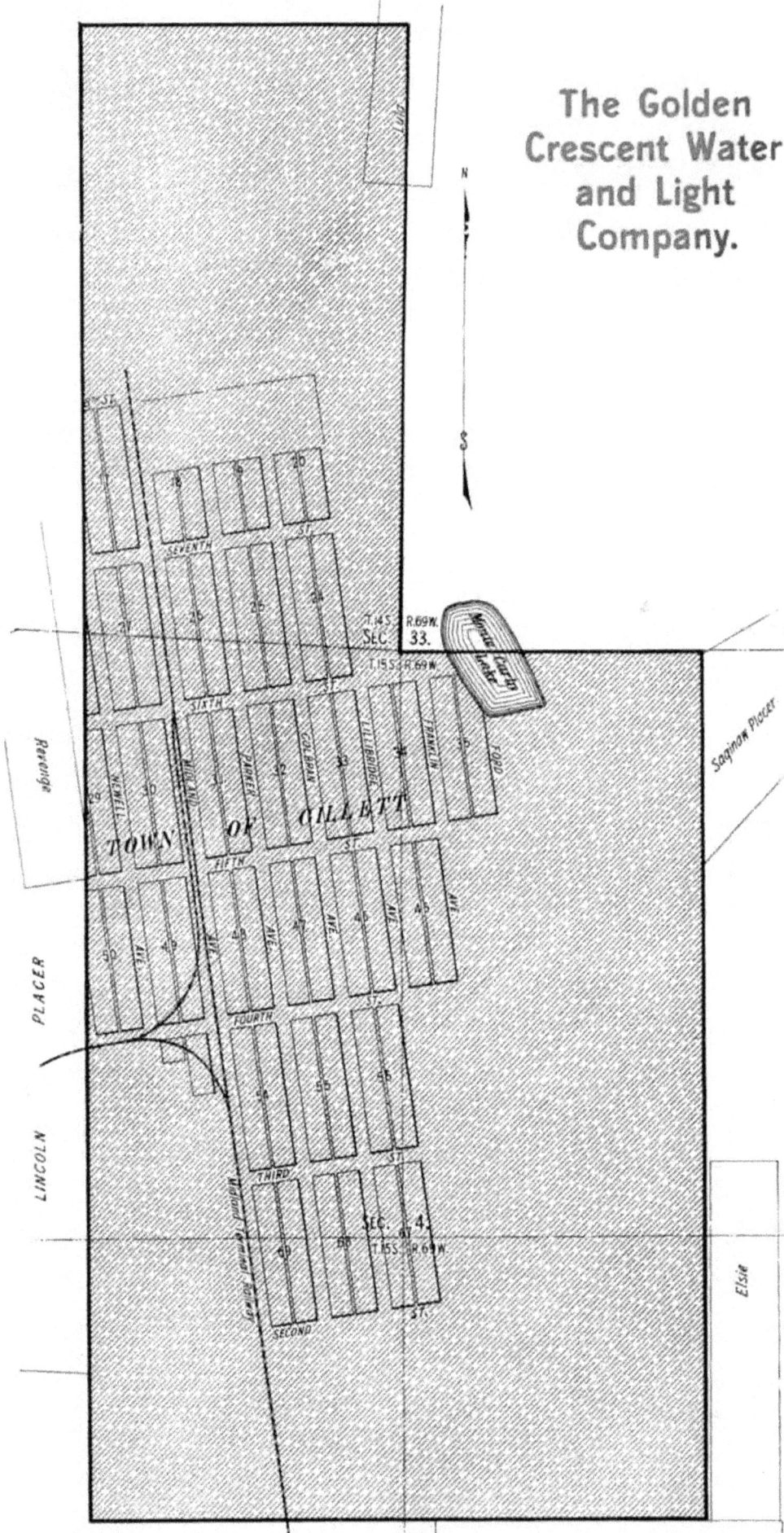

N

S

TOWN OF GILLETT

T.14 S. R.69 W.
SEC. 33.
T.15 S. R.69 W.

SEC. 4.
T.15 S. R.69 W.

Myrtle Cora'in Lode

Revenge

PLACER

LINCOLN

Saginaw Placer

Elsie

Scale of Feet
0   500   1000

Nancy E.

Copyright 1900 by Fred Hiltz

The Golden Crescent Water and Light Company.

Incorporated.

Warren Woods...........................President

H. E. Woods.......................Vice-President

F. M. Woods..............Secretary and Treasurer

Directors

Main Office—Giddings block, Colorado Springs, Colorado. Branch Offices—Gillett and Victor, Colorado.

500,000 shares. Par value, $1.00. Also $150,000 of 6 per cent. ten year gold bonds.

Capitalization

Owns the Gillett Electric Light Works, the Gillett Water Works, the townsite of Gillett, 110 acres; also franchise for water, light and sewerage at Cameron, Colorado.

Property

The Golden Cycle Mining Company.

Incorporated December 6, 1895.

Directors

D. H. Moffat...........................President
Eben Smith.......................Vice-President
R. H. Reid.................Secretary and Treasurer
Syl. T. Smith.   L. E. Campbell.

Main Office—827 Exchange building, Denver, Colorado.

Capitalization

200,000 shares. Par value, $5.00. In treasury January 1, 1900, $5,000.00 cash.

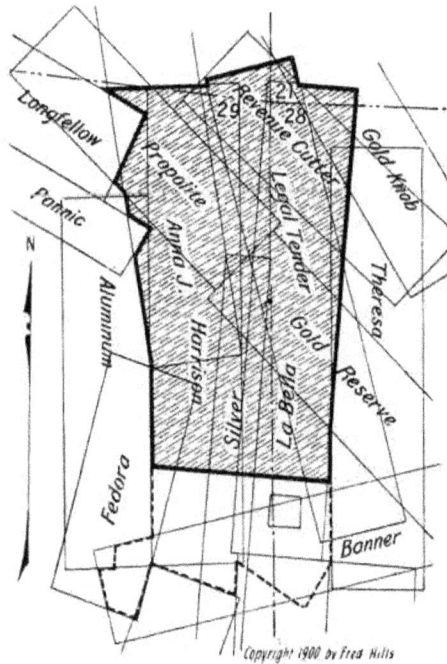

Copyright 1900 by Fred Hills

Property

Owns the La Bella, Legal Tender, Harrison and part of the Anna J., Pinkerton, Propolite and Banner, 25 acres, in the N. E. 1-4 section 29 and the N. W. 1-4 section 28, on the south slope of Bull hill. Patented.

Development

There is a complete equipment of machinery on the property for hoisting 200 tons of ore per day. There are five veins on the property and all are being worked by the company.

Production and Dividend

Production to January 1, 1900, $1,512,909.38; dividends up to January 1, 1900, $198,300.00, last being October 16, 1899, $5,000.00.

This being a close corporation the stock is not on the market, hence no prices are given.

The Golden Dale Mining and Milling Company.

Incorporated January 18, 1893.

F. H. Pettingell........................President
H. E. Gill..........................Vice-President
L. A. Civill................Secretary and Treasurer

A. McCormack.   D. Elliott.
R. Green.   Walter Scott.
H. C. Fish.   A. B. Moulder.

Main Office—11 Bank building, Colorado Springs, Colorado.

3,000,000 shares. Par value, $1.00. In treasury January 1, 1900, 142,650 shares; in treasury January 1, 1900, $242.06 cash.

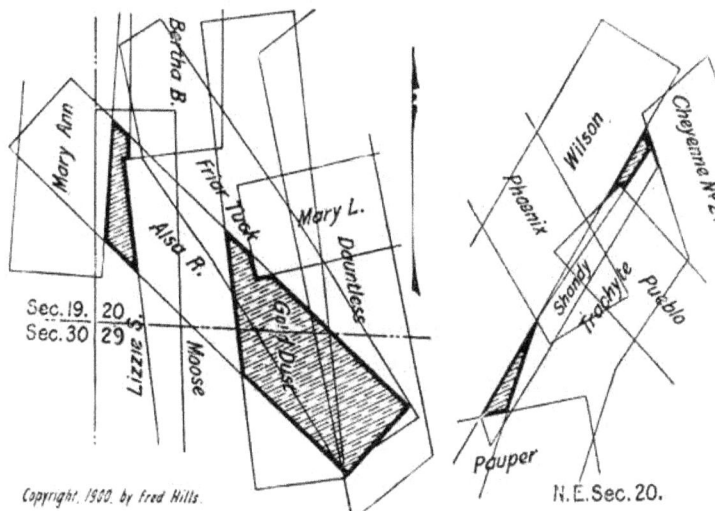

Owns the Alsa R., containing 5.082 acres, in the S. W. 1-4 section 20, on Raven hill, and the Shandy, 1 acre, in the N. E. 1-4 section 20, on Bull hill. The Alsa R. claim is patented and the Shandy held by location.

There is a 50-foot shaft on the Shandy, with 42 feet of drifting, and about 500 feet of shafts, tunnels and winzes on the Alsa R. In 1893, or near that time, the company was a hopeless wreck. The present management redeemed the property from a sheriff's sale, made good an over-issue of nearly one-half million shares of stock made by predecessors, by an increase of capital, paid several thousand dollars debt, and patented the Alsa R., which was nearly lost through neglect.

Production to January 1, 1900, about $3,000.00.

Highest price for stock during 1899, 3 7-8 cents; lowest price for stock during 1899, $2.00 per 1,000.

The Golden Eagle Mining and Milling Company.

Incorporated 1892.

Directors

Herbert Gardner........................President

E. R. Whitmarsh..........Secretary and Treasurer

W. D. Shemwell.   H. A. Young.

Louis Sheer.   N. Leipheimer.

Main Office—No. 25 East Pike's Peak avenue, Colorado Springs, Colorado.

Capitalization

1,000,000 shares. Par value, $1.00.

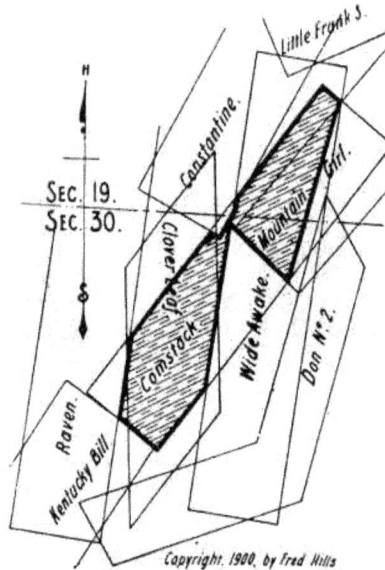

Copyright. 1900. by Fred Hills

Property

Owns the Comstock, containing 7 acres, with vein rights, situated in the N. E. 1-4 section 30, on Raven hill, east of the Raven and north of the Elkton mines. Patented.

Development

There are four shafts; two have been sunk to a depth of 45 feet each, one is 60 feet deep, and one 90 feet deep. In the 60-foot shaft 50 feet of drifting and some surface trenching has been done.

Highest price for stock during 1899, 10 cents; lowest price for stock during 1899, $5.00 per M.

The Golden Link Mining, Leasing and Bonding Company.

Incorporated July 6, 1895.

Main Office—No. 17 East Pike's Peak avenue, Colorado Springs, Colorado.

1,000,000 shares. Par value, $1.00.   Capitalization

Copyright, 1900, by Fred Hills.

Property

Owns the Allen Thurman and the Crystal Hill, containing in all 11 acres, in the N. W. 1-4 section 6, on Rhyolite mountain; also the Cream, containing 8 acres, in the N. E. 1-4 section 10, on Cow mountain. All patented.

Development

On the Allen Thurman and the Crystal Hill is a shaft of 45 feet depth, with 25 feet of drifting. On the Cream is an open cut of 30 feet and a tunnel of 25 feet. The greater part of the development work is being done on the Allen Thurman and the Crystal Hill.

History

The company has had considerable difficulty in securing patents on the Rhyolite mountain properties, but after two years' fighting the secretary of the interior has reversed the decision of the land office and given this company patents on the above claims.

Stock not on the market. It is now offered at 2 cents per share.

The Golden Treasure Mining and Tunnel Company.

Incorporated January, 1895.

Directors

Frank Ferris.............................President
W. H. Sutherland...................Vice-President
C. P. Bently...........................Secretary
R. C. Day.............................Treasurer
L. B. Spinney.

Main Office—No. 53 First National Bank building, Colorado Springs, Colorado.

Capitalization 2,000,000 shares. Par value, $1.00. In treasury January 1, 1900, 59,500 shares.

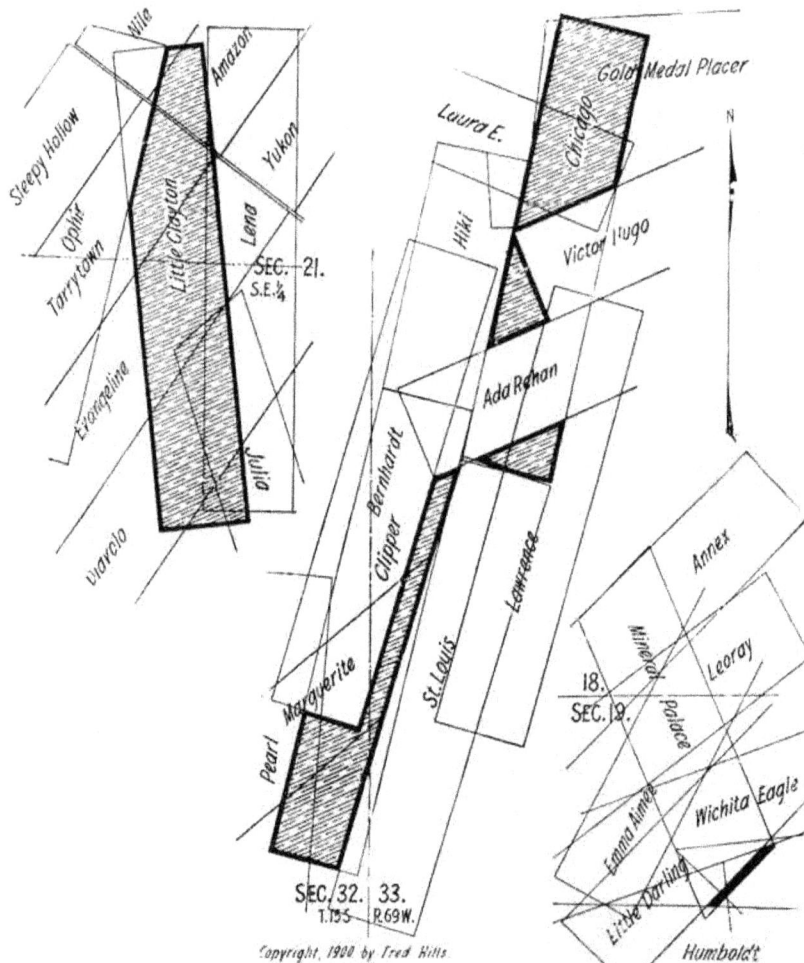

Copyright, 1900 by Fred Hills

Property Owns the Chicago and Clipper, 11 1-4 acres, in the S. W. 1-4 section 33, township 15 south, range 69 west, on Howell hill; the Little Clayton, 9 1-4 acres, in the E. 1-2 section 21, on the east slope of Bull hill; a fraction of the Mineral Palace, 0.147 acre, survey No. 10,367, in the N. E. 1-4 section 19, on Ironclad hill. Receiver's receipt held for all the above property.

Development Sufficient work, for patent only, has been done.

Highest price for stock during 1899, $8.00 per M.; lowest price for stock during 1899, $5.50 per M.

The Goldfield Mining and Leasing Company.

Incorporated April, 1895.

B. P. Anderson.........................President

F. E. Robinson.....................Vice-President

L. R. Ehrich...............Secretary and Treasurer

H. H. Melville.   F. A. Perkins.

Directors

Main Office—No. 64 Hagerman building, Colorado Springs, Colorado.

1,250,000 shares. Par value, $1.00. In treasury January 1, 1900, 100,000 shares; in treasury January 1, 1900, $5,000.00 cash.

Capitalization

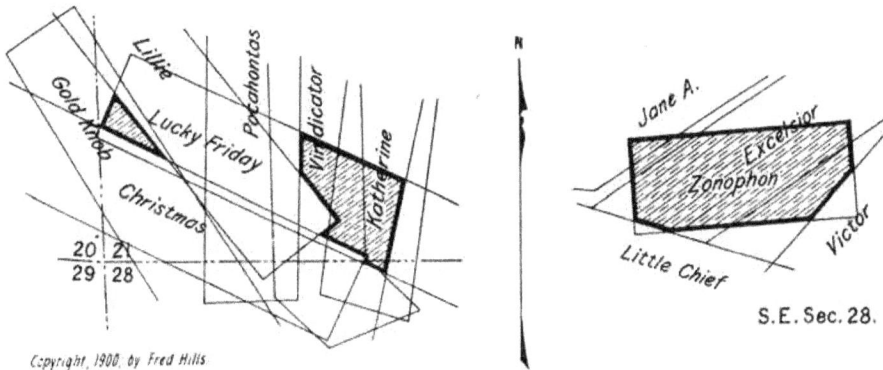

Copyright, 1900, by Fred Hills

S.E. Sec. 28.

Owns the Lucky Friday, on Bull hill, adjoining the Lillie mine; and the Zonophon, on Big Bull hill, 8 acres in all. Patented. Property is situated in sections 21 and 28.

Property

On the Lucky Friday the greater part of the development work is being done, and the company is vigorously sinking and drifting, with good indications for striking the rich Vindicator vein.

Development

Highest price for stock during 1899, 5 cents; lowest price for stock during 1899, 4 cents.

The Gold Galleon Mining Company.

Incorporated January, 1896.

Directors
Irving W. Bonbright, president; F. W. Stehr, vice-president; J. Arthur Connell, secretary and treasurer.

Transfer Office—The International Trust Company, Colorado Springs, Colorado. Main Office—Nos. 67-68 Postoffice building, Colorado Springs, Colorado.

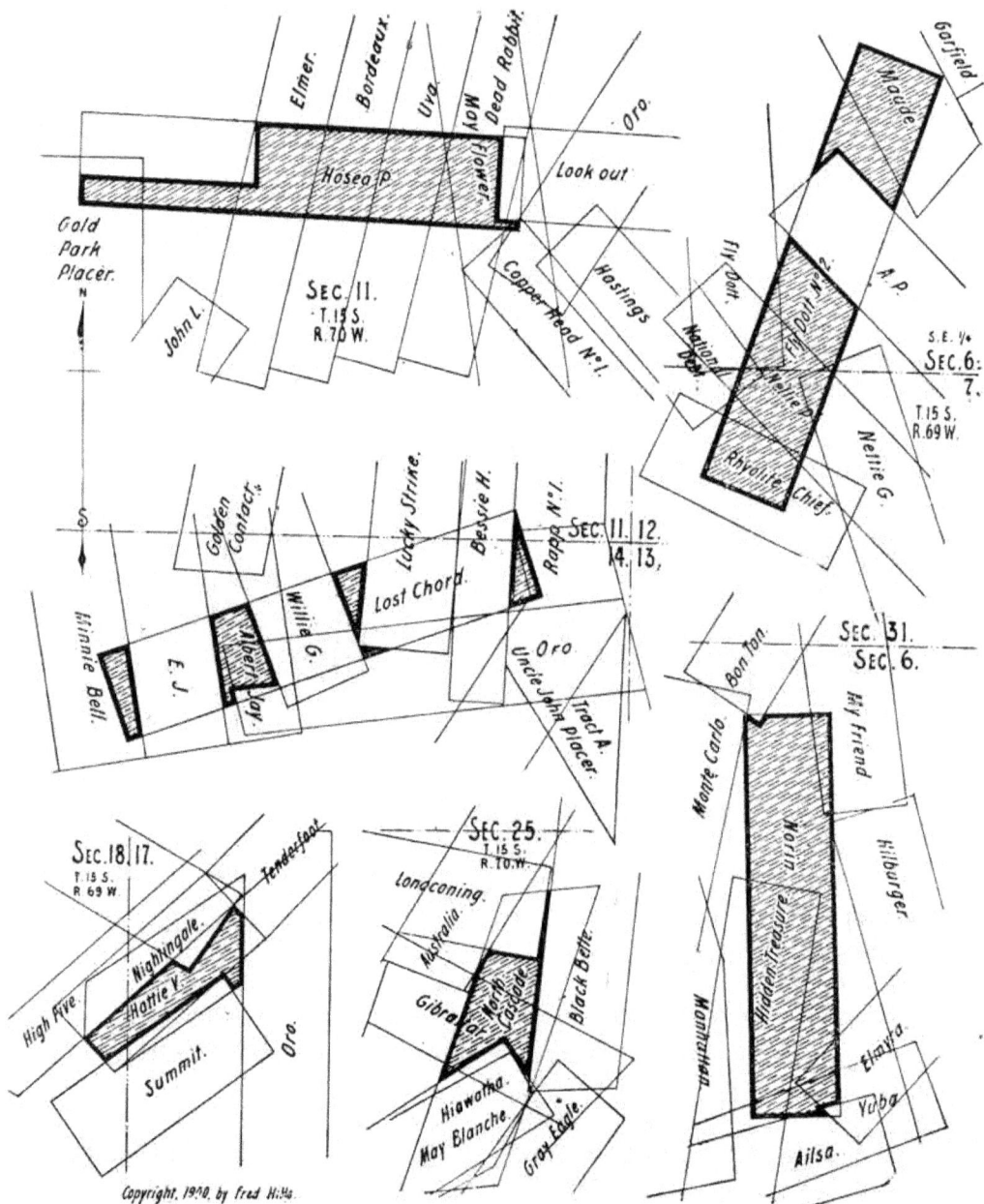

Copyright, 1900, by Fred Hills.

Capitalization
1,200,000 shares. Par value, $1.00. In treasury January 1, 1900, 200,000 shares. The company has an indebtedness of about $600.00.

Property
Owns the Lost Chord, in the S. E. 1-4 of section 11, containing 2 1-2 acres, on Mineral hill; the Hosea P., in the S. E. 1-4 of section 11, containing 7 acres, on Mineral hill; the Norin, in the N. W. 1-4 of section 6, town-

(CONTINUED ON PAGE 213.)

ship 16 south, range 69 west, containing 9 acres, on Straub mountain; the Fly Dot No. 2, in the S. E. 1-4 of section 6, 8 acres, on Tenderfoot hill; also a one-half interest in the Hattie V., in the S. W. 1-4 of section 17, containing 2 acres, on Ironclad hill; and a one-half interest in the North Cascade, in the S. E. 1-4 of section 25, and containing 3 acres, on Beacon hill. All the above property is patented except the North Cascade and the Hattie V., which are in process.

The Gold Garden Mining Company.

Incorporated February 17, 1896.

Geo. F. Batchelder, president; M. B. Carpenter, vice-president; T. J. Moynahan, secretary; E. P. Brown, treasurer; Jas. H. Blood; Thos. Hamilton. [Directors]

Main Office—No. 37 Jacobson Building, Denver, Colorado.

1,250,000 shares. Par value, $1.00. In treasury January 1, 1900, 138,000 shares; in treasury January 1, 1900, $351.50 cash. [Capitalization]

Owns the Pike's Peak placer, containing 29.31 acres; the Pike's Peak placer No. 2, containing 16.13 acres; the Moreland placer, containing 6.33 acres; the Dark Horse placer, containing 1.78 acres; and the O. B. Joyful placer, containing 7.07 acres —a total of over 61 acres, situated in the W. 1-2 of sections 12 and 13, on Mineral hill. All patented. [Property]

The only developments on the property are a few shafts of from 10 to 50 feet deep.

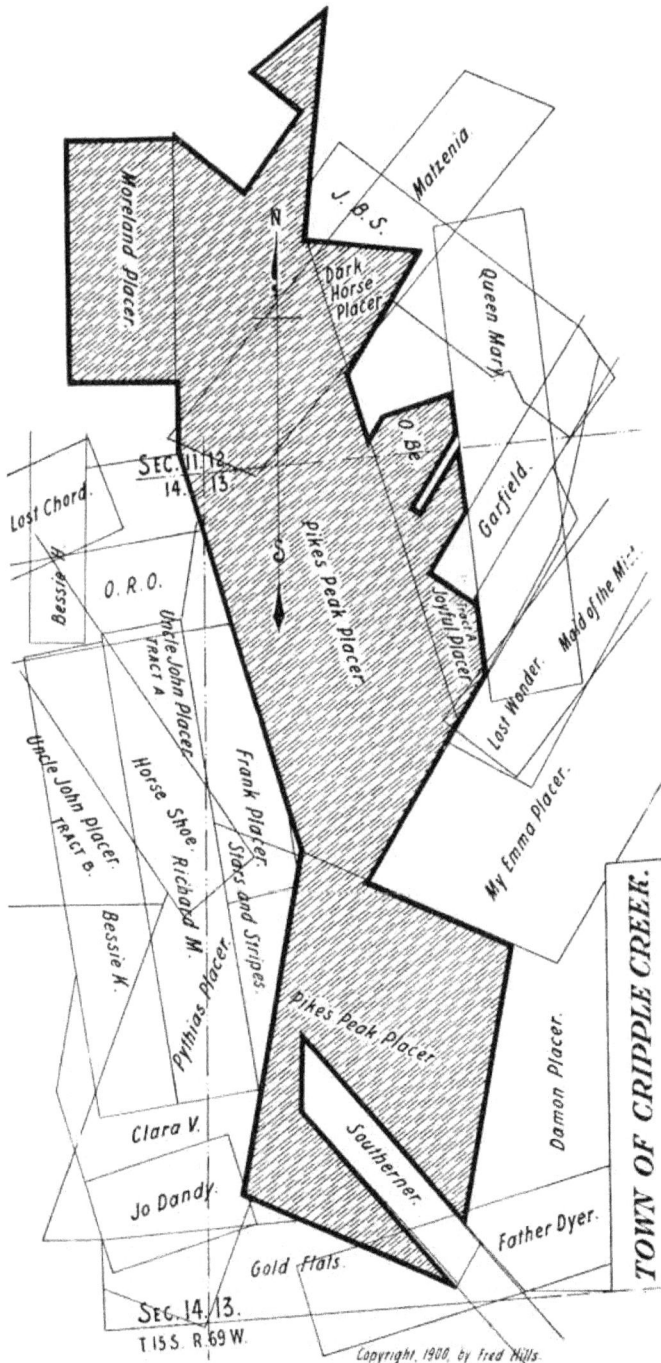

213

The Gold and Globe Hill Mining Company.

Incorporated 1891.

Directors

J. R. McKinnie.........................President

G. S. Wilson........................Vice-President

L. L. Aitken..........................Secretary

H. W. Patterson........................Treasurer

Wm. H. Gowdy.

Main Office—No. 25 East Pike's Peak avenue, Colorado Springs, Colorado.

Capitalization   750,000 shares.   Par value, $1.00.   In treasury January 1, 1900, $500.00 cash.

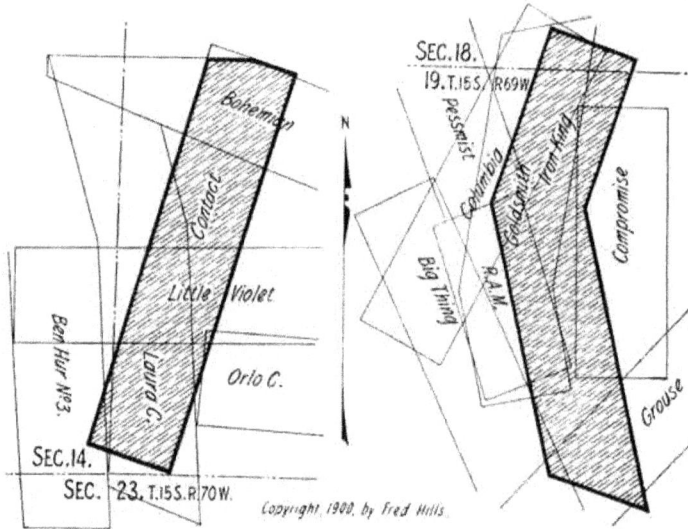

Copyright, 1900, by Fred Hills.

Property   Owns the Iron King, 10 acres, in the N. E. 1-4 section 19, on Ironclad hill; the Contact, 5.8 acres, in the S. 1-2 section 14, west of Cripple Creek; owns 200,000 shares of the Savage Gold and Copper Mining Company's stock, and 225,000 shares of the Moon Anchor Gold Mining Company. All the above property is patented.

Production and Dividend   Total dividends to January 1, 1900, $41,000.00.

Highest price for stock during 1899, 7 1-4 cents; lowest price for stock during 1899, 4 1-4 cents.

The Gold Hill Bonanza Mining Company.

Incorporated.

Irving Howbert ....President   Directors

James F. Burns.....................Vice-President

Frank G. Peck.........................Secretary

J. A. Hayes............................Treasurer

W. S. Nichols.   Chas. L. Tutt.

Spencer Penrose.

Main Office—68 Bank building, Colorado Springs, Colorado.

1,000,000 shares. Par value, $1.00.   Capitalization

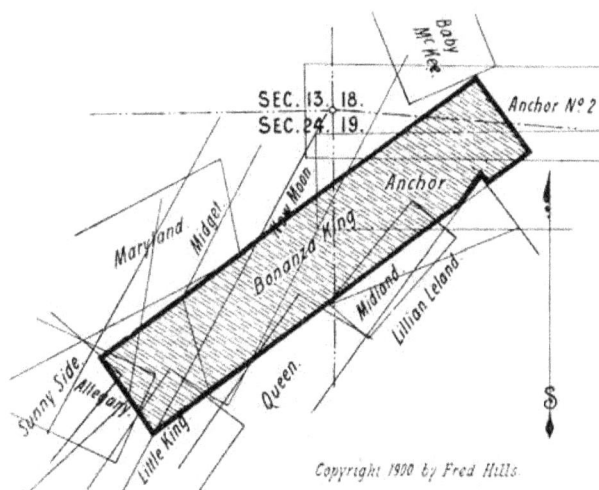

Copyright 1900 by Fred Hills.

Owns the Bonanza King and part of the Anchor and Anchor No. 2,   Property
10 acres, in the N. W. 1-4 section 19 and N. E. 1-4 section 24, on Gold hill.
Patented.

One 250-foot shaft and three 80-foot shafts, with 1,200 feet of drifts   Development
and cross-cuts.

The stock is not for sale on the general market, hence no prices can
be given.

The Gold Hill Gold Mining Company.

Incorporated February 10, 1896.

Main Office—Equitable Building, Denver, Colorado; branch office, 2 and 4 Bank Building, Colorado Springs, Colorado.

Capitalization   1,500,000 shares. Par value, $1.00. In treasury January 1, 1900, 175,000 shares; in treasury January 1, 1900, $750.00 cash.

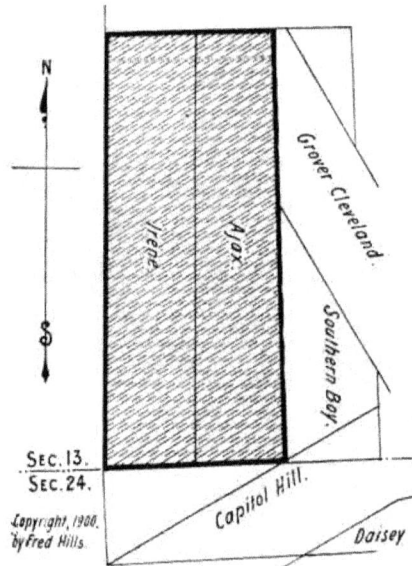

Property   Owns Irene and Ajax claims, containing 17.97 acres, situated in the S. E. 1-4 section 13, in Poverty gulch. Patented.

Development   The property is developed by lessees. There is a 105-foot double-compartment shaft on the Ajax claim, and a 130-foot shaft on the Irene claim, and the work is steadily going on. In April, 1900, a contract was let to sink the shaft to 200 feet and a new plant of machinery has been installed.

Highest price for stock during 1899, 4 5-8 cents; lowest price for stock during 1899, 2 3-8 cents.

The Gold King Mining Company.

Incorporated 1892.

E. A. Colburn...........................President

John Lennox........................Vice-President

C. H. Dudley...............Secretary and Treasurer

Wm. Lennox.   W. S. Jackson.

Directors

Main Office—14 North Nevada avenue, Colorado Springs, Colorado.

1,000,000 shares. Par value, $1.00. In treasury January 1, 1900, 63,150 shares.   Capitalization

Copyright, 1900, by Fred Hills.

Owns the El Paso, Stop Short, Rosetta, Triangle, Sundown, E. C. L. & F., Lookout and part of Jim Blaine, about 40 acres, in the W. 1-2 section 18, in Poverty gulch. All patented.   **Property**

Surface improvements consist of shaft house, ore house, blacksmith shop and boiler house, two boilers, two compressors, one 10x14 hoist and double compartment shaft; the main shaft 700 feet deep, with 5,000 feet of drifts and levels.   **Development**

Dividends up to January 1, 1900, $125,000.00; last dividend paid January 20, 1900, $30,000.00.   **Production and Dividend**

Highest price for stock during 1899, $1.04; lowest price for stock during 1899, 83 1-2 cents.

The Gold Knob Mining and Townsite Company.

Copyright, 1900, by Fred Hills

The Gold Knob Mining and Townsite Company.

Incorporated January 24, 1900.

James F. Burns.........................President
F. M. Woods.....Vice-President and General Manager
I. W. Bonbright............Secretary and Treasurer
Irving Howbert.   E. A. Colburn.
J. R. McKinnie.   Frank G. Peck.
Syl. T. Smith.

Main Office—Colorado Springs, Colorado.

3,000,000 shares. Par value, $1.00. In treasury January 1, 1900,
300,000 shares.

Owns the Aluminum lode mining claim, survey No. 9,141; the Andes, survey No. 9,166; the Alta, survey No. 9,166; the Anna, survey No. 9,221; the Banner, survey No. 8,941; the Brown Bear, survey No. 9,221; the Clara, survey No. 9,187; the Cut Diamond, survey No. 9,157; the Emma, survey No. 9,221; the Extension, survey No. 8,941; the Fedora, survey No. 9,141; the Fraction, survey No. 9,187; the Grass, survey No. 8,671; the Gold Reserve, survey No. 10,589; the Ide, survey No. 9,216; the Last Dollar, survey No. 9,187; the Lone Star, survey No. 9,141; the Lucky Friday, survey No. 9,194; the Mary Mack, survey No. 9,166; the Mary Anne, survey No. 9,216; the Old Abe, survey No. 9,221; the Oro, survey No. 9,058; the Perhaps, survey No. 9,187; the Peru, survey No. 9,187; the Plomo, survey No. 8,941; the Rough and Ready, survey No. 9,166; the Ruth D., survey No. 9,221; the Silver, survey No. 8,941; the Spring, survey No. 9,166; the Tomale, survey No. 9,216; the Texas Jack, survey No. 9,385; the Venus M., survey No. 9,058; the War Bonnet, survey No. 9,166; the Wienewurst, survey No. 9,216; the Wilson, survey No. 9,385; also an undivided one-third (1-3) interest in the surface of the Homestake lode mining claim, survey No. 8,760; a total of 150 acres, and in addition 50 acres in which the company owns surface rights only.

This recently incorporated company owns a large area of ground lying on and at the base of Bull hill and at the base of Battle mountain and Big Bull hill, comprising practically the entire townsite of Goldfield. As a treasury asset the company owns about 600 lots in this town, worth at least $200,000, which will be sold as funds are required for the vigorous development to which the property is being subjected. A 600-foot shaft has been sunk on the property at a point east of the southern portion of the Theresa claim; this will be one of the principal working shafts, and in addition a shaft is being sunk on the Aluminum claim, in the extreme northwestern portion of the property. It is known that a number of veins traverse the ground from which a heavy production is obtained in adjoining and neighboring properties; and it seems scarcely within the possibilities that development in Gold Knob territory will fail to yield results of at least equal importance and value. The management has adopted a vigorous and aggressive policy; and since this management comprises, to a very notable extent, the elements of financial strength and managerial ability which have been conspicuous in Cripple Creek's development, results of great importance both to the company and to the district at large may with confidence be expected.

The Gold Magnet Mining, Leasing and Milling Company.

Incorporated February, 1896.

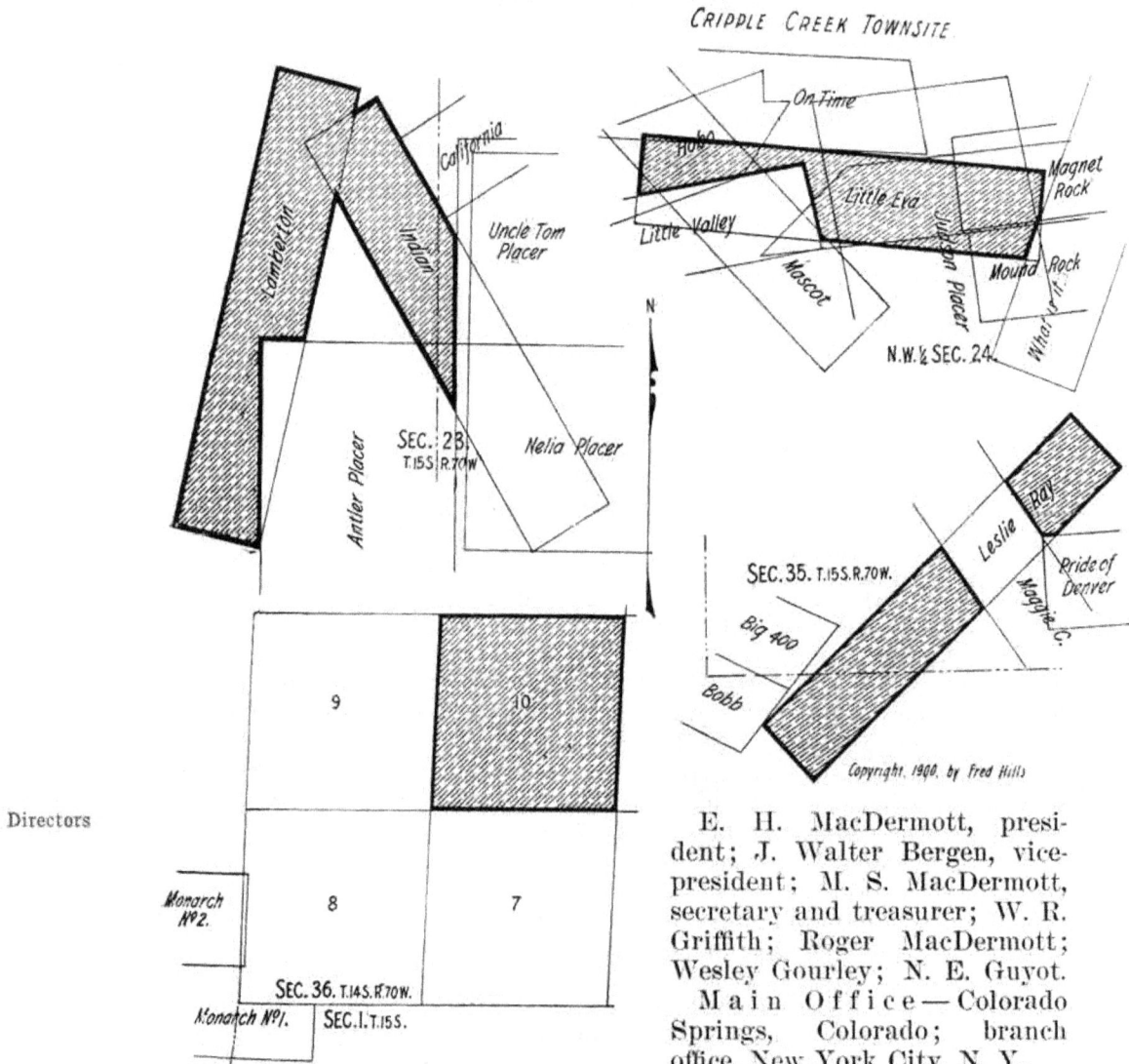

CRIPPLE CREEK TOWNSITE

On Time

California

Lamberton
Indian
Hobo
Little Eva
Magnet Rock
Uncle Tom Placer
Little Valley
Julian Placer
Mascot
Mound Rock
Wharf Ridge
N.W. ¼ SEC. 24

SEC. 23
T.15S.R.70W.
Nelia Placer
Antler Placer

N

Leslie Ray
Pride of Denver
SEC. 35. T.15S.R.70W.
Leslie
Magnet C.
Big 400
Bobb

9
10

Monarch Nº 2.
8
7
SEC. 36. T.14S.R.70W.
Monarch Nº 1.
SEC. 1. T.15S.

Copyright, 1900, by Fred Hills

Directors

E. H. MacDermott, president; J. Walter Bergen, vice-president; M. S. MacDermott, secretary and treasurer; W. R. Griffith; Roger MacDermott; Wesley Gourley; N. E. Guyot. Main Office—Colorado Springs, Colorado; branch office, New York City, N. Y.

Capitalization

1,250,000 shares. Par value, $1.00. In treasury January 1, 1900, 350,000 shares.

Property

Owns the Little Eva, survey No. 11,285, containing 7 1-2 acres, in the N. W. 1-4 section 24, on Gold hill, 300 feet south of the town of Cripple Creek and adjoining the Volcano mine; the Indian lode claim, 5 acres, in the N. 1-2 section 23, adjoining the California; the Lamberton lode claim, 9 acres, in the N. 1-2 section 23, on Gold Magnet hill; a 20 years' lease on lot No. 10, section 36, township 14 south, range 70 west, on Copper mountain; owns a one-half interest in one-half of the Leslie Ray claim, 7 acres, in the S. E. 1-4 section 35, and is negotiating for the purchase of the remaining interests. The Leslie Ray is patented. Little Eva, Indian and Lamberton in process of patenting.

Development

The Little Eva has a brick yard and buildings, which are leased. Five hundred dollars has been expended on this claim in shafts and $1,000 on other labor; $500 has been expended in shafts on the Indian

(CONTINUED ON PAGE 221.)

lode claim; $500 has been expended in shafts on the Lamberton claim. The Leslie Ray has several 10-foot prospecting shafts. Royalty has been paid on lot No. 10 up to July, 1900, and one or two shafts have been sunk for prospecting. The greater part of the development work is being done on the Little Eva, which is leased to Mr. Geo. H. Deters, until August, 1900.

Highest price for stock during 1899, 3 cents; lowest price for stock during 1899, unknown.

The Gold Retort Mining, Leasing and Bonding Company.

Incorporated December 26, 1895.

A. L. Lawton...........................President
A. J. Lawton......................Vice-President
J. H. Thedinga............Secretary and Treasurer
　　　L. C. Perkins.　　　　　　Adam Gregg.

Directors

Main Office—No. 17 East Pike's Peak avenue, Colorado Springs, Colorado.

1,500,000 shares. Par value, $1.00. In treasury January 1, 1900, 290,650 shares.

Capitalization

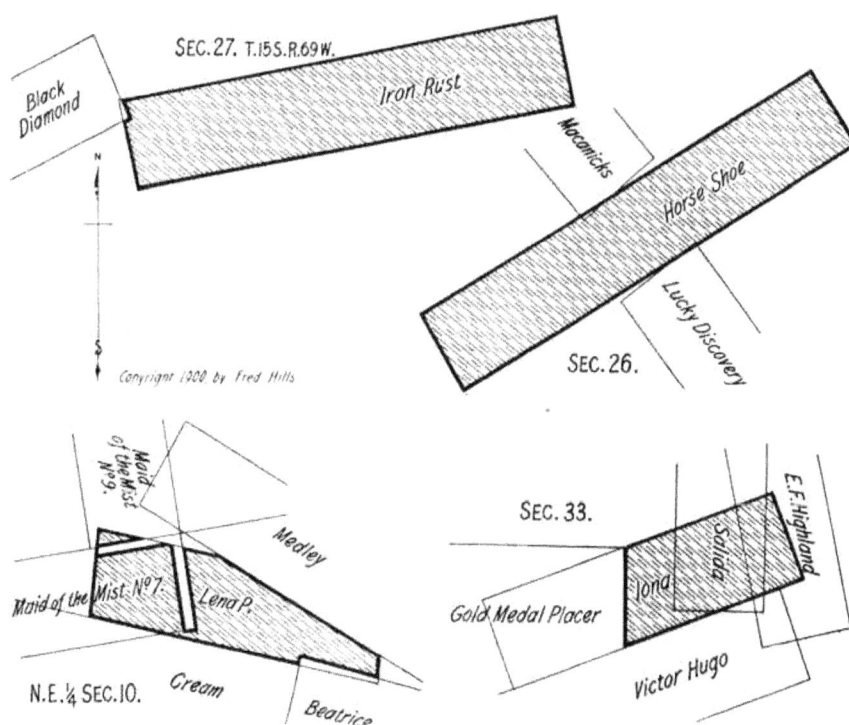

Owns the Iona, 4 1-2 acres, situated in the N. W. 1-4 section 33, on Big Bull hill; the Iron Rust, 9.687 acres, in the N. E. 1-4 section 27, on Big Bull hill; the Lena P., 6 acres, in the N. E. 1-4 section 10, on Cow mountain; and the Horse Shoe, 9.648 acres, in the E. 1-2 section 26, on Buck mountain. The Iona, Iron Rust and Lena P. are patented. Receiver's receipt for Horse Shoe.

Property

On each claim there is a shaft about 50 feet deep. The greater part of the development work is being done on the Iona.

Development

Highest price for stock during 1899, 1 cent; lowest price for stock during 1899, 1-2 cent.

The Gold Rock Mining Company.

Incorporated October, 1894.

Directors

William Clark..........................President

Fred F. Horn.............Secretary and Treasurer

Main Office—No. 23 East Kiowa street, Colorado Springs, Colorado.

Capitalization

1,000,000 shares.   Par value, $1.00.

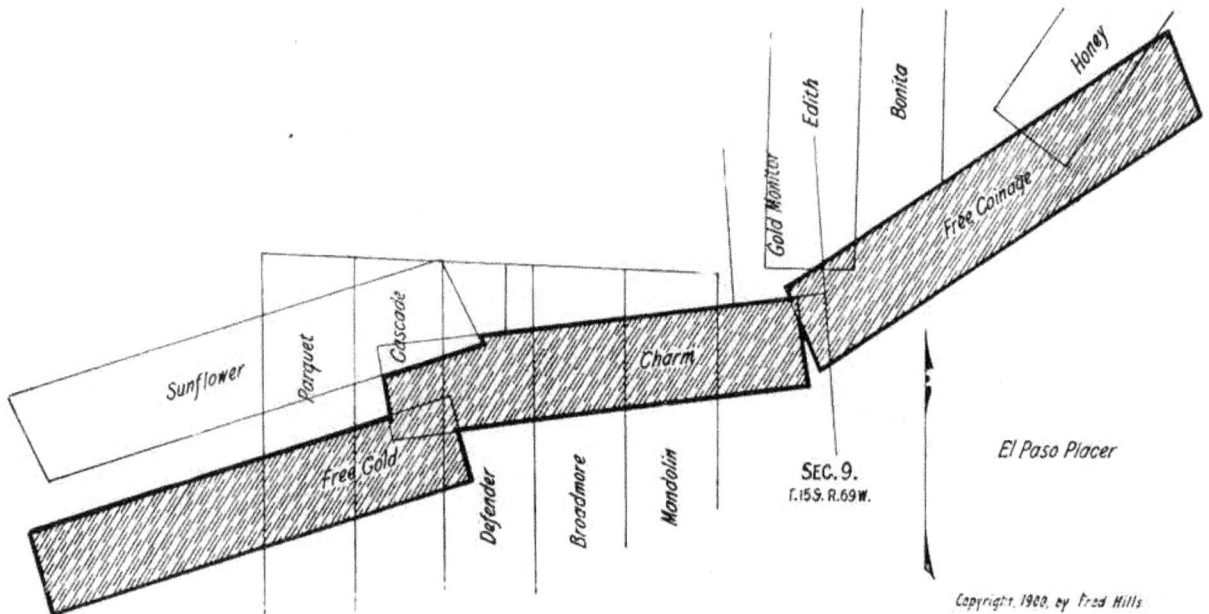

Copyright 1900, by Fred Hills.

Property

Owns the Charm, the Free Gold and the Free Coinage, containing in all 27.808 acres, situated in the S. 1-2 section 9, on Galena hill.   Property all patented.

Development

There are two 75-foot shafts, with 20 feet of drifting.   The greater part of the development work has been done on the Charm.   This claim was for a short time (some four months) under bond and lease, but at the present time work has ceased.

Highest price for stock during 1899, $17.50 per M.; lowest price for stock during 1899, $2.50 per M.

The Gold Sovereign Mining and Tunnel Company.

Incorporated November 13, 1895.

A. E. Carlton..............President and Treasurer
A. P. Mackey......................Vice-President
W. F. Rock..........................Secretary
E. M. Sharp..........................Manager
    William A. Otis.   E. C. Newcombe.
          Abram Rapp.

Directors

Main Office—365 Bennett avenue, Cripple Creek, Colorado. Transfer Office—International Trust Company, Colorado Springs, Colorado.

2,000,000 shares. Par value, $1.00. In treasury January 1, 1900, 215,212 shares; in treasury January 1, 1900, $10,000.00 cash.

Capitalization

Owns the Gold Sovereign lode, the J. G. Blaine lode and Gold Sovereign tunnel site, the surface of which amounts to 6.21 acres. All patented. Situated in section 20, on Bull hill.

Property

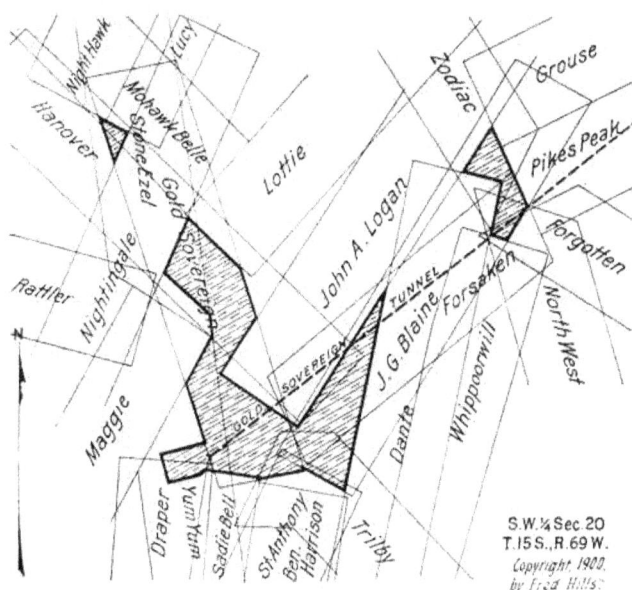

S.W. ¼ Sec. 20
T. 15 S., R. 69 W.
Copyright 1900
by Fred Hills.

There is an electric hoist on the winze of the Gold Sovereign tunnel, 200 feet deep from tunnel level, which is under contract for 200 feet more depth, and raised to the surface 100 feet, making this when completed the main working shaft of the property, located on a phonolyte and basalt dyke running parallel with each other the full length of the property. There is also another shaft 375 feet deep, known as the Whisper shaft; also a shaft 360 feet deep known as the Rawson shaft; also a shaft 211 1-2 feet deep known as the Lovett shaft; also a shaft 165 feet deep known as the Jackson shaft; also shaft 150 feet deep, known as the Barrack shaft, together with levels, drifts and winzes, making a total of about 2,000 feet of work. Tunnel driven 640 feet under Bull hill. New and valuable ore bodies are being opened at depth in several of the workings, and the property is in excellent condition. The company proposes to work the entire property after the expiration of the present leases.

Development

Production to January 1, 1900, $195,500.00.

Production

Highest price for stock during 1899, 20 cents; lowest price for stock during 1899, 4 cents.

The Goldstone Mining and Milling Company.

Incorporated.

Directors

L. C. Hall..........................President
L. Srite...............Vice-President and Treasurer
C. P. Bently...........................Secretary
J. S. Hall.   Seth Baker.

Main Office—No. 53 First National bank, Colorado Springs, Colorado.

Capitalization

2,000,000 shares. Par value, $1.00. In treasury January 1, 1900, 392,000 shares.

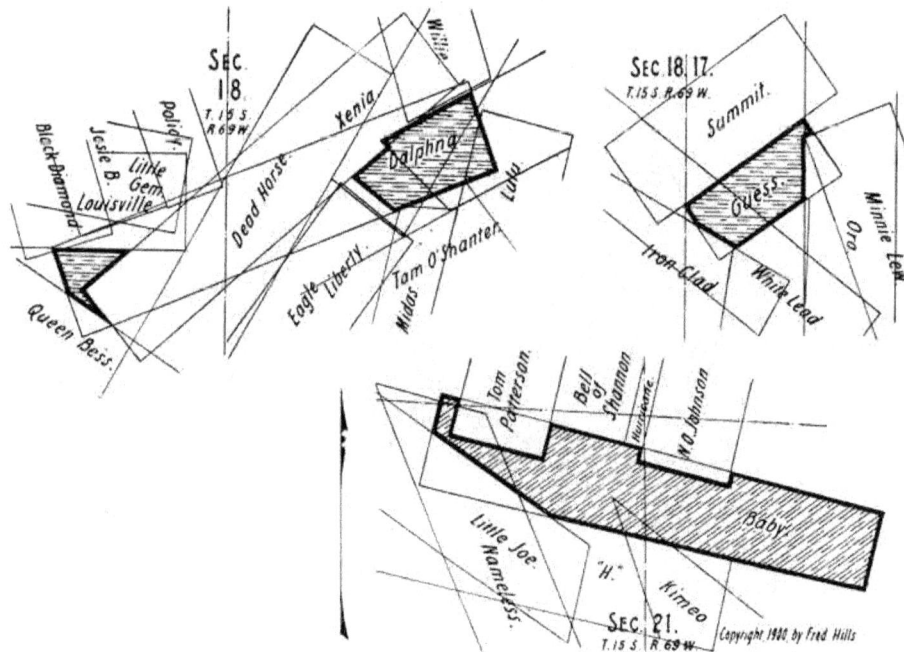

Property

Owns the Baby, 7 1-2 acres, in the center of section 21, on Bull hill; the Guess, 2 acres, in the S. W. 1-4 section 17, on Ironclad hill; the Dalphna, 3 1-2 acres, in the N. 1-2 section 18, in Poverty gulch. All patented.

Development

The Baby has 175 feet of shafts and drifts. This claim is leased for three years from March 26, 1900. The Guess has two shafts, each 80 feet deep. This claim is leased for two years from October 21, 1899. The Dalphna has 160 feet of shafts.

Highest price for stock during 1899, 3 3-8 cents; lowest price for stock during 1899, 1 cent.

The Good Hope Gold Mining Company.

Incorporated 1898.

A. S. Pope...............................President   Directors

F. H. Pettingell....................Vice-President

L. A. Civill...............Secretary and Treasurer

W. B. Boardman.

Main Office—No. 11 Bank building, Colorado Springs, Colorado.

1,000,000 shares. Par value, $1.00. In treasury January 1, 1900,   Capitalization
$500.00 cash.

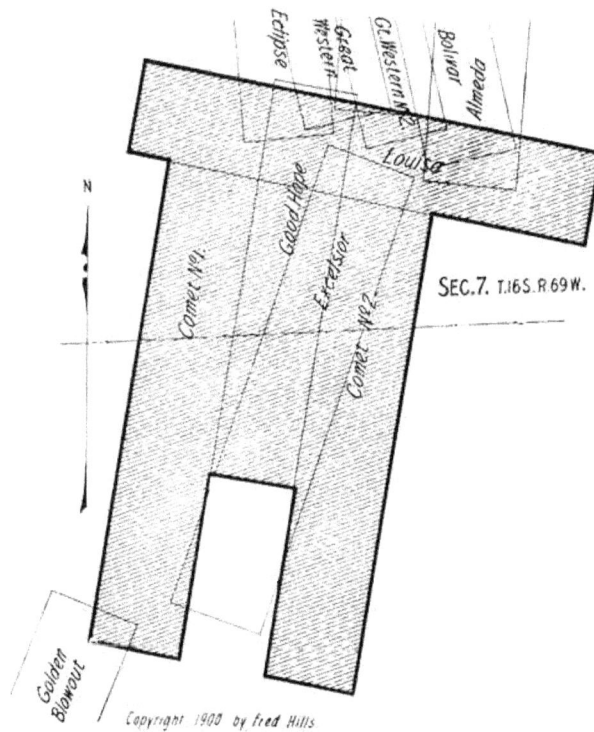

Copyright 1900 by Fred Hills

Owns the Good Hope, 8 1-2 acres; the Comet Nos. 1 and 2, the Louisa,   Property
and the Jasper tunnel site, 27 acres, in sections 7 and 18, township 16
south, range 69 west. The Good Hope is patented. Remainder in process.

Highest price for stock during 1899, $6.50 per 1,000; lowest price for
stock during 1899, $5.00 per 1,000.

The Good Will Tunnel and Mining Company.

The Good Will Tunnel and Mining Company.

Incorporated.

J. A. Hayes.............................President   Directors

Wm. P. Bonbright..................Vice-President

F. W. Stehr.............................Secretary

Main Office—With Wm. P. Bonbright & Co., Colorado Springs, Colorado.

2,000,000 shares. Par value, $1.00. In treasury January 1, 1900,   Capitalization $25,611.09 cash.

Owns the south 500 feet of the Sunnyside claim, on Gold hill; tunnel   Property will be 3,000 feet in length when completed to the Lexington mine. On April 4, 1900, the tunnel was in 1,860 feet. The company owns a power plant worth about $11,000. The Sunnyside lode is patented.

One 80-horse power boiler; one heater, 60x30; one compressor, 16x12;   Development one air receiver, 40x12; and all necessary fittings; two Rand machine drills; two Ingersoll-Sargent machine drills; powder magazine; compressor and boiler room, 34x24 and 20x24. The tunnel company has a lease on the following properties through which it passes: The Anchoria-Leland claim, the Lillian-Leland, the Missouri Company, the Progress Company, the Tom Gough and the Lexington Company. The following claims are being traversed: Queen, Little King, Lillian-Leland, Annie, Moonlight, Tom Gough, E. Porter Gold King, Clara D.

The tunnel was first started by Mr. W. Weston, M. E., in 1895. Later   History on working capital was secured by the sale of treasury stock and the tunnel was driven 1,860 feet, just entering the outer edge of the main auriferous area of the Cripple Creek district. In the autumn of 1899 further working capital was obtained sufficient to drive the tunnel to the Lexington mine and the Anchoria-Leland property, a total distance of 3,000 feet. It is entirely a mining and transportation tunnel, and no claims are made to any veins that may be discovered while driving it. If any valuable ores are cut the tunnel company will work them under the terms of the leases from the companies owning the ground through which it passes.

The Gould Mining and Milling Company.

Incorporated June 18, 1892.

Directors

D. Weyand..............................President
L. A. Keys..........................Vice-President
E. M. Purdy..............................Secretary
L. C. Weyand............................Treasurer
    D. P. Cathcart.   John Barragar.
        John Harnan.

Main Office—Room No. 7, Freeman building, Colorado Springs, Colorado.

Capitalization

1,250,000 shares. Par value, $1.00. In treasury January 1, 1900, 65,000 shares.

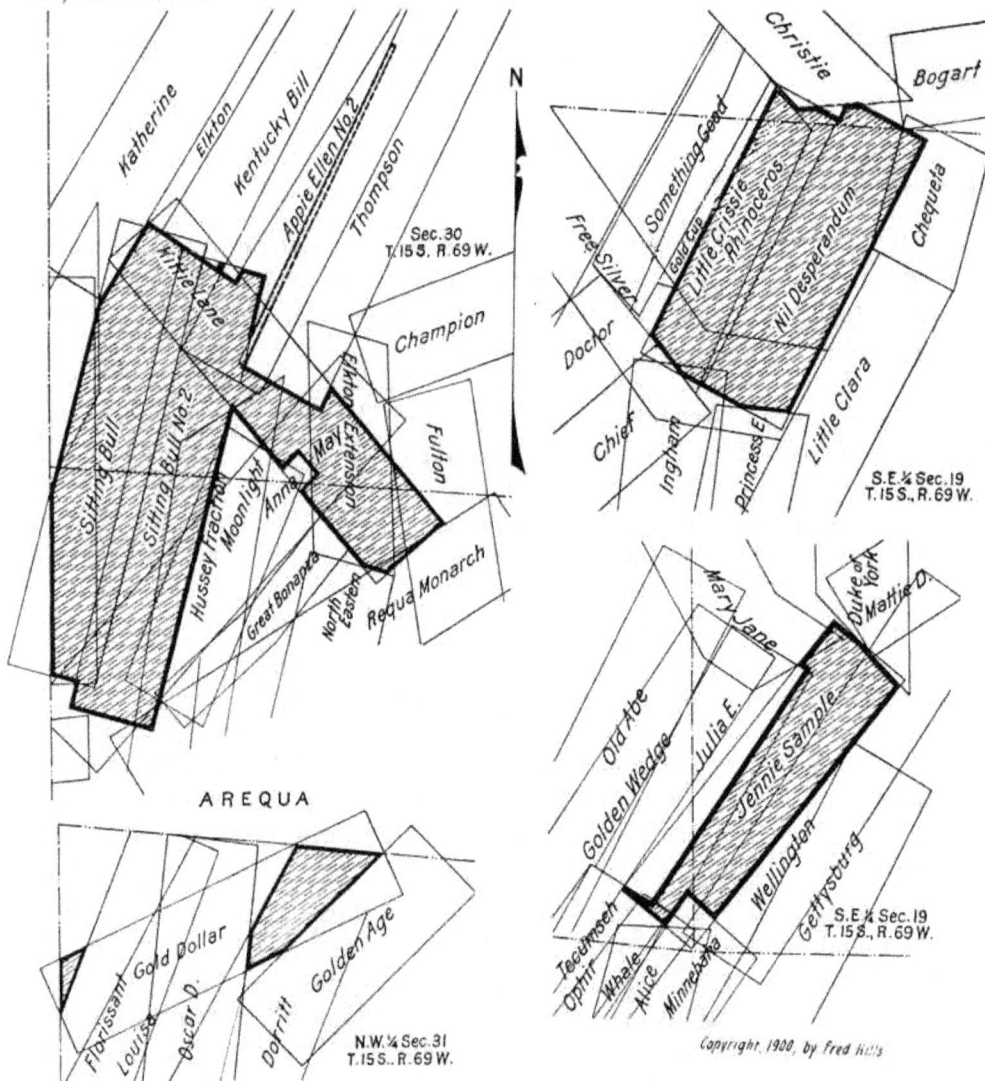

Copyright, 1900, by Fred Hills

Property

Owns the Nil Desperandum and the Rhinoceros, containing 11 1-2 acres, in the S. E. 1-4 section 19, on the northwest slope of Raven hill; the Jennie Sample, containing about 7 1-3 acres, in the S. 1-2 section 19; the Kittie Lane, Sitting Bull and Sitting Bull No. 2 lodes, containing 19 acres, in the E. 1-2 section 30, south of and adjoining the Elkton, on Raven hill; the Gold Dollar, containing 2 acres, in the N. W. 1-4 section 31; also owns the Tip Top, containing 3.5 acres in the N. E. 1-4 section 12, on Tenderfoot hill. All the above is patented with the exception of the

(CONTINUED ON PAGE 229.)

The Gould Mining and Milling Company—Continued.

Gold Dollar and the Tip Top, which are in process. The company also claims by deed a strip of the Thompson lode, as shown on the map by dotted lines. This is in controversy with the Elkton Company. The Sitting Bull Nos. 1 and 2 were purchased in March, 1900, by this company, for $75,000 cash.

There is a steam hoist and ore house on the Kittie Lane, which has a two-compartment shaft 300 feet deep, and 100 feet of drifting. The Jennie Sample has a 300-foot shaft, with 600 feet of drifting. The Nil Desperandum and the Rhinoceros have a 400-foot tunnel, with about 300 feet of cross-cutting. The greater part of the development work at present is being done on the Jennie Sample.

Development

Highest price for stock during 1899, 38 3-8 cents; lowest price for stock during 1899, 7 7-8 cents.

The Grace Gold Mining and Milling Company.

Incorporated September 5, 1894.

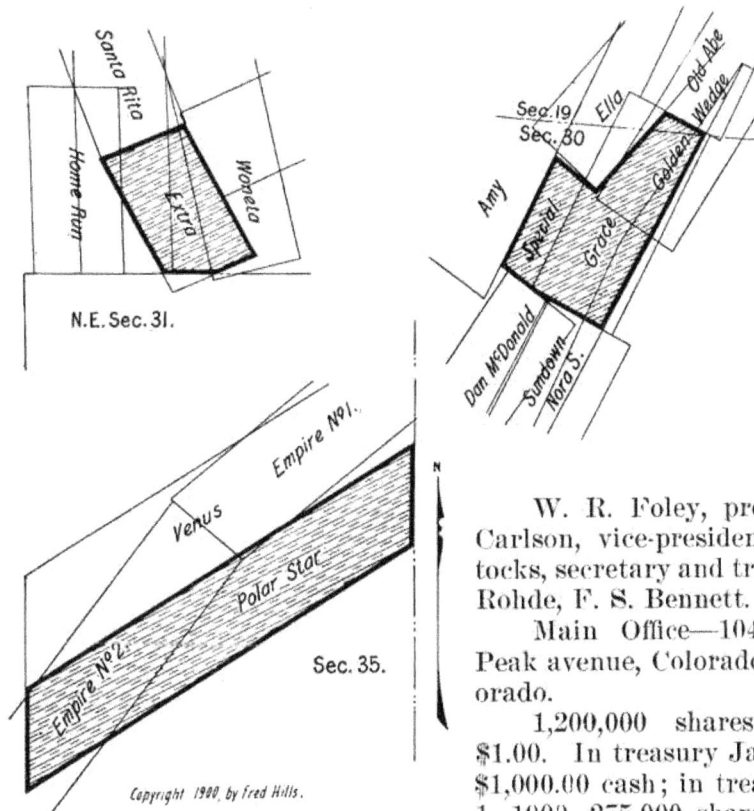

N.E. Sec. 31.

Sec. 19 / Sec. 30

Sec. 35.

Copyright 1900, by Fred Hills.

W. R. Foley, president; C. G. Carlson, vice-president; S. J. Mattocks, secretary and treasurer; W. E. Rohde, F. S. Bennett.

Directors

Main Office—104 East Pike's Peak avenue, Colorado Springs, Colorado.

1,200,000 shares. Par value, $1.00. In treasury January 1, 1900, $1,000.00 cash; in treasury January 1, 1900, 275,000 shares.

Capitalization

Owns the Grace and Special, about 6 acres, in the N. W. 1-4 section 30, on Raven hill; Extra, 5 acres, in the N. E. 1-4 section 31, south of Squaw mountain; also the Polar Star, 10 acres, in the S. E. 1-4 section 35, on Grouse mountain. All patented.

Property

A frame shaft house, boiler room and steam hoist, which cost about $5,000.00, is on the Special claim, which has a shaft 200 feet deep, with 250 feet of drifting. The company was incorporated in 1894, but no work of any amount ever done on the property until November, 1899. Considerable low grade ore is in sight.

Development

Highest price for stock during 1899, 7 1-2 cents; lowest price for stock during 1899, 2 cents.

The Grafton Gold Mining Company.

Incorporated January 3, 1896.

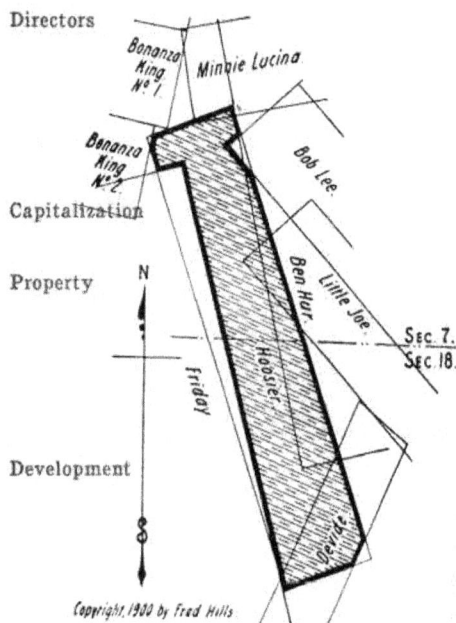

Directors

Capitalization

Property

Development

Production
and Dividend

Edmund J. A. Rogers, president; David Rubridge, vice-president and treasurer; Wm. P. Headden, secretary; O. W. Pitcher; P. H. Rogers.

Main Office—No. 1529 Lawrence street, Denver, Colorado.

1,000,000 shares. Par value, $1.00. In treasury January 1, 1900, $5,000.00 cash.

Owns the Hoosier, containing about 9 acres, situated in S. E. 1-4 section 7 and N. E. 1-4 section 18, on Tenderfoot Hill. The above property is held by patent, with the exception of .056 of an acre, which the present company acquired by deed.

Shaft house, ore house, and other machinery. The main shaft is 300 feet deep, and there is about 300 feet of drifting; 200 feet to the north of the main shaft is one 62 feet deep in process of sinking. The property is leased for three years from October 15, 1898.

Gross production to January 1, 1900, $80,500.00; net royalty on ore mined during 1899, $15,213.00; last dividend was for $10,000.00, paid October 15, 1899.

Highest price for stock during 1899, 35 cents; lowest price for stock during 1899, 10 cents.

The Granite Gold Mining Company.

Incorporated November 13, 1896.

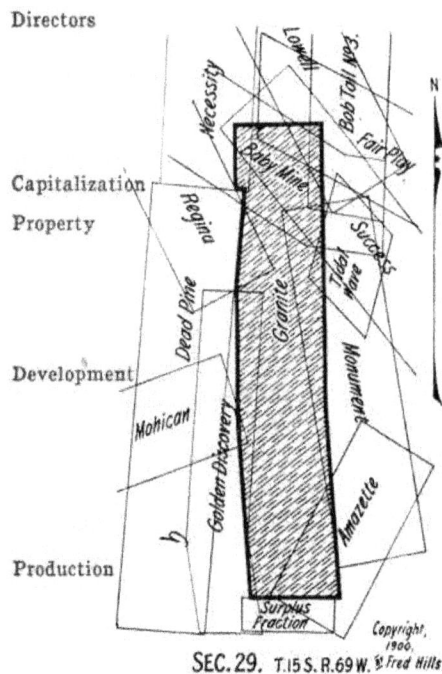

Directors

Capitalization

Property

Development

Production

W. H. Brevoort, president; Eben Smith, vice-president; R. H. Reid, secretary and treasurer; W. D. Reid, D. L. Webb.

Main Office—Equitable Building, Denver, Colorado.

1,000,000 shares. Par value, $1.00.

Owns the Granite claim, survey No. 7,861, containing 10 acres, and the Baby Mine claim, survey No. 9,523, containing a fraction of an acre, both situated in the S. W. 1-4 section 29 on Battle mountain. Patented.

The company has a complete equipment for hoisting 300 tons of ore per day The main shaft is 750 feet deep, with 5,432 feet of levels, drifts and cross-cuts. There is also an incline shaft of 155 feet. The greater part of the development work is being done on the Granite claim.

Gross production to January 1, 1900, $400,000.00.

Highest price for stock during 1899, $4.00; lowest price for stock during 1899, $3.00.

SEC. 29. T. 15 S. R. 69 W.

The Granite Hill Mining and Milling Company.

Incorporated May, 1892.

J. P. Madden, president; Jos. H. Ryan, vice-president; W. H. Gowdy, Directors
secretary; W. R. Barnes, treasurer.

Main Office—Colorado Springs, Colorado.

1,000,000 shares. Par value, $1.00. In treasury January 1, 1900, Capitalization
about 50,000 shares; in treasury January 1, 1900, $350.00 cash.

Owns the Midget, 0.6 acre, in the N. W. 1-4 section 19, on Gold hill; Property
the Granite hill, 2 1-4 acres, in the S. W. 1-4 section 18, in Poverty gulch;
the Oro, 7 1-4 acres, in the S. E. 1-4 section 11, township 15 south, range
70 west, on Rattlesnake hill. All patented.

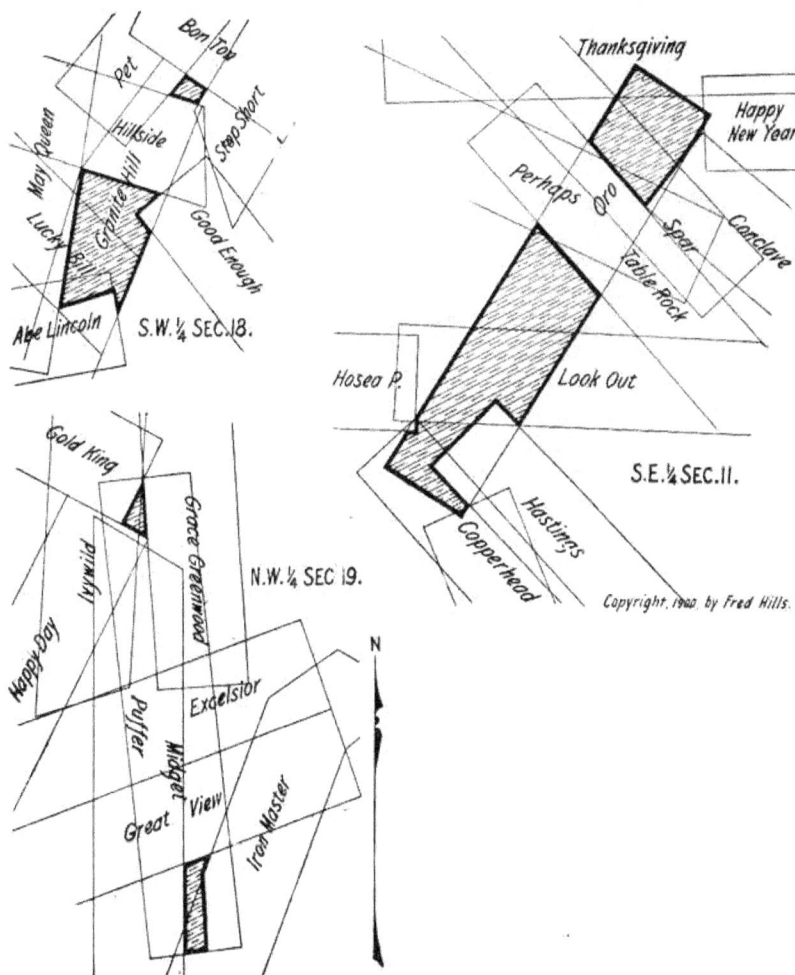

As this Manual goes to press, in May, 1900, it appears that the
Granite Hill claim had been offered to Mr. C. W. Kurie, of Colorado
Springs, and that some controversy has arisen between the directors of
the company as to whether the sale should be ratified or not. At the last
meeting, on May 11, 1900, 590,000 shares were in favor of the same.

The Granite Hill has a shaft 300 feet deep. On this claim the Development
greater part of the development work is being done. The property is
leased. The Midget claim is leased and bonded. It has been developed
by tunnels and a shaft 80 feet deep.

Gross production to January 1, 1900, about $30,000.00. Production

Highest price for stock during 1899, 10 cents; lowest price for stock
during 1899, 3 cents.

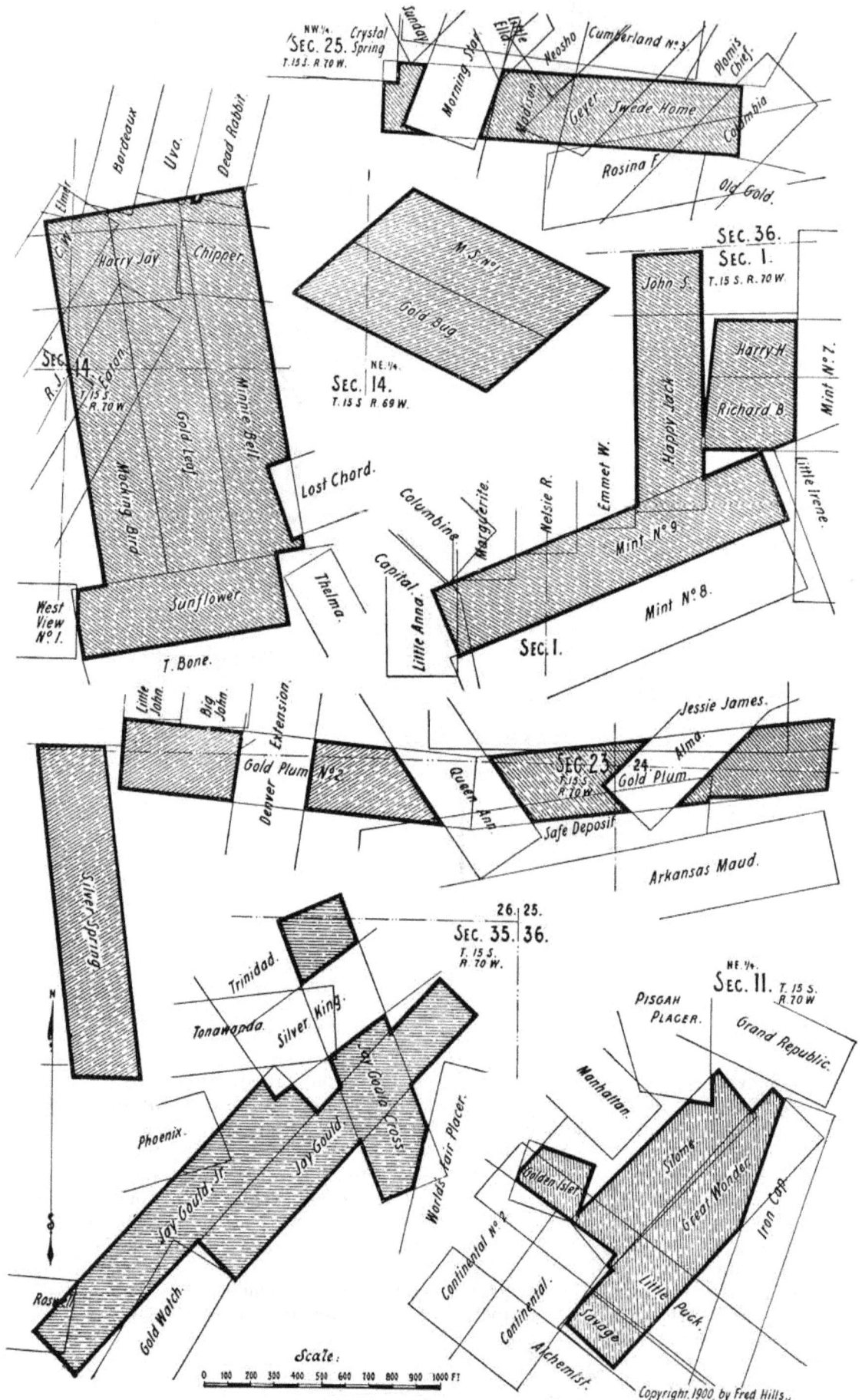

The Greater Gold Belt Mining Company.

(SEE ALSO PAGES 233 AND 234.)

The Greater Gold Belt Mining Company.

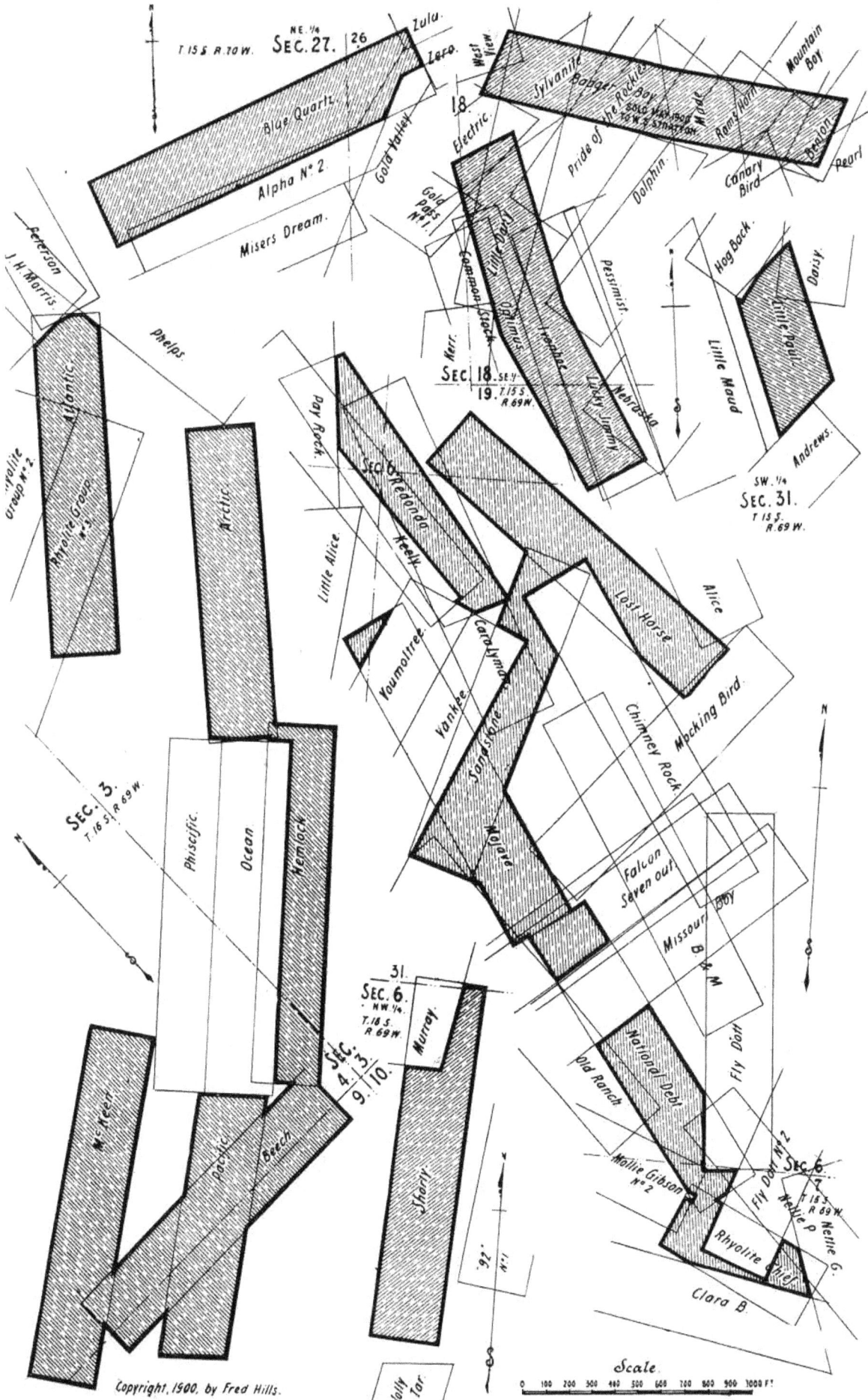

(CONTINUED ON PAGE 234.)

The Greater Gold Belt Mining Company.

(CONTINUED FROM PAGES 232 AND 233.)

Incorporated 1894.

Directors
D. V. Donaldson.........................President
A. S. Holbrook.....................Vice-President
Walter C. Frost...........Secretary and Treasurer
John S. Hall.................Assistant Secretary
John Wilson.

Main Office—No. 6 North Nevada avenue, Colorado Springs, Colorado.

Capitalization
5,000,000 shares. Par value, $1.00. In treasury January 1, 1900, 1,271,317 shares; in treasury January 1, 1900, $219.70 cash.

Property
Owns the Ivanhoe, 10 acres, on Gold hill, in the S. E. 1-4 section 18; the Badger Boy, 10 acres, on Globe hill, in the S. E. 1-4 section 18; the Gold Plum, Gold Plum No. 2 and the Silver Spring, 23 acres, on Signal hill, in the S. E. 1-4 section 23; the Redondo, Lost Horse, Sandstone, Mojave, National Debt and the Rhyolite Chief, 25 acres, receiver's receipt, on Rhyolite hill, in the S. E. 1-4 section 6; the Little Paul, 4 acres, and the Shorty, 9 acres, both in the saddle between Straub and Grouse mountains, in the S. W. 1-4 section 31 and the N. E. 1-4 section 6; the Swede Home, 8 acres, on Guyot hill, in the N. W. 1-4 section 25; the Mocking Bird, Gold Leaf, Sunflower and Minnie Bell, 36 acres, on Crystal hill, in the S. W. 1-4 section 11 and the N. 1-2 section 14; the Blue Quartz, 10 acres, on Signal hill, in the N. E. 1-4 section 27; the Jay Gould, Jay Gould, Jr., Jay Gould Cross, 20 acres, in the S. E. 1-4 section 26 and the N. E. 1-4 section 25; the Gold Bug and M. S. No. 1, 12 acres, on Cow mountain, in the N. 1-2 section 14; the Happy Jack, Harry H., Richard B. and Mint No. 9, 22 acres, on Copper mountain, in the N. E. 1-4 section 1; the Golden Islet, Silome and Great Wonder, 14 acres, on Mineral hill, in the N. E. 1-4 section 11; the McKeen, Beach, Hemlock, Arctic, Atlantic and Pacific, 50 acres, on Brind mountain, in the S. W. 1-4 section 3, township 16 south, range 69 west; and High Park City, 160 acres, 4 miles west of Cripple Creek, the latter not being shown on plat. All the above properties are patented except the Redondo et al. group, for which receiver's receipt is held.

Development
The company has been operating on its own account a portion of the Badger Boy claim, and has just completed a well-timbered compartment shaft to a depth of 500 feet. Stations have been cut at the 150, 300, 400 and 500-foot levels. At the 150-foot level a drift was run for about 150 feet on the vein; on the 400-foot level about 80 feet of drifting was done, and 170 feet on the 500-foot level. Scarcity of funds compelled the abandonment of the work at this point. There are also other shafts on the Badger Boy and the Ivanhoe claims. Five sets of lessees are working on the Badger Boy, and three sets on the Ivanhoe.

Production and Dividend
History
Small semi-monthly shipments are being regularly made from the Badger Boy. From the Ivanhoe a shipment has recently been made averaging $35.00 to the ton. On May 9, 1900, the Badger Boy claim was sold to Mr. C. W. Kurie, of Colorado Springs, supposedly for $95,000. From the money thus obtained a dividend of 2 cents will be paid and about $19,000 will be placed in the treasury of the company for development purposes.

Stock is selling readily at 5 cents per share.

BULL CLIFFS.

SHOWING ISABELLA PROPERTIES AND VICTOR MINE.

The Great Northern Gold Mining Company.

Incorporated January, 1896.

Directors

J. Arthur Connell.........President and Treasurer
Henry C. Hall......................Vice-President
D. D. Lord..............................Secretary

R. C. Thayer.

Transfer Office—The International Trust Company, Colorado Springs, Colorado.   Main Office—Postoffice building, Colorado Springs, Colorado.

Capitalization

1,500,000 shares.   Par value, $1.00.   In treasury January 1, 1900, 290,000 shares.   Present indebtedness, $7,300.00.

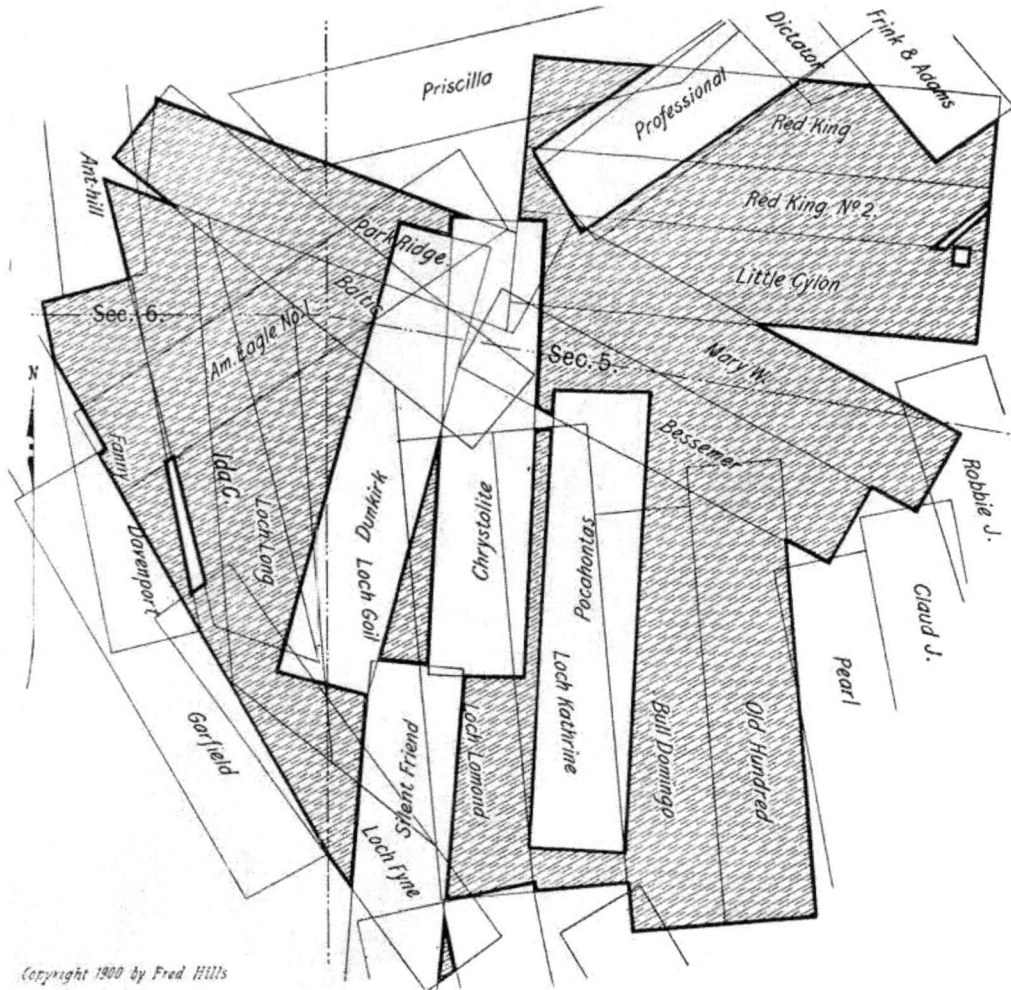

Copyright 1900 by Fred Hills

Property

Owns the Mary W., Little Cylon, Park Ridge, Ada C., Fanny Davenport, Bull Domingo, Old Hundred, Red King, Red King No. 2, Bessemer, Loch Long, Loch Goil, Loch Katrine, Loch Lomond, Loch Fyne, in all about 98 1-2 acres, lying in the E. 1-2 of section 6 and in the W. 1-2 of section 5, between Tenderfoot and Lincoln hills.   All the property is patented.

Highest price for stock during 1899, 2 cents: lowest price for stock during 1899, 1-2 cent.

The Gregory Leasing Company.

Incorporated October 24, 1896.

Ramsay C. Bogy, president; G. H. Hill, vice-president; J. D. Nieder-litz, secretary and treasurer; L. V. Bogy, E. J. Seely. Directors

Main Office—Equitable building, Denver, Colorado.

1,000,000 shares. Par value, $1.00. In treasury January 1, 1900, 45,000 shares; in treasury January 1, 1900, several hundred dollars. Capitalization

Owns the Deleware, containing about 6 acres, patented, situated in the S. E. 1-4 section 3, township 15 south, range 70 west; also about one acre of the J. A. G., adjoining the Deleware, all on Iron mountain. The company has also a five years' lease on three patented claims in Gilpin county. Property

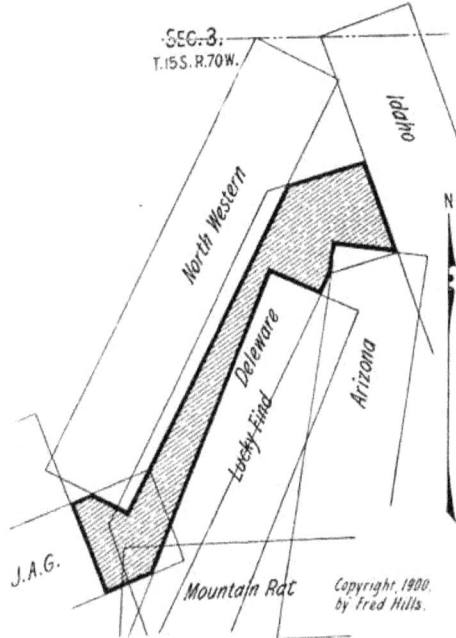

About 100 feet of work has been done on the Deleware and the J. A. G. Development

From the Gilpin county mines several hundred tons of ore have been shipped. Production

Highest price for stock during 1899, about 1 cent; lowest price for stock during 1899, about $4.50 per M.

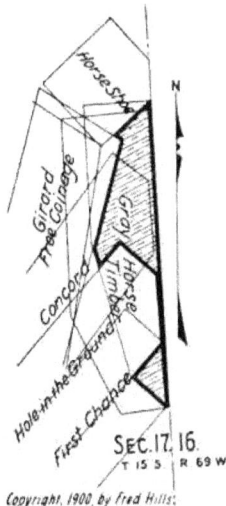

The Gray Horse Gold Mining Company.

Incorporated October 25, 1899.

Henry McAllister, Jr...............President Directors
Robt. H. Widdicombe...........Vice-President
Asa T. Jones.........Secretary and Treasurer
Joseph C. Dean.   Henry M. Blackmer.

Main Office—Colorado Springs, Colorado.

1,250,000 shares. Par value, $1.00. Capitalization

Owns the Gray Horse lode, 2 1-4 acres, in the S. E. 1-4 section 17. Patented. Enough development work only has been done to secure the patent. Property

The Hallett and Hamburg Gold Mining Company.

Incorporated 1894.

Directors
E. A. Colburn, president; C. H. Dudley, secretary and treasurer; J. L. Hallett.

Main Office—Colorado Springs, Colorado.

Capitalization
2,000,000 shares. Par value, $1.00.

Property
Owns the Hallett and the Hamburg claims, 4 1-2 acres, patented, situated in S. W. 1-4 section 29, on Battle mountain.

Development
Shaft house, steam hoist and ore bins. The company had a shaft 400 feet deep, with 400 feet of drifting and cross-cuts, when the property passed under the control of the Ajax Company.

Production and Dividend
Gross production to January 1, 1900, $50,000.00. Total amount of dividends to January 1, 1900, about $8,000.

History
The control of the stock of this company was purchased lately by the Ajax Gold Mining Company, as it adjoins their ground. It is now being delevoped by the latter company. The stock has not been placed on the market.

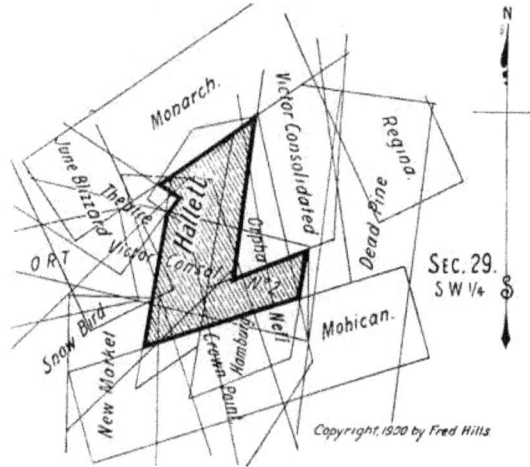

The Hanover Gold Mining Company.

Incorporated October 7, 1899.

Directors
Chas. L. Tutt, president; K. R. Babbitt, vice-president; J. McK. Ferriday, secretary; Spencer Penrose, treasurer; John Harnan.

Main Office—Room 11, El Paso Bank building, Colorado Springs, Colorado.

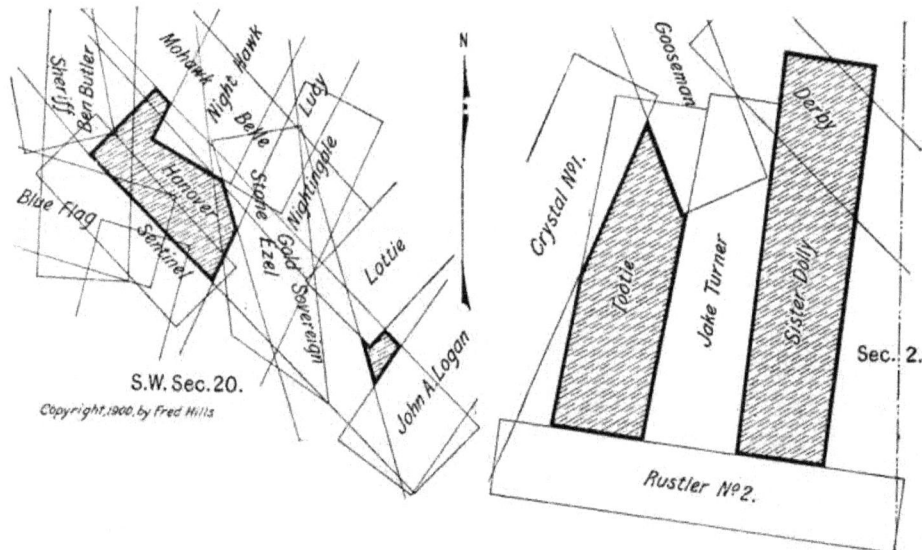

Capitalization
1,500,000 shares. Par value, $1.00. In treasury January 1, 1900, 250,000 shares; in treasury January 1, 1900, $400.00 cash.

(CONTINUED ON PAGE 239.)

The Hanover Gold Mining Company.—Continued.

Owns the Hanover, 3.11 acres, in the S. W. 1-4 section 20, on Bull hill; Sister Dolly and Tootie, about 13 acres, on Copper mountain, in the N. W. 1-4 section 2, township 15 south, range 70 west. Patented. Also the Jake Turner, adjoining the Sister Dolly, in process, consequently not shaded on plat. *Property*

There are three sets of lessees working on the Hanover claim, and as they are all responsible and competent miners, it would seem that the prospects were most flattering. *Development*

Highest price for stock during 1899, 5 cents; lowest price for stock during 1899, 3 cents.

The Hard Carbonate Gold Mining Company.

Incorporated 1895.

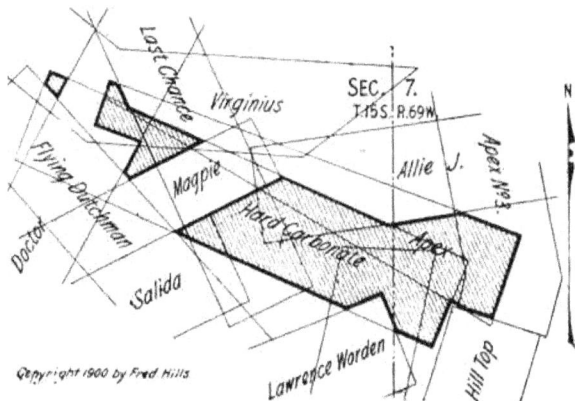

Joseph H. Smith, president; J. C. Mitchell, vice-president; L. G. Campbell, secretary; E. P. Arthur, treasurer; J. C. Johnson; J. D. Hill; W. S. Harwood, general manager. *Directors*

1,000,000 shares. Par value, $1.00. In treasury January 1, 1900, 359,000 shares; in treasury January 1, 1900, $100.00 cash. *Capitalization*

Owns the Hard Carbonate and the Apex, 9 acres, in all, situated on Tenderfoot hill, in the S. W. 1-4 section 7; also vein rights of about 1,625 feet passing through the Doctor and the Last Chance claims and the Magpie claim. All the above property is patented. *Property*

The Hard Carbonate claim has a shaft 175 feet deep, with 250 feet of levels and cross-cuts, and a shaft house and whim. The greater part of the development work is being done on this claim. The Apex has some shallow workings. A new plant of machinery is shortly to be erected, and, when the mine is in better working order, more valuable shipments of ore are expected. *Development*

The stock is closely held.

The Hart Gold Mining and Leasing Company.

Incorporated January 27, 1899.

Directors

J. R. McKinnie, president; J. W. Graham, vice-president; John E. Phillips, secretary; L. L. Aitken, treasurer; Joshua Grozier.

Main Office—No. 25 E. Pike's Peak avenue, Colorado Springs, Colo.

Capitalization

1,000,000 shares. Par value, $1.00. In treasury January 1, 1900, $6,000.00 cash.

Property

Owns the Hart, the Fresno, the Jessie G., and the Brooklyn, containing in all 30.247 acres, situated in the S. 1-2 section 17, on the north slope of Bull Hill. All the above claims are patented.

Development

Shaft house and electric power plant capable of sinking to a depth of 600 feet.

A double compartment shaft is now down 325 feet. Size of the shaft, 4 1-2x9 1-2, timbered with square sets. Cross-cutting is being done to the east and west of this shaft. Considerable prospecting has been done to the north and east by the company and by lessees. A number of veins are known to cross this property.

Highest price for stock during 1899, about 25 cents; lowest price for stock during 1899, 10 cents.

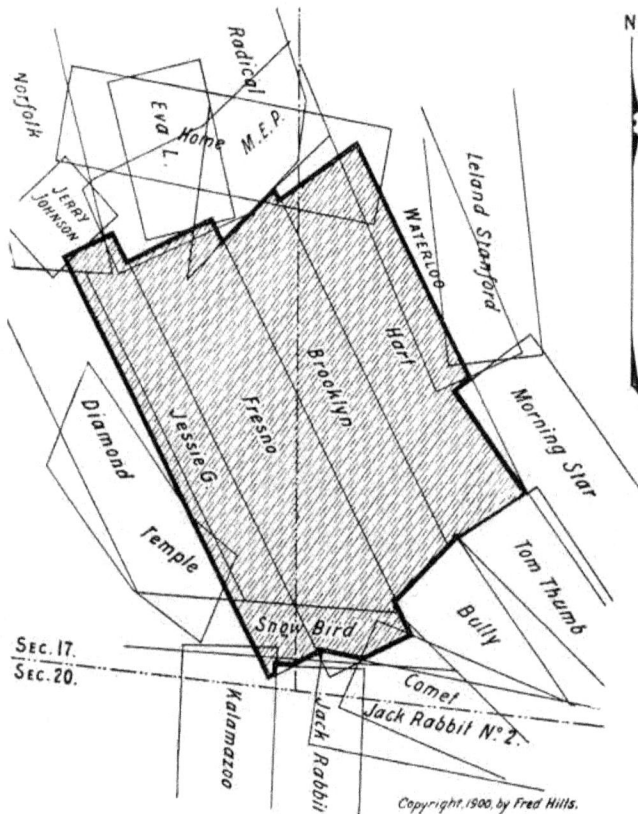

240

The Hartford Gold Mining Company.

Incorporated December 19, 1895.

F. J. Baker...........................President   Directors

J. C. Spicer......................Vice-President

H. A. Scurr................Secretary and Treasurer

Wm. Laws   C. G. Carlin.

Main Office—20 South Tejon street, Colorado Springs, Colorado.

1,500,000 shares. Par value, $1.00. In treasury January 1, 1900 Capitalization 544,000 shares; in treasury January 1, 1900, $130.00 cash.

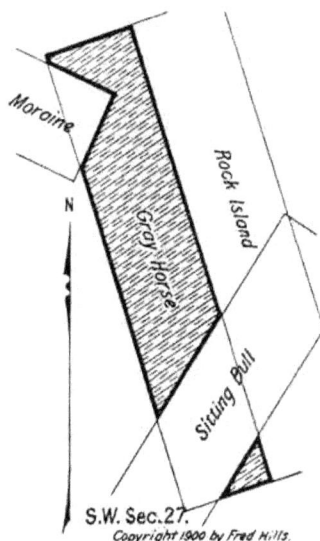

Owns the Gray Horse lode, 7 1-2 acres, in the S. W. 1-4 section 27, Property township 15 south, range 69 west, on Big Bull mountain. The company also owns the Little Fred lode, 900 feet west of the Fanny Rawlings mine, in Leadville, and the Hartford, on Blue mountain, west of Florissant. These latter properties being outside the Cripple Creek district are not shown on plat.

100 feet of sinking and drifting on the property. Assessment work Development was completed on the Gray Horse in 1899, and assessment work in shaft and some drifting on the Little Fred.

Highest price for stock during 1899, $8.00 per 1,000; lowest price for stock during 1899, $5.00 per 1,000.

The Hawkeye Gold Mining Company.

Incorporated June 17, 1898.

Directors   Joseph St. Martin...........President
R. Theroux ....Vice-President
Horace Granfield.Secretary and Treasurer
J. C. McCormack.   P. A. Primeau.

Main Office—Colorado Springs, Colorado; branch office, St. Paul Building, New York City, N. Y.; branch office, room 1, B. & M. Block, Denver, Colorado.

Capitalization   1,500,000 shares. Par value, $1.00.
In treasury January 1, 1900, 240,000 shares; in treasury, January 1, 1900, $5,-000 cash.

Property   Owns the Springion and the Valley, about 16 acres in all, situated in the N. E. 1-4 of section 4, township 16 south, range 69 west, about 1 3-4 miles southeast of the town of Victor. Property is all patented.

Development   There is a shaft house on the Springton mine, with a steam hoisting plant, dwelling house, etc., and a shaft, 4x8 in the clear, 125 feet deep.

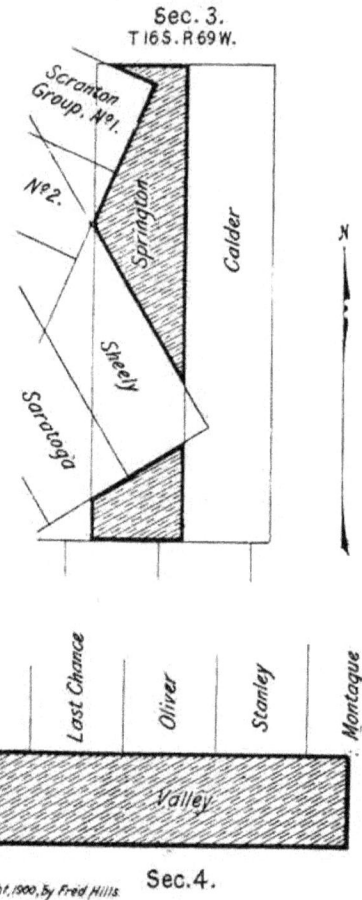

Copyright, 1900, by Fred Hills.

The Hayden Gold Mining Company.

Incorporated November 18, 1895.

Directors   Chas. L. Tutt...........................President
Spencer Penrose....................Vice-President
H. L. Shepherd.............Secretary and Treasurer
George East.

Main Office—No. 267 Bennett avenue, Cripple Creek, Colorado. Transfer Office—International Trust Company, Colorado Springs, Colorado.

Capitalization   5,000,000 shares. Par value, $1.00. In treasury January 1, 1900, 400,000 shares; in treasury January 1, 1900, $2,572.95 cash.

Property   Owns the Royal Arch and Louisa, 4 acres, in the N. W. 1-4 section 30, on Guyot hill; the Robert H., 2.5 acres, in the N. E. 1-4 section 25; Chance of '94, 1 acre, in the N. W. 1-4 section 30, on Beacon hill; Red Jacket Nos. 1, 2, 3, 4 and 5, south of Cripple Creek, 31 acres, in the N. W. 1-4 section 25. All patented. Also owns the Flora Fountain, Josephine, April Fool, Diamond Joe, Coxey No. 2, Robert Emmet and Luella S., south of Victor, in process of patenting; also 125 acres, patented mineral rights, under the Hayden placer and 204 lots in Cripple Creek, in all about 250 acres. The Diamond Joe and the Luella S. are not shown on plat.

Development   Several small plants, belonging to lessees. Seven sets of lessees are working the property. About 1,500 feet of shaft work and 400 feet of tunnel work has been done. The greater part of the development work is being done on the Louisa and the Hayden placer.

Highest price for stock during 1899, 3 cents; lowest price for stock during 1899, $7.00 per M.

242

The Hayden Gold Mining Company.

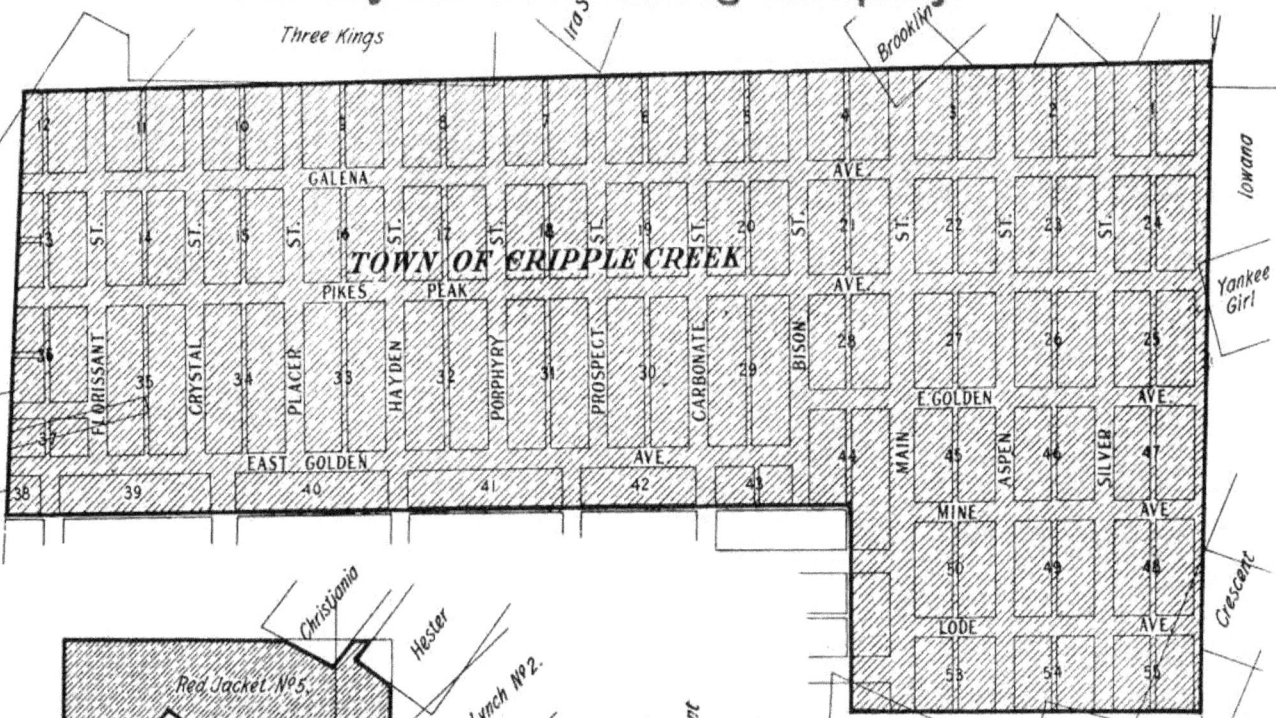

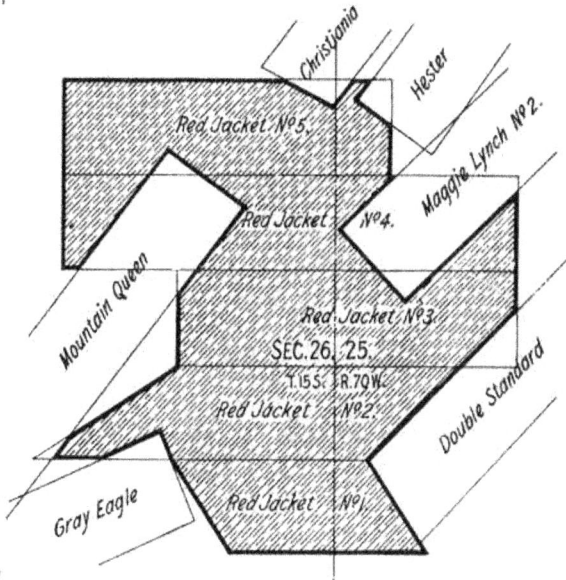

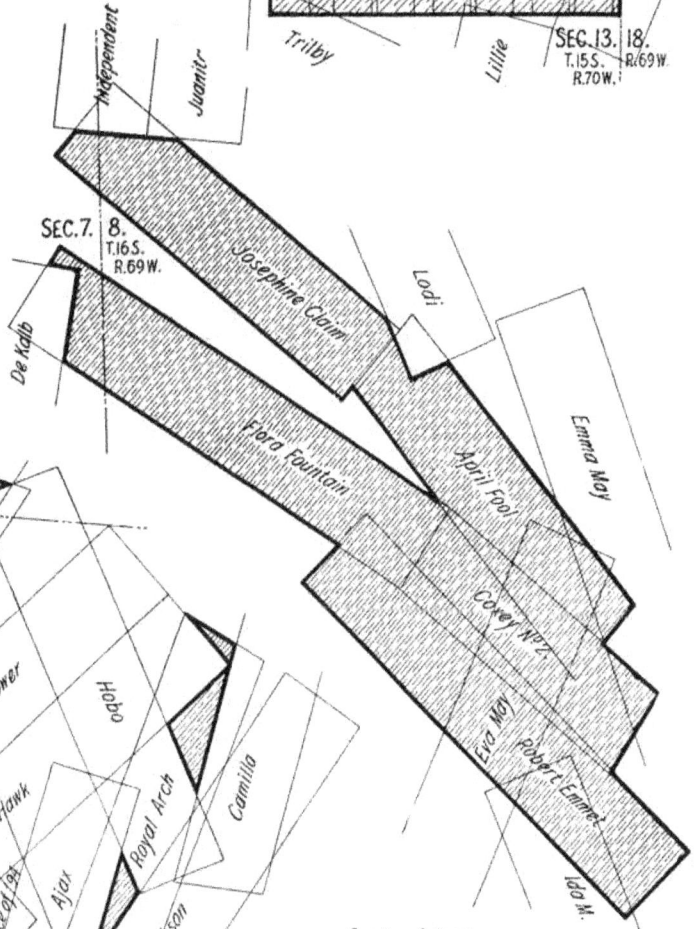

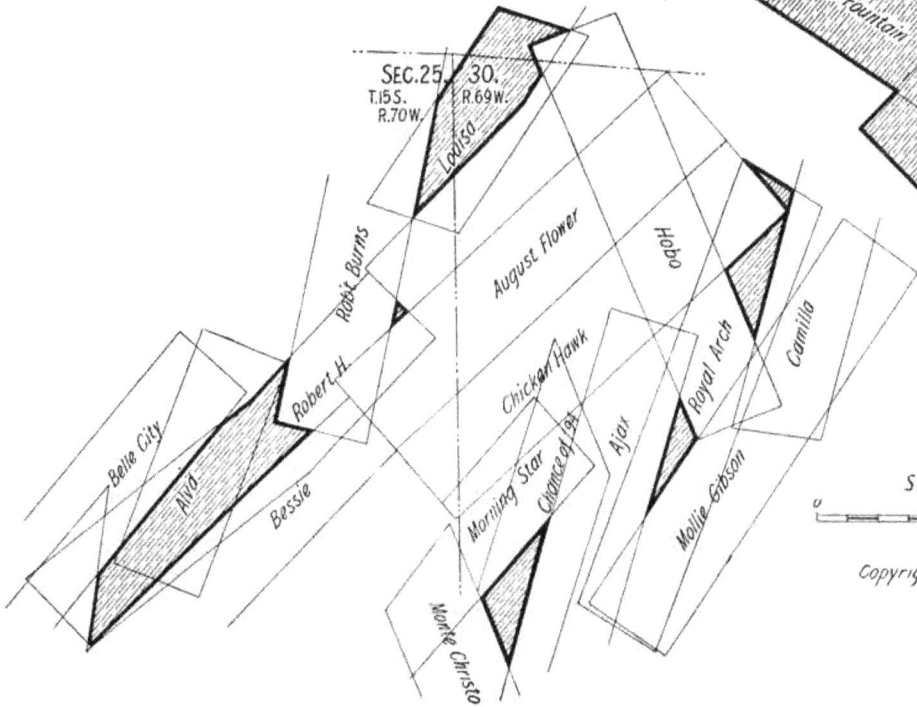

TOWN OF CRIPPLE CREEK

Scale of Feet

Copyright, 1900, by Fred Hills.

The Henrietta Gold Mining Company.

Incorporated 1895.

Directors

J. K. Miller..............................President
S. N. Nye..........................Vice-President
Phil S. Delany.............Secretary and Treasurer
J. W. Miller.   C. E. Titus.
Wm. A. Delany.   F. W. Isham.

Main Office—35 and 37 Hagerman building, Colorado Springs, Colorado.

Capitalization   1,500,000 shares.   Par value, $1.00.

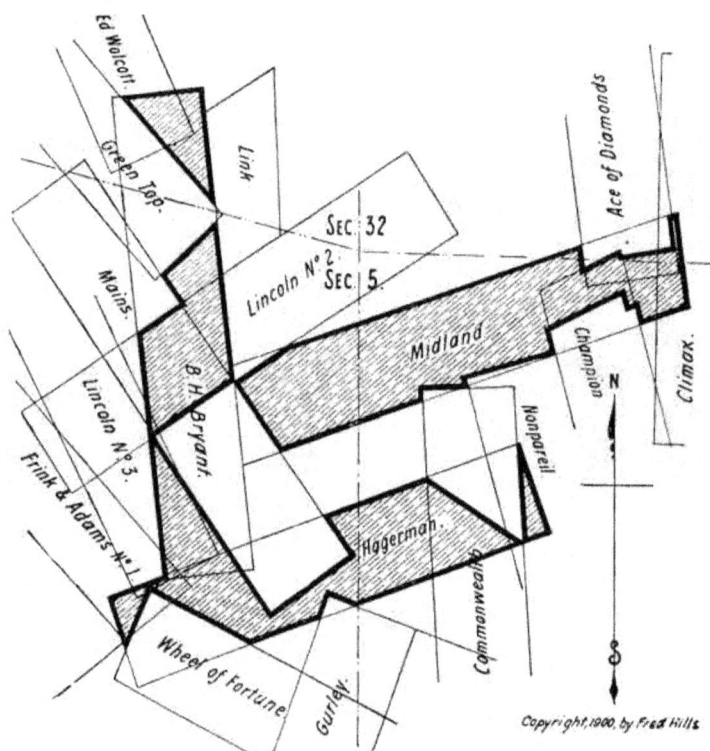

Copyright, 1900, by Fred Hills

Property   Owns the B. H. Bryant, Midland and Hagerman claims, 23 acres, in the N. 1-2 section 5, on the north extension of Tenderfoot hill, 15.5 acres patented and 8 acres held by deed.

Development   A shaft house and gallows frame; one shaft about 65 feet deep, and two or three from 40 to 60 feet.

Highest price paid for stock during 1899, 2 cents; lowest price paid for stock during 1899, $6.00 per M.

244

The Hercules Gold Mining and Milling Company.

Incorporated 1893.

S. S. Hatfield.............................President

R. H. Magee.......................Vice-President

F. M. Keeth...............Secretary and Treasurer

J. J. Grier.   T. G. Horn.

Directors

Main Office—19 Postoffice building, Colorado Springs, Colorado.

500,000 shares.   Par value, $1.00.   In treasury January 1, 1900, Capitalization 60,000 shares; in treasury January 1, 1900, $24.78 cash.

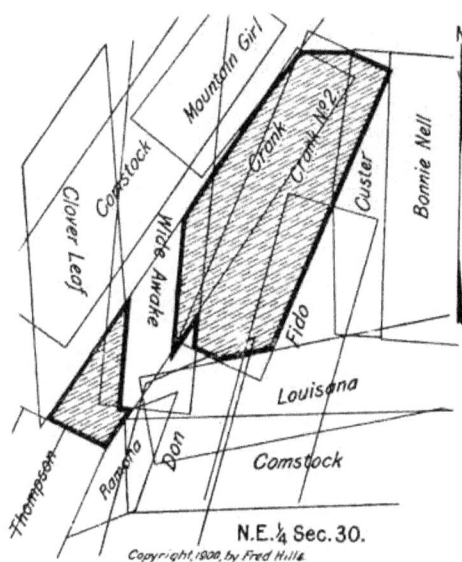

N.E. ¼ Sec. 30.
Copyright, 1900, by Fred Hills.

Property — Owns the Crank and Crank No. 2, containing 10 acres, in the N. E. 1-4 section 30 and S. E. 1-4 section 19, on the south side of Raven hill, between the Elkton and Moose mines.

Development — One good shaft house, one double compartment shaft 300 feet deep, two shafts 60 feet each, tunnel 125 feet, about 125 feet of cross-cutting in the deepest shaft.   Patent survey No. 8,435 is among the first patents on the hill and is in no danger of any litigation; stock almost all held by the directors.   Property well located and considered a good prospect.

Highest price for stock during 1899, 6 cents; lowest price for stock during 1899, 5 cents.

The Hermosa Gold Mining Company.

Incorporated 1895.

Directors

N. S. Gandy, president; Geo. E. Hasey, vice-president; J. W. Wright, secretary; R. H. Widdicombe, treasurer; Geo. Kirby.

Main Office—Hagerman building, Colorado Springs, Colorado.

Capitalization

1,500,000 shares. Par value, $1.00. In treasury January 1, 1900, about 280,000 shares; in treasury January 1, 1900, $500.00 cash.

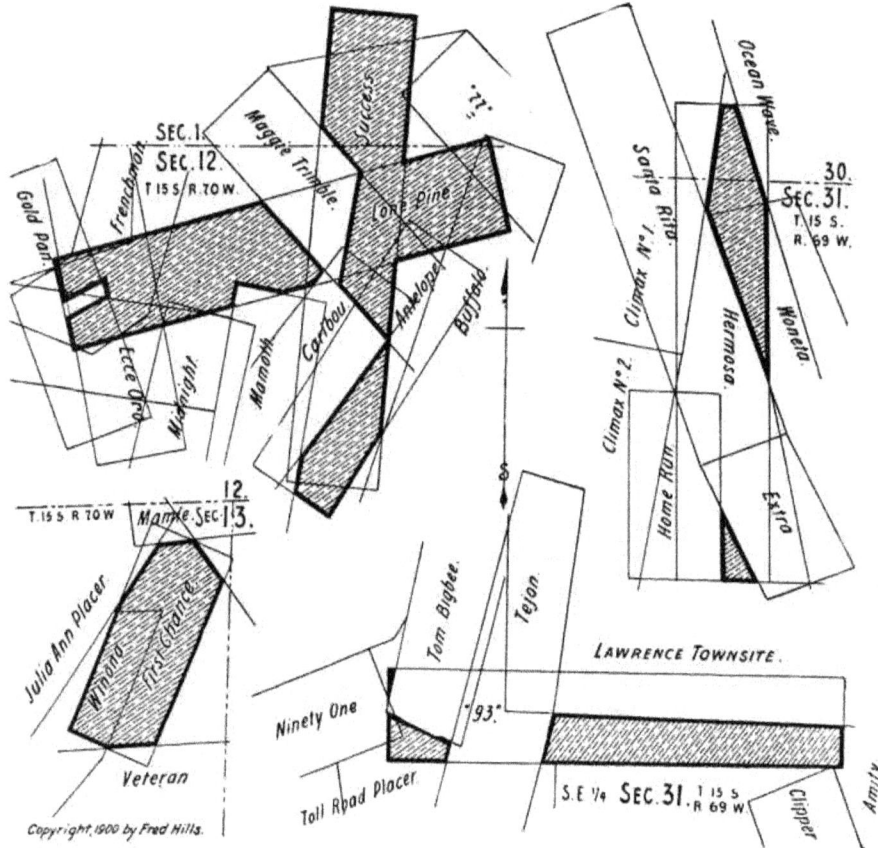

Copyright, 1900 by Fred Hills.

Property

Owns the '93, in the S. E. 1-4 section 31, containing about 4 1-2 acres, on Squaw mountain, adjoining the Tom Bigbee; also the vein rights through the Tom Bigbee and the Tejon; the Hermosa, in the N. E. 1-4 section 31, adjoining the property of the Little Puck Company; the First Chance, in the N. W. 1-4 section 13, containing about 5 acres; the Lone Pine and the Success, in the N. W. 1-4 section 12, containing about 16 acres, between Mineral hill and Copper mountain. Total acreage owned by this company is about 28 acres. The '93, Hermosa and the First Chance are patented. Receiver's receipt held for the Lone Pine and the Success.

History

The First Chance is leased and bonded for $25,000 to the Metropolitan M. & M. Co.; the Hermosa is leased and bonded for $15,000 to the same company; terms of both, 18 months, 15 feet per month, 20 per cent. royalty. This company has recently passed into new hands, and it will be their policy to push active developments.

Highest price for stock during 1899, 4 cents; lowest price for stock during 1899, 2 1-2 cents.

The Hillsdale Gold Mining Company.

Incorporated February 3, 1896.

John E. Lundstrom, president; Reginald H. Parsons, secretary and treasurer; C. F. Schneider, Fred Weeber, Louis W. Oakes. **Directors**

Main Office—Colorado Springs, Colorado.

1,250,000 shares. Par value, $1.00. In treasury January 1, 1900 **Capitalization** 230,000 shares.

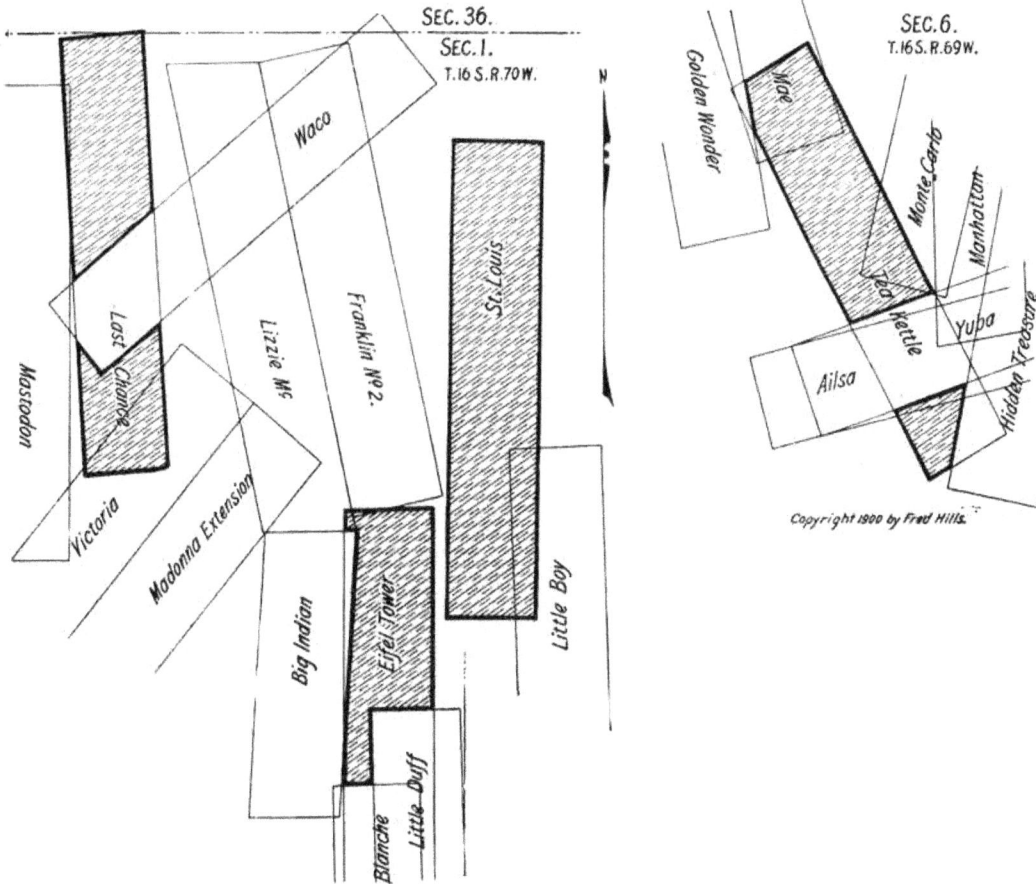

Copyright 1900 by Fred Hills.

Property Owns the Tea Kettle, 6.426 acres, in the N. E. 1-4 section 6, township 16 south, range 69 west, on Straub mountain; the St. Louis, 10.327 acres, in the N. E. 1-4 section 1, township 16 south, range 70 west, on Straub mountain; the Last Chance, 6.656 acres, in the N. 1-2 section 1, township 16 south, range 70 west, on Grouse mountain; and the Eifel Tower, 4.567 acres, in the N. E. 1-4 section 1, township 16 south, range 70 west, on Straub mountain. Receiver's receipts have been received for all the above properties.

Development There are good tunnels and shafts on all the properties. On the Eifel Tower, at the bottom of the shaft, a good vein is disclosed, from which $20 assays have been obtained. On the St. Louis the tunnel has not yet reached the vein. In a new shaft on the Tea Kettle some fine looking ore has been discovered. The Last Chance tunnel has not yet come to any ore, but the indications for doing so are good.

The stock is not on the market.

The Hobart Gold Mining Company.

Incorporated February 17, 1896.

Main Office—121 East Pike's Peak avenue, Colorado Springs, Colorado.

Capitalization

1,500,000 shares. Par value, $1.00. In treasury January 1, 1900, 109,000 shares.

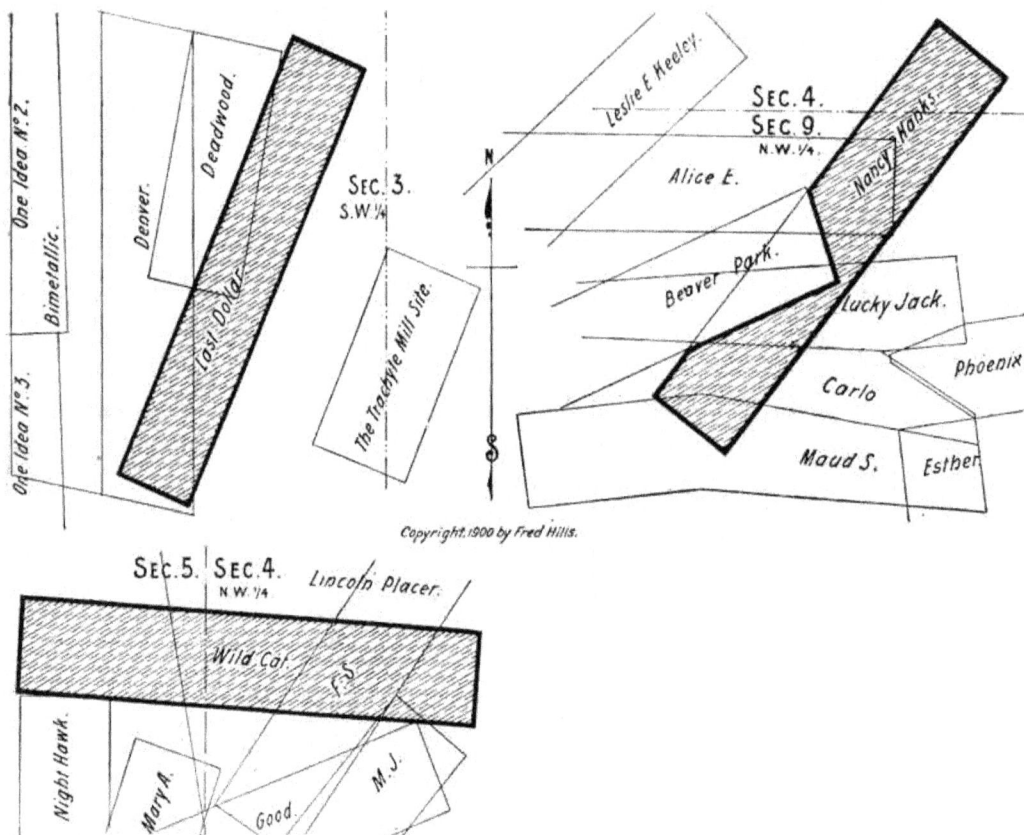

Copyright, 1900 by Fred Hills.

Property

Owns the Last Dollar, 9.03 acres, on Trachyte mountain, in the W. 1-2 section 3; Nancy Hanks, 8.900 acres, in the N. W. 1-4 section 9, on Galena hill, both patented; also the Wild Cat, 10.311 acres, in the N. E. 1-4 section 4, on Lincoln hill, in process of patenting. On April 30, 1900, this company purchased the Alice E. claim, 10 acres, adjoining the Wild Cat, on Lincoln hill, from the Silver State Consolidated Gold Mining Company. For plat of latter see Silver State C. G. M. Co.

Development

Shaft house on the Nancy Hanks claim. This claim has a shaft about 60 feet deep, and has made an excellent showing. The Last Dollar has a tunnel about 200 feet in length, with good showing.

Highest price for stock during 1899, 1 cent; lowest price for stock during 1899, 1 cent.

The Homestake Gold Mining and Milling Company.

Incorporated.

T. G. Horn..............................President

G. W. Logan..........Vice-President and Treasurer

T. Tribe................................Secretary

Chas. E. Richards.

Directors

Main Office—No. 27 North Tejon street, Colorado Springs, Colorado.

1,000,000 shares. Par value, $1.00. In treasury January 1, 1900, 175,000 shares.

Capitalization

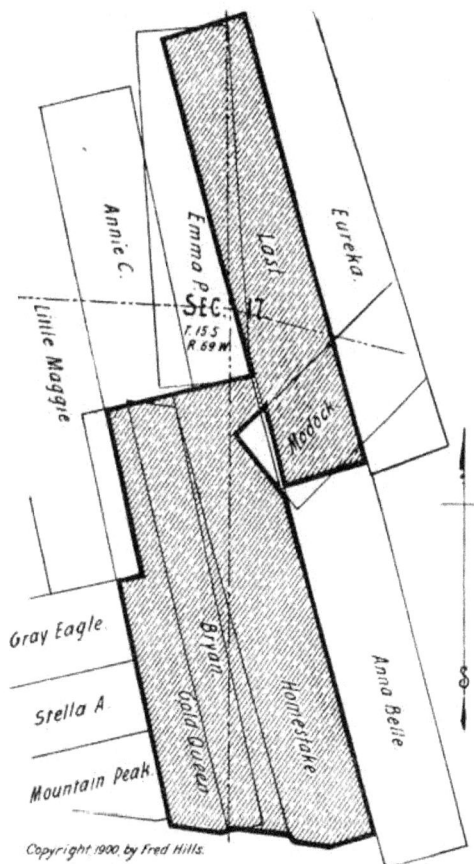

Copyright 1900 by Fred Hills.

Owns the Homestake, 10 acres; the Gold Queen, 8 1-2 acres; the Last, 10 acres; the Bryan, about 2 1-2 acres; all in the N. 1-2 section 17, near Cameron; a total of 28 acres. All patented.

Property

On all the claims there is a total of 10,000 feet of shafts and drifts. The greater part of the development work is being done on the Gold Queen and the Homestake. Nearly all the claims are leased. Good ore has been found around the property and prospects are good, but no shipping ore has as yet been produced. Assays range from $12 to $160 per ton.

Development

Highest price for stock during 1899, 5 cents; lowest price for stock during 1899, 4 cents.

The Hoosier Boy Mining and Tunnel Company.

Incorporated December 17, 1895.

Directors

C. W. Kurie..........................President
Matthew Kennedy..................Vice-President
T. A. Sloane............................Secretary
C. H. White............................Treasurer

Hugh R. Steele.   A. M. Ripley.

M. W. Kurie.

Main Office—No. 9 North Tejon street, Colorado Springs, Colorado.

Capitalization   1,250,000 shares.   Par value, $1.00.   In treasury January 1, 1900, 100,000 shares; in treasury January 1, 1900, about $1,100.00 cash.

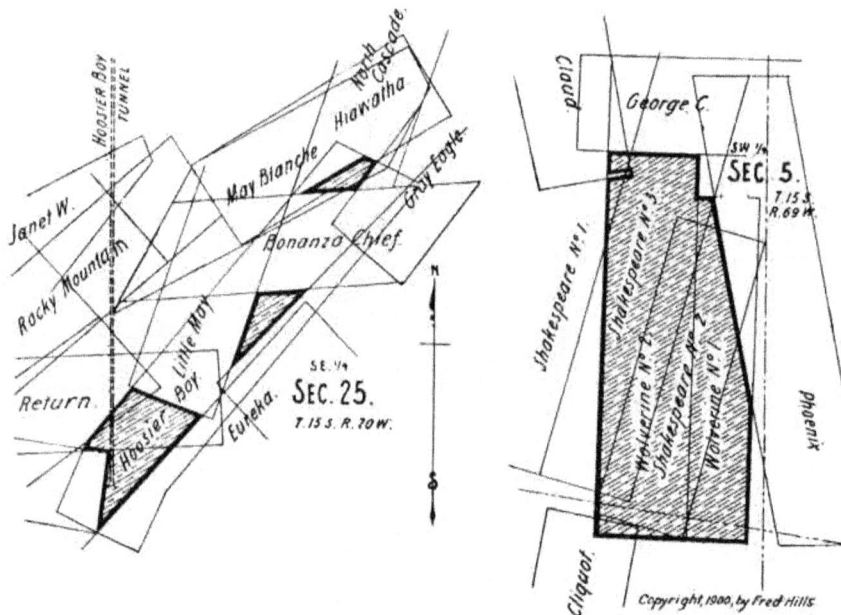

Property   Owns the Hoosier Boy claim, 3 acres, the Hoosier Boy tunnel and tunnel site, in the S. E. 1-4 section 25, on Beacon hill; the Wolverine Nos. 1 and 2, 15 acres, in the S. W. 1-4 section 5, township 15 south, range 69 west, on Tenderfoot hill.   All the above property is patented.

Development   The Hoosier Boy has a tunnel and shafts, the total development work on this claim costing over $7,000.00.   The Wolverine Nos. 1 and 2 have about $2,000 worth of development work.

Highest price for stock during 1899, 3 3-4 cents; lowest price for stock during 1899, 2 cents.

The Humboldt Gold and Silver Mining and Milling Company.

Incorporated July 15, 1892.

W. B. Pullin.........................President
W. R. Foley..........Vice-President and Treasurer
S. J. Mattocks........................Secretary
W. R. Snyder.   William Barber.

Directors

Main Office—104 East Pike's Peak avenue, Colorado Springs, Colorado.

1,000,000 shares. Par value, $1.00. In treasury January 1, 1900, 400,000 shares; in treasury January 1, 1900, $500.00 cash.

Capitalization

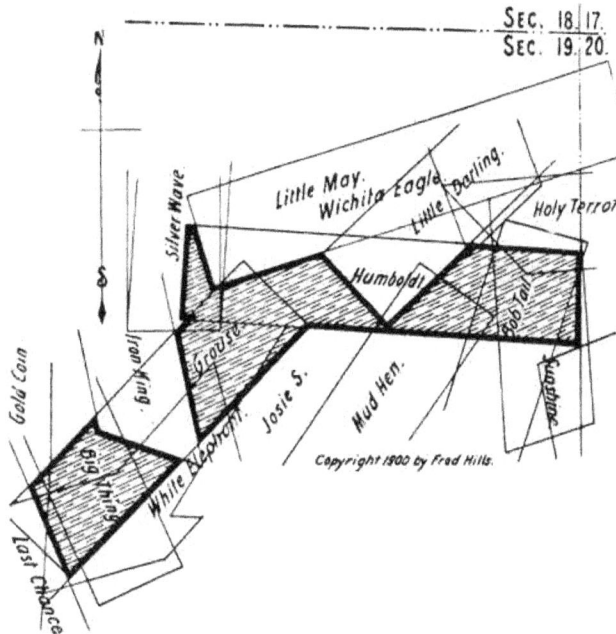

Copyright 1900 by Fred Hills.

Owns Grouse and Humboldt, containing 8 acres, patented, located in the N. E. 1-4 section 19, on Ironclad hill.

Property

A steam hoist and boiler room are on the property, together with considerable improvements. A shaft 200 feet deep has been sunk on the Humboldt and will be continued to a depth of 250 feet, from which point a main drift will be started. The property lies close to and adjoins some of the best properties; has been worked by lessees at different times and a good deal of low grade ore opened up. There has been a good deal of very rich float found, running as high as $17,000.00 a ton, but the main ore shoot has never yet been opened up.

Development

Highest price for stock during 1899, 10 cents; lowest price for stock during 1899, 2 1-2 cents.

The Ida May Gold Mining Company.

Incorporated.

Directors   J. A. Hayes, president; E. P. Shove, vice-president; L. R. Ehrich, secretary; J. G. Shields, treasurer.

Main Office—No. 64 Hagerman building, Colorado Springs, Colorado.

Capitalization   50,000 shares. Par value, $1.00. In treasury January 1, 1900, $10,000 cash.

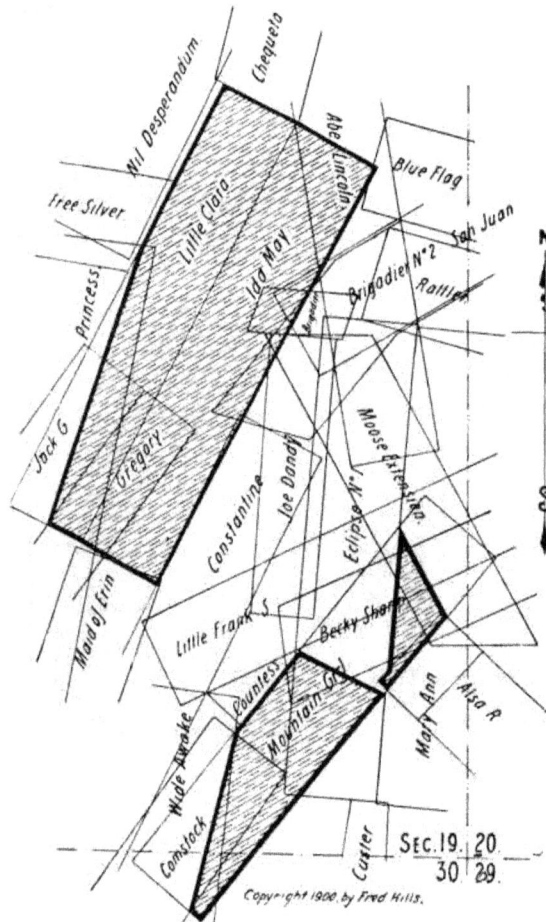

Copyright 1900, by Fred Hills.

Property   Owns the Little Clara, the Ida May, the Mountain Girl, 25 acres in all, situated in the S. E. 1-4 of section 19, on Raven hill. All patented.

Development   The property is leased. One shaft house, 40x100 feet; one 25-horse power hoister; one 25-horse power boiler; whim on 70-foot shaft of the Little Clara. The Little Clara has five shafts, one of 60 feet with an incline of 70 feet, and 110 feet of drifting; one of 85 feet, with 40 feet of drifting; one of 70 feet, one of 40 feet and one of 45 feet, with 20 feet of drifting. The Ida May has three shafts, one 375 feet deep with 800 feet of drifting, one 200 feet deep, with 300 feet of drifting; one 125 feet deep, with 160 feet of drifting; also three other shafts each 60 feet deep. The Mountain Girl has three shafts, one 60 feet deep with 80 feet of drifting, one 45 feet deep with 50 feet of drifting, and one 30 feet deep with 10 feet of drifting.

Production   Gross production to January 1, 1900, over $40,000.

Highest price for stock during 1899, 25 7-8 cents; lowest price for stock during 1899, 24 cents.

The Idlewild Gold Mining Company.

Incorporated January, 1896.

F. E. Robinson...........................President   Directors
Chas. Zobrist.......................Vice-President
James W. Coffey...........Secretary and Treasurer
F. E. Brooks.   L. H. Shafer.

Attorneys for the Company—F. E. Brooks and A. T. Gunnell.

Main Office—Colorado Springs, Colorado.

1,500,000 shares.   Par value, $1.00.   In treasury January 1, 1900, Capitalization
113,000 shares.

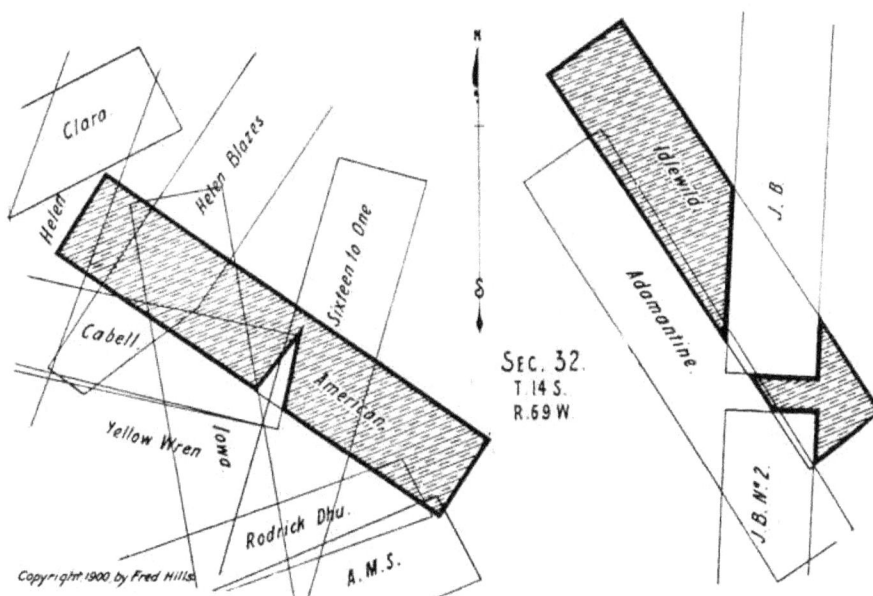

Owns the American, containing 9.906 acres, in the center of Property
section 32, township 14 south, range 69 west; the Idlewild, containing 8
acres, situated in the N. E. 1-4 section 32; also hold by location the Maggie
May, containing 10 acres. All three claims are on Lincoln hill. This last
named claim does not appear on map. The American is patented. Patent
of the Idlewild applied for.

The American has a tunnel of 110 feet. Two shipments of ore have Development
been made, but when the smelters passed a law refusing to receive low
grade ore, shipments were stopped, as the ore ran only about $17.00.
The greater part of the development work is being done on the American.

Highest price for stock during 1899, 2 cents; lowest price for stock
during 1899, 1 cent.

The Illinois Gold Mining and Milling Company.

Incorporated December 5, 1895.

Directors

M. H. Insley..............................President
Kendrick Hughes..................Vice-President
Chas. H. Millar..........................Secretary
H. E. Insley............................Treasurer
John Vories.   George S. Prentice.

Main Office—No. 206 Mining Exchange, Denver, Colorado.

Capitalization

1,250,000 shares. Par value, $1.00. In treasury January 1, 1900, 103,900 shares; in treasury January 1, 1900, $106.00 cash.

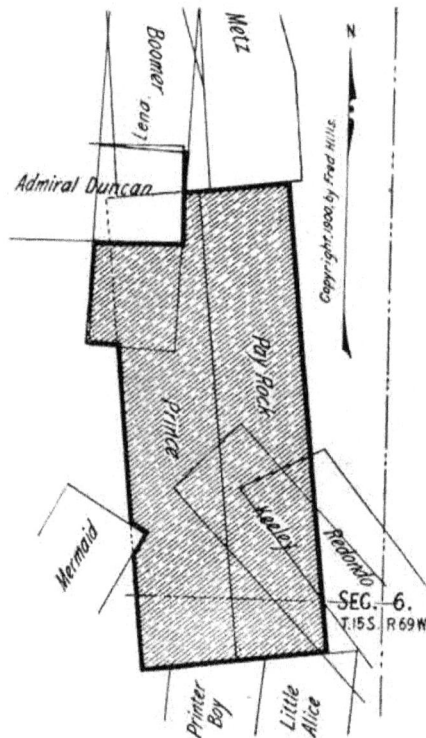

Property

Owns the Pay Rock, 10 1-3 acres; the Pearl, about 3 acres; the Lena, about 2 acres; and the Prince, about 7 acres; all in one group, in the N. W. 1-4 section 6, on the east side of Rhyolite mountain. The Pay Rock and the Lena are patented. The Pearl is held by location and does not show on map.

Development

On the Pay Rock there is a shaft house, 24x62; blacksmith shop, stable, frame dwelling house. The above claim has a shaft 200 feet deep, with 300 feet of cross-cuts. This claim, which is leased, is receiving the greater part of the development work. On the other claims only assessment work has been done.

Highest price for stock during 1899, 2 cents; lowest price for stock during 1899, 1 cent.

Copyright, 1900, by Fred Hills.

The Independence Town and Mining Company.

Incorporated October 30, 1894.

Directors — W. S. Montgomery, president; R. P. Davie, vice - president; James E. Gregg, secretary; A. D. Craigue, treasurer; James F. Burns; James F. Smith.

Main Office — Giddings Block, Colorado Springs, Colorado.

Capitalization — 1,250,000 shares. Par value, $1.00. In treasury January 1, 1900, $65,000.00 cash.

Property — The property owned by the company consists of the Hull City placer, containing 38.894 acres, and the Reindeer lode, containing 0.184 acre in the S. E. 1-4 of section 20, near the Independence townsite. United States patent is held for the Hull City placer property.

Development — The early policy of the company was to develop its ground through lessees. During the year 1898 more than $700,000 was extracted by lessees, the company receiving nearly $200,000 in royalties. In January, 1899, the producing leases were terminated, since which time the company has been operating the property. A three-compartment shaft, 5 feet by 14 feet, has been sunk to a depth of 750 feet, and stations cut at 300 feet, 400 feet, 475 feet, 550 feet, 650 feet and 750 feet. During the year ending December 31, 1899, 2,240 feet of levels were run, 138,847 cubic feet of stoping done, producing about 9,000 tons of ore. During the year two new principal veins were discovered, one to the east, and the other to the west of the old vein known as the "Fox" vein, and both of these veins have been extensively developed, proving fully as strong and as rich as the old vein. The average value of the ores from all three veins is about $65.00 per ton. The production is now from 50 to 100 tons per day. Quite a revenue is also obtained from surface rentals of this property. During the past seven years litigation has been pending before the interior department at Washington, and also in the civil courts of Colorado, affecting the title of this property. In February last a suit was tried at Cripple Creek, wherein the Wilson Creek Consolidated Mining and Milling Company, as plaintiff, had instituted a suit on behalf of the "Minnie Belle" lode mining claim, which suit involved all the questions affecting the title to the property. Judgment was rendered in favor of The Independence Town and Mining Company. An appeal has been taken by The Wilson Creek Company, which is now pending in the Supreme Court of Colorado. The officers of The Independence Company

(CONTINUED ON PAGE 256.)

are confident that the judgment of the District Court will be sustained and that all litigation materially affecting their title to the property is practically disposed of.

Production and Dividend

Production to December 1, 1899, $900,193.00. Dividend No. 1, paid March 1, 1900, $12,500.00; dividend No. 2, paid April 2, 1900, $12,500.00; dividend No. 3, paid May 1, 1900, $12,500.00; dividend No. 4, paid June 1, 1900, $12,500.00.

Highest price for stock during 1899, 65 cents; lowest price for stock during 1899, 47 1-2 cents.

The Ingham Consolidated Gold Mining Company, Limited.

Incorporated October 15, 1895.

Directors

Baron William del Marmol, president; Wm. P. Bonbright, vice-president; F. Scheffler, secretary; F. W. Stehr, treasurer; Theodor Posno, Godefroid Victor Meer, Irving W. Bonbright.

Main Office—Colorado Springs, Colorado. Branch Office—Brussels, Belgium.

Capitalization

1,500,000 shares. Par value, 50 cents. In treasury January 1, 1900, 140,331 shares; in treasury January 1, 1900, $11,635.67 cash.

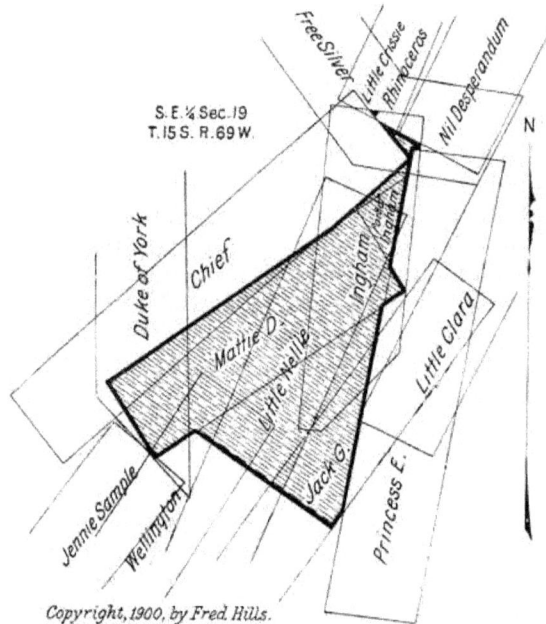

Copyright, 1900, by Fred Hills.

Property

Owns the Ingham, Survey No. 8,410; the Mattie D, Survey No. 7,758, and portions of the Jack G, Survey No. 7,866; the Little Nellie, Survey No. 7,866, and the Wellington, Survey No. 7,866, about 15 acres in all. All in the S. E. 1-4 of section 19, on Raven hill. All patented.

Development

Office, assay office, ore houses, blacksmith and carpenter shop; all at the mouth of the Mattie D. tunnel. There are also several other shaft houses on the Mattie D. and the Ingham claims. Several sets of lessees are working on the Ingham claim. Three sets of lessees are working the Mattie D. claim. On these two claims the greater part of the development work is being done.

Production and Dividend

Gross production to January 1, 1900, about $250,000.00. Net profit to company on ore mined during 1899, $9,479.52.

Highest price for stock during 1899, 15 cents; lowest price for stock during 1899, 7 1-2 cents.

The Inverness Gold Mining Company.

Incorporated 1893.

Angus Gillis..........................President

A. J. Gillis...........Vice-President and Treasurer

Con Schott...........................Secretary

Directors

Main Office—Exchange place, Colorado Springs, Colorado.

300,000 shares.   Par value, $1.00.   Capitalization

Owns the Xenia, comprising 3 3-4 acres, patented, in the N. 1-2 of   Property
section 18.   The property is in Poverty gulch, adjoining Gold King.

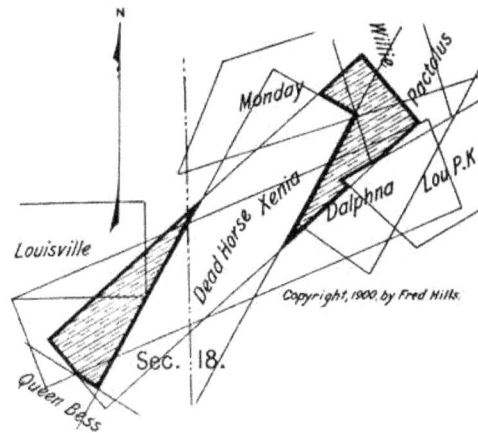

The company have on the property a steam hoist, 500 feet of shaft-   Development
ing and 200 feet of drifts.   On these shafts and drifts $6,000 have been
expended.   The directors consider the indications for a paying mine to
be good.   Lessees are now pushing developments, and, as their success
means an interest in the company, they are doing their utmost to forward the work.   Close corporation.

Highest price for stock during 1899, 15 cents.

The Iowa-Colorado Mining and Milling Company.

Incorporated Under the Laws of Iowa in Colorado, March, 1900.

Directors

Clinton S. Fletcher.........President and Treasurer
H. J. Fletcher......................Vice-President
M. W. Tilney.........................Secretary
D. S. Crain.   E. L. Ernhout.

Main Office—No. 362 Bennett avenue, Cripple Creek, Colorado.

Capitalization

1,500,000 shares.   Par value, $1.00.   In treasury March 31, 1900, 1,077,796 shares; in treasury March 31, 1900, $400.00 cash.

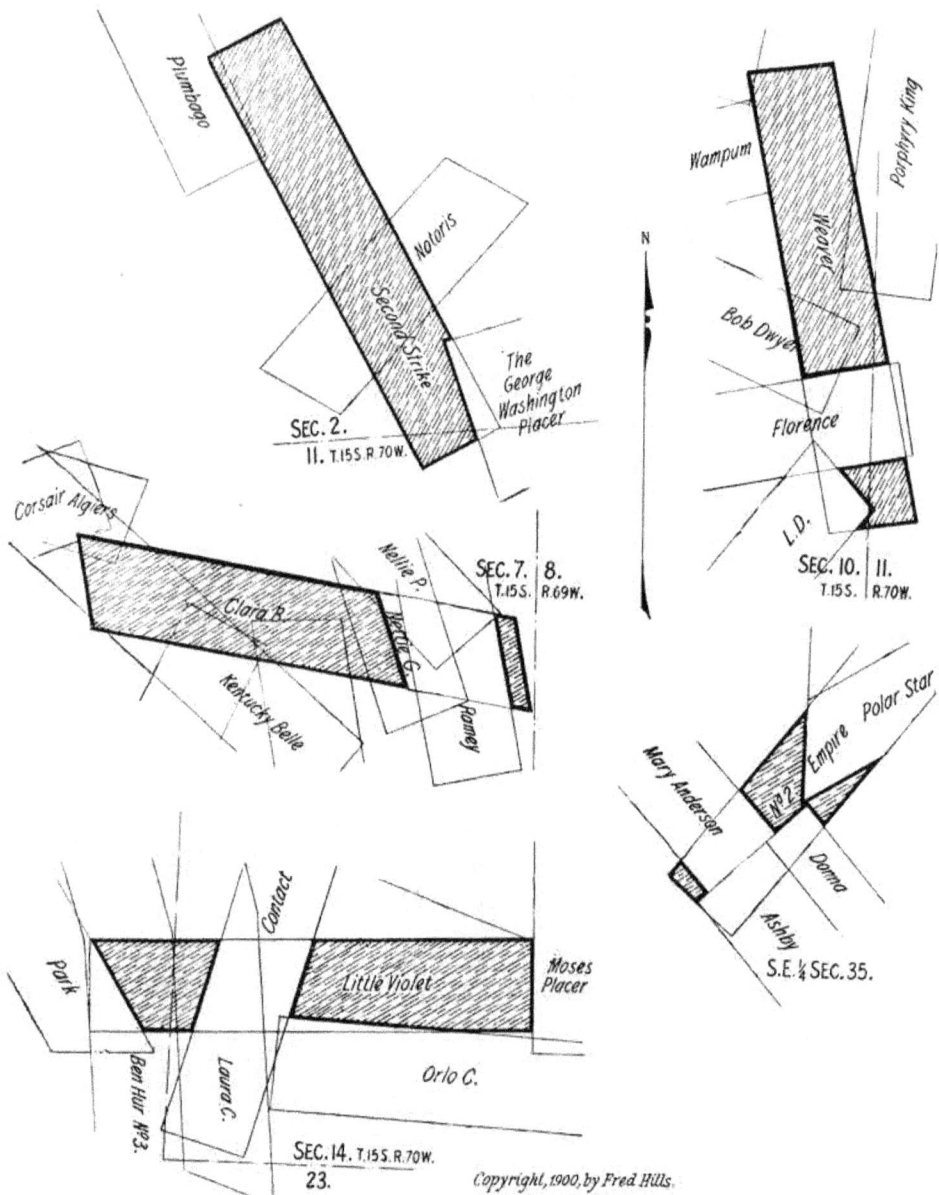

Copyright, 1900, by Fred Hills.

Property

Owns a three-eighths interest in the Little Violet lode, in the S. 1-2 section 14, containing 7.147 acres; a one-half interest in the Second Strike, 9.61 acres, in the S. W. 1-4 section 2, on Red mountain; a one-third interest in the Weaver lode, 7.3 acres, in the N. E. 1-4 section 10, on

(CONTINUED ON PAGE 259.)

Iron mountain; a one-fourth interest in the Clara B., 7.62 acres, in the N. E. 1-4 section 7, on Tenderfoot hill; the whole of the south one-third of the Empire No. 2 lode, 1.7 acres, in the S. E. 1-4 section 35, on Grouse mountain; a one-half interest in the Home Run lode, 7.5 acres, in the S. W. 1-4 section 23, township 14 south, range 69 west. All the above property is patented. The Home Run lode, being outside the Cripple Creek district, is not shown on map. The company also has leases on school lots Nos. 20, 21, 25, 26, 27, 28 and 37, in section 16, township 16 south, range 69 west, a total of 70 acres; also interests in several locations in Park and Chaffee counties.

Sufficient work to obtain a patent has been done. The Clara B. is History leased but not bonded. The policy of the company is to develop these properties by leasing and to use the proceeds of the sale of treasury stock to secure other property while it can be had at present prices.

The Ironclad Mining and Milling Company.

Incorporated.

Joseph H. Smith......................President   Directors
Walter C. Frost...................Vice-President
J. Matthews...............Secretary and Treasurer
     James Blood.   Charles D. Hoyt.

Main Office—Colorado Springs, Colorado.

1,000,000 shares. Par value, $1.00. In treasury January 1, 1900, Capitalization about $500.00 cash.

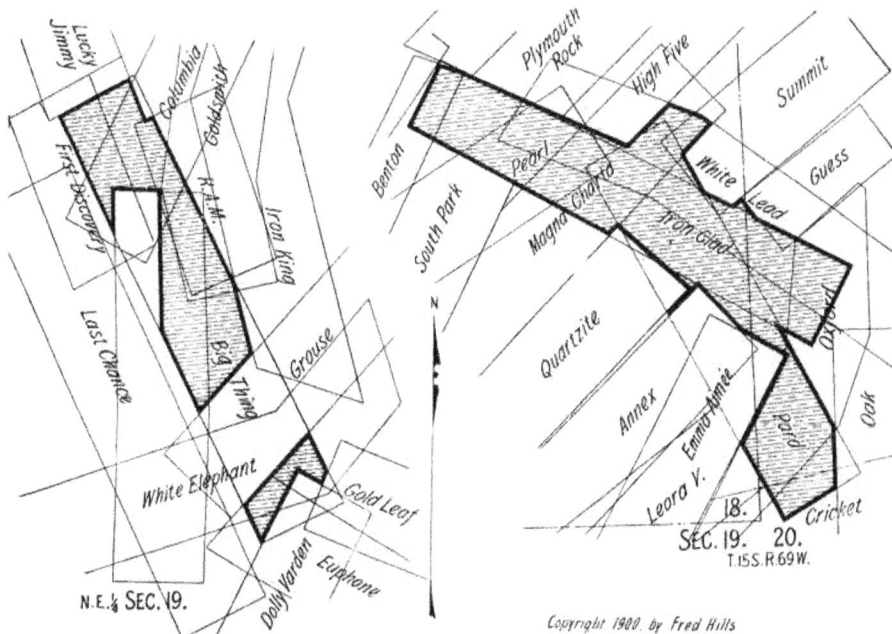

Copyright 1900. by Fred Hills

Owns the Iron Clad, Pard and White Lead, 12 1-2 acres, situated in Property S. E. 1-4 of section 18, on Ironclad hill; the Big Thing, 5 1-2 acres, in N. E. 1-4 section 19, on south slope of Globe hill. Iron Clad and Pard are patented. White Lead and Big Thing in process of patenting.

The Iron Clad claim is leased. On this property the greater part of the development work is being done.

Highest price for stock during 1899, 8 cents; lowest price for stock during 1899, 3 cents.

The Iron Mask Gold Mining Company.

Incorporated November 19, 1895.

Directors — M. D. Swisher, president; B. Whitehead, vice-president; J. M. Harden, secretary and treasurer; E. B. Sprague, R. C. Harden.

Main Office—Colorado Springs, Colorado.

Capitalization — 1,500,000 shares. Par value, $1.00. In treasury January 1, 1900, 350,000 shares.

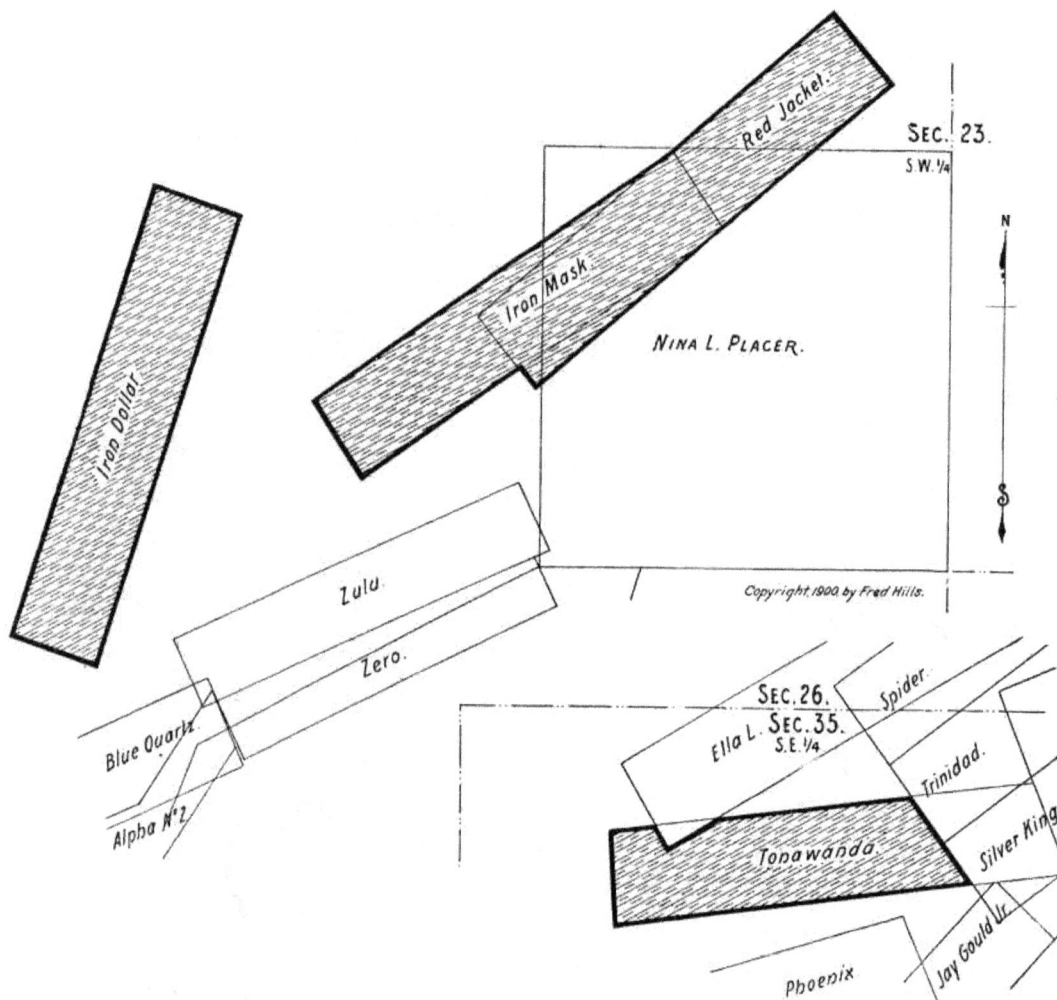

Copyright, 1900, by Fred Hills.

Property — Owns the Iron Dollar, 10 1-3 acres; Iron Mask, 5 acres, and the Red Jacket, 10 acres, in the S. W. 1-4 section 23, township 15 south, range 70 west; also the Tonawanda, 7 acres, in the N. E. 1-4 section 35. The Iron Mask, Iron Dollar and Red Jacket are located one mile southwest of Cripple Creek, while the Tonawanda is on the south slope of Beacon hill. The Iron Dollar and Iron Mask are held by location; balance in process.

Development — 70 feet of shaft work on the Tonawanda, 50 feet on Iron Dollar, 50 feet on Iron Mask, and 40 feet on the Red Jacket. As stated, the Tonawanda is on the south slope of Beacon hill and adjoins the Spider and Trinidad. The deepest shaft is 43 feet and shows a vein 7 feet wide that gives an average assay for the 7 feet of $11.00 to the ton.

The Iron Mountain Mining and Milling Company.

Incorporated February, 1893.

W. J. Murray..........................President   Directors
Jno. V. Swift......................Vice-President
E. N. Bement...............Secretary and Treasurer
H. H. Clark.   F. Warlaumont.

Main Office—Cripple Creek, Colorado.   Branch Office—1420 Logan avenue, Denver, Colorado.

700,000 shares. Par value, $1.00.   Capitalization

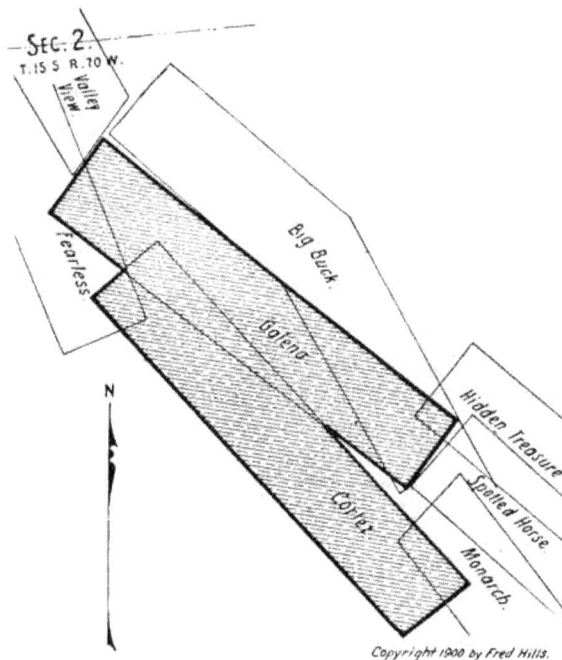

Copyright 1900 by Fred Hills.

Owns the Galena and Cortez, containing about 17.50 acres, in the   Property
S. W. 1-4 of section 2, township 15 south, range 70 west, on Red mountain;
the Hidden Treasure and the Monitor, on Iron mountain; the Little Wonder, on Little Red mountain. The Galena and the Cortez are patented.
Remainder held by location and are not shown in plat.

The Galena, on which the greater part of the development work is   Development
being done, has a shaft 155 feet deep, a tunnel of about 1,200 feet, and
from 300 to 500 feet of cross-cuts. On this claim there is an engine house
and ore house. The Cortez has two shafts, each 50 feet deep. The Monitor has three 10-foot shafts and a 20-foot open cut. The Hidden Treasure
has a 40-foot tunnel and a 30-foot shaft. The Little Wonder has two
shafts, one 30 feet deep, and one 20 feet deep.

Gross production to January 1, 1900, about $6,000.00. Average   Production
assays of ore mined during the past year ran from 2 to 8 ounces gold per
ton. No quotations on stock.

The Isabella Gold Mining Company.

Scale of Feet

0   500   1000

Copyright, 1900, by Fred Hills

The Isabella Gold Mining Company.

Incorporated 1892.

Nelson B. Williams, president; F. H. Morley, vice-president; J. F. Sanger, secretary; W. T. Doubt, treasurer; Geo. D. Kilborn, J. A. Hayes, W. S. Jackson. *Directors*

Main Office—P. O. building, Colorado Springs, Colorado.

2,250,000 shares. Par value, $1.00. In treasury January 1, 1900, *Capitalization* $476,796.06 cash.

Owns the Buena Vista, Lee, Smuggler, Cheyenne No. 1, Cheyenne *Property* No. 2, Bully, Tom Thumb, Snow Bird, Aspen, Jack Rabbit Nos. 1, 2, 3 and 4, Hope No. 2, Bouncer, Old Hickory, Valeria, Two J's, Hobo and Comet, about 160 acres in all, situated on Bull hill, in the N. E. 1-4 of section 20, and the N. W. 1-4 of section 21. All patented. A few small fractions held by location. Some leases under operation all the time.

The operations of this company are confined to the Lee shaft; all *Development* other shafts have been abandoned with a view to the erection, at once, of a shaft house at the north end of the block similar to that at the south end, the Lee shaft being situated at the southern end of the property. This latter shaft house is very commodious and is equipped to do 18,000 feet of development work per annum, which is three times the capacity of the plant it superseded. During the last year there has been added to the property a large engine room, 35x36; boiler room extended 36x36; large addition to the compressor room; assay office 15x34; ore house with 16 screens complete and another cage placed in the Lee shaft, besides the following machinery: 20x32 hoisting engine, 22x24 Norwalk air compressor, 100-light electric plant, two 100-horse power boilers, two 150-horse power boilers. During the same period the Lee shaft has been sunk 200 feet; drifts have been driven 6,110 feet; cross-cuts 1,759 feet; raises 1,233 feet. During the year 1899 the following machinery was placed underground on the property: A 15x23 and a 39x10x24 Duplex Triple Expansion and condensing pumping engine, capacity 1,000 gallons per minute, with 1,000 feet head.

This company was incorporated in 1892 by the consolidation of sev- *History* eral companies owning various tracts of land, and as that year was the 400th anniversary of the visit of Christopher Columbus to these shores, the new company was named "Isabella" in honor of the queen of Spain, the reigning queen at the time of the discovery of this continent. The new management of this company has been very successful in the saving expense attached to this mine, as can be shown by the following: The number of feet of development in 1897 was 7,138, and the expense for the same year was $258,211.90; in 1898, under the present management, the number of feet of development work was 9,134, and the expenses were $215,308.47, showing a saving in one year of $42,903.43, besides doing 2,000 feet more development than in the preceding year.

Production during 1892 and 1893, $80,944.88; 1894, $131,752.14; *Production and Dividend* 1895, $361,905.12; 1896, $568,040.65; 1897, $541,076.80; 1898, $565,279.49; 1899, $968,011.24; total, $3,217,010.32. Dividends paid from 1892 to 1897, $270,000.00. On February 1, 1898, the present management took charge and in 1899 paid $270,000.00 in dividends, besides having a large surplus in the bank, and no debts. A dividend was paid December 16, 1899, amounting to $67,500.00. Last dividend, paid March 24, 1900, amounted to $67,500.00. Total dividends to date, $607,500.00.

Highest price for stock during 1899, $1.51 per share; lowest price for stock during 1899, 38 1-8 cents.

The Jack Pot Mining Company.

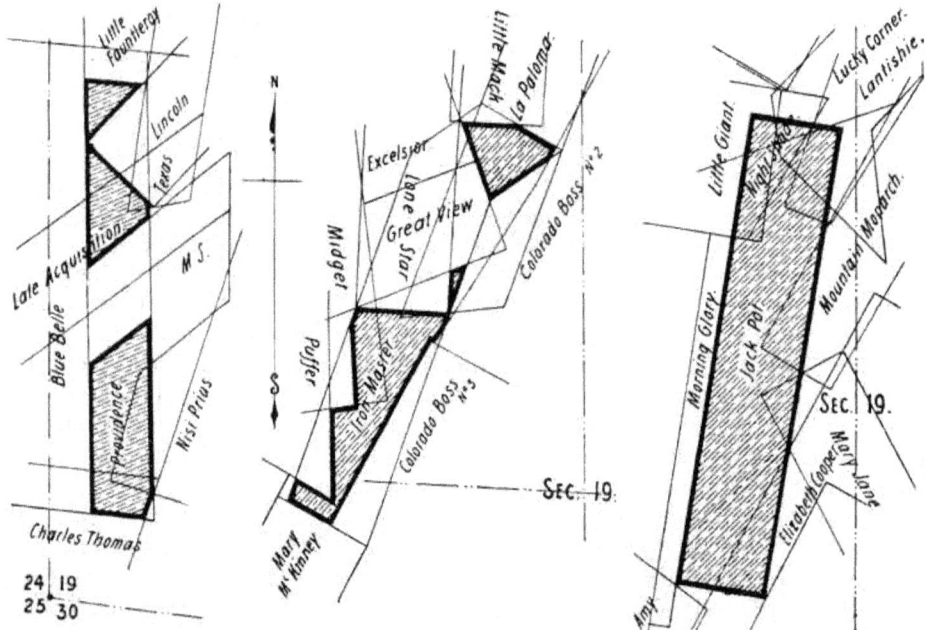

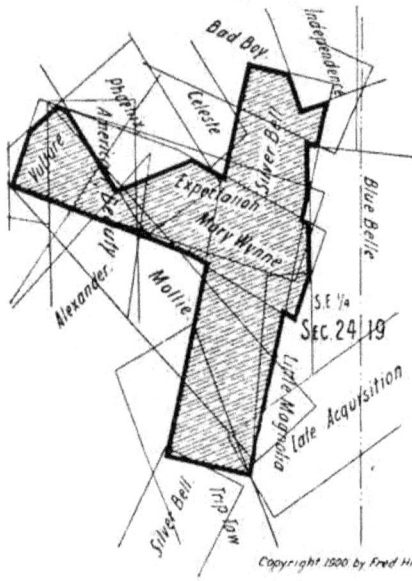

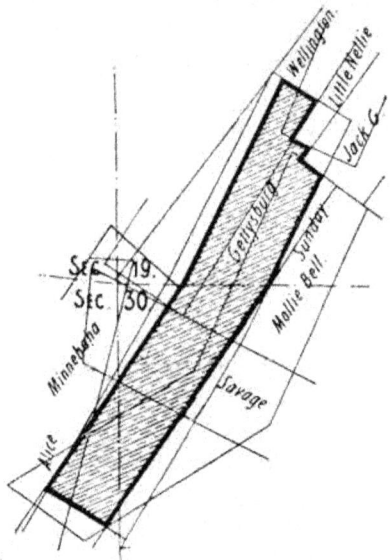

Copyright 1900 by Fred Hills.

The Jack Pot Mining Company..

Incorporated April 30, 1892.

H. E. Woods......................President   Directors

J. B. Cunningham..................Vice-President

Louis R. Ehrich......................Secretary

F. M. Woods.........................Treasurer

Samuel G. Porter.   J. W. Davenport.

Main Office—No. 64 Hagerman building, Colorado Springs, Colorado.

1,250,000 shares.   Par value, $1.00.   In treasury January 1, 1900,   Capitalization
$55,000.00 cash.

Owns the Jack Pot, containing 10 acres; the Gettysburg, containing   Property
7 acres, both situated in the S. W. 1-4 of section 19; the Ironmaster, containing 5 acres, situated in the N. W. 1-4 of section 19; the Providence, containing 5 acres, situated in the S. W. 1-4 of section 19; the Mary Wynne and the Silver Bell, 14 acres in all, situated in the S. E. 1-4 of section 24. All patented and situated on Gold and Raven hills.

There are two engine houses, two ore houses and several plants of   Development
machinery on the Jack Pot. A number of leases are in operation. The greater part of work has been done on the Jack Pot.

Total dividends amount to $150,000.00, amounting to 12 cents per   Production and Dividend
share. Last dividend, paid December 23, 1899, one of 6 cents, was for $75,000.00.

Highest price for stock during 1899, 74 cents; lowest price for stock during 1899, 37 1-8 cents.

The Jefferson Mining Company.

Incorporated May, 1892.

Directors

George McLean.........................President

William Barth.................. ....Vice-President

W. J. Chamberlain..........Secretary and Treasurer

R. S. Morrison..........................Attorney

W. N. Gourlay.   Fred Walsen.

Geo. L. Goulding.   Herbert Gardner.

Main Office—No. 114 Boston building, Denver, Colorado.

Capitalization

1,100,000 shares. Par value, $1.00. In treasury January 1, 1900, $2,500.00 cash.

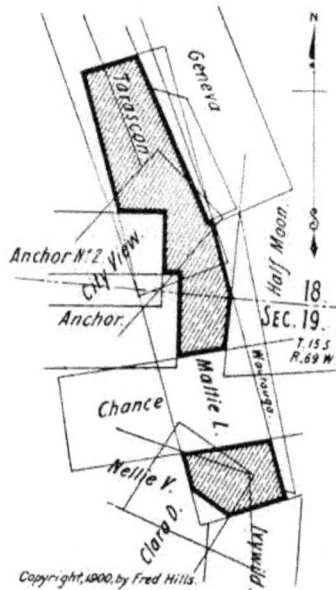

Property

Owns the Mattie L., survey No. 8,910, containing 7 acres, situated in the N. W. 1-4 section 19, on Gold hill. Patented.

Development

Shaft house. There is a shaft 800 feet deep and about 2,000 feet of drifting, etc. At present the property is being worked by three sets of lessees.

Highest price for stock during 1899, 12 cents; lowest price for stock during 1899, 6 cents.

The Jerry Johnson Mining Company.

Incorporated September 1, 1899.

A. P. Mackey, president; Dr. J. W. Graham, vice-president; Chas. F. Potter, secretary and treasurer; Wm. A. Otis, A. B. Whitmore. *Directors*

Main Office—No. 619 Ernest & Cranmer building, Denver, Colorado. Branch Office—Wm. A. Otis & Co., Colorado Springs, Colorado.

1,500,000 shares. Par value, $1.00. In treasury January 1, 1900, *Capitalization* $13,000.00 cash.

Owns thirteen-sixteenths of the Jerry Johnson Nos. 1 and 2, seven- *Property* ninths of the Arapahoe, and the Little Pedro, containing 19 acres, situated in the S. W. 1-4 section 17, on Ironclad hill. The Jerry Johnson Nos. 1 and 2 and the Arapahoe are patented. Receiver's receipt held for the Little Pedro.

There are 14 sets of lessees at work on the property. There are 6 *Development* plants of machinery; the other shafts are worked by whims. Over $20,000 worth of work has been done on the property.

Gross production to January 1, 1900, $100,000.00. *Production*

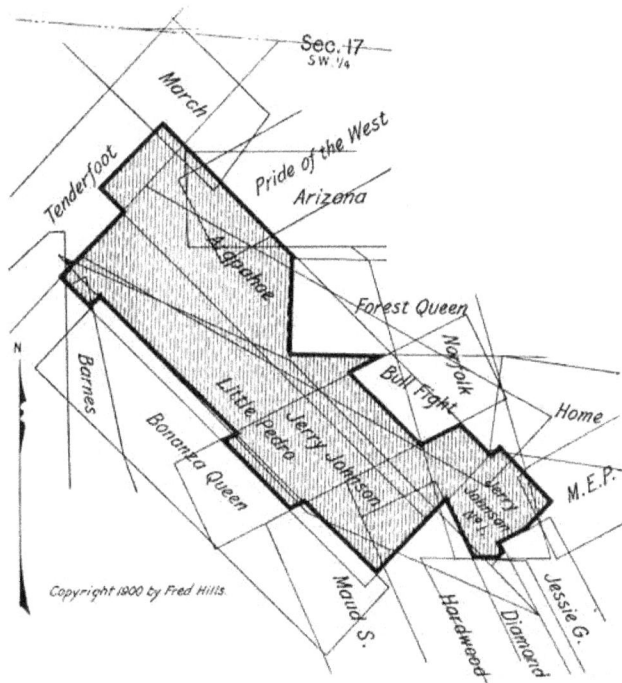

Copyright 1900 by Fred Hills.

This company claims title to large and valuable ore bodies under the *History* Damon ground, by virtue of apex rights, and litigation is now pending in the District Court of Teller county to determine the ownership of said ore bodies. These ore bodies were reached by following the vein down from the Jerry Johnson territory. This litigation relates only to the apex rights claimed by virtue of the Jerry Johnson No. 1 vein, and affects only the ore bodies under the Hardwood and Diamond claims, lying to the west of the Jerry Johnson No. 1 property. The Arapahoe lode, Jerry Johnson No. 2 lode and Little Pedro lode claims are not in any manner affected by this litigation. The Jerry Johnson No. 1 vein, as well as several other known veins, intersect and extend through the properties last named.

Highest price for stock during 1899, 38 1-2 cents; lowest price for stock during 1899, 25 cents.

The Johanna Gold Mining Company.

Incorporated 1895.

Directors

C. E. Palmer..................President
W. J. Cox...............Vice-President
E. L. Ogden.......Secretary and Treasurer
   F. M. Taylor.   J. B. Glasser.

Main Office—Aspen, Colorado.

Capitalization

1,000,000 shares. Par value, $1.00. In treasury January 1, 1900, 370,157 shares; in treasury January 1, 1900, $150.00 cash.

Property

Owns the Ogden, Little Dan, and the Fraction; also the Baltic and the Black Iron claims, containing about 20 1-2 acres, all patented, in section 21, on the south slope of Bull hill.

Development

There is a shaft house, gallows frame, 6x8 friction hoist, with a 20 H. P. vertical boiler. A 300-foot shaft is contracted for, and is now down 120 feet (4 1-2x7 ft.). The Ogden has a 65-foot shaft, with 200 feet of cross-cuts. A 35-foot shaft is on the Fraction, with 125 feet of cross-cuts. The Ogden has also a 160-foot shaft, with 25 feet of drifting. There are also several shallow shafts, open cuts, etc.

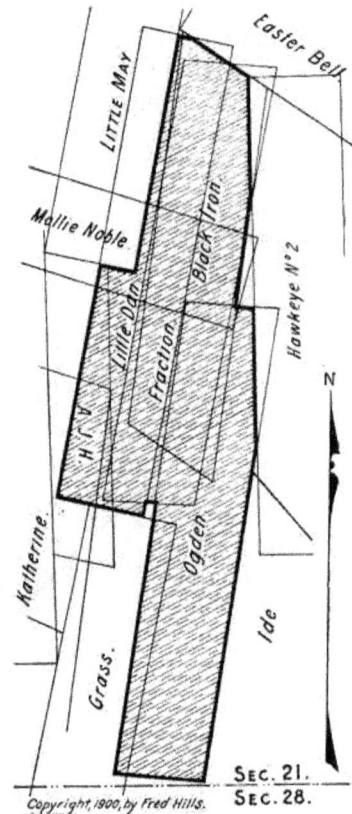

This company is the same as the Bull Hill Gold Mining Company, the name having been changed to avoid confusion with that of other corporations.

As the stock is held largely by the directors and their friends, it is not dealt in on the stock exchanges, so prices can not be quoted.

The Josephine Gold Mining Company.

Incorporated December 7, 1895.

H. H. Dorsey..........................President   Directors

J. A. Sill...........................Vice-President

F. H. Dunnington......................Secretary

E. R. Whitmarsh.......................Treasurer

John Pedersen.   O. R. Dunnington.

W. N. Gibson.

Main Office—2 N. Nevada avenue, Colorado Springs, Colorado.

1,250,000 shares.   Par value, $1.00.   In treasury January 1, 1900,   Capitalization
1,000 shares.

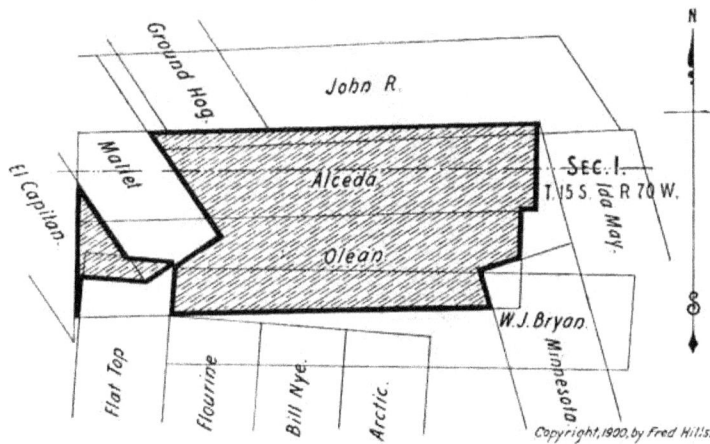

Owns the Oleon and Alceda, 17 acres, in the W. 1-2 of section 1, on   Property
Copper mountain, near the Anna C. of the Copper Mountain Company.
Patented.   There are several shafts on the property, the deepest being 85
feet deep, with 200 feet of drifting; in all, over 500 feet of shafts and 275
feet of drifting, costing about $4,500.00.   Several veins have been found
and systematically developed, but no value above $8.00 has ever been dis-
covered.

Highest price for stock during 1899, 5 cents per share; lowest price
for stock during 1899, $8.00 per thousand.

The Kaffirs Gold Mining Company.

Incorporated 1895.

Directors

T. C. Delany..............................President

H. A. Young.......................Vice-President

Phil. S. Delany............Secretary and Treasurer

Main Office—35 and 37 Hagerman building, Colorado Springs, Colorado.

Capitalization

1,250,000 shares. Par value, $1.00. In treasury January 1, 1900, 70,000 shares; in treasury January 1, 1900, $300.00 cash.

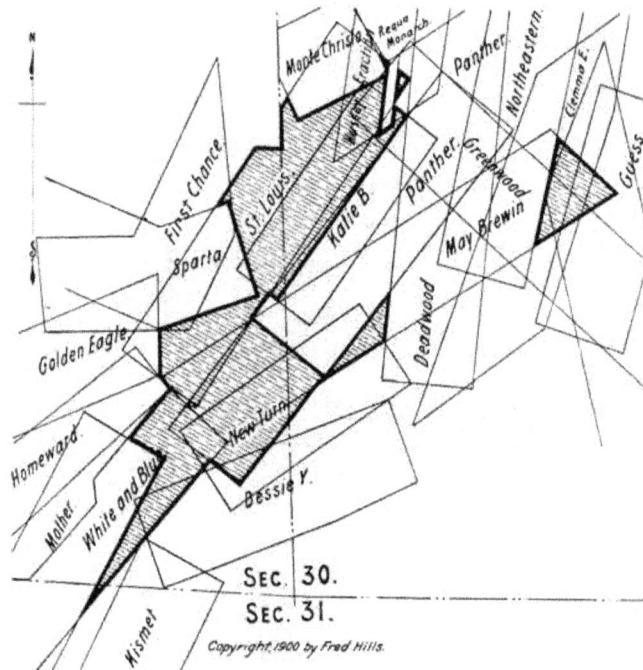

Copyright, 1900 by Fred Hills.

Property

Owns White & Blue, May Brewin, St. Louis and Katie B., containing 12 1-2 acres, all patented, situated in section 30, on Squaw mountain. On March 17, 1900, this company purchased the New Turn claim, a fraction on Squaw mountain, adjoining the May Brewin, for $2,250.

Development

The company have expended about $5,000.00 in the development of the property on various shafts, and the work will be pushed vigorously.

Highest price for stock during 1899, 6 1-4 cents; lowest price for stock during 1899, $7.00 per thousand.

270

The Katinka Gold Mining Company.

Incorporated August 27, 1896.

William Lennox.........................President   Directors

E. W. Giddings, Jr.....Vice-President and Treasurer

N. H. Partridge........................Secretary

J. S. Jones.   W. P. Sargeant.

Main Office—10, 11 and 12 Giddings block, Colorado Springs, Colorado.

1,250,000 shares. Par value, $1.00. In treasury March 15, 1900, **Capitalization**
220,000 shares; in treasury March 15, 1900, $6,000.00 cash.

Copyright, 1900, by Fred Hills.

Owns the Katinka, August Flower, Chicken Hawk and Hobo, in the **Property**
N. W. 1-4 of section 30, about 26 acres, on Guyot hill.

Shaft houses, ore bins and steam hoist on the property. There is **Development**
one shaft 500 feet deep and several other shallow shafts.

Production to January 1, 1900, $5,000.00.   **Production**

Highest price for stock during 1899, 22 cents; lowest price for stock
during 1899, 20 cents.

The Kentucky Belle Gold Mining Company.

Incorporated January, 1896.

Directors

N. E. Guyot.................President
Chas. E. Cherrington.....Vice-President
Jno. F. Bishop..Secretary and Treasurer
Wm. Barber.   Jos. E. Collier.
T. H. Devine.   R. J. W. Payne.

Main Office—217 1-2 S. Union Avenue, Pueblo, Colo.

Capitalization

1,500,000 shares. Par value, $1.00. In treasury January 1, 1900, 132,500 shares; in treasury January 1, 1900, $200.00 cash.

Property

Owns the Kentucky Belle, containing 6 acres. Patented. Situated in S. W. 1-4 section 20, on the east slope of Raven hill.

Development

There are two shaft houses and a steam hoist on the property.

The south six hundred feet of the claim is leased, and it is the policy of the company to develop the property in this way.

Production

Production to January 1, 1900, $200.00.

Highest price for stock during 1899, 9 cents; lowest price for stock during 1899, 1 1-2 cents.

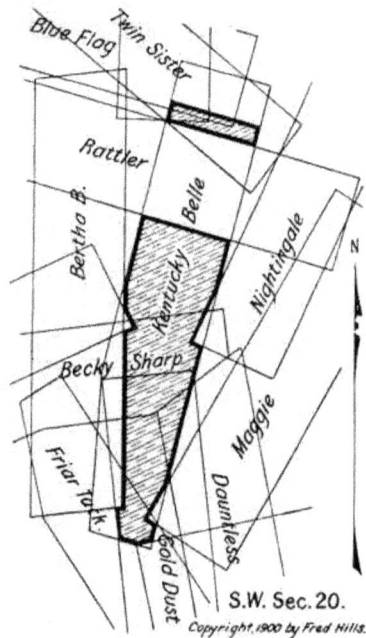

272

STRATTON'S INDEPENDENCE MINE.

SHOWING TOWN OF VICTOR.

The Keystone Mining and Milling Company.

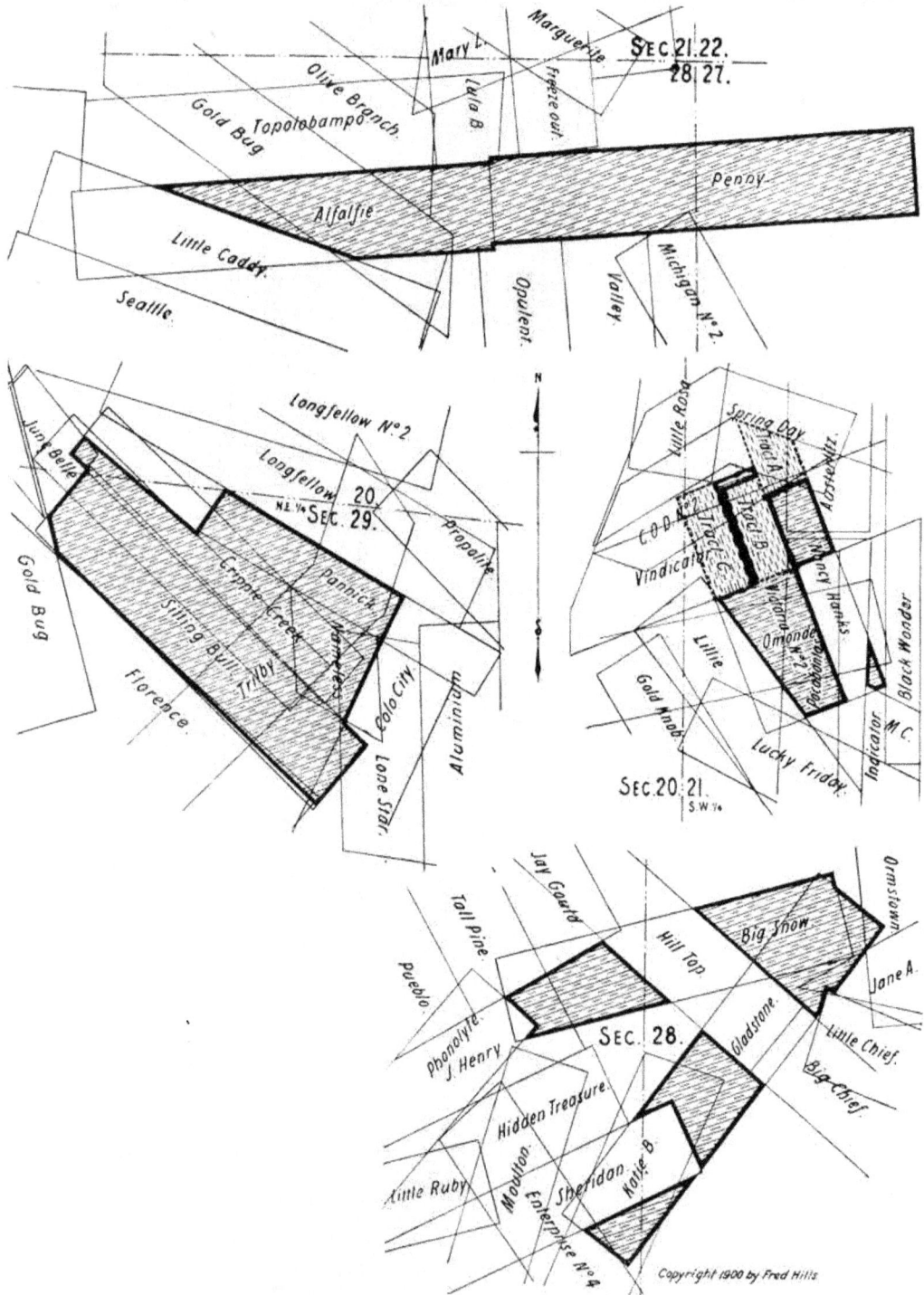

Copyright 1900 by Fred Hills

The Keystone Mining and Milling Company.

Incorporated 1894.

J. W. Graham.........................President

D. P. Cathcart....................Vice-President

J. R. Talpey...............Secretary and Treasurer

Robert Davis.   H. L. Voss.

Directors

Main Office—Room No. 8, Bank building, Colorado Springs, Colorado.

1,500,000 shares. Par value, $1.00. In treasury January 1, 1900, 21,000 shares.

Capitalization

Owns the Sitting Bull, the Cripple Creek and the Pannick, containing in all 18 acres, situated in the N. E. 1-4 section 29, on Bull hill; the Victoria No. 2 and the Nancy Hanks, containing 6 acres, in the S. W. 1-4 section 21, on Bull hill; the Alfalfie and the Penny, in the N. E. 1-4 section 28, containing 16 1-2 acres, on the saddle, southeast of the Victor mine; also the Big Snow and the Gladstone, in the S. 1-2 section 28, containing about 9 acres, on Big Bull mountain. All the above property is patented. The ownership of the tracts A. and B. is in controversy between this company and the C. O. D. Mining Company; this ground is shown in dotted lines. Tract C. is also in controversy between this company and the Vindicator Cons. G. M. Co.

Property

On the Sitting Bull is a double compartment shaft, 250 feet deep, with buildings and machinery. This is under lease. This shaft has already produced about $13,000.00. The Victoria No. 2 has a shaft about 125 feet deep. The Victoria No. 2 and the Nancy Hanks are situated beside the Vindicator and the Lillie mines, two of the big producers of the camp; consequently this portion of their holdings is considered very valuable.

Development

Highest price for stock during 1899, 21 cents; lowest price for stock during 1899, 10 cents.

The Key West Gold Mining Company.

Incorporated July 10, 1899.

Directors

J. R. McKinnie......................President
E. D. Marr.........................Vice-President
R. P. Davie................Secretary and Treasurer
J. L. Lindsay.   Theo. P. Day.

Main Office—25 East Pike's Peak avenue, Colorado Springs, Colorado.

Capitalization

1,250,000 shares. Par value, $1.00. In treasury January 1, 1900, 100,000 shares.

Property

Owns the Key West, 7 1-2 acres, in the S. E. 1-4 section 13, on Gold hill; also the May Queen and a fraction of the Fairview, 3 1-2 acres, in the S. W. 1-4 section 18, on the southwest slope of Globe hill. All the above property is patented. In April, 1900, the east end of the Key West claim, containing 1.52 acres, and adjoining the Abe Lincoln, was sold to Mr. W. S. Stratton, of Colorado Springs. This portion can be seen by referring under "S" to "Stratton's Group." Just as this Manual goes to press the company has purchased the Nest Egg and the Consolidated Virgina claims, in the N. E. 1-4 section 13, on Mineral hill, for $11,250 cash, from the Southern Boy Gold Mining Company. (See Southern Boy.)

The greater part of the property is being worked under lease.

It is surrounded by good producing mines, whose directors seem to think Key West stock a good investment.

Highest price for stock during 1899, 6 1-8 cents; lowest price for stock during 1899, 3 1-2 cents.

The Kimberly Gold Mining Company.

Incorporated 1895.

Phil. S. Delany.........................President   Directors

Wm. A. Delany....................Vice-President

T. C. Delany..............Secretary and Treasurer

Wm. Whalen.   J. Seligman.

Main Office—35 to 37 Hagerman building, Colorado Springs, Colorado.

1,500,000 shares. Par value, $1.00. In treasury January 1, 1900,   Capitalization
15,495 shares; in treasury January 1, 1900, $500.00 cash.

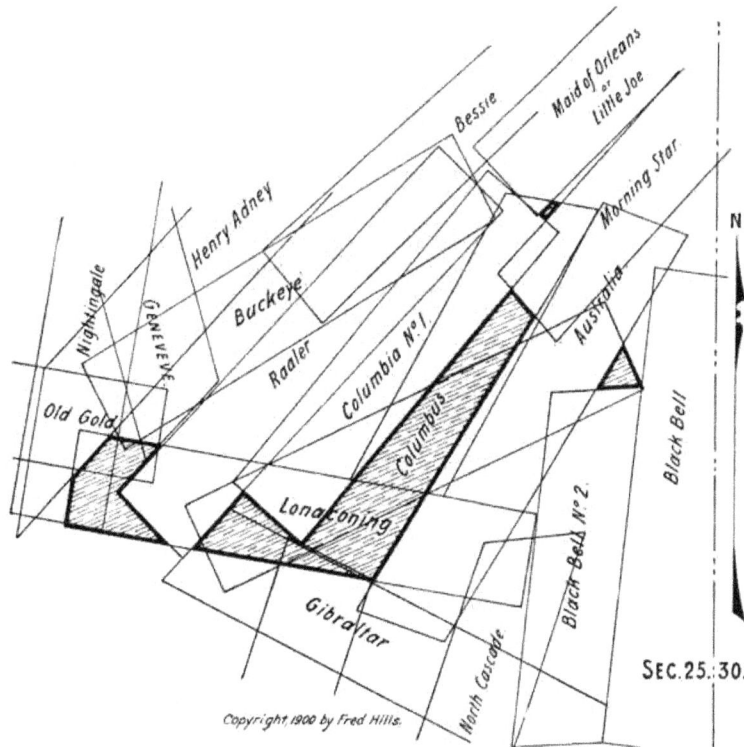

Copyright, 1900 by Fred Hills.

Owns Lonaconing, Old Hickory and Columbus, containing about   Property
7 1-2 acres, all patented, in section 25, township 15 south, range 70 west,
on the west slope of Beacon hill.

This property is being developed by lessees, of which there are four   Development
sets. It has three hoisting plants, including four shaft and ore houses.
About $30,000.00 has been expended in shafts, drifts, etc., and the
gross production from the mine to January 1, 1900, is about $50,000.00.
The lessees have paid 25 per cent. royalty after deducting freight and
treatment charges.

Production to January 1, 1900, $52,111.27.

Highest price for stock during 1899, 15 1-4 cents; lowest price for
stock during 1899, 2 1-2 cents.

The King Gold Mining Company.

Incorporated March, 1896.

Directors   H. A. Clapp.............................President
Jas. A. Howze.......................Vice-President
A. C. Labrie...............Secretary and Treasurer
A. H. Rex.   J. K. Lothridge.

Main Office—Cripple Creek, Colorado.

Capitalization   1,250,000 shares. Par value, $1.00. In treasury January 1, 1900, 382,500 shares; in treasury January 1, 1900, $200.00 cash.

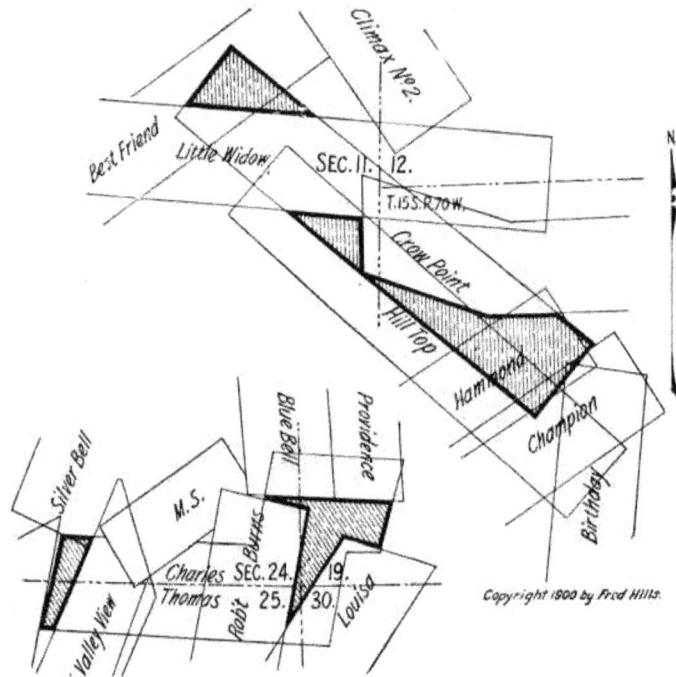

Copyright 1900 by Fred Hills.

Property   Owns the Charles Thomas lode, 1.9 acres, Survey No. 13,487, in the N. E. 1-4 of section 25, on Raven hill; the Crown Point, 4 acres, Survey No. 9,167, in the W. 1-2 of section 12, on Mineral hill; also owns the Alameen and Jock O'Dreams, Survey No. 13,538; the Magnes and Good Hope, Survey No. 13,566; the Rex No. 1, the Kid, and the Anything lodes; these latter comprise in all about 14 acres, on Tenderfoot hill. The Crown Point is patented. Remainder is held by location, and is not shown on plat.

Development   The Charles Thomas has a shaft 50 feet deep. The Magnes and Good Hope, have two shafts, 45 and 50 feet, respectively. On the Alameen is a tunnel of 30 feet and a 20-foot shaft. On the other claims only assessment work has been done. The company is now doing development work on the Magnes and the Alameen at the junction of two veins. Here they expect to sink a new shaft and place a plant of machinery by June 1, 1900. The greater part of the development work is now being done on the Magnes and the Charles Thomas.

Highest price for stock during 1899, 2 1-4 cents; lowest price for stock during 1899, 2 cents.

The King Gold Mining and Milling Company.

Incorporated October 26, 1899.

Chas. A. Johnson.........................President
H. King Johnson.....................Vice-President
Howard M. Johnson.....................Secretary
J. S. Danser............................Treasurer
E. J. Scott.

Directors

Main Office—Room No. 16, De Graff building, Colorado Springs, Colorado.

1,500,000 shares.   Par value, $1.00.   In treasury January 1, 1900, 500,000 shares.   **Capitalization**

Owns the Devil's Dream, the Last Chance, Hap-Hazard and Morning Glory, for which receiver's receipts are held, and the Norris Mayhugh mining claim, in process of patenting; all situated in the N. E. 1-4 of section 2 and containing about 31 acres, all in one group on the northwest side of Copper mountain, running from the saddle at the top over the northwest slope.   **Property**

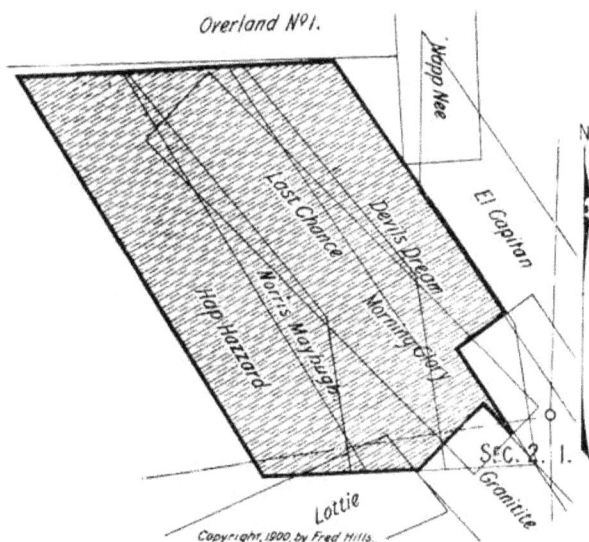

Copyright, 1900, by Fred Hills.

There is a blacksmith shop on the property. About 500 feet of shafting and drifting have been done. The company will commence work on their own account in April, 1900, with a plant of machinery, and will sink a shaft 200 feet deep and also connect with tunnel. The greater part of the work already done is on the Devil's Dream, the Last Chance and the Hap-Hazard. Work is being done on the Morning Glory and the Norris Mayhugh claims.   **Development**

These claims were purchased from the locators in 1897 by individuals now in the company. Patents will be perfected as receiver's receipts are now held. The property is located about 900 feet northwest of the celebrated Fluorine claim, from which, in 1897, about $150,000 was taken from near the surface within six months.   **History**

Highest price for stock during 1899, 7 cents; lowest price for stock during 1899, 7 cents.

The Kittie Wells Gold Mining Company.

Incorporated December 7, 1899.

Directors

C. S. Wilson, president; J. C. Johnston, vice-president; W. P. Sargeant, secretary; Wm. A. Otis & Company, treasurer; C. E. Titus, assistant secretary; S. M. Diltz, W. A. Thompson.

Main Office—Colorado Springs, Colorado.

Capitalization

1,250,000 shares. Par value, $1.00. In treasury January 1, 1900, 250,000 shares.

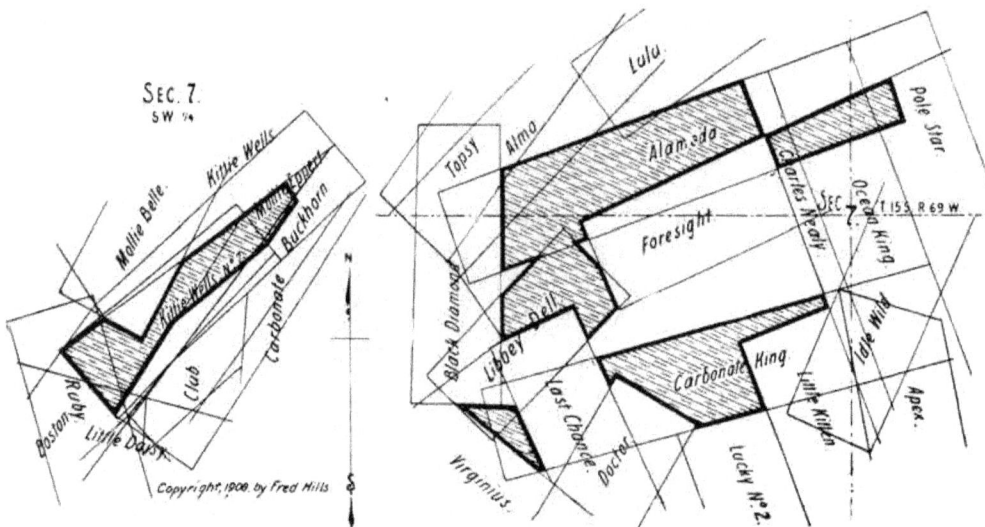

Property

Owns the Kittie Wells No. 2, 2 acres; the Libbey Dell, 1 acre; the Carbonate King, 3 acres; and the Alameda, 5 acres; a total of 11 acres, situated in the S. W. 1-4 of section 7, on Tenderfoot hill. Receiver's receipt held for the above property. Only sufficient development has been done to secure patent.

The Kitty Gold Mining Company.

Incorporated September 8, 1899.

E. S. Bach, president; Wm. Shemwell, vice-president; A. J. Bendle, *Directors* secretary; R. P. Davie, treasurer; John M. Harnan, J. A. Himebaugh.

Main Office—25 E. Pike's Peak avenue, Colorado Springs, Colorado.

1,500,000 shares. Par value, $1.00. In treasury January 1, 1900, *Capitalization* 190,000 shares; in treasury January 1, 1900, about $1,500 cash.

Owns the Katie Hollis and the Missouria placer, containing 12 *Property* acres, situated in the N. W. 1-4 of section 20, on Bull and Raven hills, between the Ramona and Ironclad and adjoining the Sheriff and Central Consolidated Mines Company. Missouria placer is patented. Katie Hollis in process.

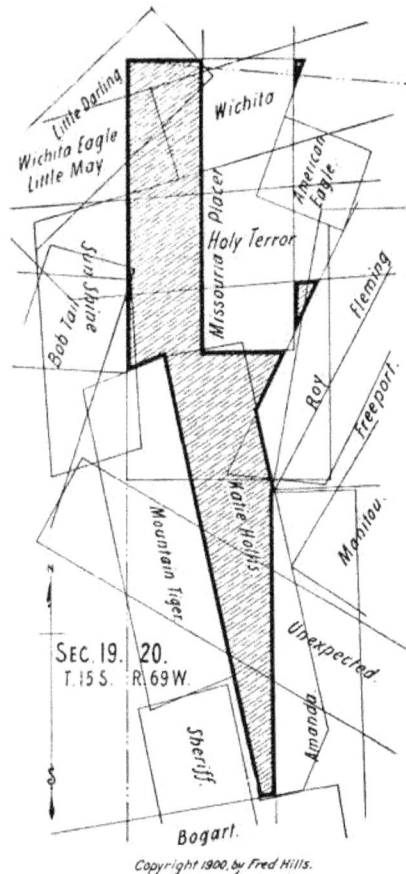

Copyright 1900, by Fred Hills.

On the Katie Hollis are shafts of 65 feet, 15 feet, 300 feet and 45 *Development* feet, respectively. One tunnel of 90 feet. Drifts of 130 feet from the N. Y. tunnel. Winzes of 25 feet. Cross-cut from shaft of 45 feet; drifts from shaft of 40 feet. The greater part of the development work has been done on this claim. On the other property there has been but superficial surface work done.

The Katie Hollis lode was located in January, 1892. The other *History* property has been recently acquired. Since location development work has been spasmodic. There are good indications in the way of float and by panning, the surface shows good values. Assays range from $2.60 to $680 per ton, but no ore in quantity has been found.

Highest price for stock during 1899, 3 5-8 cents; lowest price for stock during 1899, 2 1-4 cents.

The La Cota Gold Mining Company.

Incorporated 1895.

Directors

Wm. Munro, president; Geo. R. Bently, vice-president; P. S. Delany, secretary and treasurer.

Main Office—35 and 37 Hagerman Building, Colorado Springs, Colo.

Capitalization

1,250,000 shares. Par value, $1.00. In treasury January 1, 1900, 250,000 shares.

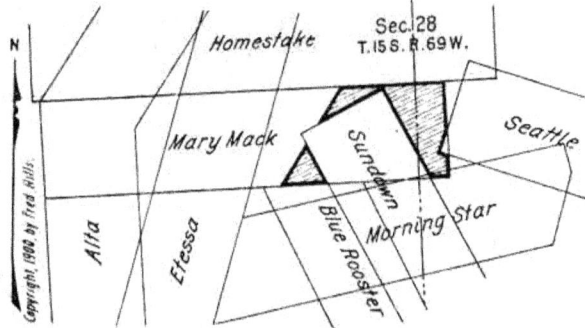

Property

Owns Mary Mack, containing about 6 acres, located in N. W. 1-4 section 28, on Big Bull mountain, near Goldfield townsite.

Development

There is a 90-foot shaft on this property, and the ground will all be leased and development continued. The surface ground owned by the company is shown on plat by shaded lines. The company also owns the mineral rights under the west end of the Mary Mack, except that portion where the Etessa crosses it.

Highest price for stock during 1899, 1 1-2 cents; lowest price for stock during 1899, $5.00 per 1,000.

The Ladessa Gold Mining and Milling Company.

Incorporated 1894.

Directors

H. S. Sommers.........................President

E. S. Cohen...............Secretary and Treasurer

Main Office—Colorado Springs, Colorado.

Capitalization

1,500,000 shares. Par value, $1.00.

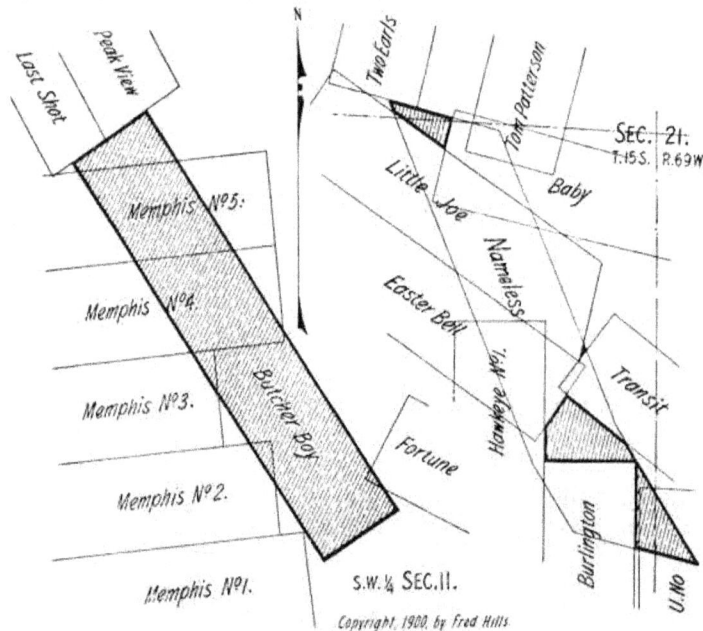

Copyright, 1900, by Fred Hills.

Property

Owns the Butcher Boy, comprising 10 1-3 acres, in the S. W. 1-4 section 11, township 15 south, range 69 west, on Cow mountain; also the Nameless, comprising 1.907 acres, in the S. W. 1-4 section 21, on Bull hill. Both claims are patented.

Highest price for stock during 1899, 1 cent; lowest price for stock during 1899, $4.00 per 1,000.

The Lady Campbell Gold Mining Company.

Incorporated April, 1897.

J. Maurice Finn, president; A. Bourquin, vice-president; E. C. Bab- Directors
bitt, secretary and treasurer; A. B. Whitmore, G. C. Blakey.

Main Office—Cripple Creek, Colorado.

1,000,000 shares. Par value, $1.00. In treasury January 1, 1900, Capitalization
50,000 shares.

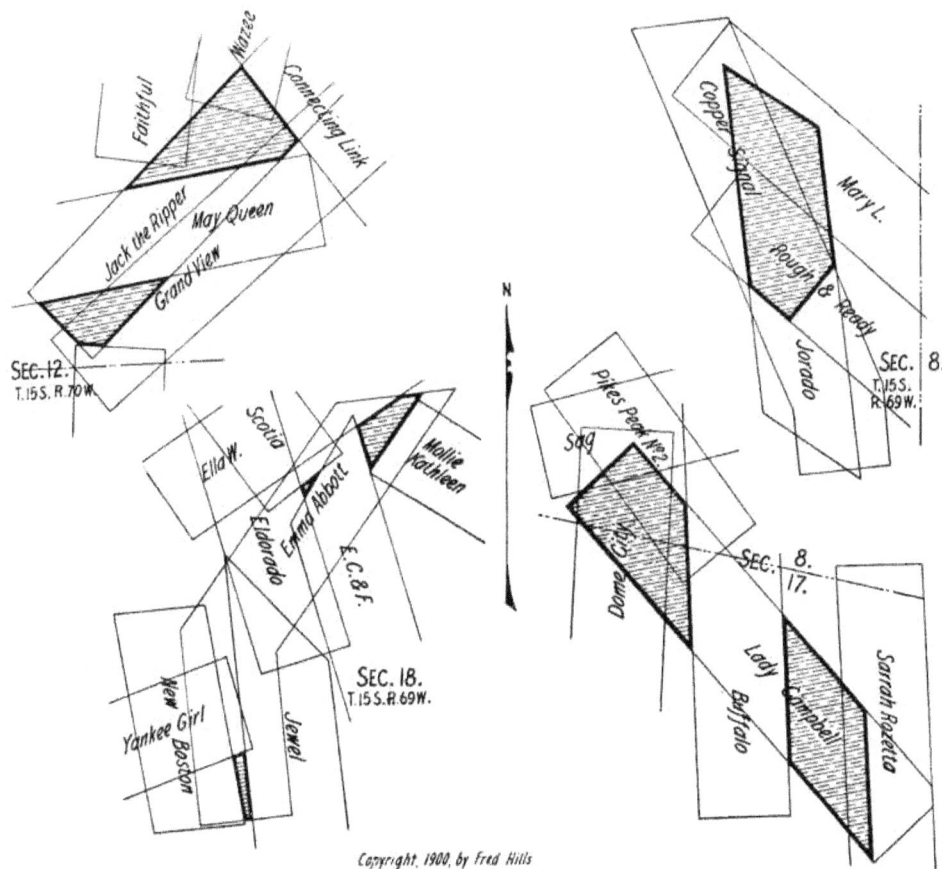

Copyright, 1900, by Fred Hills

Owns the Lady Campbell, in the N. E. 1-4 of section 17, containing Property
5 1-2 acres, on Galena hill; the Jorado, in the S. W. 1-4 section 8, containing 5 acres, on Galena hill, the ground of the latter being deeded from
the Mary L. and the Rough and Ready; also the Emma Abbott, in the
N. W. 1-4 of section 18, containing 1 acre, on Womack hill; and Jack the
Ripper, in the N. E. 1-4 of section 12, containing 2 1-2 acres, on Mineral
hill. The Jorado is patented. Receiver's receipt held for all the remainder.

On the Lady Campbell there is a 130-foot shaft, 175 feet of drifts Development
and a tunnel. Jack the Ripper has a 100-foot shaft. On the Emma
Abbott there is a 125-foot shaft. On the Jorado one of 65 feet. The
Lady Campbell is leased. On this claim the greater part of the development work has been done.

Highest price for stock during 1899, 5 cents.

The Lake Shore Gold Mining Company.

Incorporated January 28, 1896.

Directors

A. F. Seymour..President and Treasurer
John H. Mortimer........Vice-President
W. S. Heath................Secretary
Owen Millay.   James C. Duncan.

Main Office—P. O. block, Colorado Springs, Colorado.

Capitalization

1,000,000 shares. Par value, $1.00. In treasury January 1, 1900, 88,000 shares; in treasury January 1, 1900, $100.00 cash.

Property

Owns the Maceo claim, containing 10 acres, situated in the E. 1-2 section 3, on Red mountain. Patented.

A shaft, 80 feet deep, comprises the development.

Highest price for stock during 1899, $4.00 per 1,000; lowest price for stock during 1899, $2.00 per 1,000.

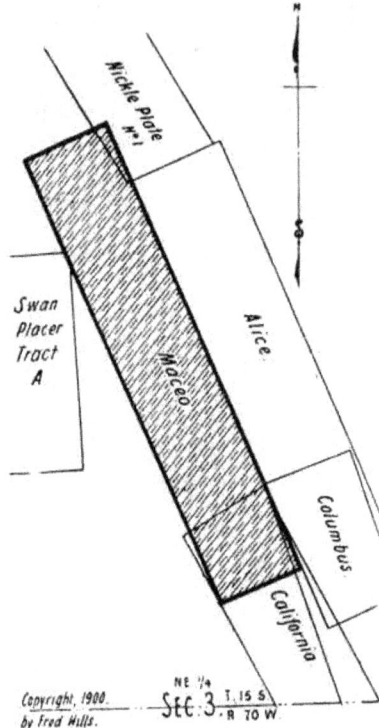

The Lamertine Gold Mining Company.

Incorporated 1896.

Directors

John W. Proudfit.....................President
R. C. Thayer......................Vice-President
C. H. Bryan...............Secretary and Treasurer
S. T. Miller.   E. W. Adams.

Main Office—No. 10 Hagerman building, Colorado Springs, Colorado.

Capitalization

1,500,000 shares. Par value, $1.00. In treasury March 1, 1900, 250,000 shares. In treasury March 1, 1900, $10,000 cash.

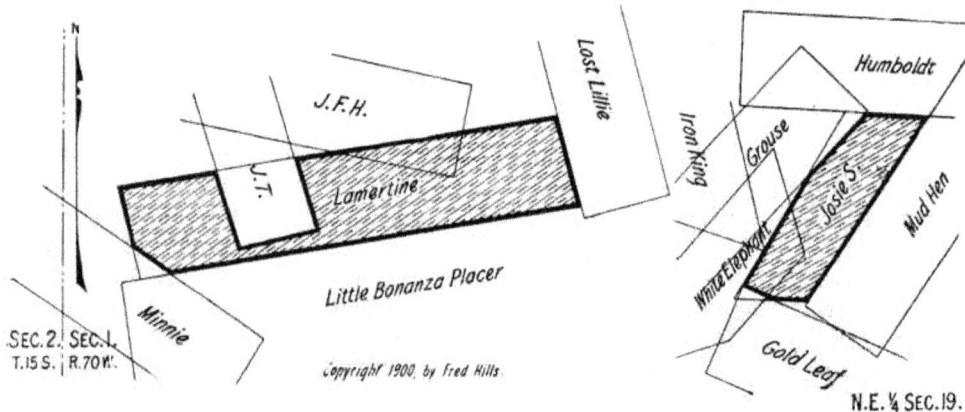

Property

Owns the Lamertine, comprising 8 acres, in the S. W. 1-4 of section 1, on Copper mountain; also the Josie S., containing 3 1-2 acres, in the N. E. 1-4 of section 19, on Raven hill; both patented.

Development

On the Josie S. a shaft 50 feet deep has been sunk and the work is steadily progressing. The greater part of the development work is being done on this claim.

Highest price for stock during 1899, 4 1-8 cents; lowest price for stock during 1899, 4 cents.

284

The Lasca Gold Mining Company.
Incorporated 1897.

J. K. Vanatta, president and treasurer; J. H. Ryan, vice-president; *Directors*
M. C. Meek, secretary; H. J. Newman, W. J. Burke, W. G. Newman.

Main Office—No. 16 North Nevada avenue, Colorado Springs, Colorado.

1,500,000 shares. Par value, $1.00. In treasury January 1, 1900, *Capitalization*
88,000 shares.

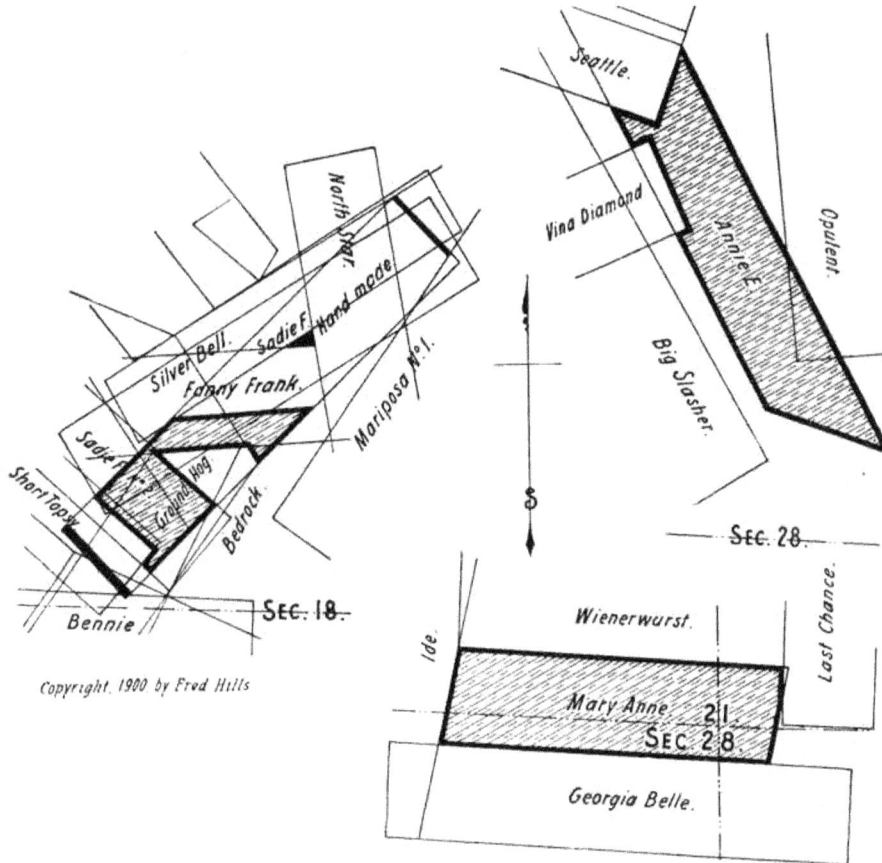

Copyright, 1900 by Fred Hills

Owns the Hand Made, comprising three acres, in the N. E. 1-4 of *Property*
section 18, on Globe hill; also the Mary Anne and the Annie E., comprising 17 acres, in the N. E. 1-4 of section 28, on Bull hill. The Annie E. and the Mary Anne are patented. The Hand Made is in process of patenting; advertising and settlements having been completed.

There are about 350 feet of shafts in all. The greater part of the *Development*
development work is being done on the Hand Made by the company. The company is now cross-cutting at a depth of 115 feet to cut a large phonolite dyke, which penetrates the claim from the Gold Hill property. A 15 months' lease at a 20 per cent. royalty has been granted on the Annie E.

Highest price for stock during 1899, 4 cents; lowest price for stock during 1899, 2 1-2 cents.

The Last Dollar Gold Mining Company.

Incorporated April, 1896.

Directors

R. P. Lounsbery, president and treasurer; A. Eilers, vice-president; H. Hanington, Jr., secretary; Willard P. Ward; Henry Seligman; B. Y. Frost; F. E. Brooks; Henry R. Wolcott.

Main Office—204 Boston Building, Denver, Colorado.

Capitalization

1,500,000 shares. Par value, $1.00. In treasury January 1, 1900, $40,000.00 cash.

Property

Owns the Last Dollar and Combination, consisting of 18 acres, in the S. 1-2 section 20, on Bull hill. Patented.

Development

There is a large and very complete shaft house on this property. Two 12"x30" Cylinder First Motion engines, 150 H. P. with a 4-foot drum; blacksmith shop, store house, large and complete ore house, dwelling houses, office and assay office. All of the above on D shaft; also small engine and shaft house on B shaft; thoroughly equipped with electric lights and electric drills.

There are two shafts on the Last Dollar, 750 feet deep, with a distance of about 800 feet between them, with about 6,000 feet of workings. Combination claim is comparatively undeveloped.

The property was originally worked under lease and bond, which was subsequently taken up.

Production and Dividend

Production to January 1, 1900, $700,000.00; dividends up to January 1, 1900, $60,000.00; last dividend, February 1, 1900, $30,000.00.

Highest price for stock during 1899, $1.03 per share; lowest price for stock during 1899, 89 cents.

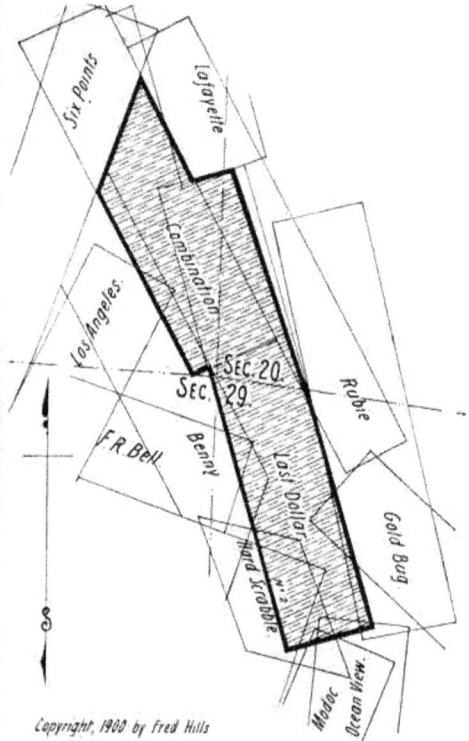

Copyright, 1900 by Fred Hills

The Legal Mining and Milling Company.

Incorporated November 25, 1895.

Directors

F. A. Vorhees, president; Fred A. Young, vice-president; Horace H. Mitchell, secretary and treasurer.

Main Office — No. 45 Bank Building, Colorado Springs, Colorado.

Capitalization

1,000,000 shares. Par value, $1.00. In treasury January 1, 1900, 52,000 shares.

Property

Owns the Saturday, 3 3-8 acres, on Rosebud hill, in section 25, township 15 south, range 70 west. Patented. Also owns the Coniago, 5 acres, in the N. W. 1-4 section 8, on North Tenderfoot; receiver's receipt. The Saturday claim is being developed by lessees. The policy of the company will be to develop its property by the leasing system.

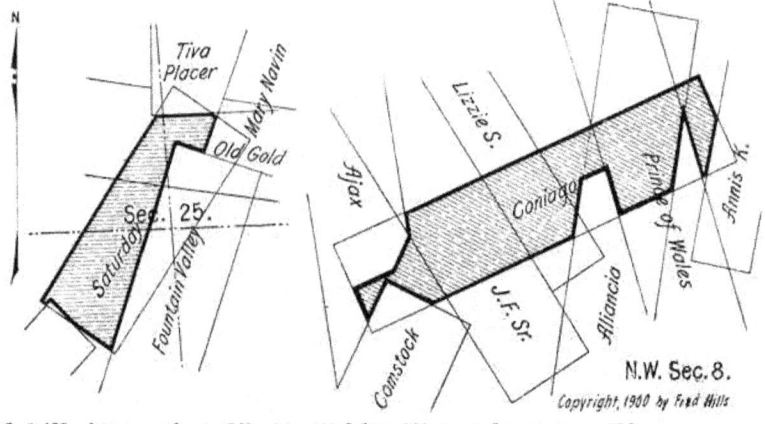

N.W. Sec. 8.

Copyright, 1900 by Fred Hills

The Leland Stanford Mining Company.

Incorporated May, 1899.

Chas. F. Potter.........................President   Directors
John J. O'Donnell...................Vice-President
A. O. Downs..............Secretary and Treasurer

Main Office—129 North Tejon street, Colorado Springs, Colorado.

350,000 shares. Par value, $1.00. In treasury January 1, 1900, Capitalization
36,500 shares; in treasury January 1, 1900, $30.00 cash.

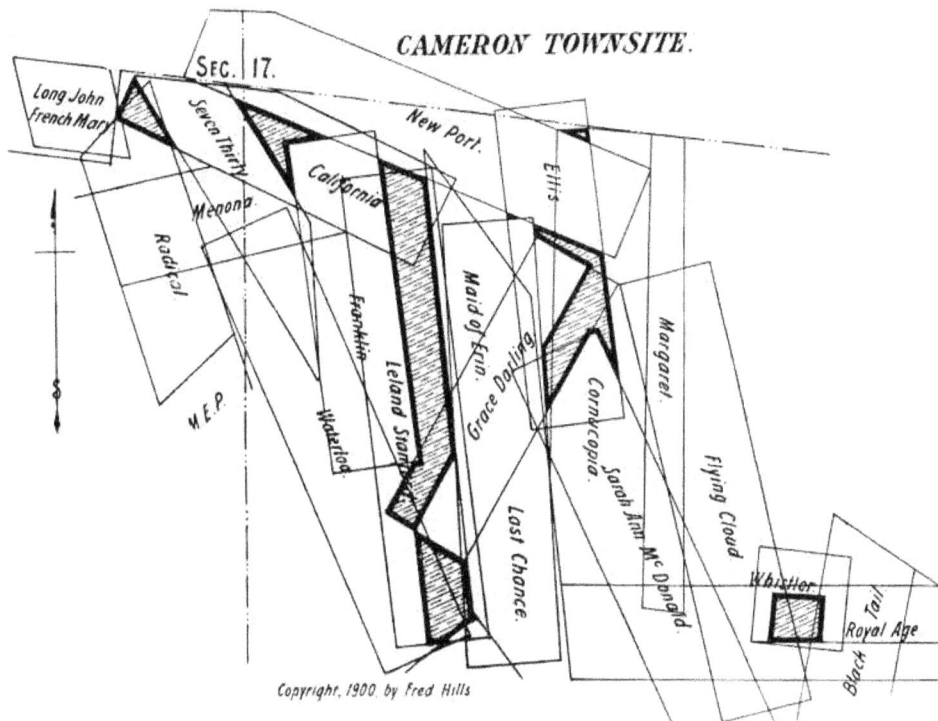

CAMERON TOWNSITE.

Copyright, 1900, by Fred Hills

Owns the Leland Stanford, California, Ellis and Whistler claims, Property
containing about 9 acres, in the S. 1-2 section 17, on the north slope of
Bull hill, adjoining the Acacia and Pinnacle properties. The Leland
Stanford and Ellis claims are patented; balance held by location.

One-half of the Leland Stanford and the Ellis are being worked by Development
lessees. The Leland Stanford claim has one shaft 130 feet deep and one
60 feet deep, both timbered, with 50 feet of drifts. The Whistler claim
has one shaft 45 feet deep, with a drift of 15 feet, and one shaft 50 feet
deep. The Ellis claim has one shaft 50 feet deep, with 40 feet of drifting,
and one shaft 50 feet deep. The properties were formerly owned by the
Chimborazoo Mining Company, and came to this company by sheriff's
deed.

Highest price for stock during 1899, 12 cents; lowest price for stock
during 1899, 5 1-2 cents.

287

The Lenora Mining and Milling Company.

Incorporated May, 1895.

B. P. Anderson.........................President

A. L. New........................Vice-President

F. A. Perkins..............Secretary and Treasurer

J. E. Hundley.

Main Office—108 North Tejon street, Colorado Springs, Colorado.

1,000,000 shares.   Par value, $1.00.   In treasury January 1, 1900, 250,000 shares.

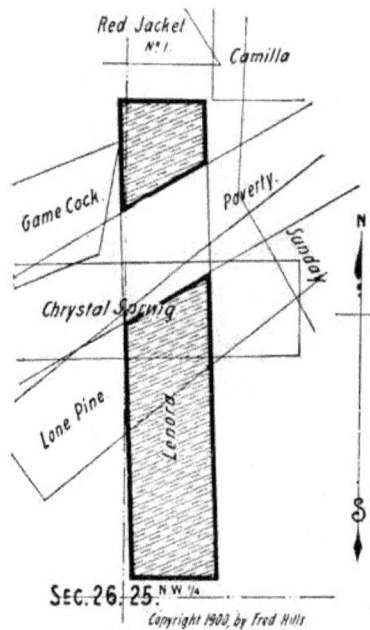

Owns the Lenora claim, in the N. W. 1-4 section 25, containing 8 acres.   Patented.

This is a close corporation and consequently no quotations on stock, as it is mostly held by the officers of the company.

The Lexington Gold Mining Company.

Incorporated December 2, 1895.

J. Stanley Jones..........................President   Directors
C. C. Butler.........................Vice-President
N. H. Partridge...........Secretary and Treasurer
E. C. Fletcher.   W. B. Storer.

Main Office—10, 11 and 12 Giddings block, Colorado Springs, Colorado.

1,500,000 shares. Par value, $1.00. In treasury January 1, 1900,   Capitalization
284,000 shares; in treasury January 1, 1900, $20,000.00 cash.

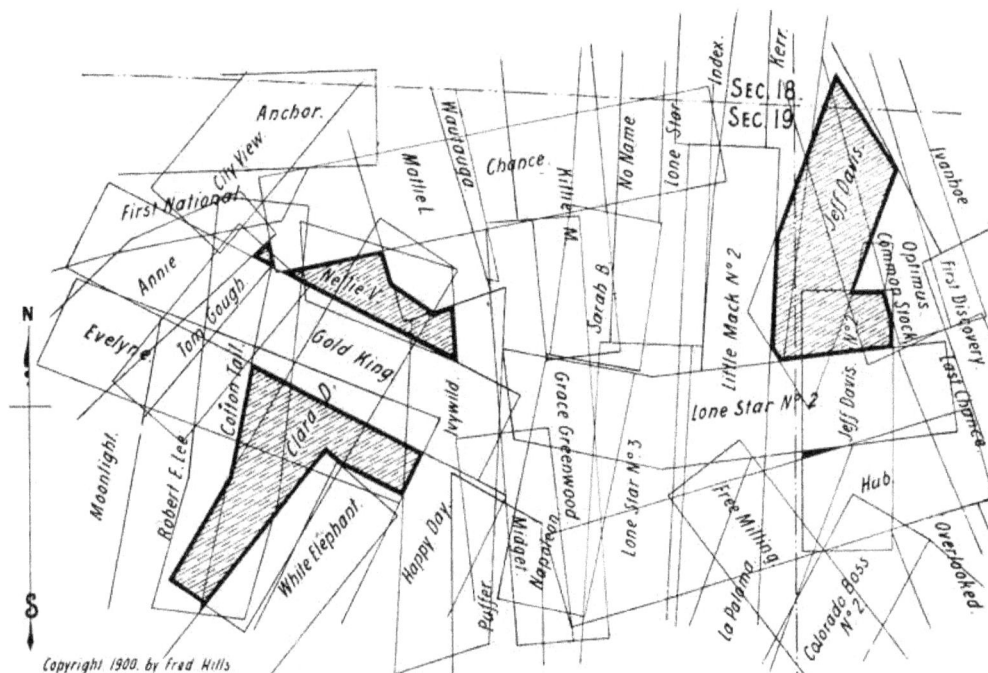

Copyright 1900. by Fred Hills.

Owns the Clara D., 4.560 acres; Nellie V., .641 acre; Cotton Tail,   Property
.741 acre; Evelyne, .831 acre, and Jeff Davis Nos. 1 and 2, 5.808 acres,
making a total of 12.58 acres, situated in the N. W. 1-4 section 19, on
Gold hill. Patented.

There are shaft houses, ore bins, trestle and machinery on the prop-   Development
erty. Lease on the south 300 feet of the Clara D., also lease on the Jeff
Davis Nos. 1 and 2, and portions of Evelyne and Nellie V.

Production to January 1, 1900, $97,000.00.   Production

Highest price for stock during 1899, 32 cents; lowest price for stock
during 1899, 6 cents.

The Liberty Bell Gold Mining Company.

Incorporated January 17, 1896.

Directors

James Bell........................President

Wm. Vineyard......................Vice-President

W. J. Hendrickson..................Secretary

J. W. Campbell....................Treasurer

W. T. Doubt.

Main Office—Room 25, First National Bank building, Colorado Springs, Colorado.

Capitalization

1,250,000 shares. Par value, $1.00. In treasury January 1, 1900, 200,000 shares.

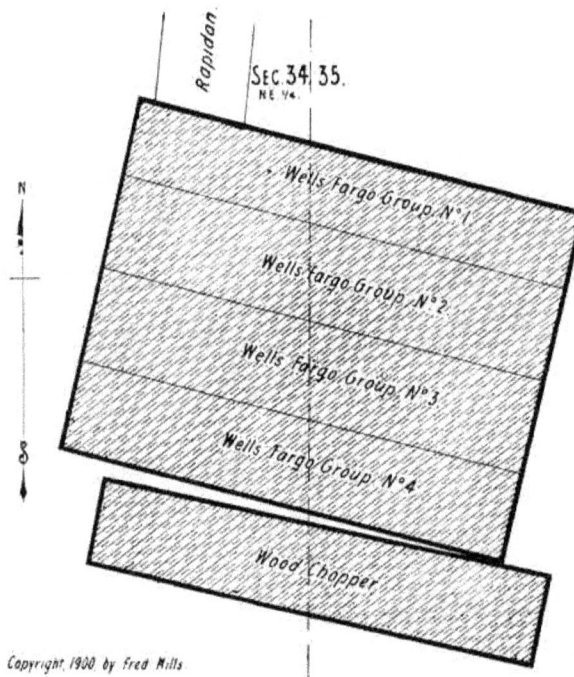

Copyright, 1900 by Fred Hills.

Property

Owns the Wells Fargo group, Nos. 1, 2, 3 and 4, and the Wood Chopper, a total of 49.297 acres, situated between sections 34 and 35, on Big Bull hill. All patented.

Development

There are 200 feet of shafts and tunnels. The greater part of the development work is being done on the Wells Fargo group, Nos. 1, 2 and 3, which claims are leased. No sales of stock reported.

The Lillie (C. C.) Gold Mining Company, Limited.

Registered in London in 1898.

F. H. Morley, chairman; Wm. A. Otis, secretary and treasurer; A. P. Mackey, R. J. Preston, Wm. F. Fisher, J. P. Pomeroy. — *Directors*

LONDON BOARD.

W. F. Fisher, chairman; W. Weston, director; H. A. G. Lewis, secretary.

Main Office (American)—With Wm. A. Otis & Company, Colorado Springs, Colorado. London Office—15 George Street, Madison House, London, E. C.

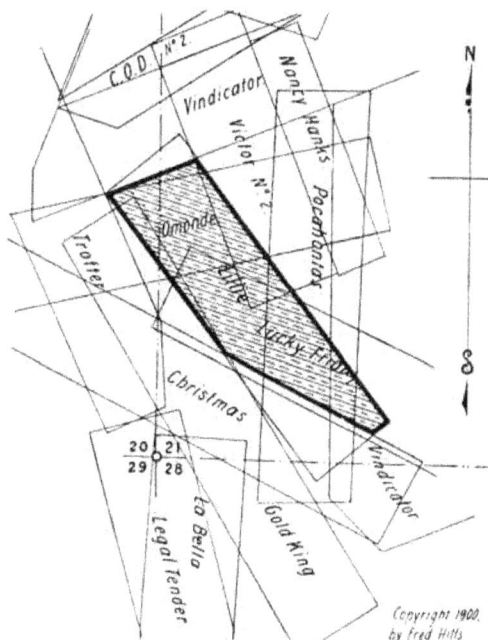

Copyright 1900.
by Fred Hills

225,000 shares. Par value, one pound sterling. In treasury January 1, 1900, $125,000 cash. — *Capitalization*

Owns the Lillie claim, containing 7 acres, in the S. W. 1-4 of section 21, on Bull hill. — *Property*

The main shaft is 1,000 feet deep. It is a double compartment shaft (cages, etc.), has two hoisting engines, air compressor run by electricity, and has, in all, one of the best equipped plants in the district. The workings are being largely developed. — *Development*

Gross production to January 1, 1900, $1,268,610.62. Total dividends to April 1, 1900, $361,433.68. Last dividend paid April 1, 1900, $11,367.19. Stock is closely held. — *Production and Dividend*

The Lipton Gold Mining Company.

Incorporated October 16, 1899.

Directors
F. F. Dennis............President
H. S. Sommers......Vice-President
C. P. Bently...Sec'y and Treasurer
John R. Newby.   Geo. Neuer.
Main Office—No. 53 Bank block,
Colorado Springs, Colorado.

Capitalization
1,000,000 shares. Par value, $1.00.
In treasury January 1, 1900, 200,000
shares.

Property
Owns the M. J. lode and the Lura
May lode, containing in all about 8
acres, located in the S. 1-2 section 1,
on the east slope of Copper mountain. The M. J. is patented. Receiver's receipt held for the Lura
May.

Development
Enough work has been done to obtain a patent. The greater part of
the development work is at present being done on the M. J.
The stock is not on the market.

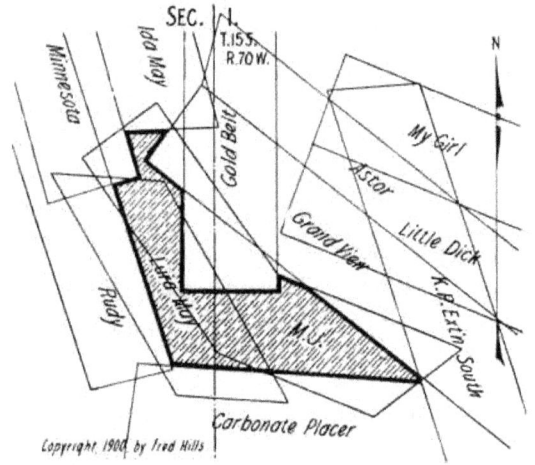

The Little Alice Mining and Milling Company.

Incorporated January, 1896.

Directors
John J. Grier, president;
T. F. McCarthy, vice-president and treasurer; J.
Maurice Finn, secretary;
J. M. Parker, J. L. Lindsay.
Main Office — Cripple
Creek, Colorado.

Capitalization
1,000,000 shares. Par
value, $1.00. In treasury
January 1, 1900, 40,000
shares; in treasury January 1, 1900, $190.00 cash.
Indebtedness January 1,
1900, $1,000.00.

Property
Owns the Little Alice,
Alise Mc., Keystone, Good
Luck, Vindicator and Good
Hope, a total of 31.4 acres,
in the S. 1-2 section 32,
township 14 south, range
69 west. Property held by
receiver's receipt.

Sufficient development work for patent only has been done.
Highest price for stock during 1899, 2 1-2 cents.

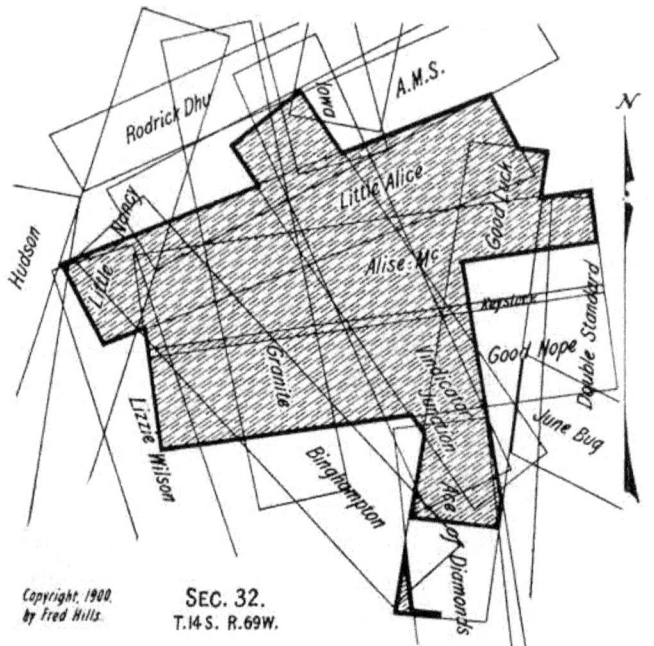

The Little Bessie Gold Mining Company.

Incorporated October 19, 1899.

J. E. Hundley............................President   Directors
C. C. Hamlin.......................Vice-President
O. H. Shoup...............Secretary and Treasurer
    J. W. Miller.   E. P. Shove.

Main Office—With the Reed & Hamlin Investment Company, Bank block, Colorado Springs, Colorado.

1,250,000 shares. Par value, $1.00. In treasury January 1, 1900, Capitalization 99,995 shares; in treasury January 1, 1900, about $4,600.00 cash.

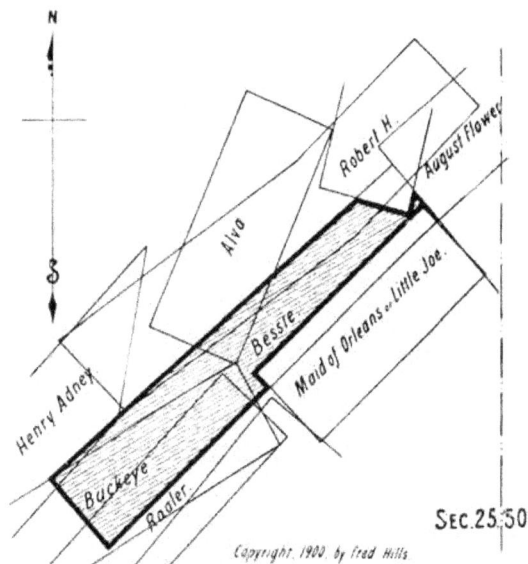

Owns the Bessie, containing about 8 acres, located in section 25, on Property Beacon hill. Patented.

This being a newly incorporated company, there are no surface im- Development provements and very little development has been done up to January 1, 1900, except sufficient to patent the claim. The company expects to develop its property by lessees. The showing on the Columbia, adjoining this company's claim, is such as to warrant the company expecting an extension of the dykes passing through the claim.

Highest price for stock during 1899, 3 3-4 cents; lowest price for stock during 1899, 3 cents.

The Little Corporal Gold Mining Company.

Incorporated December 4, 1895.

Directors

Geo. E. Lindley..........................President

J. E. Phillips.........Vice-President and Treasurer

Chas. H. Peters...........................Secretary

J. G. Tucker.   J. R. McKinnie.

Main Office—307 Mining Exchange, Denver, Colorado.

Capitalization   1,250,000 shares.   Par value, $1.00.   In treasury January 1, 1900, 63,000 shares; in treasury January 1, 1900, about $250.00.

Copyright, 1900, by Fred Hills

Property   Owns the Comstock and the Kearney, in the N. W. 1-4 of section 8, an area of about 16.87 acres, located on Tenderfoot hill, about 1,000 feet due north of the Hoosier mine. The property is fully patented.

Development   There are about 140 feet of shafts and drifts, also some other shafts and cuts. Probably $2,000 have been expended. The policy of the company is to work its property by leasing.

Highest price for stock during 1899, 2 1-4 cents; lowest price for stock during 1899, 3-4 of a cent.

The Little Cut Diamond Mining and Milling Company.

Incorporated 1896.

S. S. Bernard, president; J. Bishoff, vice-president; J. M. Jordan, Directors secretary; Jos. H. Ryan, G. D. Kennedy, Jennie E. Lauth.

Main Office—115 E. Pike's Peak avenue, Colorado Springs, Colorado.

1,500,000 shares. Par value, $1.00. In treasury January 1, 1900, Capitalization 230,000 shares; in treasury January 1, 1900, $100.00 cash.

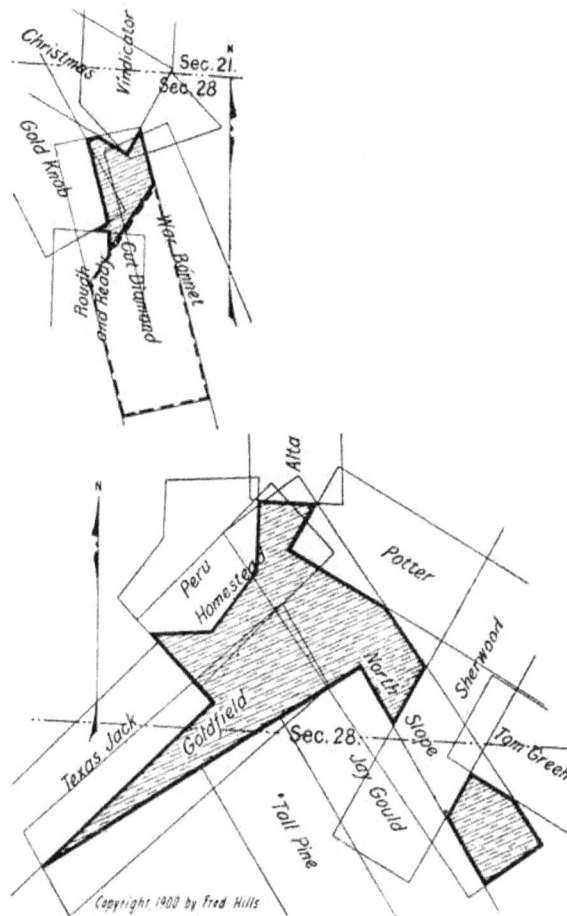

Owns the mineral and surface rights of the Cut Diamond, 2 acres, on Property Bull hill; the mineral rights only, 4 acres, of the Cut Diamond lying under the town of Goldfield; also owns the Homestead, Goldfield, and the North Slope, containing 15 acres, in the N. W. 1-4 of section 28, in a group on Big Bull hill.

Two shafts on the property, one 42 feet deep, the other 122 feet deep; Development 118 feet of drifting at the bottom of the deepest shaft. The company is now drifting in a N. E. direction so as to cross-cut at right angles the formations which run through the Vindicator and Christmas properties.

Highest price for stock during 1899, 4 cents; lowest price for stock during 1899, 1 1-2 cents.

The Little Joan Mining Company.

Incorporated July, 1892.

Directors

Warren Woods, president; W. F. Crosby, vice-president; Louis R. Ehrich, secretary; F. M. Woods, treasurer; H. E. Woods.

Main Office—No. 64 Hagerman building, Colorado Springs, Colorado.

Capitalization

1,000,000 shares. Par value, $1.00. In treasury January 1, 1900, 18,000 shares; in treasury January 1, 1900, $699.10 cash.

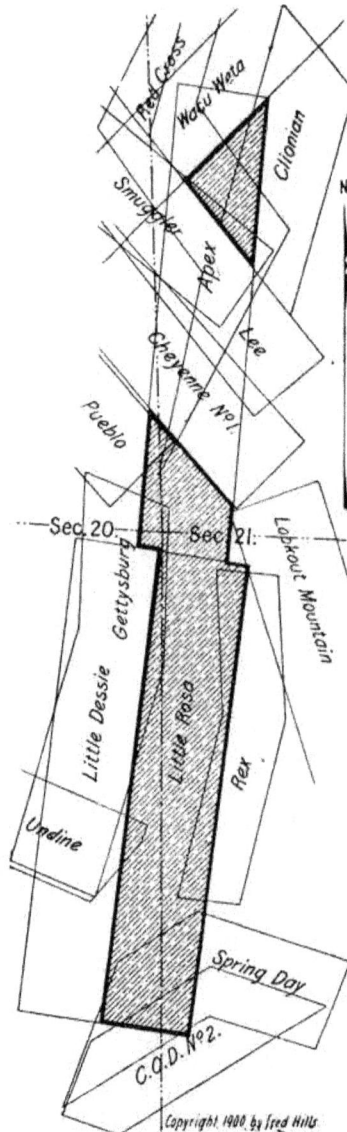

Copyright 1900, by Fred Hills.

Property

Owns the Apex and the Little Rosa, situated between sections 20 and 21, containing 15.044 acres, on Bull hill. All patented. Also owns the vein rights passing through the Apex, Smuggler, Lee and Wacu Weta claims. The property is leased. The greater part of the development work has been done on the Little Rosa.

Highest price for stock during 1899, 12 cents; lowest price for stock during 1899, 6 cents.

The Little May Mining Company.

Incorporated November 14, 1892.

Josiah Hughes, president; E. J. Seeley, vice-president; H. J. English, secretary; R. W. English, treasurer; F. G. White, R. J. Coleman. — Directors

Main Office—51 Jacobson building, Denver, Colorado.

1,500,000 shares. Par value, $1.00. In treasury February 10, 1900, $8,000.00. — Capitalization

Owns Little May No. 1 lode, and 6-10 of Little Blanche No. 1 lode, containing 13 1-3 acres, in the S. W. 1-4 section 21, on Bull hill, Cripple Creek district, adjoining the Victor mine. All patented. — Property

Improvements consist of shaft house and plant of machinery. On the Little May there is one incline shaft 275 feet deep. Shaft on Little Blanche 65 feet deep. — Development

The Victor vein has been developed to the northeast corner of Little Blanche. The shaft on the Little May is sunk on the continuation of this vein, but more depth seems necessary to find the ore shoot. Quantities of low grade ore exist in the present workings and probably could be made to pay with a mill. The company will immediately commence to sink the shaft deeper.

Highest price for stock during 1899, 10 cents; lowest price for stock during 1899, 10 cents.

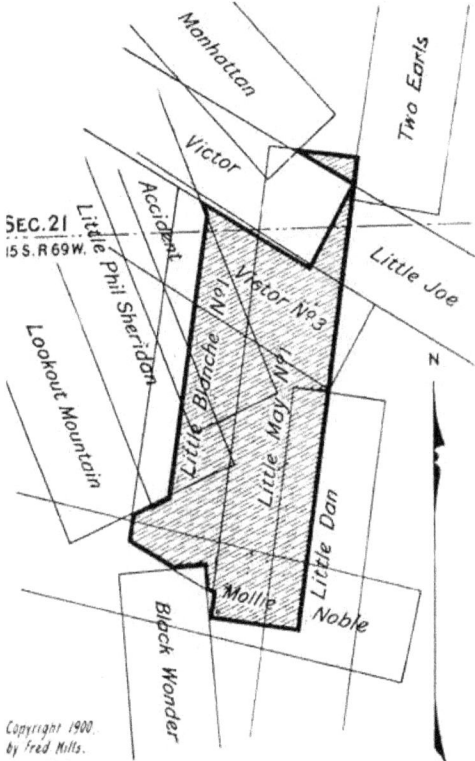

The Little Nell Gold Mining Company.

Incorporated January, 1900.

Julius A. Myers, president; Ramsay C. Bogy, vice-president and treasurer; Hiram E. Hilts, secretary; E. V. Haughwout, E. E. Quentin. — Directors

Main Office (Transfer)—International Trust Company, Colorado Springs, Colorado.

1,500,000 shares. Par value, $1.00. In treasury April 23, 1900, 150,000 shares; in treasury April 23, 1900, $6,000.00 cash. — Capitalization

Owns over 15 acres in the town of Arequa, in the S. W. 1-4 of section 30, on Raven hill. Patented. — Property

Three or four hundred feet of work on the property. There are four sets of lessees working and others will commence soon. The company is also working. Company organized recently and stock floated at 5 cents. It is listed on the Colorado Springs Exchange. The company has for its neighbors such well-known paying mines as the Elkton, Katinka, Gould, Raven, Jack Pot and Monarch, and therefore its prospects are good. — Development

Highest price for stock in 1900, 6 1-2 cents; lowest price for stock in 1900, 5 cents.

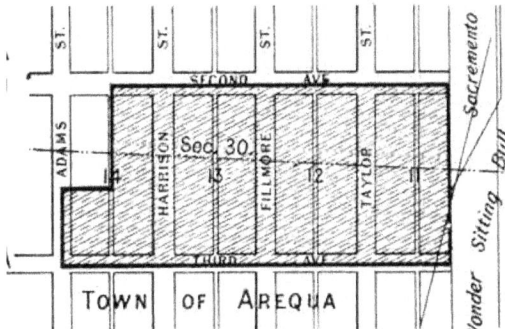

The Little Puck Gold Mining Company.

Incorporated January 11, 1896.

Directors   E. A. Colburn, president; D. N. Heizer, secretary.

Main Office—No. 16 North Nevada avenue, Colorado Springs, Colorado.

Capitalization   2,000,000 shares. Par value, $1.00. In treasury January 1, 1900, 17,168 shares.

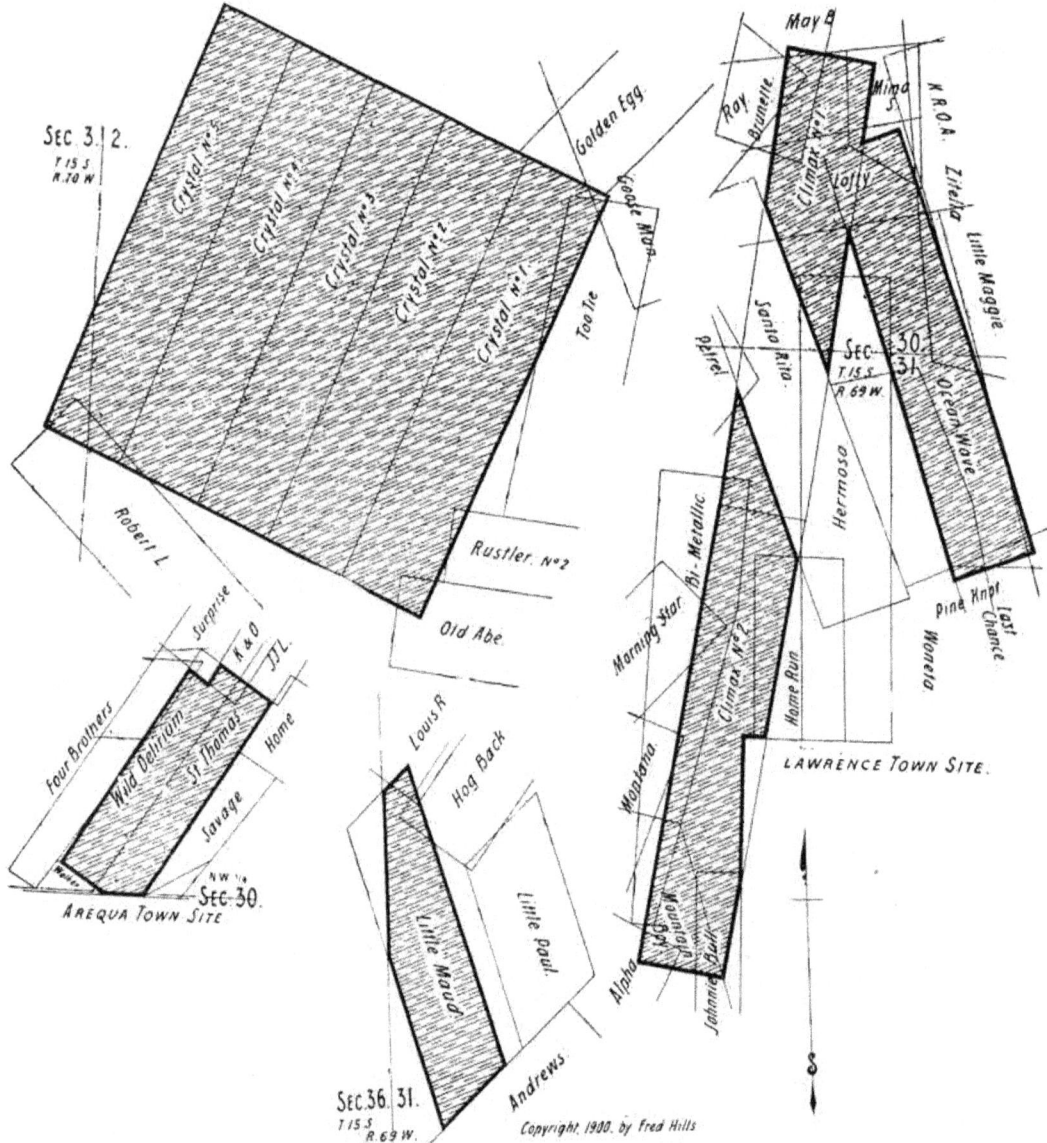

Copyright, 1900, by Fred Hills.

Property   Owns the Ocean Wave, 9.142 acres, in the S. E. 1-4 section 30, on Squaw mountain; holds contract giving the company the right to follow its vein across the Santa Rita; owns an undivided two-thirds interest in the Climax Nos. 1 and 2, 15.97 acres, in the S. E. 1-4 section 30 and the N. E. 1-4 section 31, on Squaw mountain; an undivided one-half interest in the St. Thomas, 5.031 acres, and bond on the other one-half for $40,000, situated in the N. W. 1-4 section 30; the Little Maud, 5.56 acres, in the S. W. 1-4 section 31, on Grouse mountain; the Crystal group, Nos. 1, 2, 3, 4 and 5, 46.766 acres, in the W. 1-2 section 2, on Red mountain; the Little Josie, McKinley, Billy Sherman and Robert W., 36.894 acres, in

(CONTINUED ON PAGE 299.)

The Little Puck Gold Mining Company—Continued.

section 20, township 14 south, range 70 west, on the south slope of Phonolite hill. The latter group not shown on plat, being outside of map limits. The total acreage owned by this company is 119.649 acres. All the above property is patented.

Steam hoist and ore bins on the St. Thomas. Two steam hoists, ore *Development* bins, etc., are now being set up on the Climax No. 1. The St. Thomas has a shaft 300 feet deep, and at this point the vein has been cut and is being drifted on. The Climax No. 1 has a shaft 100 feet deep. A vein has been struck and is being opened up by open cuts on the surface for a length of 300 feet north. This claim is a shipper. The Climax No. 2 is leased. On the north end of this claim a tunnel begins and runs for a distance of 600 feet through the Climax No. 1. The Ocean Wave, which is leased, has a shaft 85 feet deep. The greater part of the development work is being done on the Climax No. 1 and on the St. Thomas.

Gross production to January 1, 1900, $25,000.00. *Production*

Highest price for stock during 1899, 15 cents; lowest price for stock during 1899, 2 cents.

The London and Colorado Gold Mining Company.
Incorporated 1896.

Otto Fehringer, president; H. T. Cooper, vice-president; K. Mac- *Directors* Millan, secretary and treasurer; Wm. Lake, J. W. Kriger.

Main Office—No. 28 Midland block, Colorado Springs, Colorado.

1,250,000 shares. Par value, $1.00. In treasury January 1, 1900, *Capitalization* 175,000 shares; present indebtedness of the company, $100.00.

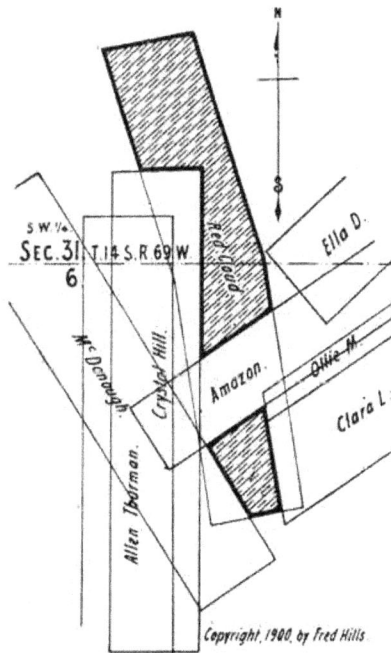

Owns the Red Cloud, containing about 6 acres, situated on Rhyolite *Property* mountain, in the N. W. 1-4 section 6, township 15 south, range 69 west, and the S. W. 1-4 of section 31, township 14 south, range 69 west. In process of patenting.

All work necessary to obtain a patent has been done. The property is leased for 18 months at a 15 per cent. royalty.

Stock not on the market.

The Longfellow Gold Mining Company.

Incorporated.

Directors

W. S. Stratton.... President

W. A. Ramsay.... Secretary

Property

Owns the Longfellow and the Longfellow No. 2, about 18 acres, patented, in section 20, on Bull hill.

This company is owned and controlled by Mr. W. S. Stratton, of Colorado Springs. No information obtainable.

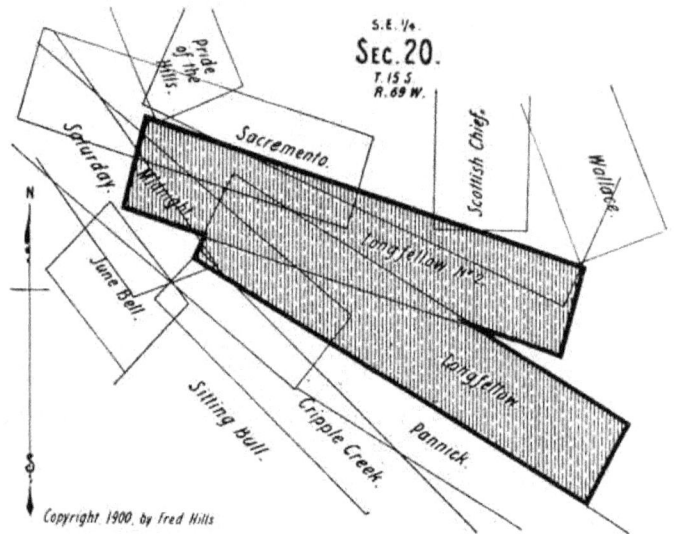

The Loraine Gold Mining Company.

Incorporated December 6, 1895.

Directors

E d w i n Arkell, president; Chas. A. Brooker, secretary and treasurer; Jos. P. Walsh, assistant secretary; Nathan Oakes.

Main Office—No. 37 Postoffice building, C o l o r a d o Springs, Colorado.

Capitalization

1,500,000 shares. Par value, $1.00. In treasury January 1, 1900, 175,000 shares.

Property

Owns the Bally-clare, the Bally-more, the Alabama, the Atlanta, the Elba, the Pensacola, the Tuscumbia, the Vendetta, in all about 78 acres, situated in the E. 1-2 of section 11, township 16 south, range 70 west, on Little Pisgah Peak; in process of patenting. Merely the amount of work required to secure a patent has been done in the way of development.

Highest price for stock during 1899, $7.50 per thousand; lowest price for stock during 1899, $2.50 per thousand.

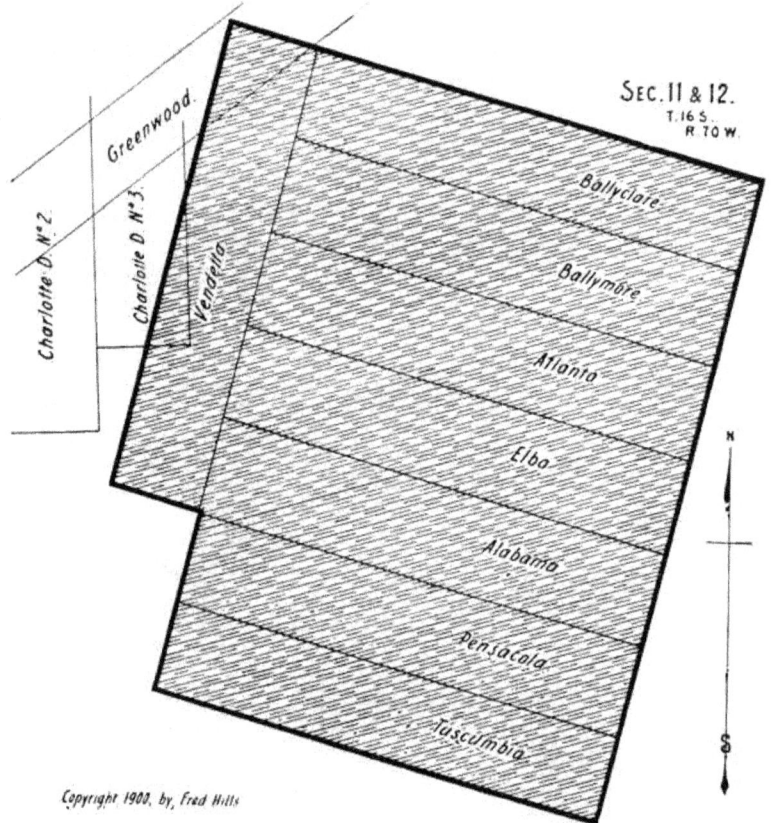

The Los Angeles Gold Mines Company.

Incorporated December 31, 1895.

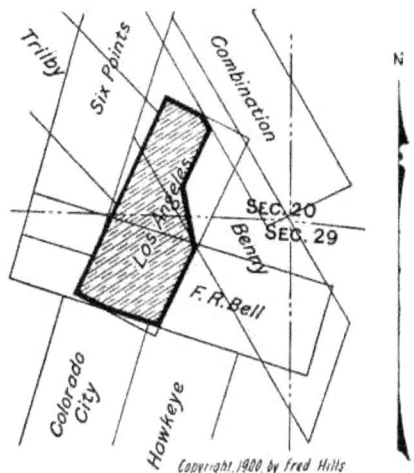

Copyright. 1900. by Fred Hills

Earl B. Coe...............President — Directors
T. J. Moynahan........Vice-President
W. N. McBird.Secretary and Treasurer
Chas. W. Babcock. De Putron Gliddon.
M. B. Carpenter.

Main Office—Suite No. 811-814 Ernest & Cranmer Building, Denver, Colorado.

1,500,000 shares. Par value, $1.00. — Capitalization
In treasury January 1, 1900, 50,000 shares.

Owns the Los Angeles lode claim, — Property
5 acres, on the line between sections 20 and 29, in the saddle between Bull hill and Battle mountain, adjoining the Portland property on the north. Property is patented and leased to B. Clark Wheeler for 3 years from February, 1900. Royalties 10 per cent. to 40 per cent.

There is a shaft house, blacksmith shop, assay office, as well as general offices, ore house, ore bins, etc., on the property. A shaft has been sunk to a depth of 560 feet, with about 900 feet of levels. — Development

Gross production to January 1, 1900, about $100,000.00. — Production

Highest price for stock during 1899, 20 cents; lowest price for stock during 1899, 7 1-2 cents.

The Lucky Guss Gold Mine, Limited.

W. P. Henderson, chairman, Copthall House, London, England.

This company is an English corporation and it has been impossible — Description
to secure particulars in regard to the same. As this book goes to press, May, 1900, it is reported that the property has been sold for $300,000.00,

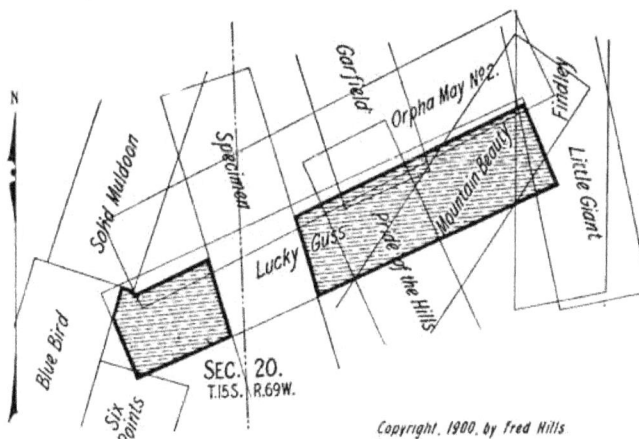

Copyright. 1900. by Fred Hills

to Mr. W. S. Stratton, of Colorado Springs. The Lucky Guss Company bought the property from the Wilson Creek G. M. Company some three years ago and it has produced over 6,000 tons of ore of a valuation of about $327,000.00. No stock has been sold.

The Mabel M. Consolidated Mining Company.

Incorporated 1893.

Directors

Horace W. Bennett......................President
Julius A. Myers......................Vice-President
F. M. Woods............................Secretary
H. E. Woods............................Treasurer
Warren Woods.   Dr. J. A. Whiting.
Thomas Murray.

Main Office—Giddings block, Colorado Springs, Colorado.

Capitalization

1,500,000 shares.   Par value, $1.00.   In treasury January 1, 1900, 500,000 shares.

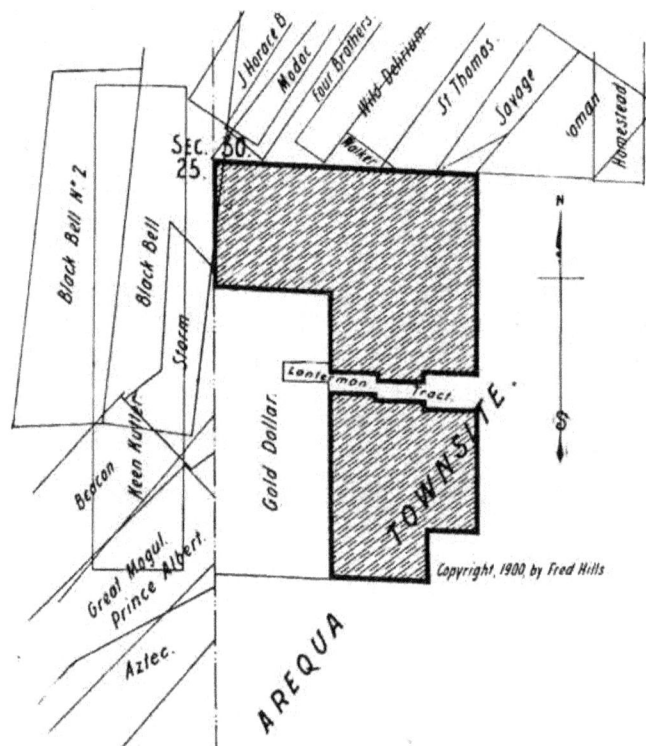

Copyright, 1900, by Fred Hills

Property

Owns a portion of the Arequa townsite, containing 18 acres of patented ground, in the W. 1-2 of section 30, adjoining the Gold Dollar and the Prince Albert groups, which are big producers of high grade ore.

Development

There is a main double-compartment shaft, 255 feet deep and 500 feet of drifting.   There are also 24 other shafts ranging in depth from 35 to 175 feet.   The property is leased, three leases being in operation at a 25 per cent. royalty.

Production

Gross production to January 1, 1900, over $100,000, and a steady output is now being made by the company.   The stock is not being offered on the market.

The Madison Gold Mining Company.

Incorporated August 29. 1899.

J. R. McKinnie..........................President

R. P. Davie................Secretary and Treasurer

A. J. Bendle.   J. L. Lindsay.

T. P. Day.

Directors

Main Office—No. 25 Pike's Peak avenue, Colorado Springs, Colorado.

1,250,000 shares.   Par value, $1.00.   In treasury January 1, 1900,   *Capitalization*
200,000 shares; in treasury January 1, 1900, $600.00 cash.

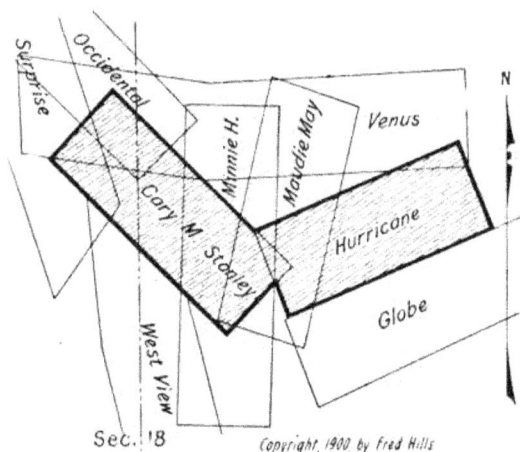

Copyright, 1900 by Fred Hills

Owns the Cary M. Stanley, containing 5 1-4 acres, and the Hurricane,   *Property*
containing 5 acres—both situated in the centre of section 18, on Globe
hill. Both patented.

Tunnel house on the Cary M. Stanley. Shaft house on the Hurricane.   *Development*
The Cary M. Stanley has also 150 feet of underground track, cars, etc. A
tunnel has been driven for 200 feet on the Stanley. Shafts and drifts on
the Hurricane. On this claim the greater part of the development work
has been done. Just as this Manual goes to press (May, 1900), the property of the above company was sold, for $35,000 cash, to C. W. Kurie,
trustee. A 3-cent dividend will be paid and the balance in the treasury
will be used by the company to purchase more property and develop it.

Highest price for stock during 1899, 4 cents; lowest price for stock
during 1899, 3 1-2 cents.

The Maggie Gold Mining Company.

Incorporated 1899.

Directors   J. M. Beaty, president; L. E. Sherman, vice-president; W. W. Williamson, secretary; John A. Thatcher, treasurer; W. G. McLean; W. A. MacWhorter.

Main Office—No. 25 1-2 N. Tejon street, Colorado Springs, Colorado.

Capitalization   1,500,000 shares.   Par value, $1.00.   In treasury January 1, 1900, 225,000 shares; in treasury January 1, 1900, $773.00 cash.

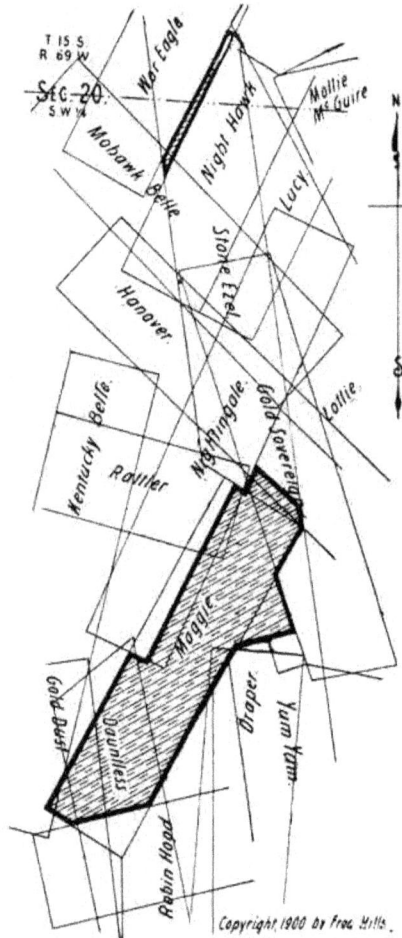

Property   Owns the Maggie, Yum Yum, Stone Ezel, a total of 8.737 acres; also a 1-2 conflict between the Maggie and the Nightingale, 0.737 acre, and the vein rights underground in conflict between the Stone Ezel and the Hanover, containing 1.168 acres; situated in the S. W. 1-4 section 20. Also owns all vein rights under the Gold Sovereign dump. All patented. Leased until January 1, 1902.

Development   Shaft house. Several shafts have been sunk and a number of drifts worked. About $15,000 worth of work has been done and a number of veins disclosed bearing low grade ore.

Gross production to January 1, 1900, about $1,500.00.

The Magna Charta Mining and Milling Company.

Incorporated.

C. B. Seldomridge......................President
L. W. Ralston.......................Vice-President
A. J. Smith...........................Secretary
M. F. Stark...........................Treasurer
F. M. Woods.

Main Office—No. 3 E. Huerfano street, Colorado Springs, Colorado; transfer office, No. 7 Barnes building, Colorado Springs, Colorado.

1,000,000 shares. Par value, $1.00. In treasury January 1, 1900, 103,000 shares; in treasury January 1, 1900, $2,500.00 cash.

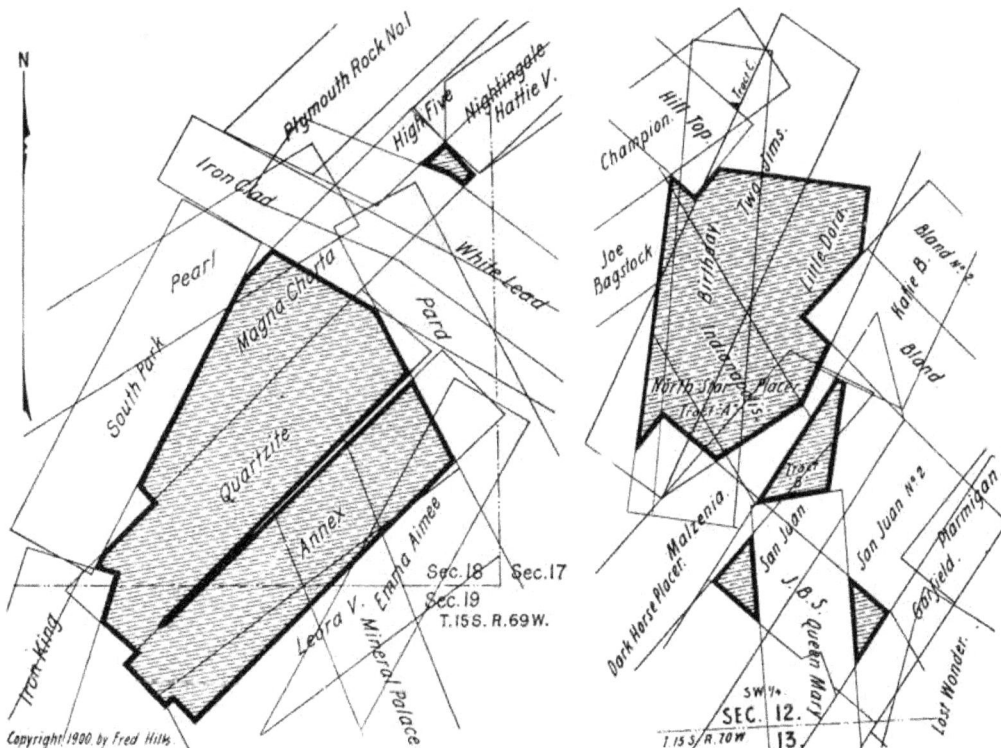

Copyright 1900 by Fred Hills

Owns the Magna Charta, Quartzite, and Annex, 22.53 acres, in the S. E. 1-4 section 18, on Ironclad hill; the North Star placer, and the J. B. S. lode, about 10 acres, on Mineral hill. All patented except the J. B. S. lode, which is held by location.

The Quartzite has a shaft 160 feet deep, and a tunnel of about 150 feet. The Annex has two shafts, one 100 feet deep, and one 500 feet deep. The greater part of the devlopment work is being done on the Quartzite.

Highest price for stock during 1899, 6 7-8 cents; lowest price for stock during 1899, 2 3-8 cents.

The Magnet Rock Gold Mining Company.

Incorporated November 29, 1895.

Directors

Clarence Edsall.........................President

Geo. R. Buckman....................Vice-President

Bertram N. Beal...........Secretary and Treasurer

John J. Key.   Wm. E. Jones.

Main Office—Hagerman building, Colorado Springs, Colorado.

Capitalization   1,250,000 shares. Par value, $1.00. In treasury January 1, 1900, 6,000 shares; in treasury January 1, 1900, $100.00 cash.

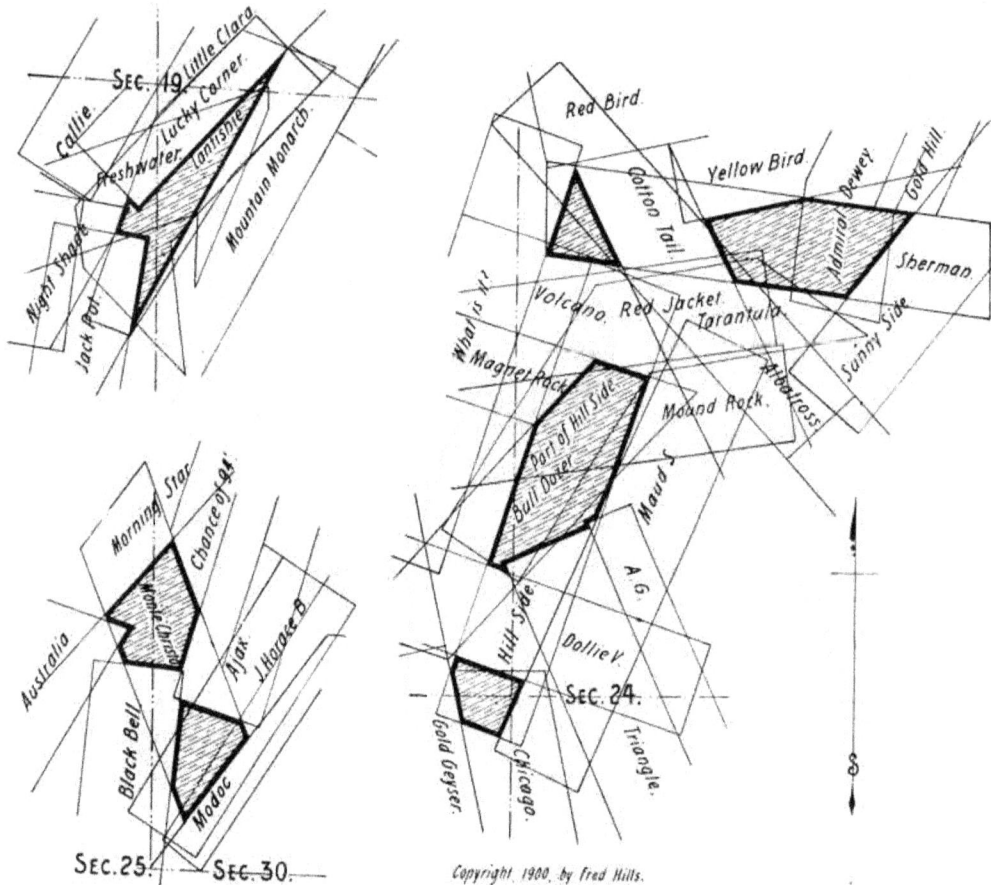

Copyright 1900, by Fred Hills.

Property   Owns the Lucky Corner, 2 1-2 acres, on north slope of Raven hill, in the center of section 19; the Monte Christo, 2 1-2 acres, on Beacon hill, in the N. W. 1-4 of section 30; the Hillside, 4 acres, in the N. E. 1-4 of section 24, and a two-thirds interest in the Sherman, 4 1-2 acres, on Gold hill, in the N. E. 1-4 of section 24. All patented.

The greater part of the development work has been done on the Sherman, which is leased for two years.

Highest price for stock during 1899, 5 3-4 cents; lowest price for stock during 1899, 1 3-4 cents.

The Magnolia Gold Mining Company.

Incorporated 1899.

S. R. Bartlett, president; Warren Woods, vice-president; F. M. Directors Woods, secretary and general manager; H. E. Woods, treasurer; J. M. Allen.

Main Office—The Woods Investment Company, Giddings building, Colorado Springs, Colorado.

1,250,000 shares. Par value, $1.00. In treasury January 1, 1900, Capitalization 150,000 shares; in treasury January 1, 1900, $10,000.00 cash.

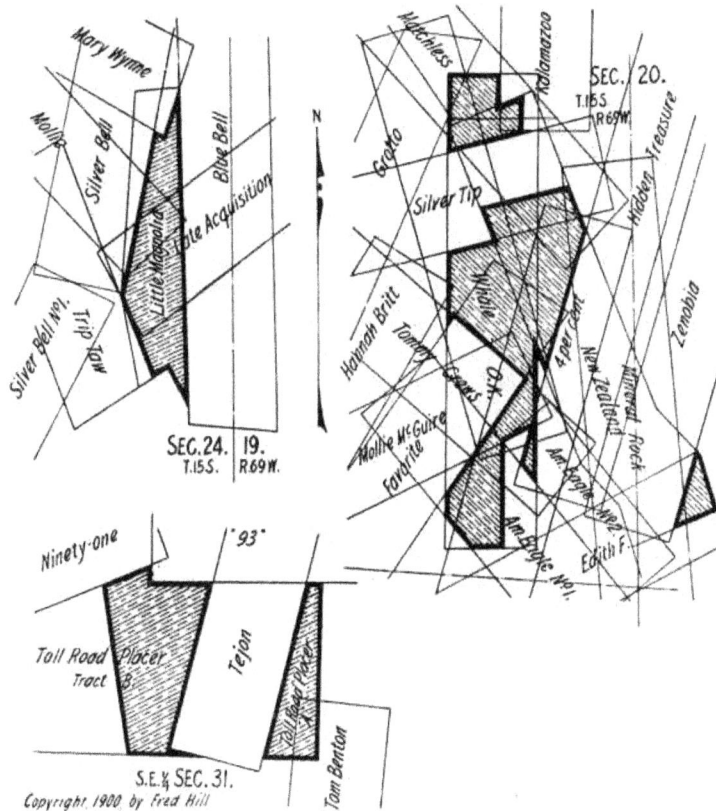

Owns the O. K., New Zealand, a portion of the Favorite and the Property Mollie McGuire, containing, in all, 7.154 acres, on Bull hill, in N. E. 1-4 of section 20; also the Little Magnolia, containing 2.94 acres, on Guyot hill, in S. E. 1-4 of section 24, and the Toll Road placer, containing 5 1-4 acres, on Straub mountain, in S. E. 1-4 of section 31, near the Lawrence townsite. All patented.

Shaft house and hoist on the New Zealand. Two small shaft houses Development on the O. K., besides several cabins and a blacksmith shop. The New Zealand has a shaft 500 feet deep and a small amount of drifting. The O. K. has a shaft 200 feet deep, one 50 feet deep, and several minor shafts. The Magnolia has a tunnel 285 feet in length. Winze in tunnel on this property. The greater part of the development work is being done on the O. K. and the New Zealand group.

On March 31, 1900, stock sold at 15 1-2 cents.

The Mahoning Gold Mining Company.

Incorporated September 14, 1898.

Directors — T. H. Whiteside, president; B. F. Rebman, vice-president; S. R. Frazier, secretary; F. G. Whiteside, treasurer and assistant secretary; Harry Bonnell, J. B. Adamson, A. D. Thomas.

Main Office—108 W. Wood street, Youngstown, Ohio.

Capitalization — 125,000 shares. Par value, $1.00. In treasury January 1, 1900, 28,500 shares; in treasury January 1, 1900, $500.00 cash.

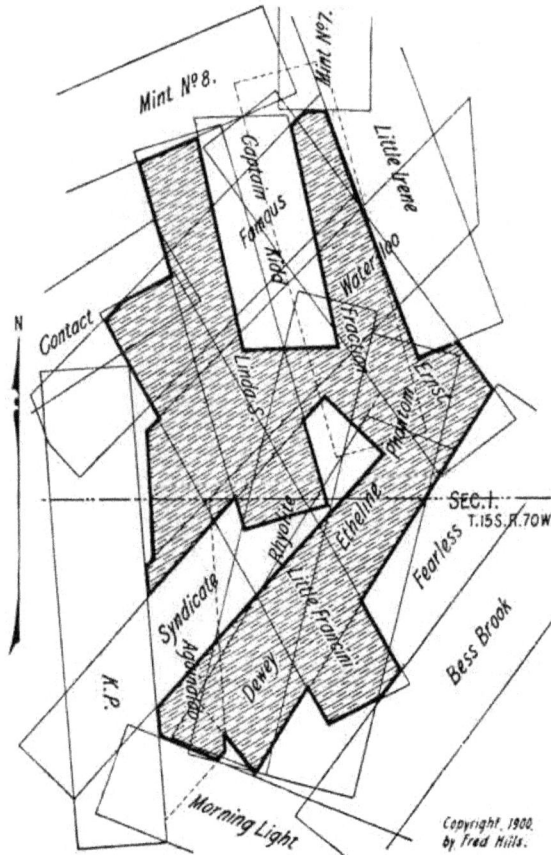

Copyright, 1900, by Fred Hills.

Property — Owns the Little Francini and the Ernst, containing 11.511 acres; the Linda S. and the Waterloo, containing 11.749 acres; and the Dewey, the Aguinaldo, and Fractions Nos. 1 and 2, containing about 10 acres; all situated on Rhyolite mountain, in the E. 1-2 of section 1, township 15 south, range 70 west. The Little Francini and the Ernst are patented. Receiver's receipts held for the Linda S. and the Waterloo. The Dewey, Aguinaldo and the Fractions Nos. 1 and 2 in process of patenting. The latter is not shown on plat.

Development — The Ernst has a shaft house and whim; shaft house also on the Francini. On the Ernst is a shaft 100 feet deep and 30 feet of drifting. The Little Francini has a shaft 110 feet deep and one 50 feet deep. The Linda S. has two shafts, one of 75 feet depth and one of 50 feet depth. The Waterloo has two shafts, one of 30 feet, one of 25 feet depth. On the other properties are several shafts from 10 to 20 feet deep.

History — The most of these claims were staked for some years, but no active work towards patenting was done until 1898, and the result is as above indicated.

Highest price for stock during 1899, 30 cents; lowest price for stock during 1899, 30 cents.

The Major-Loughrey Gold Mining and Milling Company.

Incorporated September 24, 1898.

John Loughrey.........................President
Edward Major.....................Vice-President
T. E. Major...............Secretary and Treasurer
Jos. O. Major............................Manager

Mrs. Emma Major.

Directors

Main Office—At the mine, Copper mountain. Branch Office—Room 217, No. 300 Bennett avenue, Cripple Creek, Colorado.

500,000 shares. Par value, $1.00.

Capitalization

Owns the Violet, Sans Pareil, Loughrey, Grand Review and the Last Chance placer, containing in all 25 acres, in the S. E. 1-4 section 2, on Copper mountain, adjoining the Fluorine, Fort Wilcox and Copper mountain properties. Receiver's receipt held for the above property.

Property

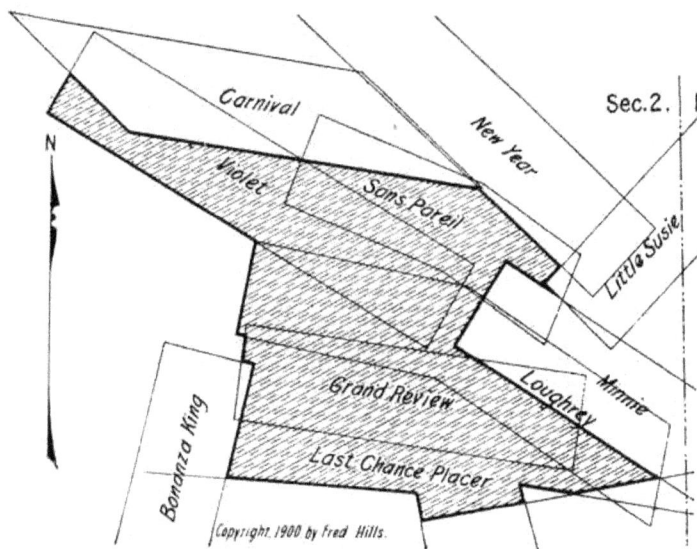

Copyright. 1900 by Fred Hills.

Three log cabins on the property, 18x20-foot blacksmith shop, tracks, cars and tools to work 25 men. Four shafts on four separate veins on the top of Copper mountain, No. 1, 90 feet; No. 2, 70 feet; No. 3, 50 feet; No. 4, 40 feet. These are all on the Violet. Also 200 feet of drifts on veins; two pay ore shoots in shafts No. 1 and No. 4. Tunnel at the base of the mountain is in 500 feet and requires only 120 feet more to cut the veins opened on the property. The four main shafts, as above mentioned, are on the Violet. First-class timbering, with a view to putting in No. 1 machinery. The tunnel is being continuously worked, and, when completed, will have cut not only the two pay ore shoots, but seven well-defined veins, all carrying pay ore. The tunnel will be completed within 90 days, or by the 1st of June, 1900. Thus ore can be taken out through the shafts and the tunnel, giving an immense output.

Development

Highest price for stock during 1899, 25 cents; lowest price for stock during 1899, 5 cents.

The Margaret Gold Mining Company.

Incorporated July 10, 1899.

Directors   H. J. Newman, president; John C. Mitchell, vice-president; H. L. Shepherd, secretary and treasurer; F. H. Dunnington, assistant secretary; J. E. Jones.

Main Office—2 North Nevada avenue, Colorado Springs, Colorado.

Capitalization   1,250,000 shares. Par value, $1.00. In treasury January 1, 1900, 100,000 shares; in treasury January 1, 1900, $1,800.00 cash.

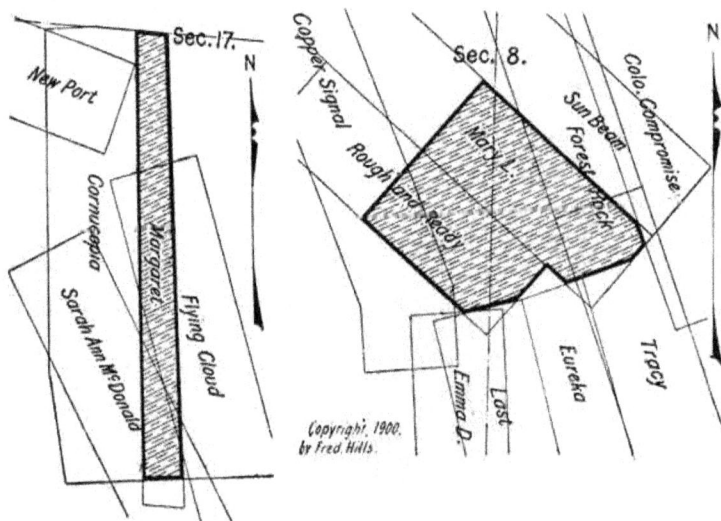

Property   Owns the Margaret, 4 acres, in the S. E. 1-4 section 17, township 15 south, range 69 west, and a portion of the Mary L. and Rough and Ready, 8.28 acres, on Galena hill, in the S. E. 1-4 section 8. All patented.

Development   Four shafts on the property, with but very little drifting. Two sets of lessees developing the Margaret claim. A very good vein has been opened up on the Margaret and a little pay ore found, but not in paying quantities. The outlook is very good. A leasing system is followed by the company.

Highest price for stock during 1899, 5 3-8 cents; lowest price for stock during 1899, 3 cents.

The Margery Gold Mining Company.

F. A. Williams, president; J. L. Lindsay, vice-president; W. H. MacIntyre, secretary; H. A. MacIntyre.
Directors

Main Office—Bank block, Colorado Springs, Colorado.

1,500,000 shares. Par value, $1.00. In treasury January 1, 1900, 300,000 shares; in treasury January 1, 1900, about $2,000.00 cash.

Capitalization

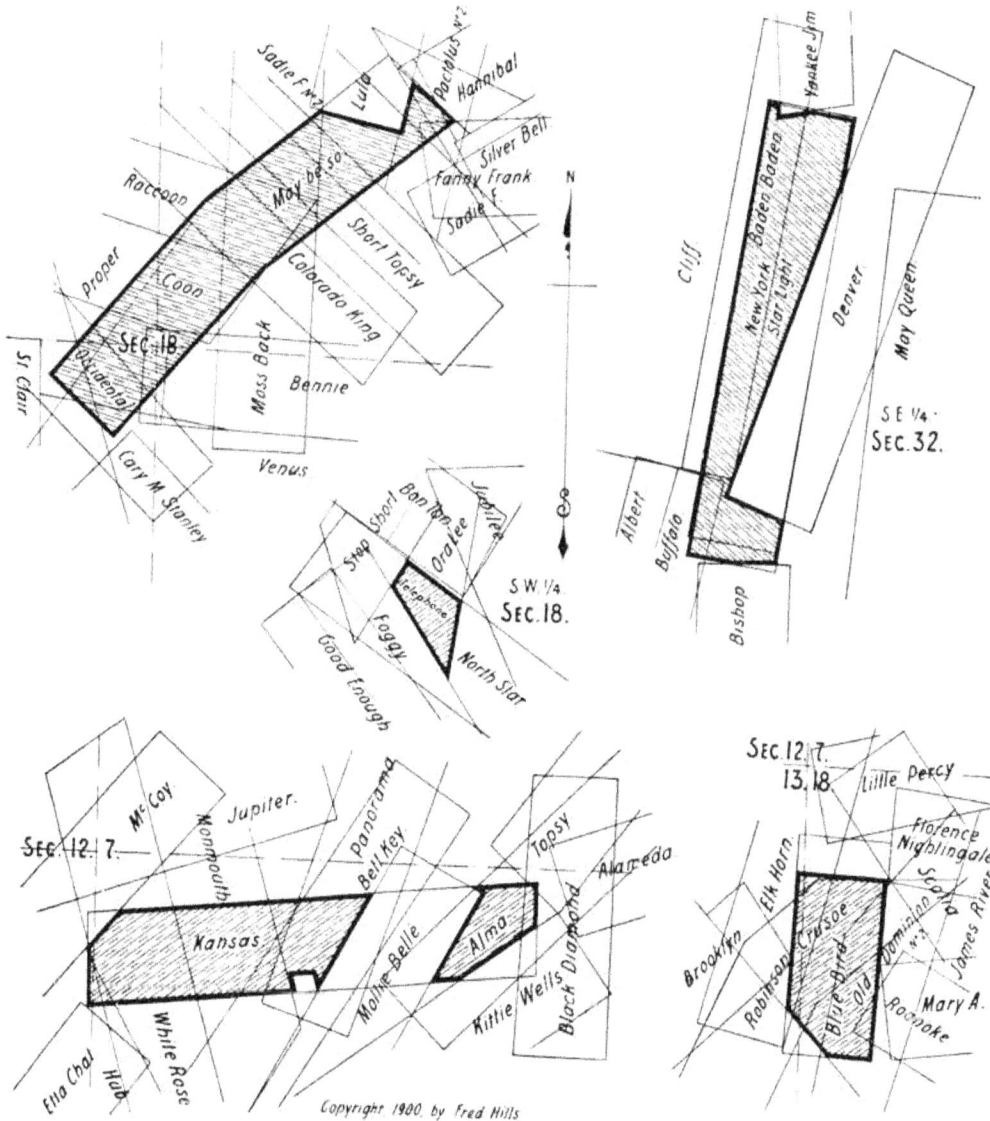

Copyright 1900, by Fred Hills

Owns the May-Be-So, 9 acres, in the N. E. 1-4 section 18; the Telephone, 1 acre, in the S. W. 1-4 section 18; both on Globe hill; the Kansas, 7 acres, on Carbonate hill, in the S. W. 1-4 section 7; the Blue Bird, 4 acres, in the S. W. 1-4 section 18, on Mineral hill; the Starlight, 6 acres, in the S. E. 1-4 section 32, south of Victor. All the above property is patented. In April, 1900, a cash offer of $50,000 was made for the May-Be-So claim, which offer was refused by the company.

Property

A good shaft house and steam plant of machinery is on the property; also one shaft 225 feet deep has been sunk and some drifting. The greater part of the development work has been prosecuted on the May-Be-So claim.

Development

Highest price for stock during 1899, 6 1-8 cents; lowest price for stock during 1899, 4 7-8 cents.

The Maria A. Gold Mining Company.

Incorporated December, 1895.

Directors

J. R. McKinnie...........................President

B. C. Joy.............................Vice-President

O. H. Shoup...............Secretary and Treasurer

Verner Z. Reed.   Frank G. Peck.

W. S. Jackson.

Main Office—Bank block, Colorado Springs, Colorado.

Capitalization

1,250,000 shares. Par value, $1.00. In treasury January 1, 1900, 249,500 shares.

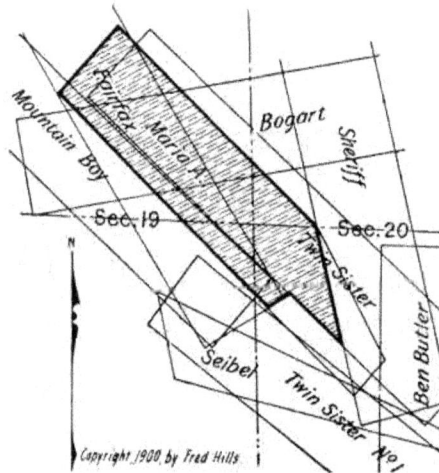

Copyright 1900, by Fred Hills.

Property

Owns the Maria A., containing 7.89 acres, in the S. W. 1-4 section 20, on Raven hill. Patented.

Development

The company has always pursued an active leasing system of their territory, and for the past year three sets have been actively engaged in development of the claim. As yet no shipments have been made, but the indications are very favorable for the property to become a mine if development is properly done.

Highest price for stock during 1899, 6 cents; lowest price for stock during 1899, 3 1-8 cents.

The Marinette Gold Mining Company.

Incorporated.

Property

Owns a part of the Abe Lincoln, containing about 1 3-4 acres, in section 13, in Poverty gulch. This property is now owned by Mr. W. S. Stratton, of Colorado Springs, and can be seen by referring to "Stratton's Group" in this Manual.

The Mariposa Mining and Tunnel Company.

Incorporated January 17, 1896.

W. R. Foley, president; W. M. Dutton, vice-president; S. J. Mattocks, Directors secretary and treasurer; C. H. Mattocks; D. Weyand.

Main Office—104 E. Pike's Peak avenue, Colorado Springs, Colorado.

1,500,000 shares. Par value, $1.00. In treasury January 1, 1900, Capitalization 350,000 shares; in treasury January 1, 1900, $7,000.00 cash.

Owns the Yellow Bird and Cotton Tail claims, containing about 14 Property acres; all patented. Situated in N. 1-2 of section 24, on Gold hill. Also Cripple Creek claims 1, 2, 3, 4, 5, 6, 7, 8 and 9, 1 1-4 miles S. W. of Cripple Creek, containing about 60 acres, situated in section 22.

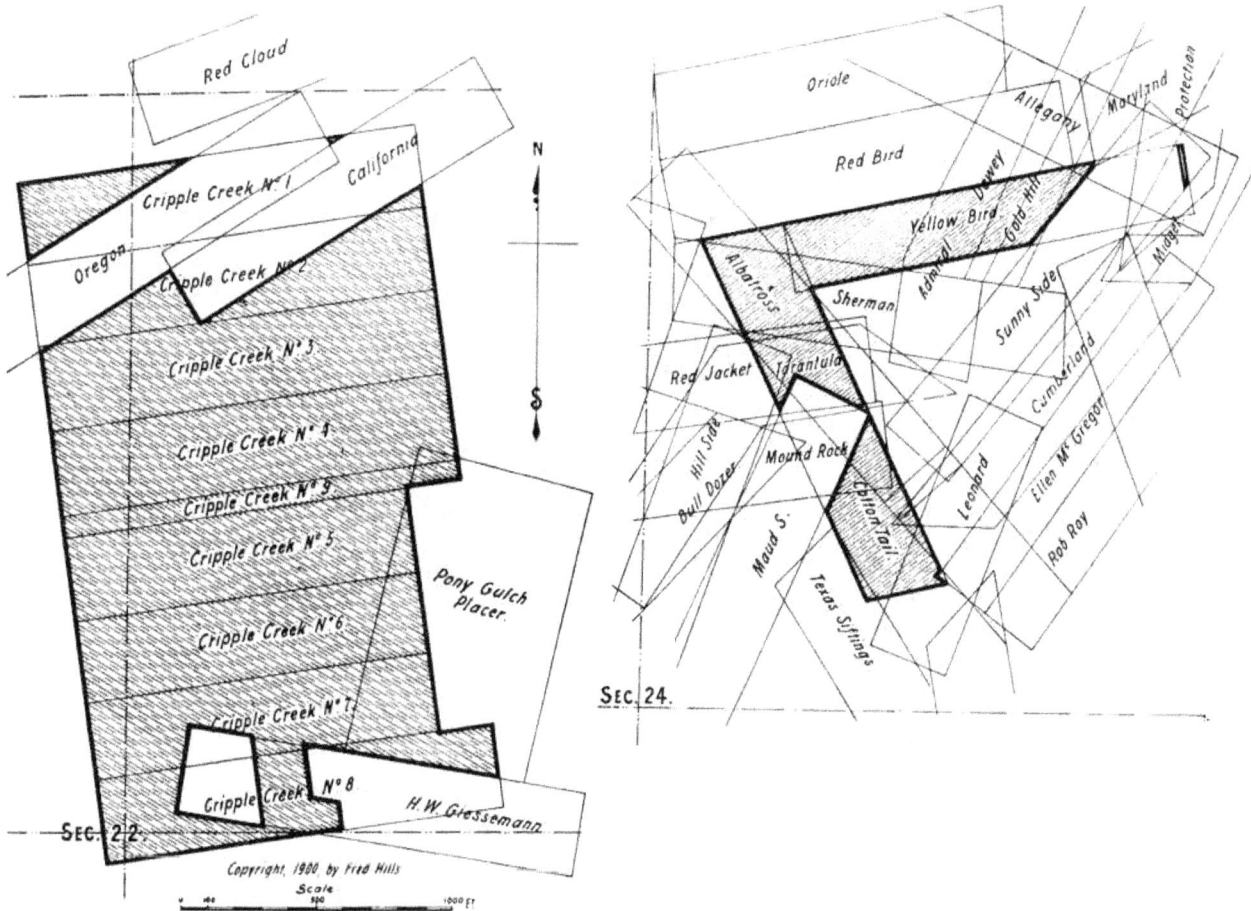

Improvements consist of gallows frames over two shafts on the Development Yellow Bird claim, each 150 feet deep. The north 200 feet of the Yellow Bird claim is leased for two years. The property side lines with the Red Bird claim of the National Mining, Tunnel and Land Company, and the company have a working contract to work the property through the shaft of the National Company. The Yellow Bird and Cotton Tail claims have been in litigation for five years, and the same has just been settled by the purchase of the claims by the Mariposa Company, and development is to be vigorously pushed. There is a good body of ore opened up in the second level, and also ore opened up in the third level. It is the intention to open up the property from the National shaft, which is now down 400 feet.

Production to January 1, 1900, $10,000.00; highest price for stock Production during 1899, 8 cents; lowest price for stock during 1899, 8 cents.

The Maroon Tunnel and Mining Company.

Incorporated November 25, 1889.

Directors

A. L. Lawton..............................President

J. A. Leech.........................Vice-President

T. G. Horn.............................Secretary

J. H. Thedinga..........................Treasurer

D. H. Rice.

Main Office—No. 30 S. Tejon street, Colorado Springs, Colorado.

Capitalization
1,000,000 shares. Par value, $1.00. In treasury January 1, 1900, 27,-000 shares; in treasury January 1, 1900, $175.00 cash.

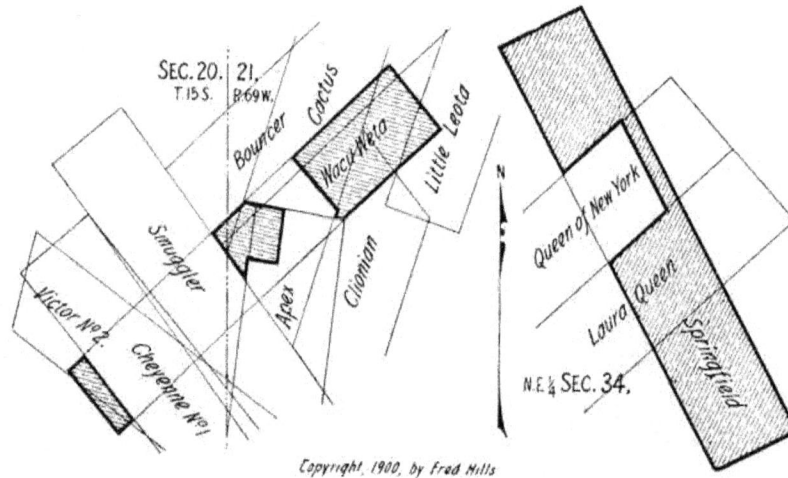

Copyright, 1900, by Fred Hills

Property
Owns the Wacu Weta, 4 3-4 acres, in the N. W. 1-4 section 21, on Bull hill; and the Springfield, 9 acres, in the N. E. 1-4 section 34, on the east slope of Big Bull hill. All patented. The company also own the Voice and the Expectation, each containing 10 acres, in the Aspen district. Patented. The latter can not be shown on plat.

Development
The Wacu Weta, which is leased, has a shaft 185 feet deep, and 100 feet of drifting. A steam hoist is on this claim. The greater part of the development work is being done on the Wacu Weta claim. The shaft on this claim will be sunk 200 feet deeper under the present lease. This property lies near the Isabella and other producing properties, hence the owners consider it very valuable property. There is an 80-foot tunnel on the Aspen property.

Highest price for stock during 1899, 5 cents; lowest price for stock during 1899, 2 cents.

The Marquette Gold Mining Company.

Incorporated September, 1899.

John M. Harnan........................President

Wm. H. Powell....................Vice-President

Geo. A. Powell.............Secretary and Treasurer

L. L. Aitken.   Edw. S. Kelley.

Directors

Main Office—4 Mining Exchange, Colorado Springs, Colorado.

1,250,000 shares. Par value, $1.00. In treasury January 1, 1900, 250,000 shares.

Capitalization

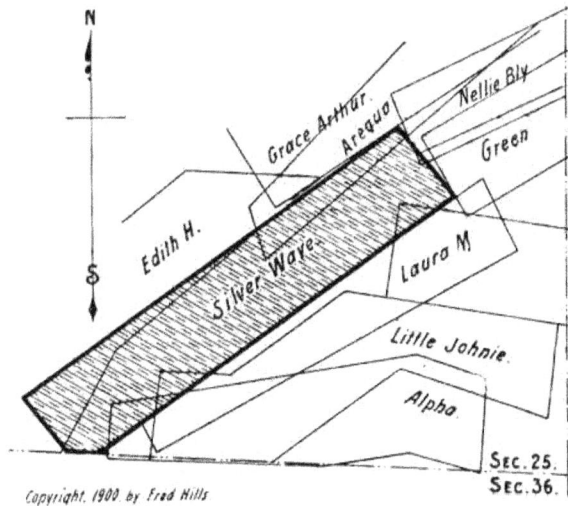

Copyright, 1900 by Fred Hills

Owns the Silver Wave, 9.83 acres, in the S. W. 1-4 section 25, on Beacon hill. Patented.

Property

One shaft 108 feet deep, one 40 feet deep, with tunnel 210 feet and about 100 feet of drifting.

The Mars Consolidated Gold Mining Company.

Incorporated.

Directors

C. S. Wilson, president; J. C. Staats, vice-president; E. A. Meredith, secretary; Wm. A. Otis, treasurer.

Main Office — Giddings Building, Colorado Springs, Colo.

Capitalization

1,500,000 shares. Par value, $1.00.

Property

Owns the Steuben, Pulaski, Henry, Mars and part of the Lone Star No. 1, 13 acres in Poverty Gulch, S. E. 1-4 section 13.

Development

There is a 120-foot shaft on the Henry, with 300 feet of drifts and levels. There are seven shafts on the Steuben, Pulaski and Lone Star No. 1, varying in depth from 40 to 80 feet, as also numerous shallow cuts and trenches. There is also a 65-foot shaft on the north end of the Pulaski, which is now being worked on a good vein.

History

The largest portion of this property has been leased to F. O. Woods of Colorado Springs. As this Manual goes to press there is an impression that Mr. W. S. Stratton has bought control of this company, but it is impossible to state accurately that this is a fact.

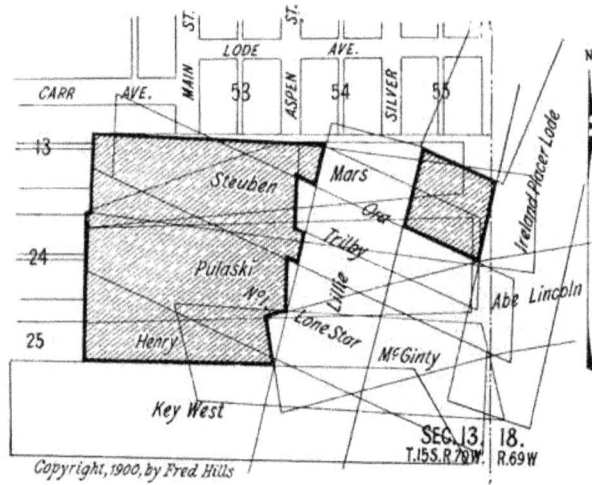

The Mary Ann Mining Company.

Incorporated January 10, 1898.

Directors

M. Kennedy .... President
C. F. Bryant .... Vice-President
Wm. A. Otis ...... Secretary and Treasurer
J. C. Connor.   G. Kissell.

Main Office—Wm. A. Otis & Co., Giddings Block, Colorado Springs, Colorado.

Capitalization

1,000,000 shares. Par value, $1.00. In treasury January 1, 1900, 200,000 shares.

Property

Owns the Mary Ann, containing 7 acres, situated in the S. E. 1-4 section 19, on Raven hill.

The claim is patented.

On the property there is an electric hoist, and surface plant.

The company have leased the claim.

The stock has been very inactive and was selling at about 6 cents in May, 1900.

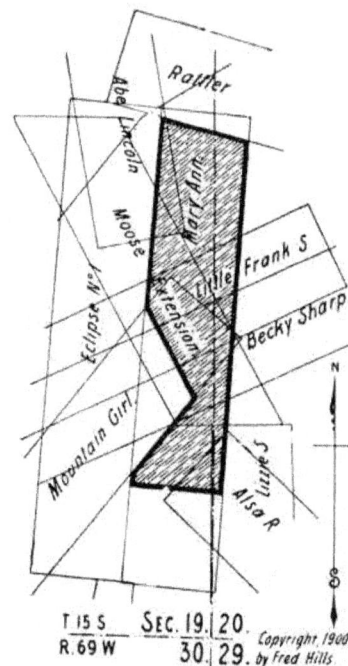

The Mary Cashen Mining Company.

Incorporated November, 1899.

J. P. Pomeroy, president; B. P. Waggener, vice-president; Clarence **Directors** Edsall, secretary and treasurer; J. P. Sweeney, second vice-president; W. T. Bland, B. N. Beal.

Main Office—Hagerman building, Colorado Springs, Colorado.

1,500,000 shares. Par value, $1.00. In treasury January 1, 1900, **Capitalization** 150,000 shares; in treasury January 1, 1900, $7,000.00 cash.

Owns the mineral rights to all that portion of the Spicer claim north **Property** of Victor avenue, containing about 3 acres, and a portion of the northeast part of the Mt. Rosa placer, including the Spicer Extension claim, in the town of Victor, in section 29, containing about 7 acres. The company also owns the surface of lots Nos. 5, 6, 7, 8, 9 and a portion of No. 4 of block 7, in said town of Victor.

This property had formerly been worked under bond and lease for **Development** $100,000 from the Mt. Rosa M. & M. Co., for two years, when it was purchased by this company. The former lessees had been cross-cutting to the west, but failed to find pay ore; in cross-cutting to the east, however, this company has opened up a large fissure vein, varying from 8 to 12 feet wide, and containing several ore chutes. This has been opened up at the 350-foot and 425-foot levels. Besides three ore houses, the company has a commodious building, containing a 30-horse power electric hoist and compressor and five air drills, the air being supplied by the La Bella Mill, Water and Power Company.

The mine is now producing and is expected to pay dividends in a **Production** very short time. **and Dividend**

As the company was only incorporated in November, 1899, the price of stock during 1899 can not be given. At the time of the incorporation the price of stock was 20 cents per share, and on February 23, 1900, was 39 1-2 cents.

The Mary McKinney Mining Company.

Incorporated March 30, 1892.

Directors — Frank F. Castello, president; W. S. Nichols, vice-president; A. C. Van Cott, secretary and treasurer; Charles Thurlow; P. J. Ryan; W. H. Ellice.

Main Office — Room 11, P. O. Block, Colorado Springs, Colorado.

Capitalization — 1,000,000 shares. Par value, $1.00. In treasury January 1, 1900, 100,000 shares; in treasury, January 1, 1900, a very substantial surplus in cash.

Property — Owns the Mary McKinney, the Mary McKinney No. 2, the Republic, the Mayflower, the Le Claire, and the Thurlow lodes, containing 34 1-2 acres, situated in section 19, on Gold hill, and including the townsite of Anaconda. All the above property is patented.

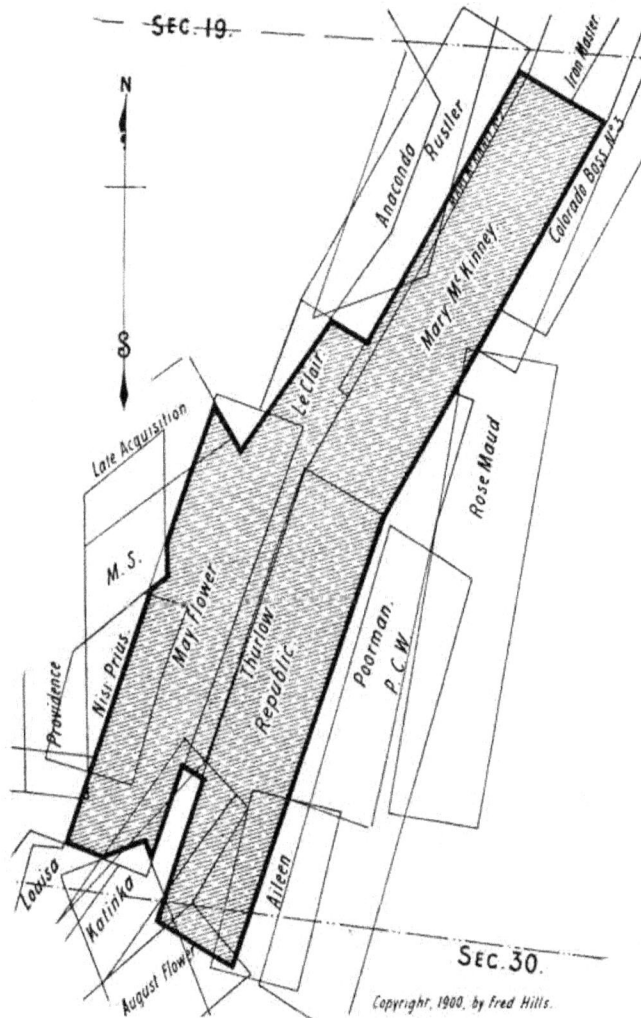

Copyright, 1900, by Fred Hills.

Development — This is, and has been, a very close corporation, and the property was operated and developed under the leasing system with varying success until 1898. Early in 1899 the company took hold, built a commodious shaft house and installed a fine hoisting plant, costing in the neighborhood of $50,000. Since then the company has carried on its own mining operations. A large gross production of ore has been made, and two quarterly dividends have been paid of 3 cents each, amounting to $30,000, the last being paid January 10, 1900. A substantial surplus is at all times in the treasury and, in fact, the company is one of the most prosperous in the camp. As practically no stock changes hands, no quotations can be given. Being a close corporation, it is impossible to obtain the amount of the gross production.

Production and Dividend — Total dividends up to January 1, 1900, $60,000.00. Amount of dividend, paid January 10, 1900, $30,000.00; amount of last dividend, paid April 10, 1900, $60,000.00.

The Matoa Gold Mining Company.

Incorporated April 27, 1892.

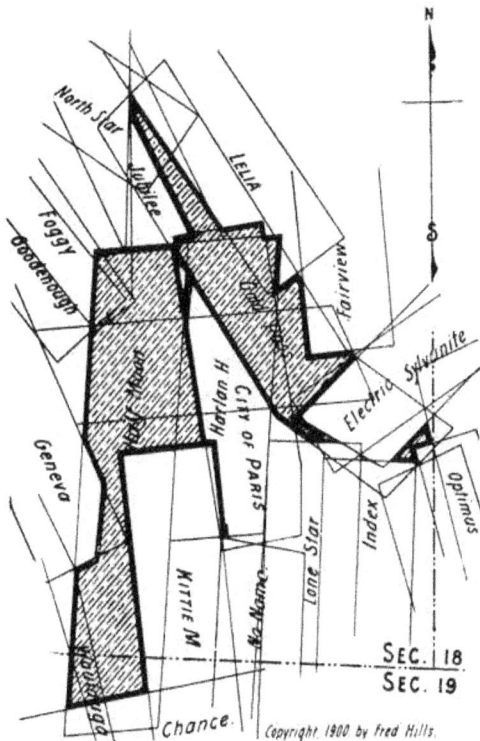

Copyright 1900 by Fred Hills.

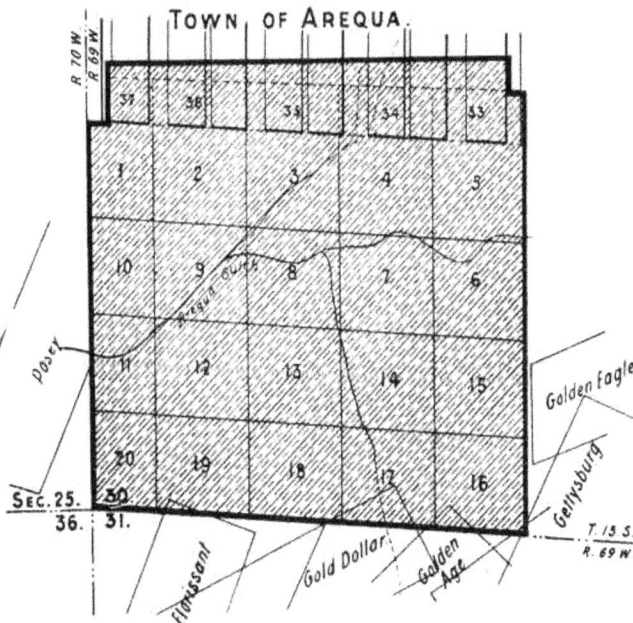

TOWN OF AREQUA.

H. P. Lillibridge, president; E. F. Smith, vice-president and treasurer; W. S. Reynolds, secretary; W. M. Lillibridge; James F. Burns. — **Directors**

Main Office — No. 51 Hagerman Building, Colorado Springs, Colorado.

1,000,000 shares. Par value, $1.00. — **Capitalization**

Owns the Half Moon, the Harlan H., and the Gold Pass No. 1, all on Gold hill and containing 12.44 acres; also the territory embraced in the S. W. 1-4 of the S. W. 1-4 of section 30, township 15 south, range 69 west, reaching from Beacon hill downwards through Arequa gulch and on to Grouse mountain, and containing 47.794 acres. — **Property**

Titles of all are based on U. S. patents, except about 1-3 of one acre, to which patent has not yet been received.

There are three shaft houses on the property, equipped with steam and electric machinery; also ore houses, etc. — **Development**

Air drills working underground from surface power plant.

There are numerous shafts from 100 to 1,000 feet deep. The shaft on Block No. 1, now 835 feet deep, is being sunk to a depth of 1,000 feet. The shaft on Block No. 5 is 415 feet deep. The Gold Pass shaft is 410 feet deep. A large amount of cross-cutting, stoping and drifting has been done on these shafts. The Half Moon is receiving the greater part of the development work.

Gross production to January 1, 1900, $625,474.48. One dividend has been paid (December 24, 1898) of $25,000.00. — **Production and Dividend**

Highest price for stock during 1899, 52 cents; lowest price for stock during 1899, 27 cents.

The Matt France Mining and Milling Company.

Incorporated 1893.

Directors
J. K. Miller............................President
W. W. Williamson..........Secretary and Treasurer
Matt France (Deceased).   S. Davidson.
N. H. West.   A. Hemenway.
E. G. Davis.

Main Office—No. 25 1-2 N. Tejon street, Colorado Springs, Colorado.

Capitalization
900,000 shares. Par value, $1.00. In treasury January 1, 1900, 59,-850 shares; in treasury January 1, 1900, $250.00 cash.

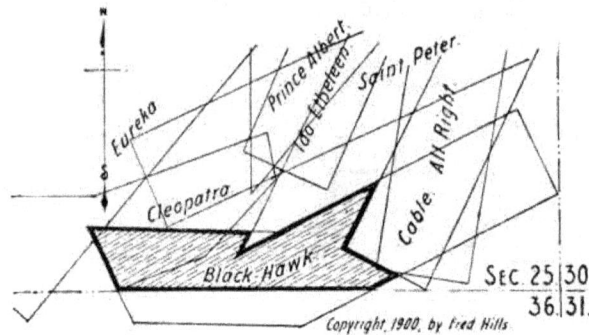

Property
Owns a part of the Black Hawk (4.736 acres), and the vein rights under the rest of the Black Hawk (3.964 acres), in conflict with the Posey and Cable. Property is situated in the S. E. 1-4 of section 25, on the north slope of Grouse mountain. The company also holds, under a 20-year lease, lots 50 and 63 in section 36, south of the Black Hawk on Grouse mountain. Not shown on plat.

Highest price for stock during 1899, $8.50 per 1,000; lowest price for stock during 1899, $4.00 per 1,000.

The Mayflower Gold Mining Company.

Incorporated October 8, 1894.

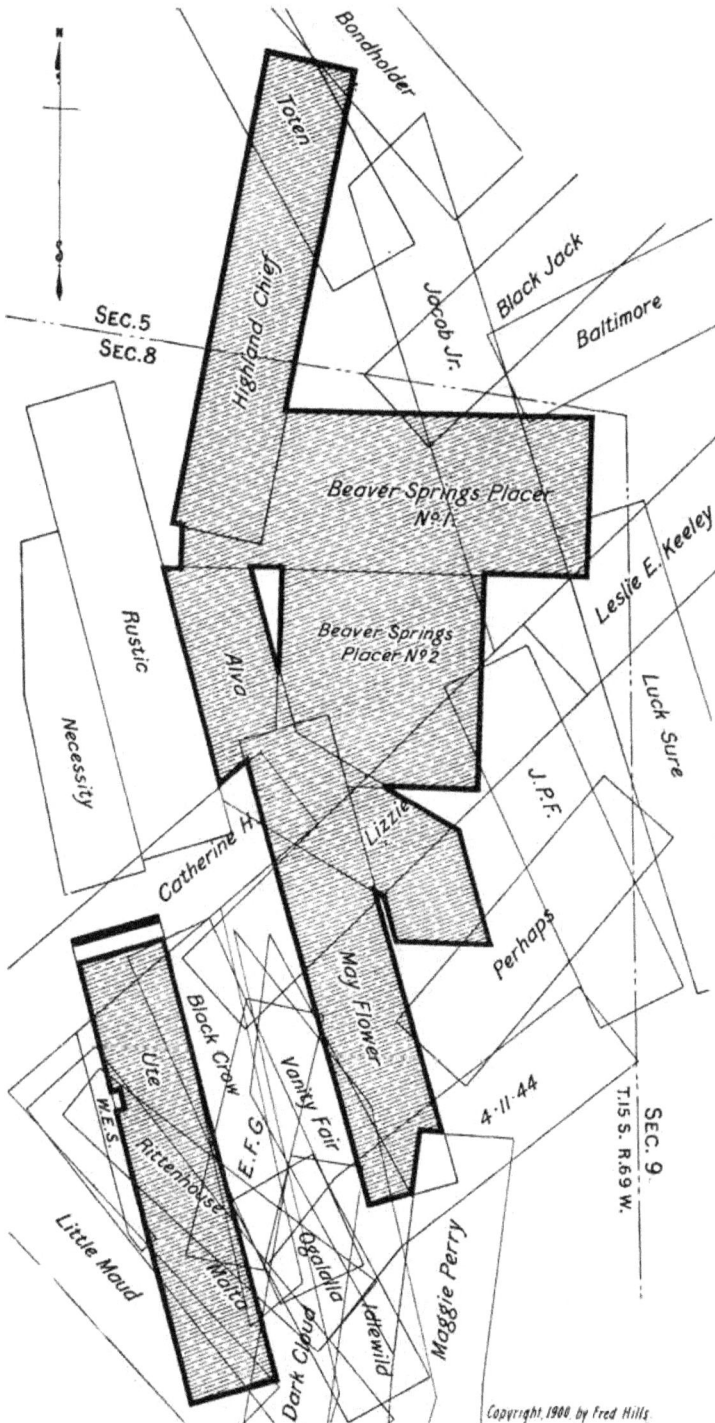

Copyright, 1900. by Fred Hills.

Directors
H. P. Lillibridge, president; E. F. Smith, vice-president and treasurer; W. S. Reynolds, secretary; S. B. Stewart; W. M. Lillibridge.

Main Office—No. 51 and 52 Hagerman building, Colorado Springs, Colorado.

Capitalization
1,000,000 shares. Par value, $1.00. In treasury January 1, 1900, 15,000 shares.

Property
Owns the May Flower, Highland Chief, Ute, Alva, Beaver Springs Placer Nos. 1 and 2, containing in all 57.248 acres, all patented except the Alva—which is in process. All property located in section 8, on Galena hill.

Development
There are numerous shafts on the property from 25 to 130 feet in depth; also one tunnel 50 feet long.

Highest price for stock during 1899, 10 cents; lowest price for stock during 1899, 3 cents.

The Memphis Gold Mining Company.

Incorporated 1895.

Directors

S. J. Burris.............................President

W. S. Sexton........................Vice-President

Wm. Barber...............Secretary and Treasurer

C. E. Miller.   Chas. E. Cherrington.

Main Office—No. 235 N. Union avenue, Pueblo, Colorado.

Capitalization   1,250,000 shares. Par value, $1.00. In treasury January 1, 1900, 75,-000 shares; in treasury April 1, 1900, $150.00 cash.

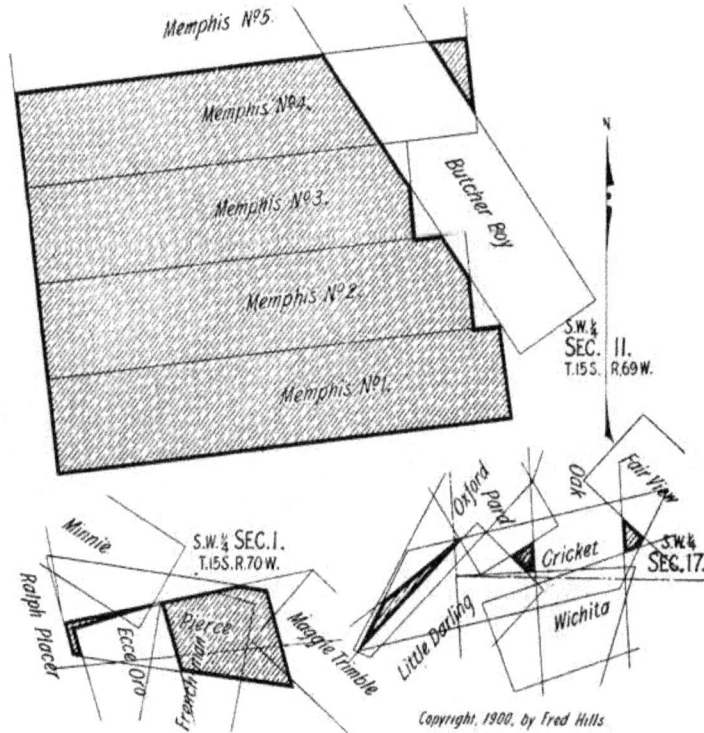

Copyright, 1900, by Fred Hills.

Property   Owns the Memphis Nos. 1, 2, 3, and 4, containing 36 acres, situated on Cow mountain, in the S. W. 1-4 section 11, township 15 south, range 69 west; the Pierce, survey No. 10,924, containing 2 1-2 acres, on Copper mountain, in the S. W. 1-4 section 1; the Cricket, survey No. 9,712, containing 0.528 acres, situated on Ironclad hill in the S. W. 1-4 section 17. All the above property is patented.

The Memphis group is leased until April 1, 1902. The stock is not quoted.

The Merrimac Consolidated Mines Company.

Incorporated January 30, 1900.

E. E. Armour..........................President   Directors
James M. Downing..................Vice-President
Jos. A. Michel.............Secretary and Treasurer
Charles M. Sumner.   A. E. Thomas.

Main Office—519 Equitable building, Denver, Colorado; branch offices, 109 East Kiowa street, Colorado Springs, Colorado; 315 Bennett avenue, Cripple Creek, Colorado.

1,500,000 shares.   Par value, $1.00.   In treasury January 1, 1900,   Capitalization 300,000 shares.

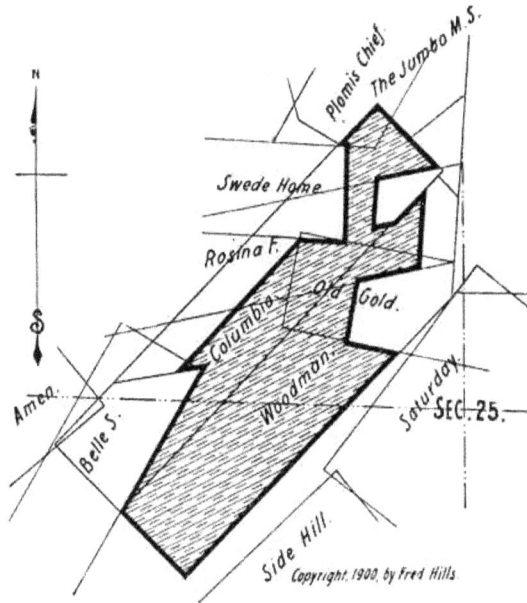

Copyright, 1900, by Fred Hills

Owns the Woodman and Columbia lodes, containing about 12.4 acres, Property located in section 25, township 15 south, range 70 west, on Beacon hill. Receiver's receipt is held for this property.

The south 500 feet of the Woodman is leased for eighteen months at a Development royalty of 15 per cent., and the property will be developed by lessees. The company consider their prospects very bright for the future, and think it only a question of more development work when they will have good ore.

Stock has been sold for 2 cents per share.

The Metropolitan Mining and Milling Company.

Incorporated April, 1900.

Directors
Franc O. Wood, president; Wm. H. Powell, vice-president and treasurer; Geo. A. Powell, secretary; Wm. Dissman; O. B. Willcox.

Main Office—With Powell Brothers, Mining Exchange building, Colorado Springs, Colorado.

Capitalization
500,000 shares. Par value, $1.00. In treasury January 1, 1900, 100,000 shares; in treasury April 1, 1900, $10,000.00 cash.

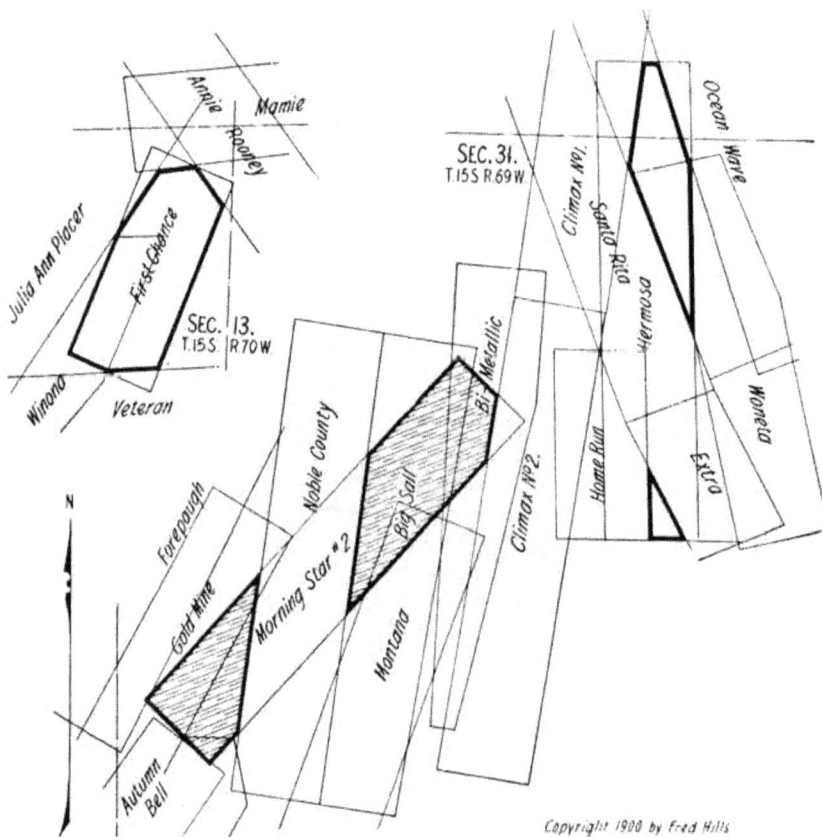

Copyright 1900 by Fred Hills

Property
Owns the Morning Star No. 2, containing 6.76 acres; also lease and bond for $25,000 for 18 months on the First Chance, containing 5.50 acres, and on the Hermosa, containing 2.59 acres, for $15,000 for the same period, both of which properties are owned by the Hermosa Gold Mining Company. The above properties are situated on Squaw mountain and Mineral hill. The Morning Star is patented. As the First Chance and Hermosa are held only by bond, the plats are not shaded. See Hermosa.

Development
On the First Chance there is a steam plant, and the company is sinking a shaft to a depth of 300 feet. The Morning Star No. 2 has two shafts, one 100 feet deep, the other 50 feet—both sunk on a strong vein. The First Chance is now receiving the greater part of the development work.

Stock not quoted on the market at the present time.

The Midget Gold Mining and Milling Company.

Incorporated February 21, 1896.

James F. Burns.........................President
Frank G. Peck.........Vice-President and Treasurer
L. F. Curtis...............................Secretary

T. F. Burns.   R. A. Trevarthen.

Directors

Main Office—Colorado Springs, Colorado.

1,000,000 shares.   Par value, $1.00.

Capitalization

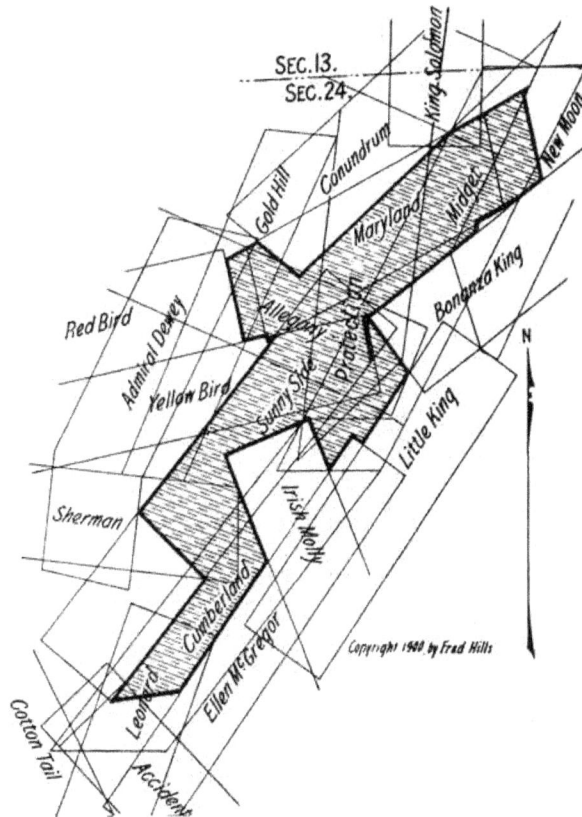

SEC. 13.
SEC. 24.

Copyright 1900 by Fred Hills

Owns the Protection, Survey No. 11,862; Maryland, Survey No. 9,474; Sunnyside, Survey No. 9,253; a part of the Alleghany, Survey No. 9,371; a part of the New Moon, Survey No. 9,048; the Midget, Survey No. 9,429; the Cumberland, Survey No. 9,571; and a part of Little Kate; all situated in the N. E. 1-4 of section 24. Total acreage owned by this company is about 20 acres. All patented.

Property

There are about $10,000 worth of surface improvements and machinery. One shaft has been sunk 600 feet and there are over 5,000 feet of drifts, etc. The greater part of the development work has been done on the Midget, which is being worked by two lessees. The company also holds a lease on a part of the Bonanza King.

Development

Stock not on the market.

The Midway Gold Mining Company.

Incorporated May 9, 1899.

Directors

J. R. McKinnie.........................President
E. P. Shove..........................Vice-President
R. P. Davie..........................Secretary
J. S. Tucker.........................Treasurer

John Harnan.

Main Office—25 E. Pike's Peak avenue, Colorado Springs, Colorado.

Capitalization

1,250,000 shares. Par value, $1.00. In treasury January 1, 1900, 190,000 shares.

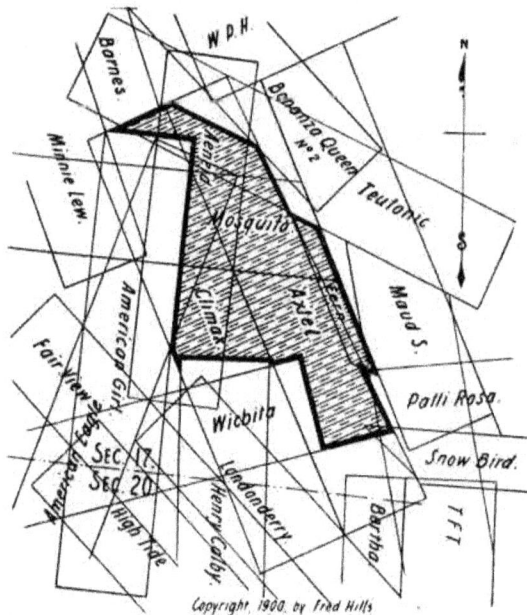

Copyright, 1900, by Fred Hills

Property

Owns the Axtel and Climax, 10 acres; the Fern, 1 acre; the Aeneid, 1 acre, and the Mosquito, 2 acres, in the S. W. 1-4 of section 17, on Bull hill. Steam and electric hoist. Two sets of lessees are working the property, with good showing for values.

Highest price for stock during 1899, 10 cents; lowest price for stock during 1899, 6 cents.

The Milwaukee and Cripple Creek Mining and Land Company.

Incorporated December 23, 1895.

Edwin Arkell........................President
Chas. A. Brooker..........Secretary and Treasurer
Jos. P. Walsh......Assistant Secretary and Treasurer
Nathan Oakes.

Directors

Main Office—37 P. O. building, Colorado Springs, Colorado.

1,500,000 shares. Par value, $1.00. In treasury January 1, 1900, 175,000 shares.

Capitalization

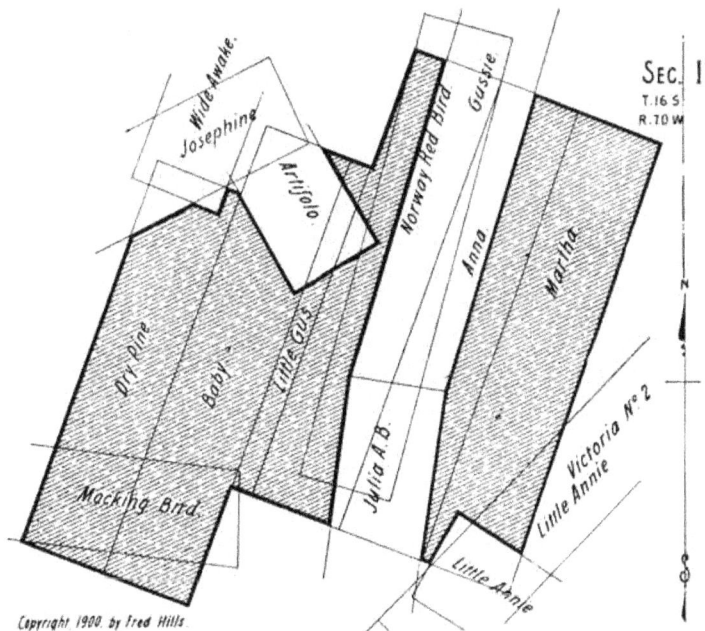

Owns the Martha, Dry Pine, Little Gus, Baby, Norway and Anna, a total of 25 acres, in the N. W. 1-4 of section 1, township 16 south, range 70 west, south of Grouse mountain. The above claims are in process of patenting. Sufficient work has been done to obtain a patent.

Property

Highest price for stock during 1899, $7.50 per thousand; lowest price for stock during 1899, $2.50 per thousand.

The Missouri Mining Company.

Incorporated 1899.

Directors

George Bernard........................President
J. E. Hundley.....................Vice-President
J. W. Miller...............Secretary and Treasurer
S. S. Bernard.   Wm. Shemwell.

Main Office—Bank building, Colorado Springs, Colorado.

Capitalization

1,250,000 shares. Par value, $1.00. In treasury January 1, 1900, 192,995 shares; in treasury January 1, 1900, $1,000.00 cash.

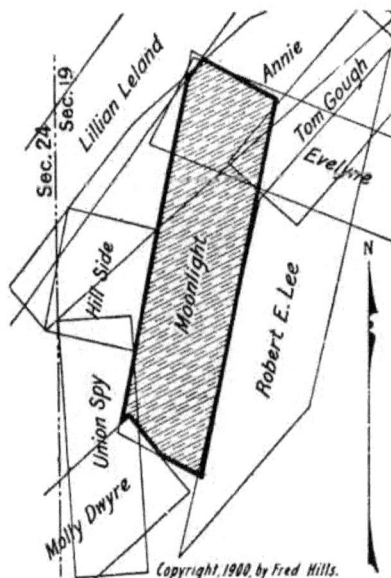

Copyright, 1900, by Fred Hills.

Property

Owns the Moonlight, containing 9.5 acres, in the N. W. 1-4 of section 19, on Gold hill. Patented.

Development

There are several shafts on the property and it is being worked by lessees. On March 17, 1900, a lease on the north one-half of the Moonlight claim was granted to the Creston Leasing Company for 18 months, at royalties of 20 per cent., lessees to sink a three-compartment shaft to a depth of 250 feet.

Highest price for stock during 1899, 7 cents; lowest price for stock during 1899, 4 cents.

The M. J. T. Gold Mining Company.

Incorporated August 26, 1899.

E. D. Marr..............................President
F. E. Brooks......................Vice-President
A. J. Bendle..........................Secretary
R. P. Davie...........................Treasurer

C. M. MacNeill.

Main Office—25 East Pike's Peak avenue, Colorado Springs, Colorado.

1,250,000 shares. Par value, $1.00. In treasury January 1, 1900, 98,495 shares; in treasury January 1, 1900, $7,502.00 cash.

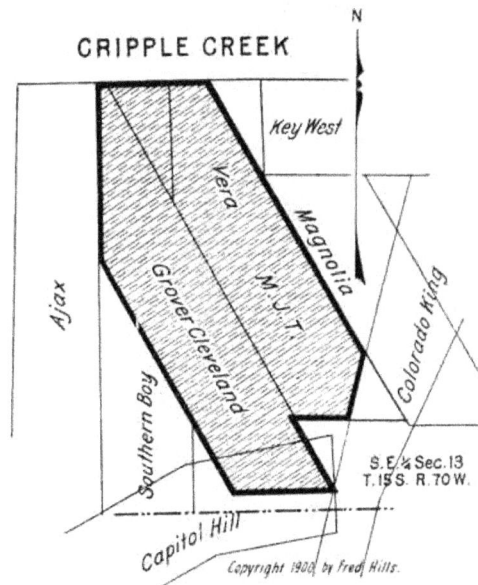

Owns the M. J. T. and Grover Cleveland, containing 18 acres, in the S. E. 1-4 section 13, on Gold hill, adjoining the Oriole and Key West properties, in the town of Cripple Creek. Patented.

The company expects soon to put in a good plant of machinery and thoroughly develop the property.

Highest price for stock during 1899, 7 cents; lowest price for stock during 1899, 3 1-8 cents.

The Mobile Gold Mining Company.

Incorporated February 3, 1896.

Directors

E. S. Bach, president; J. M. Marsh, vice-president; F. B. White, secretary and treasurer; B. H. Bryant, Edwin Arkell, E. W. Young, F. J. Parkinson.

Main Office—28 1-2 North Tejon street, Colorado Springs, Colorado.

Capitalization

1,250,000 shares. Par value, $1.00. In treasury January 1, 1900, 145,000 shares; in treasury January 1, 1900, $150 cash.

Property

Owns the Last Chance, 9.33 acres, in the N. 1-2 section 19, on Gold hill, adjoining the Lone Star and Hub of the Anaconda Gold Mining Company, also near the Dolly Varden of the Enterprise Gold Mining Company. Property is patented.

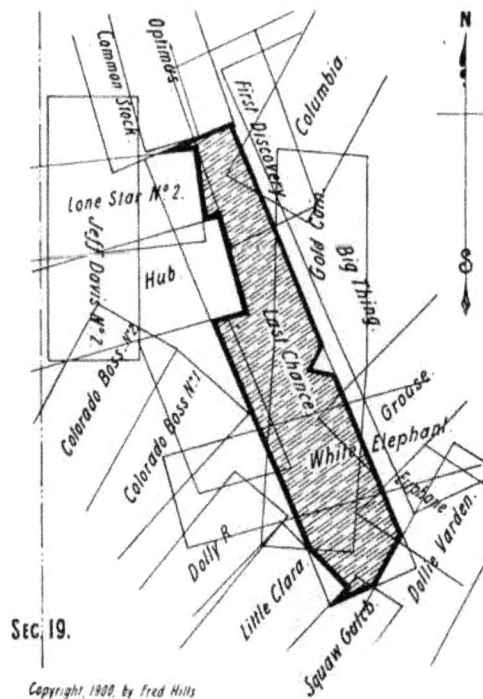

Sec. 19.

Copyright, 1900, by Fred Hills

Development

About 485 feet, with 300 feet of drifting, has been done. There are three shafts on the property, varying in depth from 40 to 300 feet deep. The main working shaft is 300 feet deep. At the present time work is being done on a drift at the 159-foot level. Electric hoist on property. The north 700 feet is leased to the Colorado Springs Mining and Leasing Company.

History

On the development of this property about $17,000 has been expended. Assays have been secured that run from $10 to $360 to the ton. As yet no paying ore in shipping quantities has been found.

Production to January 1, 1900, about $200.

Highest price for stock during 1899, 9 cents; lowest price for stock during 1899, 3 1-2 cents.

The Modoc Mining and Milling Company.

Incorporated June, 1893.

F. H. Frankenberg, Sr., president; Henry Herman, vice-president; Chas. H. Hermsmeyer, secretary; W. F. Greer, treasurer; Geo. E. Bragdon. **Directors**

Main Office—Room No. 41, Mechanics' block, Pueblo, Colorado.

500,000 shares. Par value, $1.00. In treasury January 1, 1900, **Capitalization** $30,000 cash.

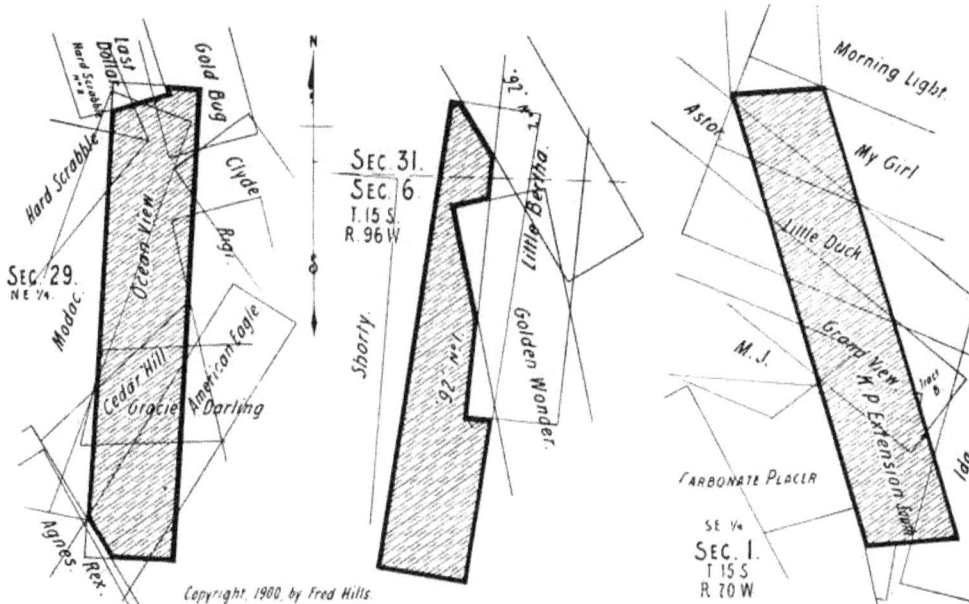

Copyright, 1900, by Fred Hills.

Owns the Ocean View, in the N. E. 1-4 of section 29, containing 10 **Property** acres, in the saddle of Battle mountain and Bull hill; the K. P. Extension South, in the S. E. 1-4 of section 1, containing 10.331 acres, on Rhyolite mountain, and the 92 No. 1, in the N. W. 1-4 of section 6, township 16 south, range 69 west, on Squaw mountain. The Ocean View and the 92 No. 1 are patented. Receiver's receipt held for the K. P. Extension South.

One ore and shaft house, 30x45 feet; one 90-horse power boiler; one **Development** 12x14 hoister, with 1,000 feet cable, 6 air drills, etc., of the most modern and latest improvements. Ocean View has a shaft 725 feet deep, with eleven levels and drifts varying from 50 to 600 feet. This claim has received the greater amount of development work. The work on the others has been merely sufficient for patent.

The Ocean View claim was worked under different leases up to **History** October 1, 1897. The first shipment, 14,450 pounds, with a value per ton of $64.24, was made by Hardten and Renshaw, lessees, April 5 or 6, 1896. Since October 14, 1897, the company has been working the property and has practically paid dividends constantly ever since, the highest being one of 4 per cent., in July, 1898.

Gross production to January 1, 1900, 10,111,788 1-2 pounds. Net **Production and Dividend** earnings on ore during 1899, $65,034.91. Dividends of $150,000 paid up to January 15, 1900. Last dividend paid March 15, 1900, of $5,000.

Highest price for stock during 1899, $1.00; lowest price for stock during 1899, 70 cents.

The Molly Dwyre Gold Mining Company.

Incorporated 1899.

Directors

J. A. Sill, president; John C. Mitchell, vice-president; A. W. Chamberlin, secretary and treasurer; F. H. Dunnington, assistant secretary; J. A. Paine.

Main Office—No. 2 N. Nevada avenue, Colorado Springs, Colo.

Capitalization

1,500,000 shares. Par value, $1.00. In treasury January 1, 1900, 200,000 shares; in treasury January 1, 1900, $1,500.00 cash.

Property

Owns the New Discovery claim, 7.239 acres, in the N. W. 1-4 section 31, on the S. E. slope of Beacon hill, and the Molly Dwyre, 7.132 acres, in sections 19 and 24 on Gold hill. All patented.

Development

There are five 50-foot shafts on the property, and a new shaft being sunk on the Molly Dwyre. A good vein has been encountered, but no ore in paying quantities. The New Discovery lode is not very promising. The Molly Dwyre claim is in a very good location, and the indications point to an ore body not very far from the present workings.

Highest price for stock during 1899, 4 3-4 cents; lowest price for stock during 1899, 2 7-8 cents.

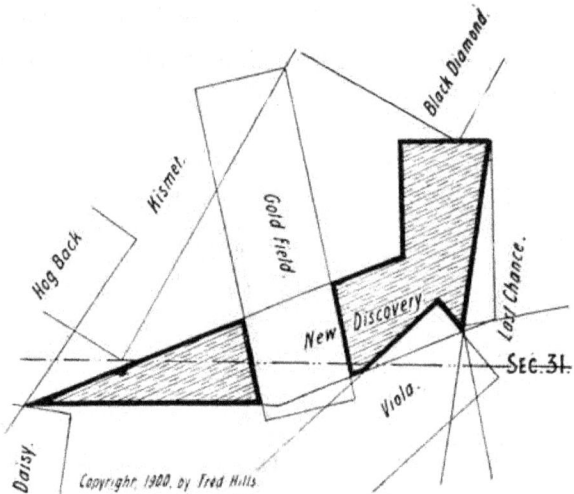

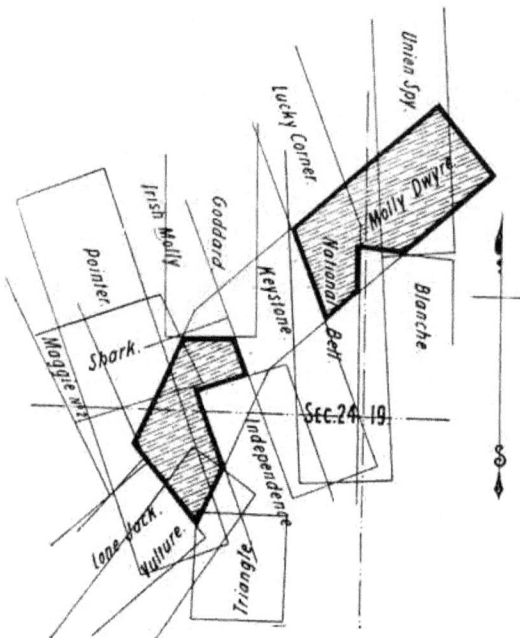

The Monarch Gold Mining and Milling Company.

Incorporated May 8, 1894.

Directors

Capitalization

Property

Development

Sam Strong, president; C. A. McLain, vice-president; F. M. Young, secretary and treasurer; H. H. Barbee.

Main Office—No. 14 El Paso Bank block, Colorado Springs, Colorado.

1,000,000 shares. Par value, $1.00. In treasury January 1, 1900, 32,000 shares; in treasury January 1, 1900, $300.00 cash.

Owns the Minnehaha, containing about 9 acres, in the N. W. 1-4 section 30, on Raven hill, west of the Elkton mines. Property is patented.

A shaft has been sunk 150 feet, and some surface prospecting is going on. Work is being actively prosecuted. The south half of this property is leased. This company formerly owned the North Star, Brown Leggings, Silver State and Monarch claims, about 30 acres, on Globe hill, which was sold in March, 1900, for $140,000. A 12-cent dividend was paid and $10,000 put in the company's treasury to be used in developing the N. 1-2 of the Minnehaha claim, the S. 1-2 being leased to the Creston Mining and Leasing Company.

Highest price for stock during 1899, 12 cents; lowest price for stock during 1899, 5 cents.

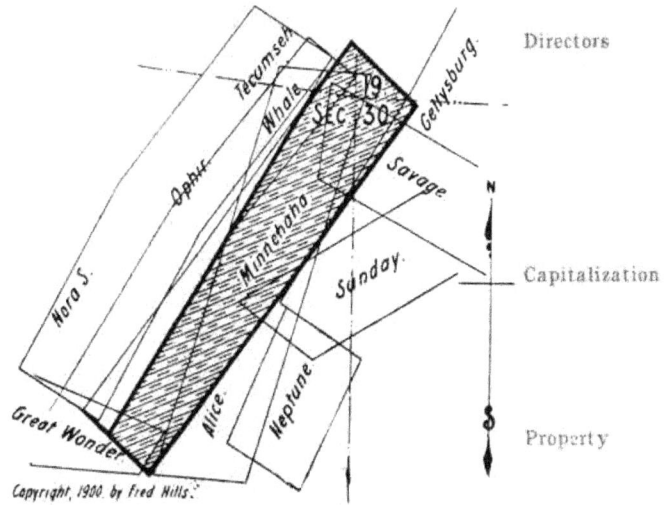

The Monroe Doctrine Gold Mining, Leasing and Bonding Company.

Incorporated.

Directors

Capitalization

Property

J. H. Thedinga, president; H. I. Bennett, secretary and treasurer; A. L. Lawton; Geo. H. Madin; Peter Crisman.

Main Office—Colorado Springs, Colo.

1,250,000 shares. Par value, $1.00. In treasury January 1, 1900, 352,000 shares.

Owns Wild Cat and Wild Cat No. 1 lodes, 6.5 acres in the S. E. 1-4, section 6, township 16 south, range 69 west, on Straub mountain. Receiver's receipt. A forty-foot shaft on Wild Cat.

No stock sold during 1899.

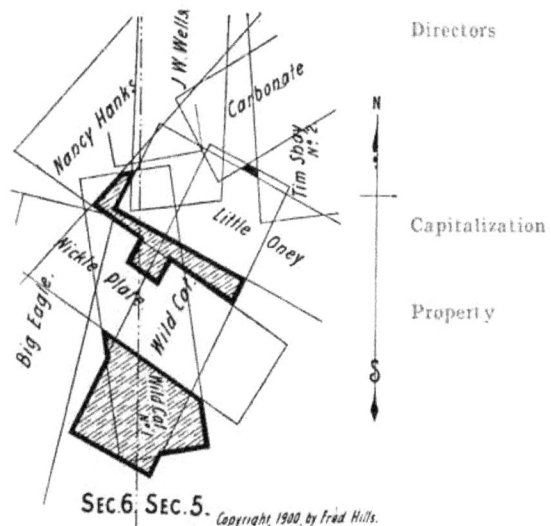

333

The Monrovia Mining and Tunnel Company.

Incorporated February, 1896.

Directors

F. A. Slosson, president; Frank Cotton, vice-president and treasurer; C. L. McKesson, secretary; G. S. Wilson; F. W. Robinson.

Main Office—Rooms No. 21 and 22 Giddings block, Colorado Springs, Colorado.

Capitalization

1,000,000 shares. Par value, $1.00. In treasury January 1, 1900, 315,000 shares.

Property

Owns the Friday, 9.948 acres, patented, survey No. 11,269; the Deer lode, 9 acres, in process; the John W., 10 acres in process, the latter not being shown on plat; also a portion of the Little Tom No. 2, containing about 1 1-2 acres, for which a bond for a deed is held. All the above property is situated in the E. 1-2 section 12 and the W. 1-2 section 11, township 15 south, range 69 west, on Cow mountain. The Dandy and Little Tom No. 2 claims, as shown on map, are not owned by this company, except the small conflicts of the same with the Deer and Friday claims.

Development

Shaft house on the Friday.

The Friday has an incline shaft of about 120 feet, with several drifts, cross-cuts, etc.; also a well timbered straight shaft, 90 feet deep, and several other shallow shafts. This claim is receiving the greater part of the development work. On the Deer lode are several shafts about 10 feet deep and a 55-foot tunnel.

History

A large amount of development work was done on the Friday in the latter part of 1895 and in 1896 with encouraging results, but the company was not able to reach a paying basis. Work was discontinued for lack of funds. At present there is a large amount of development work being done in the immediate vicinity of the company's property and indications seem good for realizing in the near future, from the sale of treasury stock, sufficient money for development purposes.

Highest price for stock during 1899, $5.20 per thousand; lowest price for stock during 1899, $5.20 per thousand.

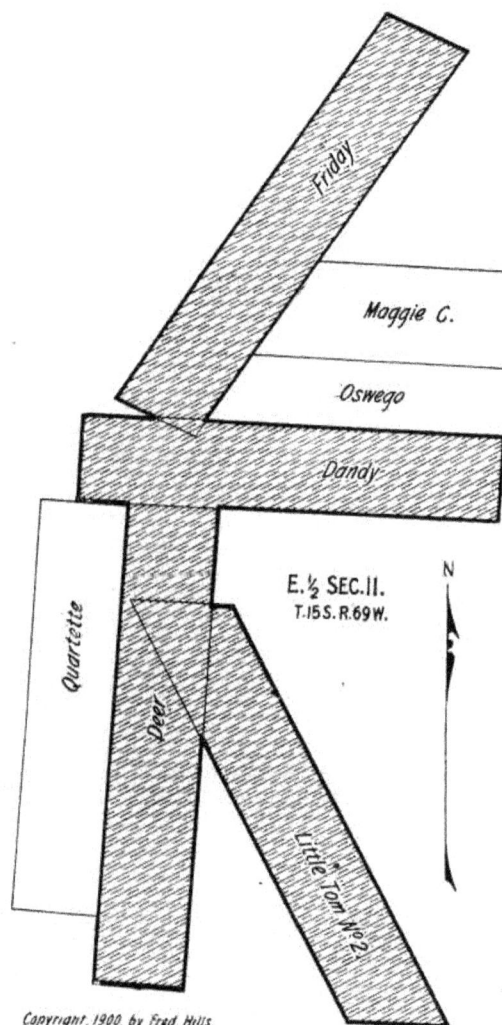

334

The Montivedo Mining and Milling Company.

Incorporated April 4, 1894.

John Pedersen..........................President   Directors
P. Ottenheimer......................Vice-President
H. C. McArthur............Secretary and Treasurer
F. N. Strong.   A. F. Woodward.

Main Office—Room 50, P. O. building, Colorado Springs, Colorado.

1,250,000 shares. Par value, $1.00.   Capitalization

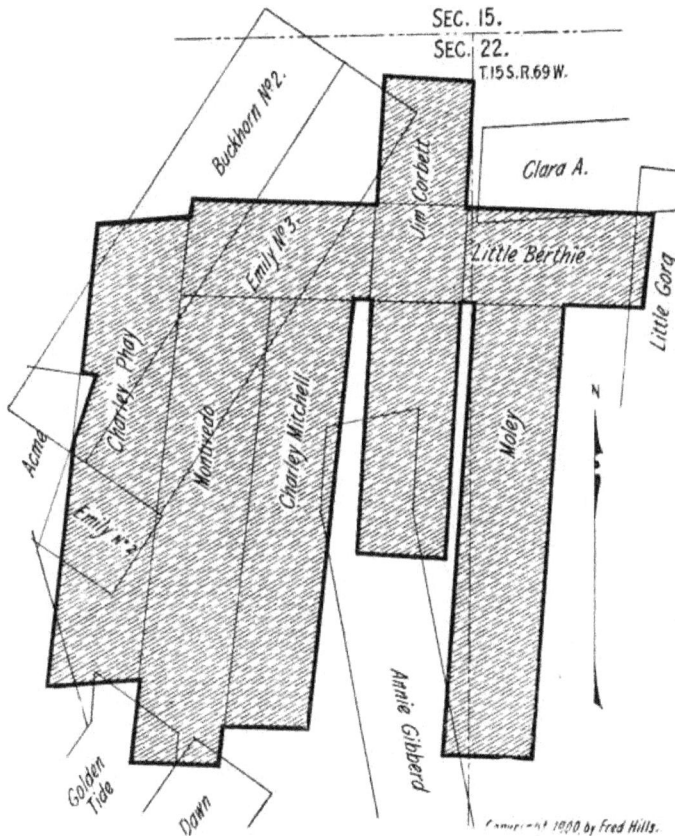

Owns the Charley Phay, the Montivedo, the Charley Mitchell, the **Property**
Jim Corbett, the Moley, and the Little Berthie, all situated in the N. 1-2
of section 22; a total of 55 acres. Receiver's receipt held for all the
above property.

Only government assessment work has been done.   **Development**

The first four stub books, containing stubs to certificates issued **History**
from No. 1 to 1,001 inclusive, were destroyed by fire in 1898, and the
books had not been posted since January, 1896. There is, therefore, no
means of telling how much stock is out. The report of stock to date
makes an over-issue of nearly 1,500,000 shares.

Highest price for stock during 1899, $2.00 per thousand; lowest
price for stock during 1899, $1.00 per thousand.

The Montreal Gold Mining and Milling Company.

Incorporated November 19, 1895.

Directors
James F. Burns, president; Irving Howbert, vice-president; L. F. Curtis, secretary; Frank G. Peck, treasurer; F. M. Woods, Theophilus Harrison.

Main Office—Colorado Springs, Colorado.

Capitalization
1,000,000 shares. Par value, $1.00.

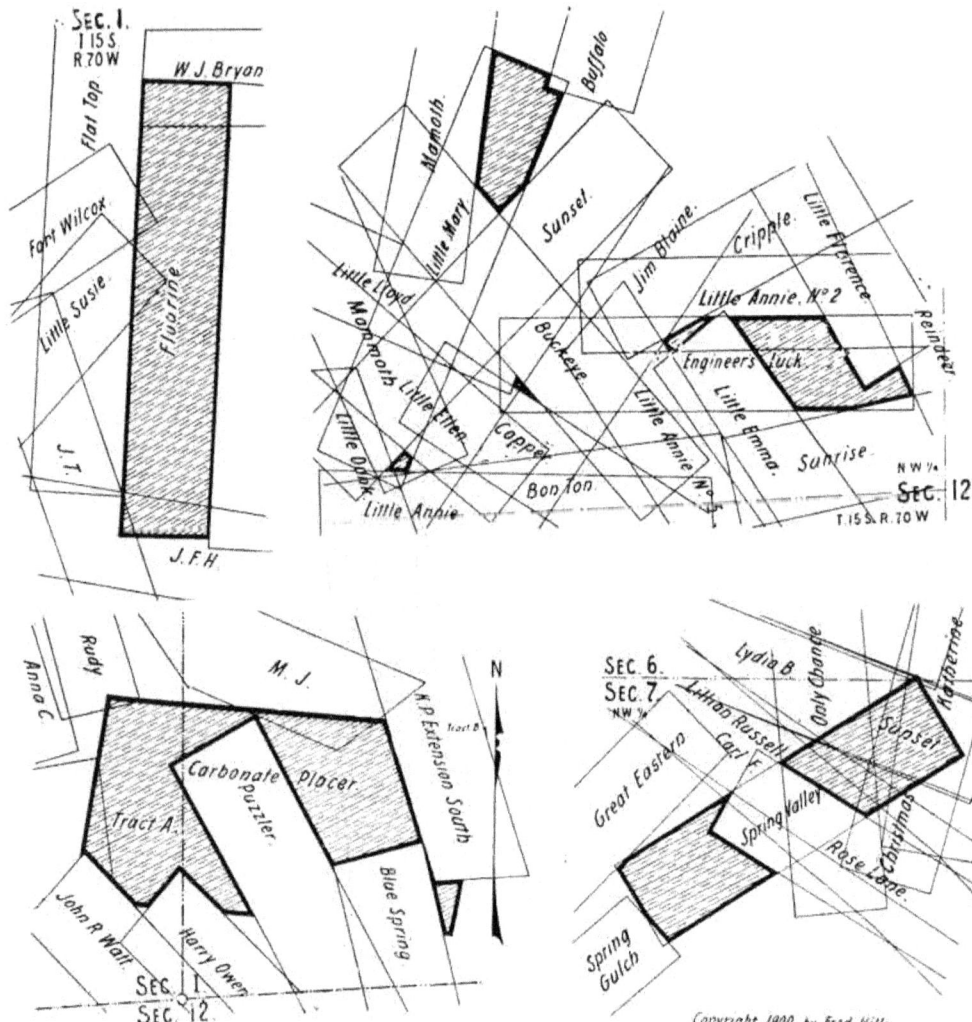

Copyright, 1900, by Fred Hills

Property
Owns the Fluorine, in the W. 1-2 of section 1, Survey No. 9,800, containing 9.61 acres; the Spring Valley, in the N. W. 1-4 of section 7, Survey No. 10,187, containing 5.455 acres; the Carbonate placer, in the S. 1-2 of section 1, containing 9.975 acres; the Little Mary, 2 1-2 acres, and the Engineer's Luck, 2 1-4 acres, both situated in the N. W. 1-4 of section 12; also a one-half interest in the Minneapolis tunnel site. Total acreage owned by this company is 34.540 acres. All patented except the Engineer's Luck and the Little Mary, which are in process.

Development
Two thousand feet of work has been done on the Fluorine, on which claim the greater part of the development work has been done. Surface improvements and machinery are insignificant. Net profit on ore mined during the year, $376.45.

Production and Dividend
Gross production to January 1, 1900, about $160,000. Dividend of $7,500 was paid November 12, 1898.

Highest price for stock during 1899, 16 cents (about); lowest price for stock during 1899, 6 cents.

The Montrose Gold Mining Company.

Incorporated September 29, 1899.

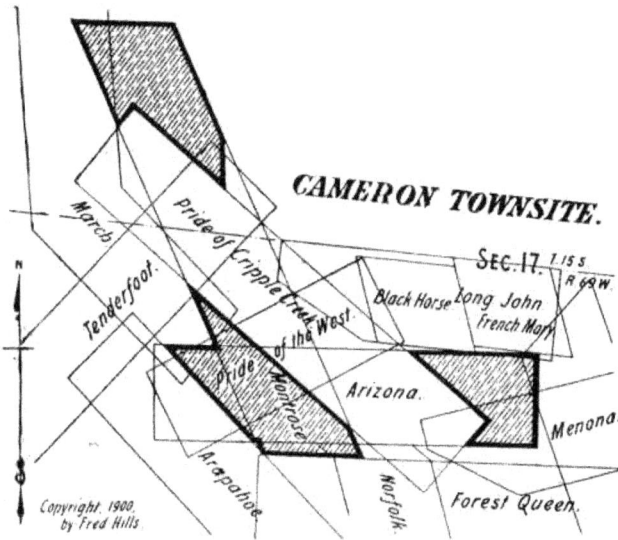

Dr. C. P. Elder, president; Warren Woods, vice-president; F. M. Woods, secretary; H. E. Woods, treasurer; W. S. Tarbell; C. M. Clinton; J. M. Allen.

Directors

Main Office—Colorado Springs, Colorado; branch offices, Denver and Victor, Colorado.

1,250,000 shares. Par value, $1.00. In treasury January 1, 1900, 140,000 shares.

Capitalization

Owns the Montrose and the Arizona, containing about 8 acres. Patented.

Property

Owns hoisting plant, etc. There is a shaft 154 feet deep on the property, with 100 feet of drifting; also some other minor shafts. The property has not been worked lately, but the company is getting ready to prosecute development work vigorously in the near future.

Development

Highest price for stock during 1899, 10 cents; lowest price for stock during 1899, 6 1-2 cents.

The Monument Gold Mining Company.

Incorporated November 23, 1898.

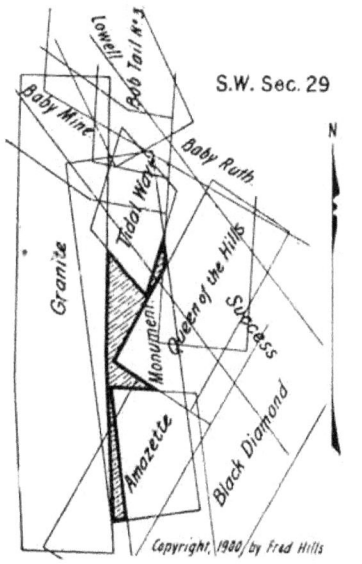

Ira Williams....................President
Verner Z. Reed.................Vice-President
A. F. Woodward.......Secretary and Treasurer
E. C. Bales.   Mrs. Ella McGovney.

Directors

Main Office — Kiowa street, Colorado Springs, Colorado.

300,000 shares. Par value, $1.00.

Capitalization

Owns the Monument claim, containing about 2 acres, in the S. W. 1-4 section 29, on Battle mountain, near the Portland mine. Patented.

Property

The property of the company is worked under lease. Although the claim is very small, it is considered very valuable, as it adjoins the Portland mine on the one side and the Granite on the other.

Development

For two years this claim has been a steady shipper.

Gross production to January 1, 1900, $97,614.83. Total dividends to January 1, 1900, $20,000.00. Amount of last dividend, $3,000.00.

Production and Dividend

Highest price for stock during 1899, 40 cents; lowest price for stock during 1899, 30 cents.

The Moon-Anchor Consolidated Gold Mines, Limited.

Incorporated December 19, 1898.

Directors

English Board—W. W. Lowe, Esq., chairman; Geo. D. Nichol; Frederick W. Baker; Henry Richards, secretary.

American Board—Verner Z. Reed, managing director; J. R. McKinnie, manager; O. H. Shoup, local secretary.

English Office—No. 3 Princes street, London, E. C.; local office, Bank block, Colorado Springs, Colorado.

Capitalization

400,000 shares. Par value, £1.

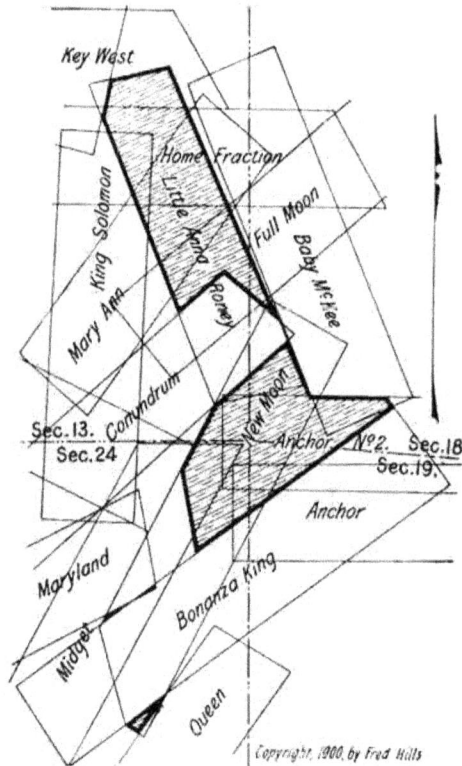

Copyright, 1900, by Fred Hills

Property

Owns the New Moon, part of Anchor and Anchor No. 2, and the Little Anna Roney, containing about 10 acres, in sections 13, 18, 19 and 24. Located on Gold hill. All the above property is patented.

Development

The property is fully equipped with pumps, hoist and all necessary machinery to work the property to a depth of 1,000 feet or more. The company has always pursued an active development policy. Early in 1899 the British company took possession of the property and fully equipped the mine with pumps and machinery for deep mining. The shaft is now being sunk for the eighth level. The seventh level is about 700 feet from the surface.

Production and Dividend

Total dividends up to January 1, 1900, $261,000.00; last dividend paid November, 1899, was $15,000.00. These dividends were paid to the stockholders of the Moon-Anchor G. M. Co. before the sale to the English company.

The Moon-Anchor Gold Mining Company.

Incorporated.

J. R. McKinnie . . . . President

Verner Z. Reed . . . . Vice-President and Treasurer

O. H. Shoup . . . . Secretary

W. S. Stratton.   L. L. Aitken.

Directors

Main Office—Bank block, Colorado Springs, Colorado.

600,000 shares.   Par value, $1.00.

Capitalization

Copyright, 1900, by Fred Hills.

Owns the Zeolite, Bloomington and Surplus Fraction, about 15 acres, in the N. E. 1-4 section 29, on Battle mountain. Patented. The company also own a large interest in the Moon-Anchor Consolidated Gold Mines, Ltd., which latter company is fully described in another page of this Manual.

Property

Dividends to January 1, 1900, $261,000.00; last dividend paid November, 1898, $15,000.00.

Production and Dividend

Highest price for stock during 1899, $1.22 7-8 cents; lowest price for stock during 1899, 69 cents.

The Moose Gold Mining Company.

Incorporated 1893.

Directors

Ward Hunt...........................President
J. B. Glasser..........Vice-President and Treasurer
John S. Hunt..........................Secretary
O. C. Townsend.   J. A. Hayes.

Main Office—Hagerman building, Colorado Springs, Colorado.

Capitalization   1,200,000 shares.   Par value, $1.00.

Property   Owns the Moose, 10 acres; the Ben Harrison, 8 acres, and the Trilby, 0.4 acre—all situated in the S. W. 1-4 section 20, and the N. W. 1-4 section 29.

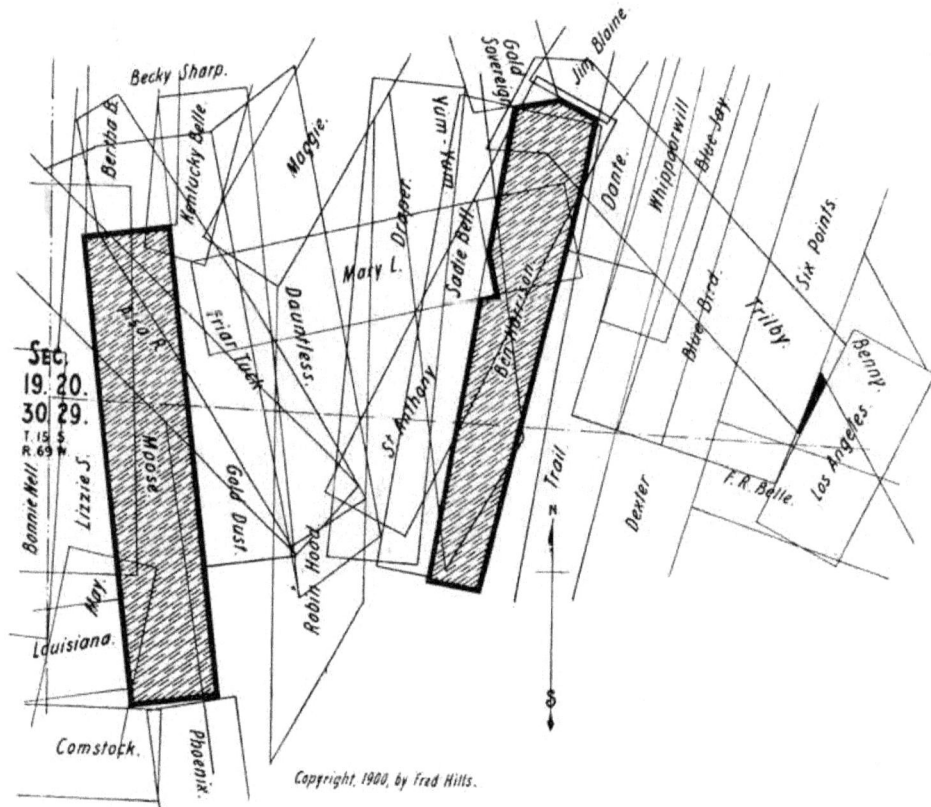

Copyright, 1900, by Fred Hills.

Development   Large and commodious plant with offices; three boilers, hoister, and compressor.   Underground machinery, large station pump between the tenth and eleventh levels.   On the Moose the main shaft is 890 feet deep, and several thousand feet of drifting.   There is another shaft 200 feet deep on this claim.   The Ben Harrison has been developed by several small shafts and tunnels.   The Trilby has a shaft about 300 feet deep, with a plant of machinery and electric hoist.

Production and Dividend   Gross production to January 1, 1900, over $500,000.00; total dividends to January 1, 1900, $145,000.00.

History   This property was one of the first shippers in the district.   As it is a close corporation, there are no stock transactions.

The Morgan Gold Mining Company.

Incorporated 1900.

John W. Proudfit.........................President   Directors

H. K. Devereux......................Vice-President

R. C. Thayer...............Secretary and Treasurer

Main Office—Hagerman building, Colorado Springs, Colorado.

1,500,000 shares. Par value, 10 cents. In treasury May 1, 1900, Capitalization 300,000 shares.

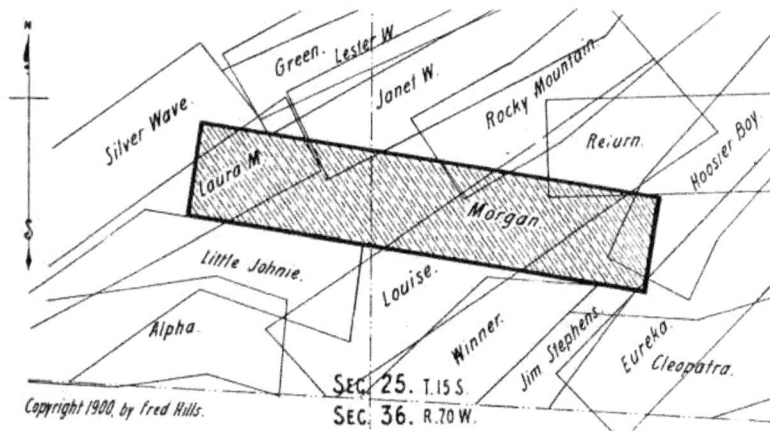

Copyright 1900, by Fred Hills.

Owns the Morgan claim, 10 1-3 acres, situated on Beacon hill, in Property section 25.

Very little development work has been done and the property is now History being developed by lessees. As this is a newly-organized company, and this data was only received just as the Manual was going to press, this property is not shown in color on the folding map herewith.

The Morning Glory Mining and Leasing Company.

Incorporated January 19, 1900.

Directors

Warren Woods, president; H. E. Woods, vice-president and treasurer; F. M. Woods, secretary and general manager; J. M. Allen; W. D. Hatton.

Main Office — Giddings building, Colorado Springs, Colorado; branch office, Victor, Colorado.

Capitalization

1,250,000 shares. Par value, $1.00. In treasury January 1, 1900, 35,000 shares; in treasury January 1, 1900, $60,000.00 cash.

Property

Owns the Aileen, containing 1.004 acres; the Little Giant and the Lantishie, 3 acres; the P. W. C., 3.71 acres; all situated on Raven hill in section 19.

Besides the above claims, this company has a lease for three years on the Morning Glory, the Morning Glory No. 2, and the Ida B. claims of the Work G. M. Co., and also a lease on the Rose Maud claim of the Rose Maud G. M. Co. (See plats of last mentioned companies.)

Development

On the Morning Glory claim there is a three-compartment shaft, 585 feet deep, for the development of the entire territory. At the 545-foot level of this shaft is 400 feet of drifting, aside from the old workings on the leased territory.

The Aileen has a shaft 250 feet deep and 100 feet of drifting. The Little Giant has three shafts of 50, 80 and 175 feet, respectively; also 125 feet of drifting. On the Morning Glory shaft is a first-class hoisting and compressor plant.

Highest price for stock during 1899, 40 cents; lowest price for stock during 1899, 25 cents.

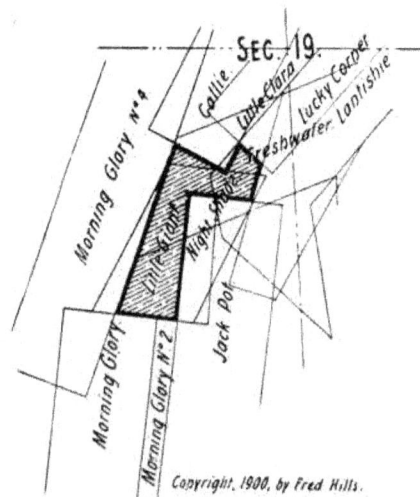

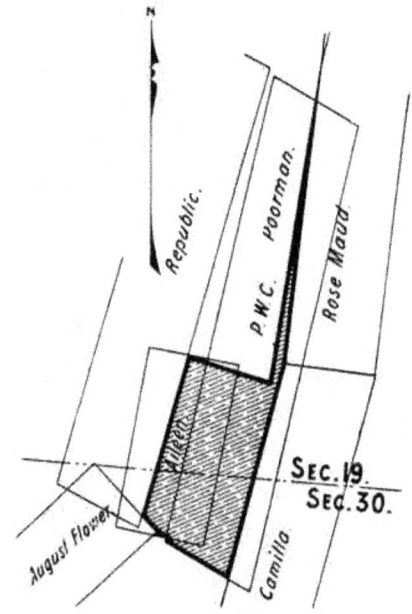

The Morning Star Gold Mining Company.

Incorporated September 8, 1899.

Main Office—Nos. 10, 11 and 12, Giddings block, Colorado Springs, Colorado.

1,250,000 shares. Par value, $1.00. In treasury January 1, 1900, Capitalization 250,000 shares; in treasury January 1, 1900, $500.00 cash.

Copyright, 1900, by Fred Hills.

Owns the Fleming, 6 acres, in the N. W. 1-4 section 20, on Bull hill; Property and the Morning Star, 6 acres, in the N. E. 1-4 section 25, on Beacon hill. All patented.

There is some surface development and shallow shafts and tunnels.

Highest price for stock during 1899, 5 1-4 cents; lowest price for stock during 1899, 4 1-2 cents.

The Mound City Gold Mining Company.

Incorporated January 23, 1896.

Directors   Frank A. Waters, president and treasurer; W. E. Frenaye, vice-president; J. C. Kimsey, secretary; D. D. Findley, M. S. Herring.

Main Office—No. 64 P. O. building, Colorado Springs, Colorado.

Capitalization   2,000,000 shares. Par value, $1.00. In treasury January 1, 1900, 30,000 shares.

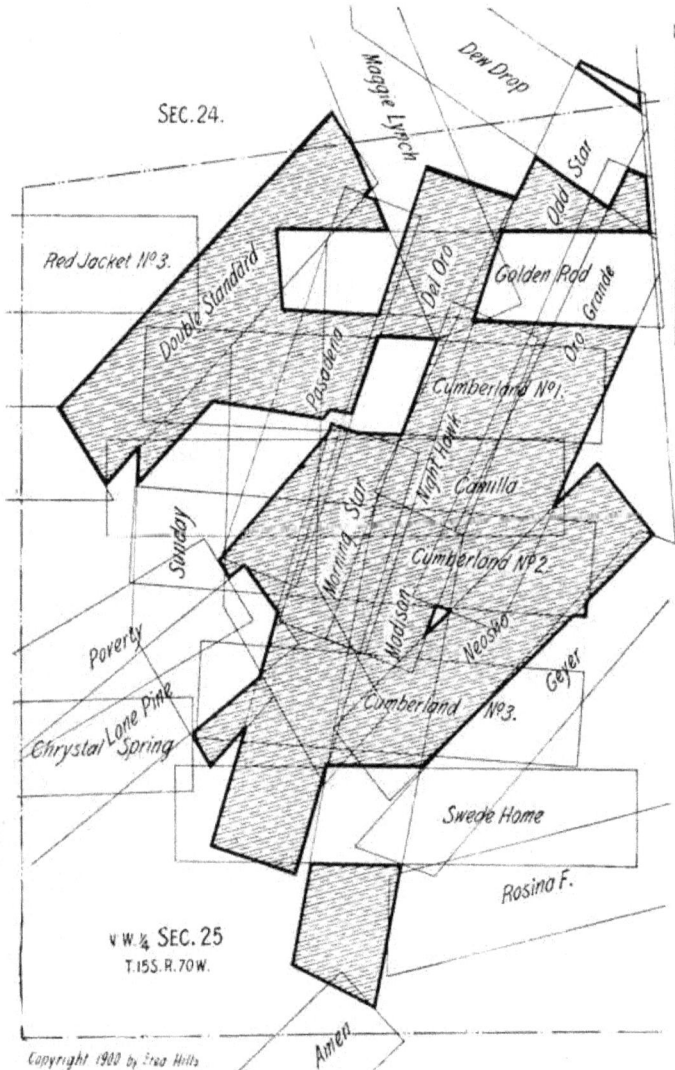

Property   Owns the Oro Grande, the Odd Star, the Del Oro, the Madison, the Morning Star, the Pasadena, the Double Standard, the Little Ella, the Sunday and the Neosho, containing in all 50 acres, in the N. W. 1-4 section 25, on Signal hill. The Morning Star is patented. Remainder in process. The Little Ella lies parallel to and east of the Sunday lode; the name was omitted in plat.

Development   There are several shafts of from 50 to 100 feet deep. No work except that necessary for a patent has been done on the property, but in doing this the company has disclosed four veins running the length of the property, and two phonolite dykes. Assays from these have run from $2 to $38 per ton. There is also a tunnel site, with a tunnel now in 250 feet.

No stock is being offered.

The Mount Rhyolite Gold Mining Company.

Incorporated January, 1896.

W. H. O'Brien..........................President

G. W. Earle..........................Vice-President

A. S. Whitaker..........................Secretary

R. C. Webster..........................Treasurer

E. P. Leech.   H. J. Dencker.

Main Office—Denver, Colorado.

1,500,000 shares.   Par value, $1.00.   In treasury January 1, 1900, Capitalization 200,167 shares; in treasury January 1, 1900, $250.00 cash.

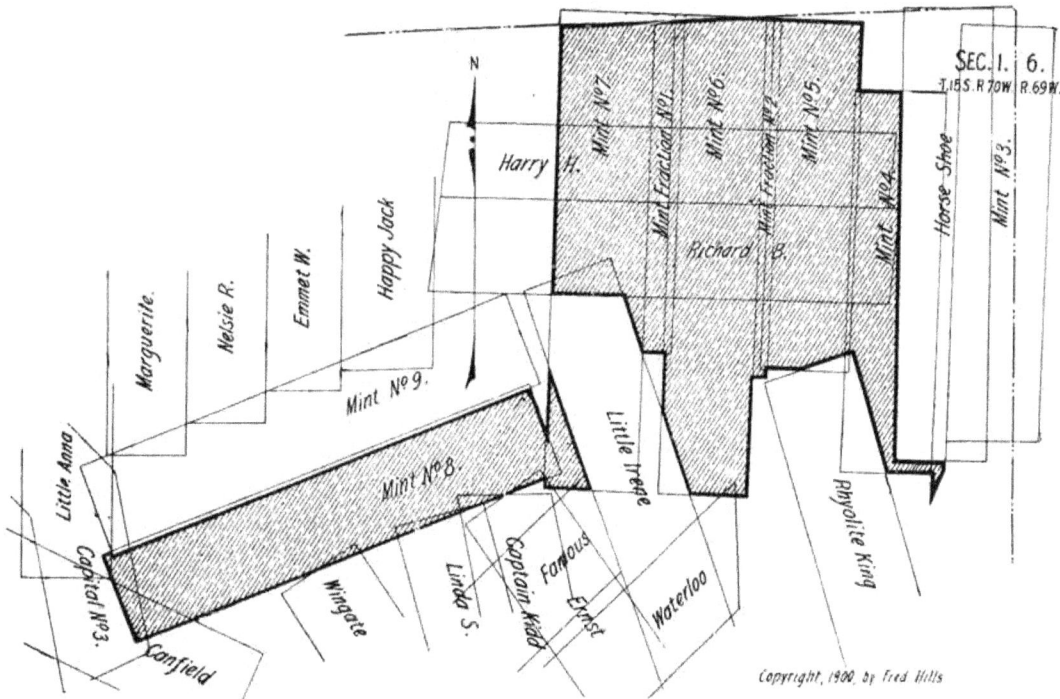

Copyright, 1900, by Fred Hills

Owns 45 acres, patented, situated in the N. E. 1-4 section 1, on Property Rhyolite mountain.

On lode No. 3 a shaft has been sunk 50 feet.   The company intends sinking a shaft 200 feet deep on this lode during the coming season.

Highest price for stock during 1899, 4 cents; lowest price for stock during 1899, 2 cents.

The Mountain Beauty Gold Mining Company.

Incorporated.

Directors

J. Maurice Finn......President and General Manager

A. E. Carlton.....................Vice-President

M. W. Levy............................Secretary

H. C. Cassidy...........................Treasurer

Geo. Rex Buckman.

Main Office—365 Bennett avenue, Cripple Creek, Colorado.

Capitalization   2,000,000 shares. Par value, $1.00. In treasury January 1, 1900, $200.00 cash.

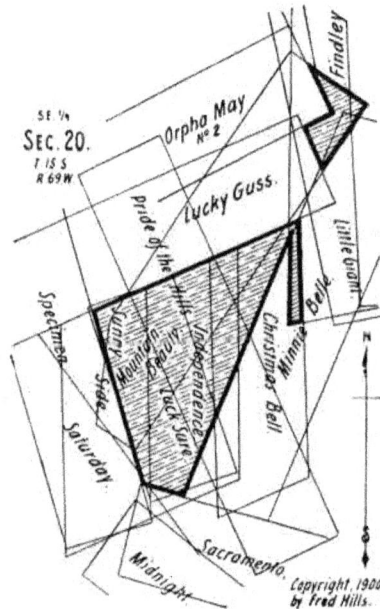

Property   Owns the Sunny Side, Independence, Luck Sure, Mountain Beauty, and the Christmas Bell, containing, in all, about 8 1-2 acres, located in a group on the south side of Bull hill. Receiver's receipt held for the above property.

Development   The splendid location of these claims has made it possible for the company to lease them on very advantageous terms. During 1899 five sets of lessees have been at work. Block No. 1, known as the Carpenter lease, has a 150-foot shaft. Block No. 2 has two shafts, the one at the west end being 250 feet deep, and the one at the east end 200 feet deep. It is proposed to eventually make this latter shaft the main working shaft of the Mountain Beauty Company. On blocks Nos. 3 and 4 a shaft has been sunk to a considerable depth. There is also a shaft on block No. 5 down to a depth of 250 feet. With such development work in progress, the prospects for this company seem very encouraging.

Production   Gross bullion value of ore mined during 1899, about $5,000. Net royalties to the company, $1,000.00.

Highest price for stock during 1899, 15 cents; lowest price for stock during 1899, 8 1-2 cents.

TOWN OF ALTMAN—1900.
SANGRE DE CRISTO RANGE IN BACKGROUND.

347

The Mt. Rosa Mining, Milling and Land Company.

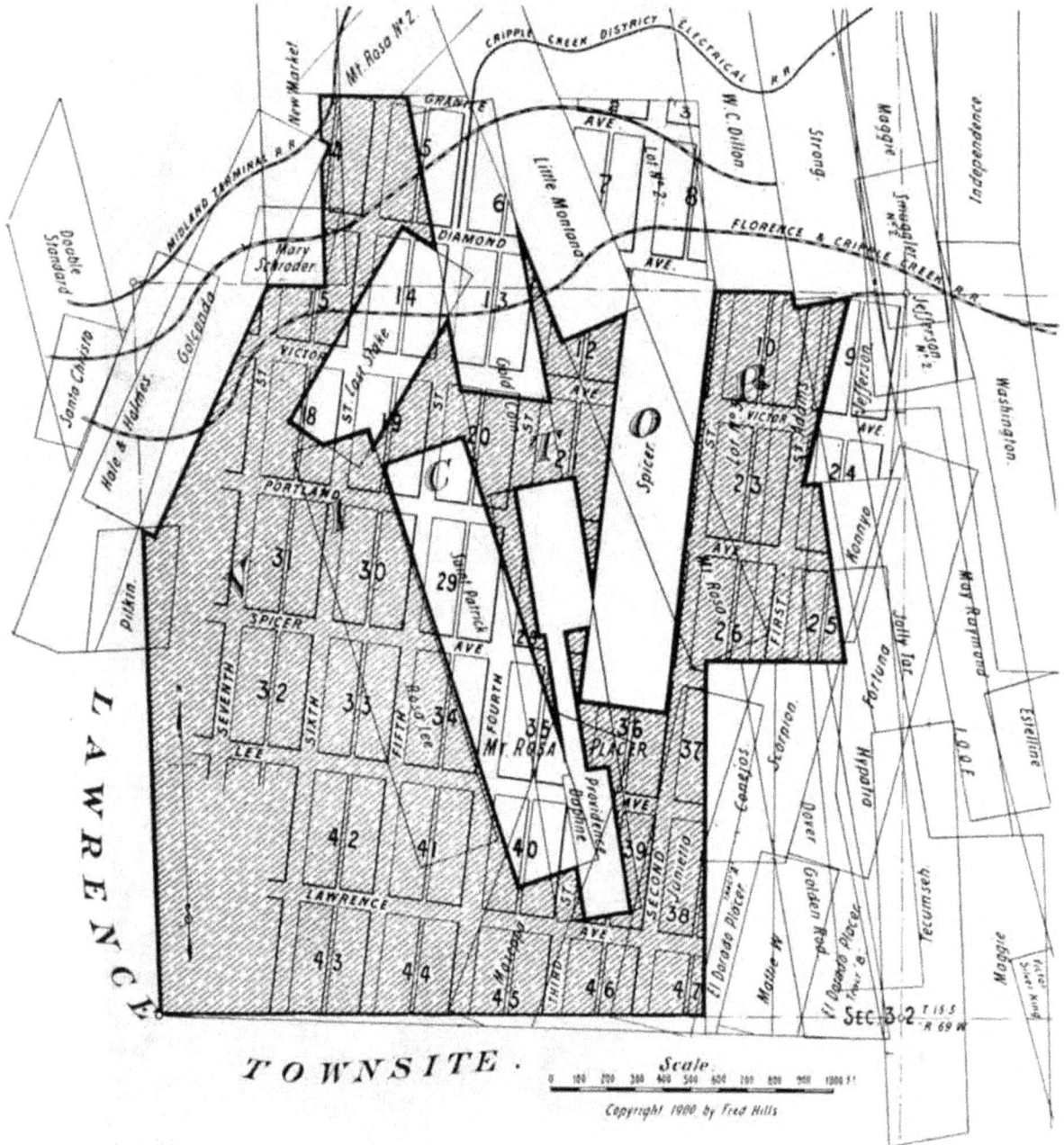

The Mt. Rosa Mining, Milling and Land Company.

Incorporated January 9, 1892.

Incorporation Amended March 21, 1893.

Warren Woods..........................President
H. E. Woods......................Vice-President
F. M. Woods..............Secretary and Treasurer
J. M. Allen.   F. H. Pettingell.

Directors

Main Office—Giddings building, Colorado Springs, Colorado. Branch Office—Victor, Colorado.

1,000,000 shares. Par value, $1.00. In treasury January 1, 1900, $43,000.00 cash.

Capitalization

Owns the Mt. Rosa placer, which is the site of the town of Victor; the Mt. Rosa, the Rosa Lee, a part of the Gold Coin, the Adams, the La Paloma, and the Professor Grubbs, containing, in all, about 106 acres, in the N. W. 1-4 of section 32. The La Paloma and the Professor Grubbs, not being patented, are not shown on the plat.

Property

The Adams has a shaft 250 feet deep. The Mt. Rosa has a shaft 200 feet deep. About 1,500 feet of work has been done in driving a tunnel and in various minor shafts.

Development

The Mundo Gold Mining Company.

Incorporated 1892.

Directors

Geo. W. Perkins.........................President

Whitney Newton.....................Vice-President

K. C. Perkins.............Secretary and Treasurer

Joseph Majors.   Robert Smith.

Main Office—106 N. Tejon street, Colorado Springs, Colorado.

Capitalization

1,500,000 shares.   Par value, $1.00.   In treasury January 1, 1900, 8,000 shares.

Property

Owns the Granitite, 5 acres, on Copper mountain, in the S. W. 1-4 of section 1, township 15 south, range 70 west, which alone is situated in the district covered by map; also owns the Iron Hand, 10 acres, on Iron hill; the Big Humbug placer, at the foot of Iron hill; the Black Diamond, on Iron hill, and the Mica Nos. 1 and 2, 20 acres, in High Park, Fremont county. The Granitite is patented. Remainder held by location.

Development

The Granitite has one 80-foot shaft and is pierced by the York tunnel. This claim is situated next to the Fluorine.

Highest price for stock during 1899, 1 cent.

The Mutual Mining and Milling Company.

Incorporated December 18, 1891.

L. E. Hawkins.........................President
J. R. McKinnie.....................Vice-President
C. W. Fairley.............Secretary and Treasurer

Geo. O. Talpey (deceased).   H. S. Hawks.

Directors

Main Office—No. 23 South Tejon street, Colorado Springs, Colorado.

700,000 shares.   Par value, $1.00.   In treasury January 1, 1900, 206,525 shares.

Capitalization

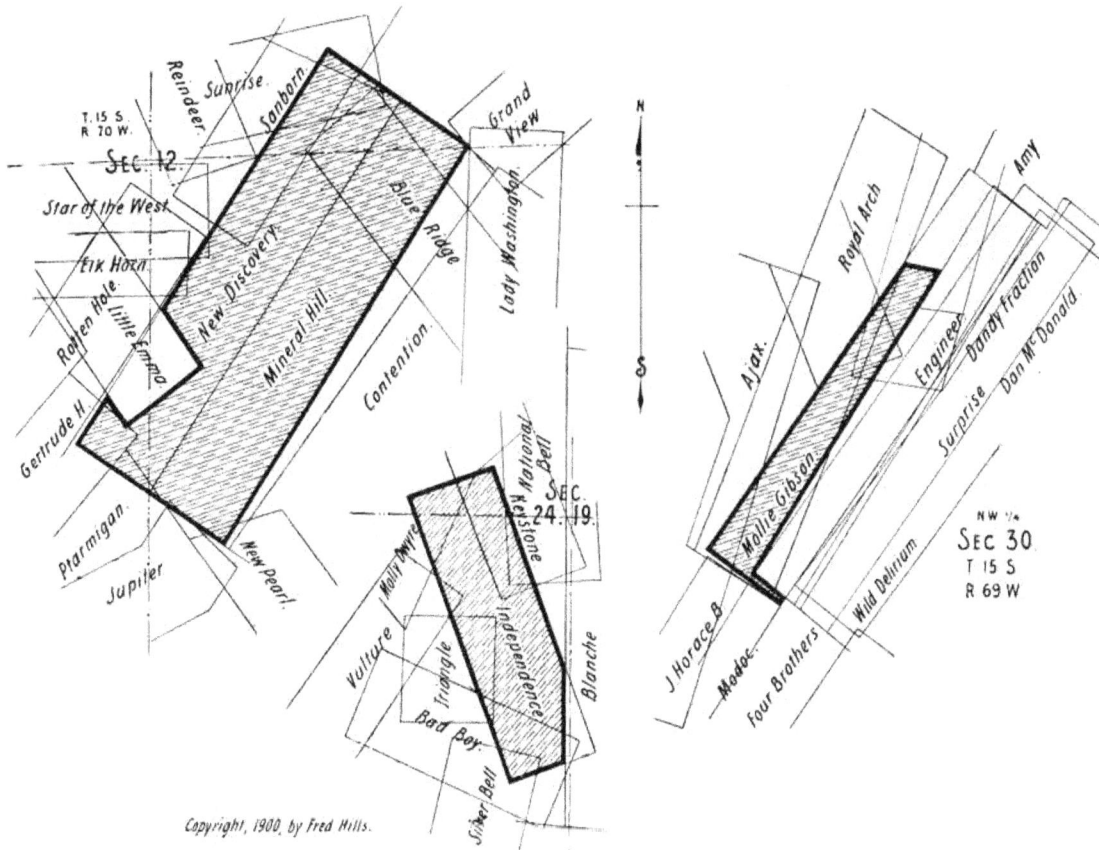

Copyright, 1900, by Fred Hills.

Owns the New Discovery and the Mineral Hill, in the S. E. 1-4 section 12, containing 19 acres, on Mineral hill; the Independence, in the eastern part of section 24, containing 6 1-2 acres, on Gold hill; and the Mollie Gibson, in the N. W. 1-4 section 30, containing 3 acres, on Guyot hill. Receiver's receipt held for the Mollie Gibson. The New Discovery, the Independence and the Mineral Hill are patented.

Property

The greater part of the development work is being done on the New Discovery and the Independence.

Highest price for stock during 1899, 13 1-4 cents; lowest price for stock during 1899, 4 cents.

The National Mining, Tunnel and Land Company.

Incorporated September 7, 1895.

Directors   W. R. Foley, president; W. M. Dutton, vice-president; S. J. Mattocks, secretary and treasurer; D. Sindlinger, W. R. Snyder.

Main Office—104 East Pike's Peak avenue, Colorado Springs, Colorado.

Capitalization   1,500,000 shares. Par value, $1.00. In treasury January 1, 1900, $3,400.00 cash.

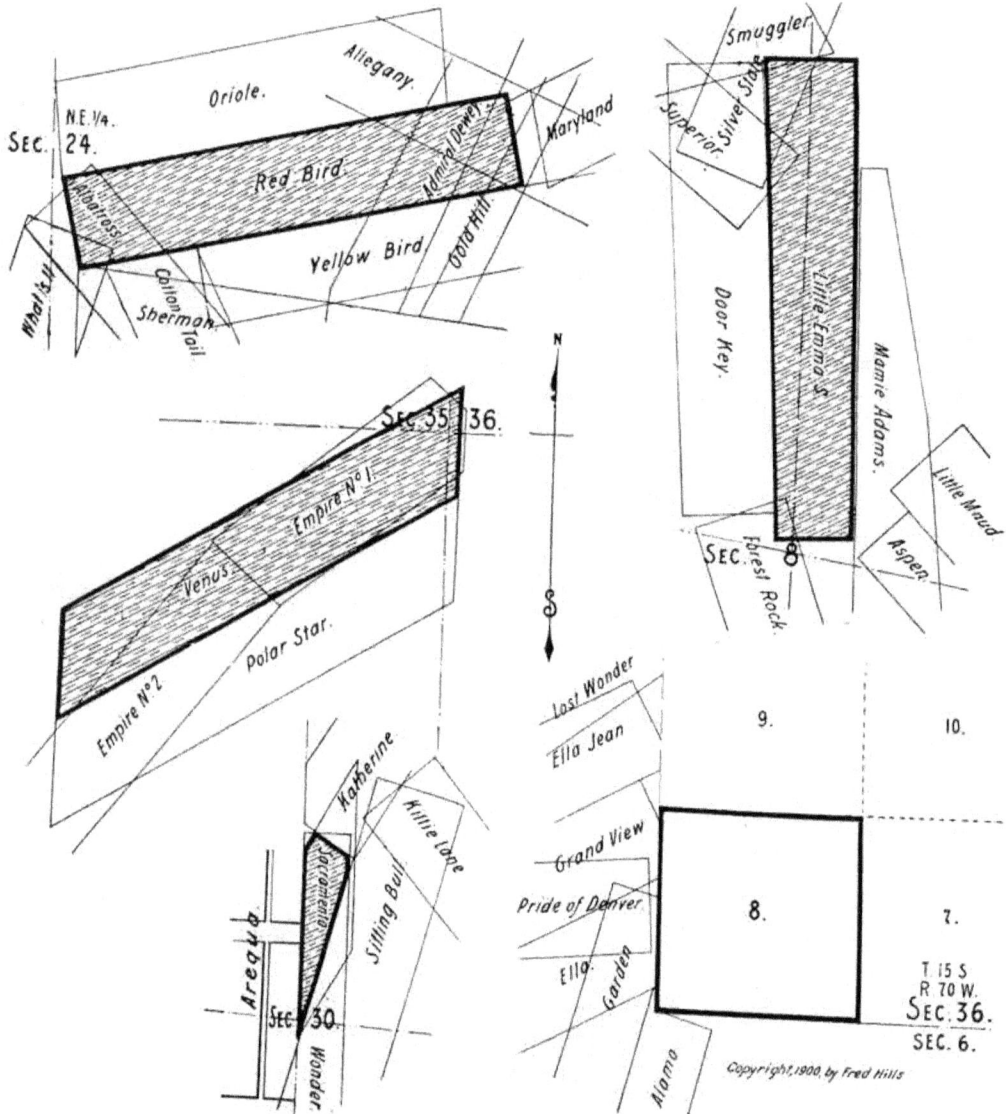

Copyright, 1900, by Fred Hills

Property   Owns the Red Bird, 10 acres, in the N. E. 1-4 section 24, on Gold hill; one-third interest in the Little Emma S., 9 1-2 acres, in the N. 1-2 of section 8, on Tenderfoot hill; both patented. Also owns a one-half interest in the Sacramento, 1 acre, in the N. E. 1-4 section 30, near the

(CONTINUED ON PAGE 353.)

352

Elkton; and the Venus, 9 acres, in the E. 1-2 of section 35; patented. The company has also a lease on block 8, school section 36, and on blocks 22, 23 and 59, in school section 16, Fremont county. The three latter blocks can not be shown on plat.

There is a large plant of machinery on the Red Bird claim, including air drills, plant, machinery and buildings, which cost $10,000.00. A 400-foot shaft on the Red Bird claim, with 600 feet of levels. The company commenced development work in April, 1899, and has sunk a shaft 300 feet and run about 600 feet of levels and drifts during the year; have opened up rich ore chute from which some very high assays have been had. The ore is increasing in width and in values as the development progresses. *Development*

Production to January 1, 1900, $15,000.00. *Production*

Highest price for stock during 1899, 20 1-8 cents; lowest price for stock during 1899, 2 cents.

The Nellie V. Gold Mining Company.

Incorporated October 5, 1899.

N. S. Gandy, president; E. M. De Lavergne, vice-president and treasurer; W. H. Allen, secretary; J. R. McKinnie; Louis R. Ehrich. *Directors*

Main Office—104 E. Pike's Peak avenue, Colorado Springs, Colorado.

1,500,000 shares. Par value, $1.00. In treasury January 1, 1900, 374,995 shares; in treasury January 1, 1900, $2,000.00 cash. *Capitalization*

Owns the Nellie V. and War Eagle claims, consisting of over 13 acres, located in section 29 on Squaw mountain. Patented. *Property*

One shaft 325 feet deep with about 1,000 feet of drifting and cross-cutting and two other shafts at a depth of about 200 feet. The Columbine-Victor tunnel crosses both properties at a depth of about 400 feet. *Development*

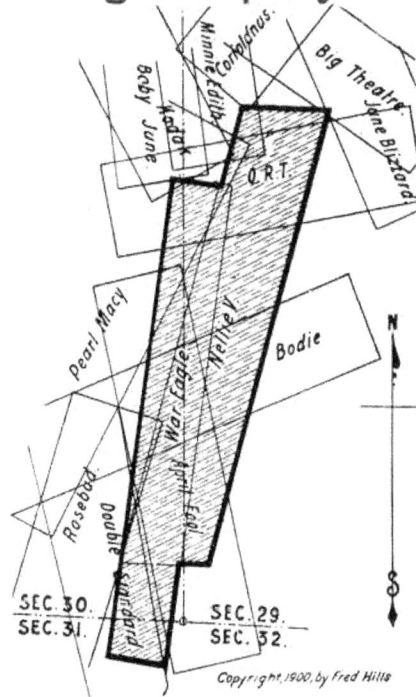

Copyright, 1900, by Fred Hills.

Between the years 1895 and 1897 the property produced something like $80,000.00 worth of very high grade ore. Values ran as high as $300,000.00 to the ton, and nothing less than $160.00. There was one car shipped which was divided into three grades; the high grade ran $1,-150.00, medium $550.00, and the low grade $190.00. At that time there was an offer of $300,000.00 cash made for the property and same was refused. There is an abundance of low grade which will be shipped at a profit. The property is being worked by lessees at present, but will soon be worked by the company. There is a very good prospect of opening up former ore chute.

Production to January 1, 1900, $80,000.00, which was all taken out by lessees prior to the incorporation of this company. *Production*

Highest price for stock during 1899, 16 1-2 cents; lowest price for stock during 1899, 11 cents.

The New Haven Gold Mining Company.

Incorporated November 30, 1895.

Main Office—First National Bank building, Colorado Springs, Colorado.

Capitalization

1,500,000 shares.   Par value, $1.00.   In treasury January 1, 1900, 235,000 shares.

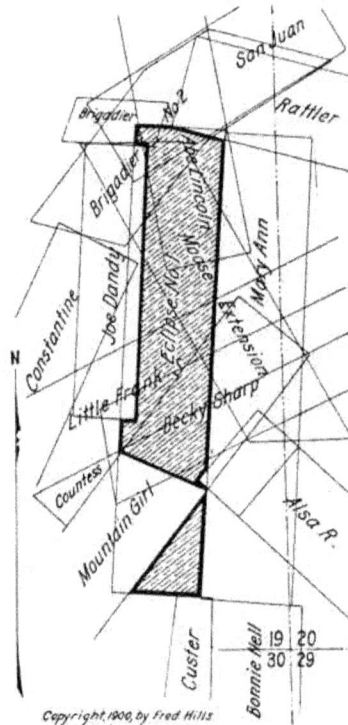

Copyright, 1900, by Fred Hills

Property

Owns the Eclipse No. 1, situated in the S. E. 1-4 of section 19, containing about 7 1-2 acres, on Raven hill.   Patented.

Development

One steam hoisting plant.   An inclined shaft has been sunk 185 feet deep and another 150 feet; some drifting has been done.   The property is being worked by one lessee.

Production

Gross production up to January 1, 1900, about $4,000.00.

Highest price for stock during 1899, 7 cents; lowest price for stock during 1899, 2 cents.

The New Zealand Mining Company.

Incorporated April 7, 1892.

H. E. Woods, president; R. R. Latta, vice-president; F. M. Woods, secretary and treasurer; J. A. Small, assistant secretary and assistant treasurer; Bruce Glidden, J. M. Allen.

Directors

Main Office—Denver, Colorado. Branch Office—Fremont, Colorado.

600,000 shares. Par value, $1.00. In treasury January 1, 1900, 1,300 shares.

Capitalization

Copyright, 1900, by Fred Hills.

Owns the New Zealand, 3 acres; the Pauper and the Deadwood, 14.4 acres, situated in section 20. This company also owns 40 per cent. of the Compromise Gold Mining Company, controls the Magnolia Company and owns 182,000 shares of Pinnacle stock.

Property

There are three shaft houses and two plants of machinery. The property is being worked under lease. The South Deadwood shaft is 450 feet deep, with 1,900 feet of workings. The North Deadwood shaft is 335 feet deep, with 900 feet of underground work. The shaft on the New Zealand is 500 feet deep, with 500 feet of drifting. Shipments have been made from all the above shafts.

Development

Gross production to January 1, 1900, $175,000.00.

Production

Highest price for stock during 1899, 52 cents; lowest price for stock during 1899, 16 cents.

The Niagara Gold Mining Company.

Incorporated November, 1895.

Directors

W. G. Fraser......................President

T. H. Devine.......................Vice-President

N. D. Hinsdale.........................Secretary

D. R. Greene...........................Treasurer

W. A. Baldwin.   E. M. Jackson.

F. E. Sage.   W. R. Grace.

Main Office—243 N. Union avenue, Pueblo, Colorado.

Capitalization 1,250,000 shares. Par value, $1.00. In treasury January 1, 1900, $9,000.00 cash.

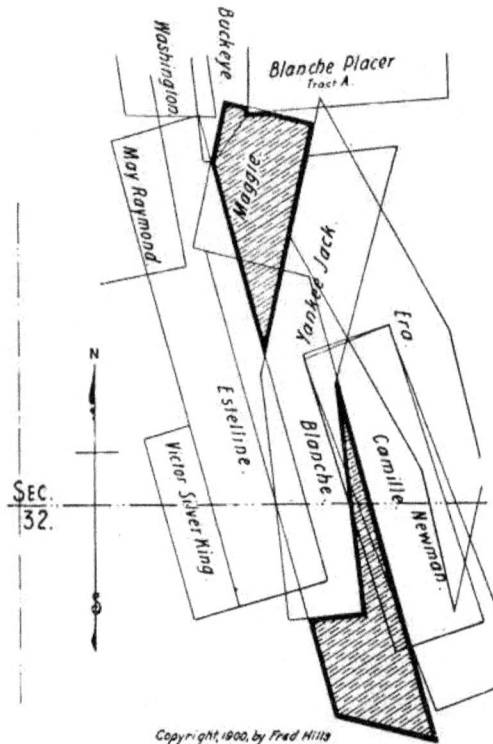

Copyright, 1900, by Fred Hills

Property Owns the Blanche and Maggie, containing about 8 acres, patented, in the E. 1-2 of section 32, 1,400 feet south of Stratton's Independence, on extension of Battle mountain.

Development Shaft house, engine and hoisting machinery. One shaft 160 feet deep with cross-cuts and drifting of 200 feet. The company has recently let contract for sinking of the shaft 200 feet more.

The North American Gold Mining and Development Company.

Incorporated November 18, 1898.

Aubert L. Eskridge......................President

Frank J. Gross......................Vice-President

Howard R. Burk...........Secretary and Treasurer

J. F. Anderson.   Dr. N. F. Brown.

Henry R. Swartley.   Chester R. Lawrence.

J. P. Reymond.   Steward B. Bontems.

Main Office—606-607 Kittredge building, Denver, Colorado.

1,000,000 shares. Par value, $1.00. In treasury January 1, 1900, 327,695 shares; in treasury January 1, 1900, $2,467.41 cash.

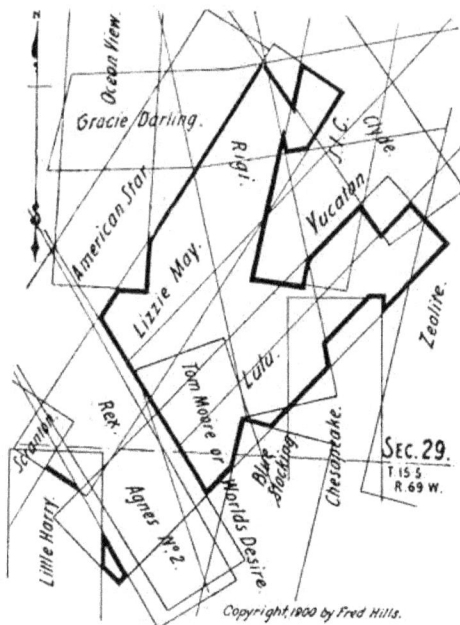

Copyright 1900 by Fred Hills.

The company holds, by lease, all that portion of the Rigi group lying south of the south side line of the American Star, on Battle mountain, adjoining the Portland. This lease runs until March 1, 1902. Their entire attention is now being devoted to the development of this property. The company also owns the Douglas lode mining claim, situated within 700 feet of the famous Caribou mine, in Boulder county, Colorado.

Three shafts have been sunk. The main shaft, No. 2, is 500 feet deep. The company are now negotiating to cross-cut the ground to the west of this shaft from the 500-foot level. Shaft No. 3 is 135 feet deep.

First shipment of ore, April 23, 1900, was 20 tons, averaging over $100.00 per ton.

Highest price for stock during 1899, 35 cents; lowest price for stock during 1899, 5 cents.

The Nugget Mining Company.

Incorporated May, 1900.

Main Office—Room 20, Hagerman building, Colorado Springs.

Capitalization   1,250,000 shares. Par value, 50 cents. Amount of stock in treasury May 15, 1900, 259,005 shares. Cash in treasury May 15, 1900, $11,000.

Property   Owns the Elizabeth Cooper, containing 7.8 acres in the south 1-2 section 19 on Raven hill, and Charles B., containing 3.9 acres, situated in the N. W. 1-4 section 1, township 16 south, range 70 west, on Grouse mountain. Patented.

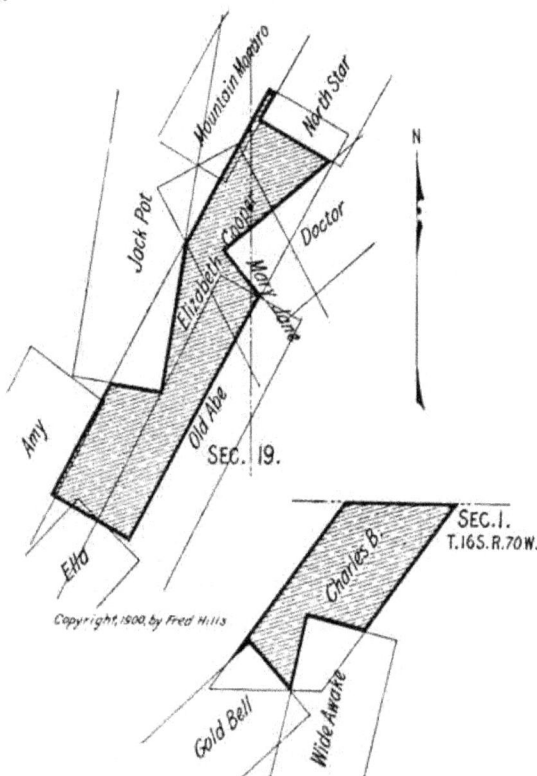

Copyright, 1900, by Fred Hills

Development   There is a 6x8 hoist on the Elizabeth Cooper and a 30 H. P. boiler. There is also a 9x10 Fairbanks hoist and 80 H. P. boiler with good shaft houses, etc. The Elizabeth Cooper has a double-compartment shaft 375 feet deep, with 400 feet of drifting. There is also a 325-foot shaft on the south end of the property with about the same amount of drifting. A lease has been granted on the Elizabeth Cooper, to Franc O. Wood, which pays 25 per cent. royalty.

History   The company formerly owned the Katherine, which was a producer and paid dividends. This claim was sold to the Elkton company. The Elizabeth Cooper lies between the Jack Pot and Doctor mines. There has been some litigation between this company and that of the Doctor, the Doctor lying on the Apex of the Jack Pot vein, and this litigation is still pending. In a cross-cut, in what is known as the Smith and Riley shaft, a body of ore has been encountered, there being 12 inches of very high grade ore and 3 inches low grade.

Highest price for stock during 1899, 25 cents; lowest, 9 1-2 cents.

The Oasis Gold Mining Company.

Incorporated December 5, 1899.

A. P. Mackey...........................President

C. S. Wilson.......................Vice-President

J. C. Connor.........................Secretary

Wm. A. Otis.........................Treasurer

C. E. Titus...................Assistant Secretary

W. H. Thompson.   W. B. Milliken.

Main Office—Colorado Springs, Colorado.

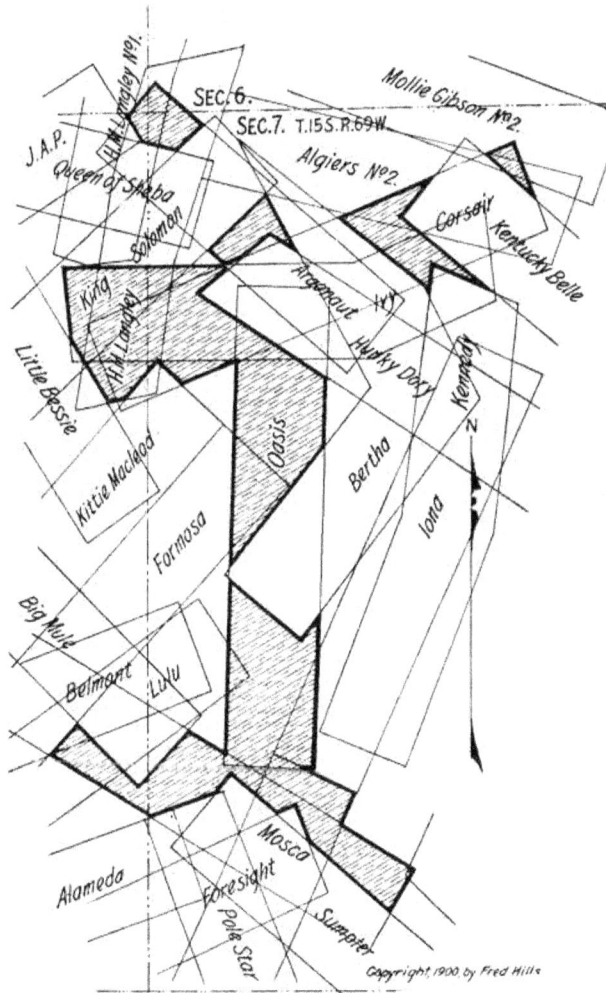

Copyright 1900 by Fred Hills

2,000,000 shares. Par value, $1.00. In treasury January 1, 1900,
500,000 shares.

Owns the Mosca, the Oasis, the Corsair, the Argonaut, the Ivy, and
part of the Idlewild, all situated in the N. E. 1-4 section 7, containing a
total estimated acreage of 15 acres. Receiver's receipt held for the above.

The O. K. Gold Mining Company.

Incorporated February 27, 1896.

Directors

J. E. McIntyre........................President

Clarence Edsall.....................Vice-President

W. C. Frost...............Secretary and Treasurer

R. D. Munson.   A. B. Whitmore.

Main Office—6 N. Nevada avenue, Colorado Springs, Colorado.

Capitalization

1,250,000 shares. Par value, $1.00. In treasury January 1, 1900, 150,000 shares; in treasury January 1, 1900, $4,500.00 cash.

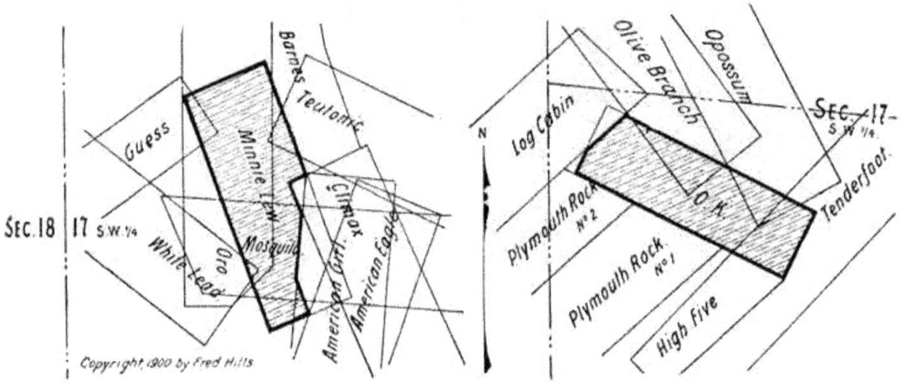

Copyright 1900 by Fred Hills

Property

Owns the O. K. and Minnie Lew claims, containing 9 acres. Patented. Located in S. W. 1-4 section 17, on top of Iron Clad hill.

Development

This property is being developed by lessees, there being four sets actively at work on the same. The company consider the prospects very flattering indeed. The ore chutes of the Damon and Jerry Johnson properties are thought to extend through the O. K. grounds.

Highest price for stock during 1899, 6 cents; lowest price for stock during 1899, 4 cents.

The Old Gold Mining Company.

Incorporated December 30, 1895.

James M. Downing.....................President
S. T. Rush.........................Vice-President
Warwick M. Downing..................Secretary
H. K. Chittenden.....................Treasurer

Owen Owen.   E. T. Jones.
J. E. Schiff.   Chas. M. Sumner.

Main Office—No. 519 Equitable building, Denver, Colorado; branch office, room 2 First National Bank building, Colorado Springs, Colorado.

1,800,000 shares. Par value, $1.00. In treasury January 1, 1900, 30,-000 shares.

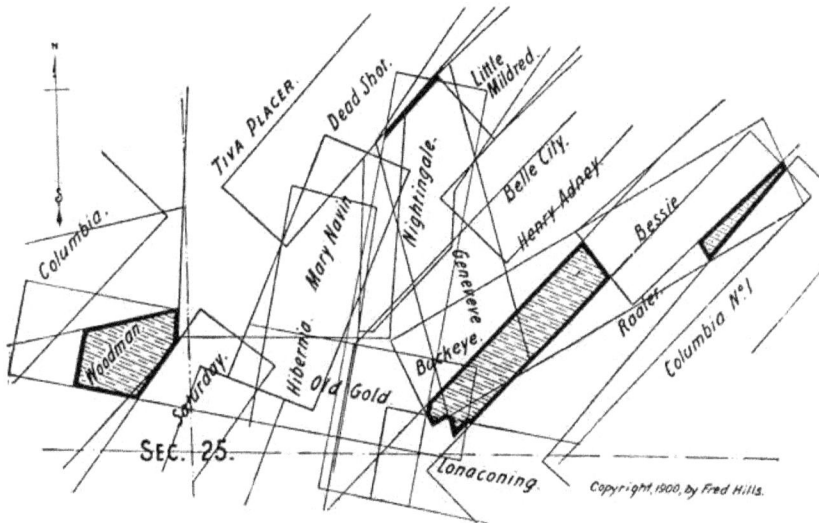

Owns the Old Gold, the Buckeye, and the Geneveve, situated in the N. E. 1-4 section 25, township 15 south, range 70 west, containing 1.147 acres, on the west slope of Beacon hill; also a small portion of the Lonaconing lode of the Kimberley company, and also the northern 400 feet of the Raaler lode belonging to the C. K. & N. Co., i. e., the north 400 feet of that part of that claim lying between the Columbia No. 1 and the Bessie lodes; also 1-2 of the Old Gold tunnel, 550 feet in Beacon hill; also the Mayflower, Puritan and America lodes, unpatented, on Straub mountain. All the Beacon hill property is patented. The Straub mountain property is held by location and not shown on plat.

On the Buckeye there is a shaft 70 feet deep, which has disclosed a phonolite dyke 12 feet wide and indicating rich mineral. High assays have been taken but no pay ore. The Buckeye has also about 100 feet of drifting. The Old Gold tunnel has been driven eastward 550 feet through Beacon hill, starting near the Buckeye shaft. This tunnel has a valid tunnel site location. Several good veins have been found but no pay ore.

Highest price for stock during 1899, 2 1-8 cents; lowest price for stock during 1899, 1 1-8 cents.

The Olive Branch Gold Mining Company.

Incorporated March 4, 1896.

Main Office—First National Bank building, Colorado Springs, Colorado.

Capitalization   1,250,000 shares.   Par value, $1.00.   In treasury January 1, 1900, 90,000 shares.

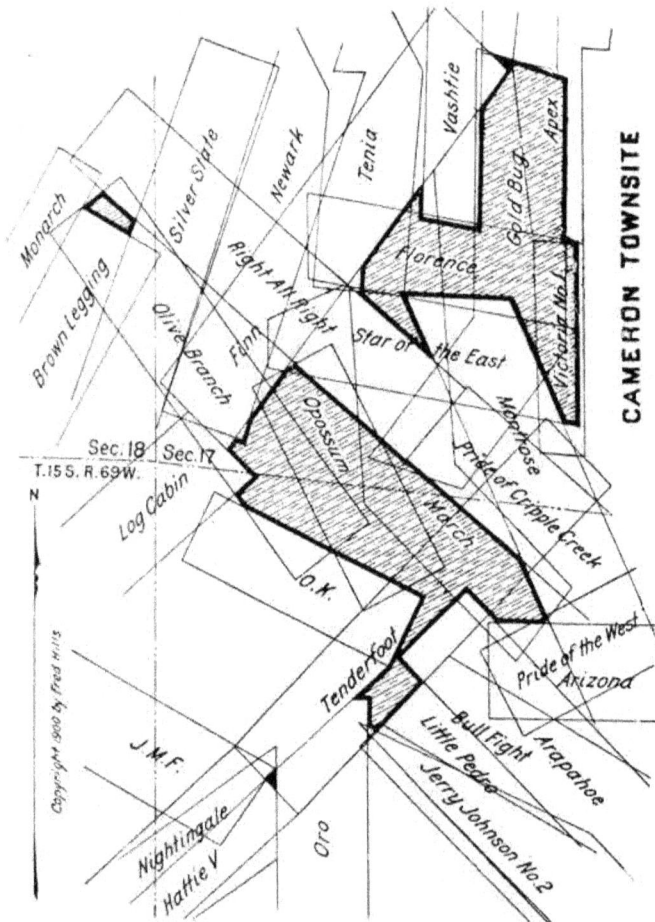

Property   Owns the Olive Branch, Opossum and Tenderfoot claims, lying in a group, containing about 8 acres; the March and Goldbug claims, lying together, containing about 8 1-2 acres; all on Ironclad hill. All this property is patented and is being worked by five sets of lessees.

Highest price for stock during 1899, 10 7-8 cents; lowest price for stock during 1899, 5 1-2 cents.

The Olympian Gold Mining Company.

Incorporated January, 1896.

E. N. Bement, president; H. H. _{Directors} Clark, secretary; O. F. Shattuck; W. A. Moore; C. W. Hawver.

Main Office—Cripple Creek, Colorado.

2,000,000 shares. Par value, $1.00. _{Capitalization} In treasury January 1, 1900, 500,000 shares.

Owns a fraction of the Forsaken _{Property} and the North West, in the S. W. 1-4 section 20, containing 1-2 acre, on Bull hill; also owns a 1-2 interest in the Score lode, in the N. E. 1-4 section 6, township 16 south, range 69 west, containing 9.03 acres, on the north end of Straub mountain.

The Score lode is patented. Forsaken and North West in process of patenting.

There is a 105-foot shaft on the _{Development} North West, and a 55-foot shaft on the Forsaken.

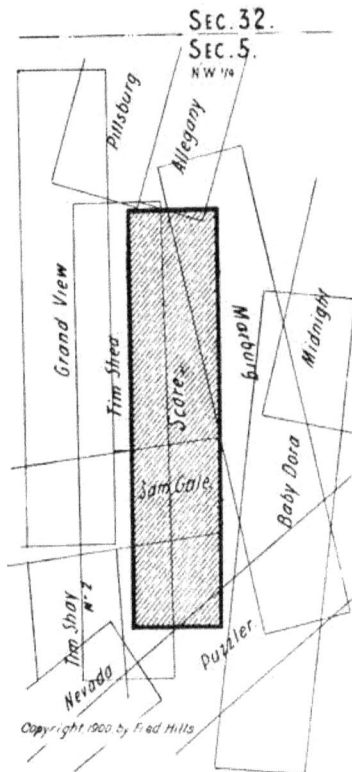

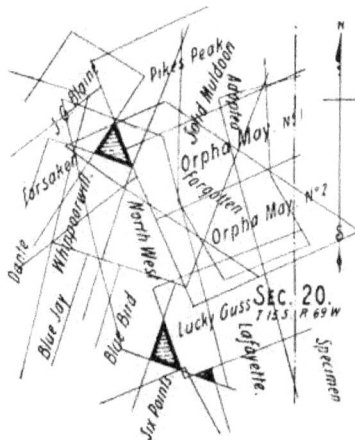

The Ontario Gold Mining and Milling Company

Incorporated August, 1895.

Directors

W. H. MacIntyre.......................President
M. S. Herring......................Vice-President
Horace H. Mitchell..........Secretary and Treasurer
L. E. Sherman.   Geo. W. Logan.

Main Office—No. 45 Bank building, Colorado Springs, Colorado.

Capitalization

1,250,000 shares.   Par value, $1.00.   In treasury January 1, 1900, 5,000 shares.

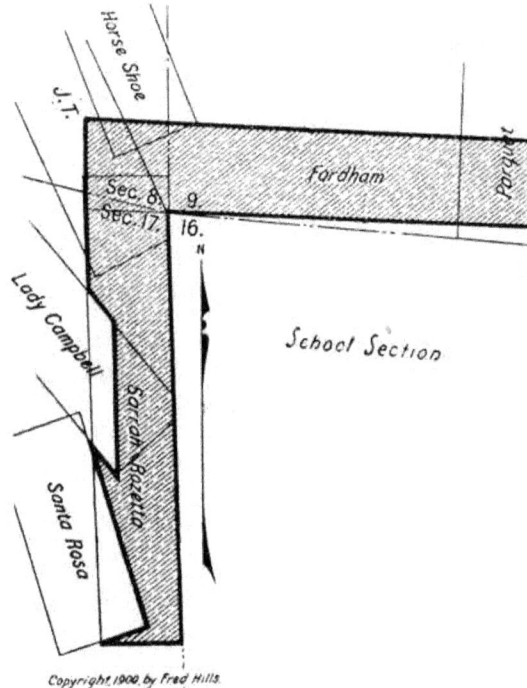

Copyright, 1900, by Fred Hills.

Property

Owns the Sarrah Rozetta and the Fordham, containing about 16 1-4 acres, in the S. W. 1-4 section 9 and the N. E. 1-4 section 17, on Grassy mountain. Receiver's receipt has been obtained for both. The policy of the company will be to develop the property by a liberal leasing system.

Highest price for stock during 1899, 2 cents; lowest price for stock during 1899, $1.20 per M.

The Ophir Mining and Milling Company.

Incorporated March 7, 1892.

J. K. Fical........................President Directors

John Maclean....................Vice-President

D. Chisholm..............Secretary and Treasurer

C. W. Waterman.   B. W. Fical.

Main Office—Room 17, Hagerman block, Colorado Springs, Colorado.

700,000 shares. Par value, $1.00. Capitalization

Owns the Dead Pine and the Carbonate Queen, containing in all Property about 20 acres, situated in the S. W. 1-4 section 29, on Battle mountain. Patented.

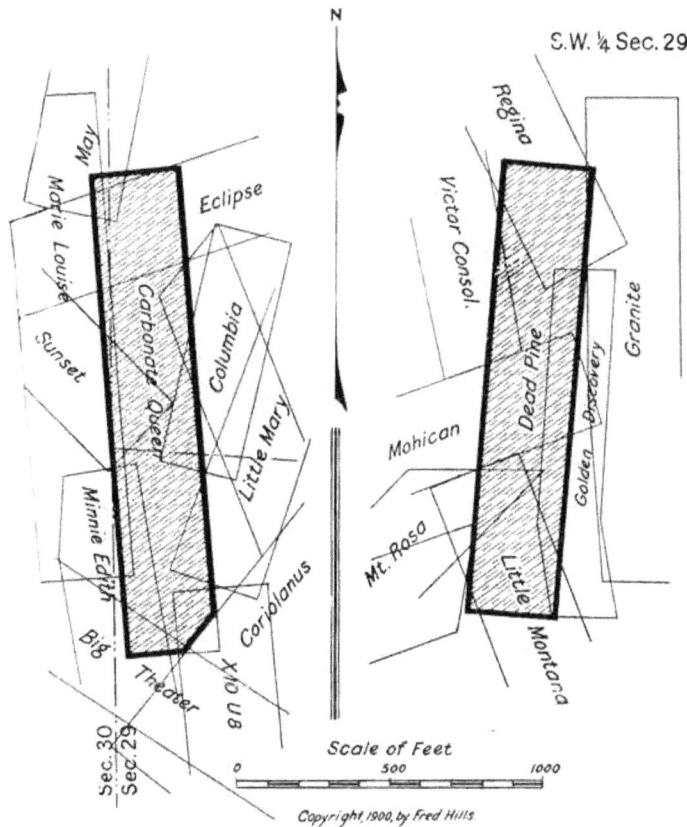

There are two shaft houses and two complete hoisting plants, with Development ore bins, on the property. This property has been largely developed by lessees, and, as the company is a close corporation, it is impossible to state the gross production. The mine is being constantly worked.

Highest price for stock during 1899, 75 cents; lowest price for stock during 1899, 15 cents.

The Ore or No Go Gold Mining and Leasing Company.

Incorporated December 23, 1895.

W. H. Sutherland......................President
Samuel P. Duff....................Vice-President
J. H. Thedinga.............Secretary and Treasurer
W. R. Mason.   L. M. Peck.
J. R. Friedline.

Main Office—No. 17 East Pike's Peak avenue, Colorado Springs, Colorado.

Capitalization

1,000,000 shares. Par value, $1.00. In treasury January 1, 1900, 149,000 shares; in treasury January 1, 1900, $981.95 cash.

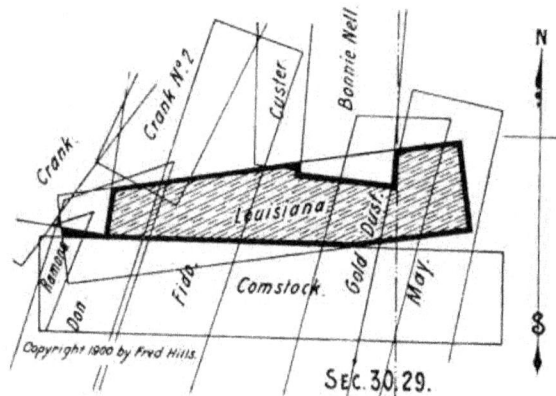

Property

Owns the Louisiana, containing 6 acres, located in the N. E. 1-4 section 30, on Raven hill. Patented.

Development

There is one shaft on the property 142 feet deep; one level running westerly 75 feet, and one level running northeasterly 90 feet; one drift on tunnel 30 feet, with winze 85 feet. A new working shaft 4x8 is now being sunk by the company. The contract for the first 50 feet was let January 1, 1900.

Highest price for stock during 1899, 3 1-2 cents; lowest price for stock during 1899, 2 cents.

The Oriole Gold Mining Company.

Incorporated November 12, 1895.

Clarence Edsall.........................President
Duncan Chisholm..............First Vice-President
John Matthew................Second Vice-President
John J. Key...............Secretary and Treasurer
B. N. Beal........Assistant Secretary and Treasurer

John S. Hunt.   W. E. Jones.

Directors

Main Office—Hagerman building, Colorado Springs, Colorado.

1,250,000 shares. Par value, $1.00. In treasury January 1, 1900, 140,000 shares; in treasury January 1, 1900, $150.00 cash.

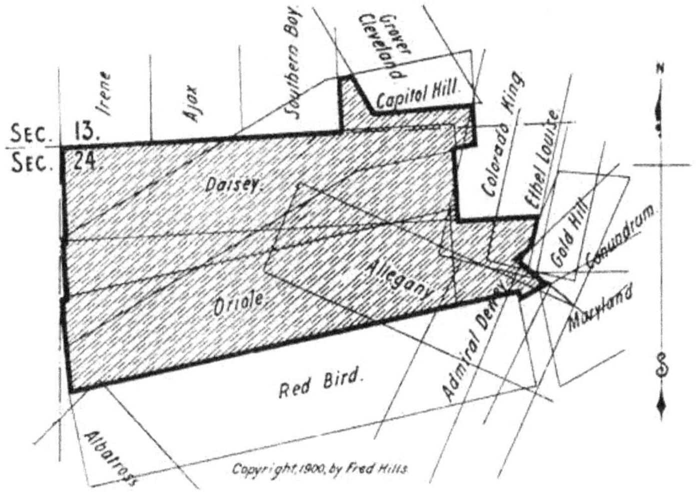

Capitalization

Copyright, 1900, by Fred Hills

Owns the Daisey, Capitol Hill, Allegany, Oriole and Pearl Cecil, containing 25 acres, in the N. E. 1-4 section 24, on Gold hill. Patented. Also owns the Oriole townsite on surface of above.

Property

There is a plant of machinery on the Oriole, and this claim is being worked by lessees.

Development

Highest price for stock during 1899, 9 1-4 cents; lowest price for stock during 1899, 2 1-2 cents.

The Orphan Bell Mining and Milling Company.

Incorporated 1892.

Directors

F. B. Gibson, president and treasurer; J. F. Sanger, vice-president and secretary; W. H. Spurgeon.

Main Office — International Trust Company, Colorado Springs, Colorado.

Capitalization

1,000,000 shares. Par value, $1.00. In treasury January 1, 1900, 21,000 shares.

Property

Owns the Phoenix, containing 0.864 acre, in the N. E. 1-4 section 20, on Bull hill.

Production and Dividend.

Gross production to January 1, 1900, about $300,000.00. Total amount of dividends to January 1, 1900, $215,000.00.

History

This company formerly owned the Orphan No. 1 and the Orphan No. 2, the Ida Belle and the Ida Bellle No. 2 claims, and produced a large amount of ore; but in 1899 these claims were sold to two companies, namely, the Orphan Gold Mining Company and the Arrow Gold Mining Company, as will be seen in another portion of this book. The company is going into liquidation as soon as some lawsuits now pending are settled.

Highest price for stock during 1899, 11 cents; lowest price for stock during 1899, 5 1-2 cents.

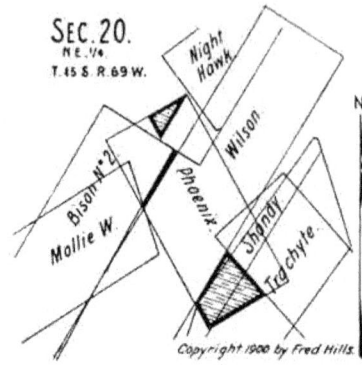

The Orphan Gold Mining Company.

Incorporated July 20, 1899.

Directors

H. M. Blackmer, president; C. E. Titus, secretary; Wm. A. Otis, treasurer and vice-president; W. P. Sargeant; P. B. Stewart.

Main Office—Wm. A. Otis & Co., Giddings Block, Colorado Springs, Colorado.

Capitalization

1,250,000 shares. Par value, $1.00. In treasury January 1, 1900, 114,995 shares; in treasury January 1, 1900, $900.00 cash.

Property

Owns the Ida Belle, Nos. 1 and 2; and the Orphan, Nos. 1 and 2, containing in all 12 acres, situated in the N. E. 1-4 section 20, on Bull hill. All patented.

Development

There has been done between 10,000 and 15,000 feet of work in shafts, drifts, etc. The Ida Belle No. 1 and the Orphan No. 2 are leased.

Highest price for stock during 1899, 23 cents; lowest price for stock during 1899, 18 cents.

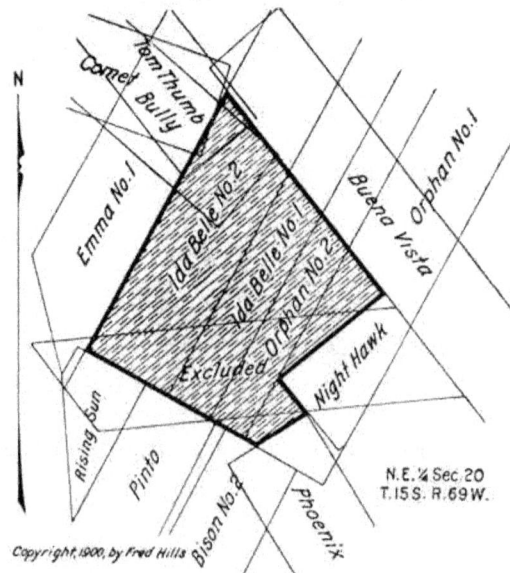

The Palace Gold Mining Company.

Incorporated August 25, 1899.

F. L. Ballard..........................President   Directors
S. G. Tucker..........Vice-President and Secretary
H. L. Shepherd.......................Treasurer
    J. E. Jones.   A. L. Houck.

Main Office—109 East Pike's Peak avenue, Colorado Springs, Colorado.

1,250,000 shares. Par value, $1.00. In treasury January 1, 1900,   Capitalization
99,995 shares; in treasury January 1, 1900, $950.00 cash.

Copyright, 1900, by Fred Hills

Property

Owns the Palace claim, 7.377 acres, and the Homeward, 3.5 acres, near the Palace, in the S. E. 1-4 of section 30, on Squaw mountain. Patented.

Development

On the Palace claim there is one shaft 90 feet deep, one 80 feet deep, one 60 feet deep, one 12 feet deep and one 40 feet deep. The north 400 feet of the Palace is leased for 18 months at 25 per cent. royalty. About $8,000.00 has been expended on the property and assays from $30.00 to $70.00 obtained from the Palace, and as high as $100.00 from the Homeward, and the prospects are excellent.

Highest price for stock during 1899, 3 cents; lowest price for stock during 1899, 1 7-8 cents.

The Pappoose Gold Mining Company.

Incorporated March 8, 1894.
Amended March 4, 1899.

Directors   John F. Hardy, president; A. L. Hardy, secretary and treasurer; Walter C. Frost, assistant secretary; L. O. Byrer, A. M. Hardy.

Main Office—6 N. Nevada avenue, Colorado Springs, Colorado.

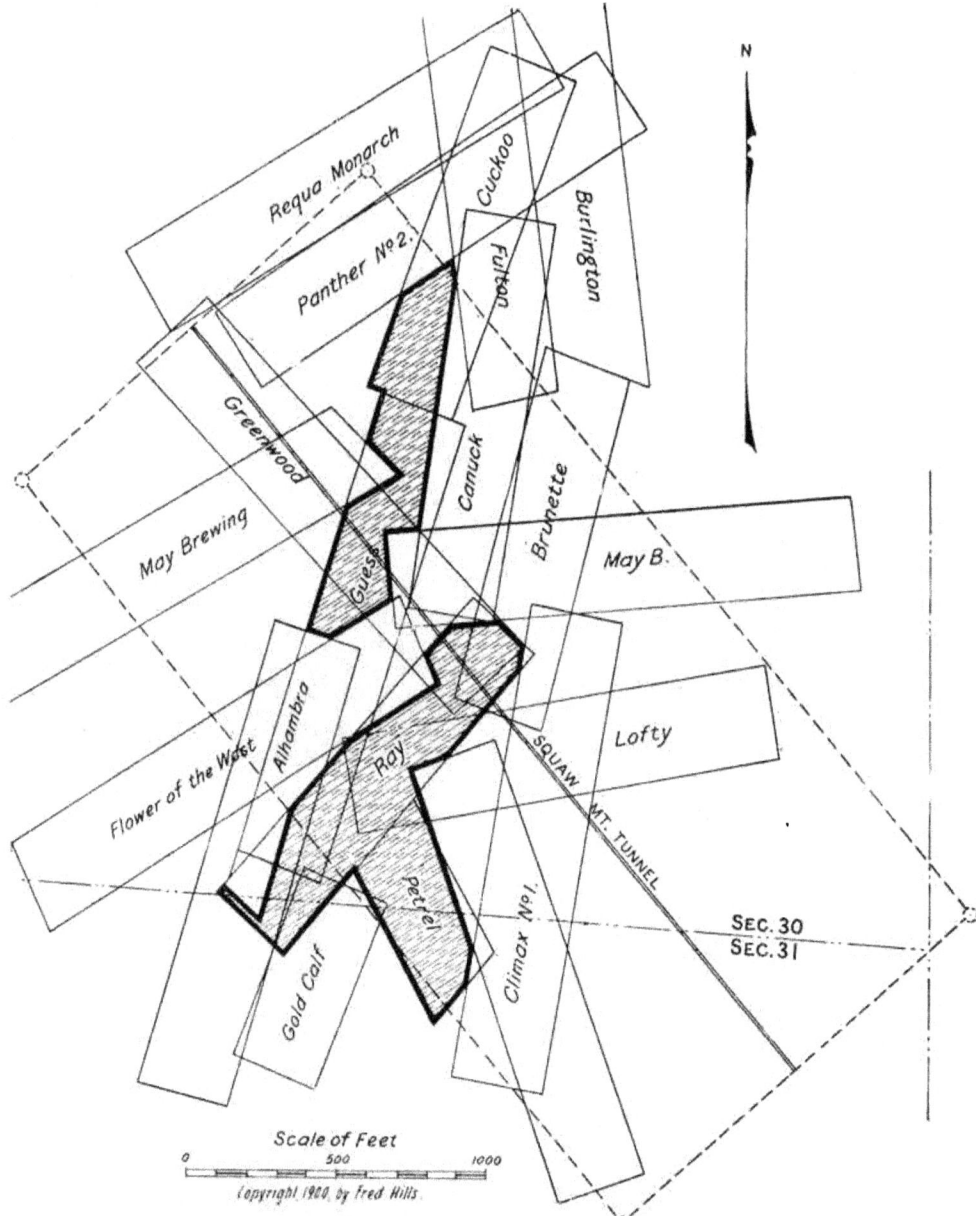

Scale of Feet

Copyright 1900, by fred Hills.

Capitalization   2,250,000 shares. Par value, $1.00. In treasury January 1, 1900, 50,000 shares; in treasury January 1, 1900, $600.00 cash.

Property   Owns the lode claims Ray, Petrel, Cuckoo and Guess, containing 21 acres, situated in the S. E. 1-4 of section 30, on the southwest slope of Squaw mountain, a short distance from the Gold Coin, Portland, Mary Cashen and other prominent shippers.

The property is being worked by lessees and it is estimated that it **Development** has already produced a gross valuation of about $60,000.00. The vein out of which this ore has been shipped is now being opened up below the 300-foot level, and the owners of a large portion of the stock are confident that the property will make a dividend payer.

Gross production to January 1, 1900, $60,000.00. **Production**

Highest price for stock during 1899, 17 1-2 cents; lowest price for stock during 1899, 3 cents.

The Park View Mining and Milling Company.
Incorporated 1893.

J. G. Medley, president; J. M. **Directors** Dorr, vice-president; R. B. Massey, secretary; M. L. Dorr, treasurer; R. C. Hart.

Main Office—No. 52 P. O. building, Colorado Springs, Colorado.

1,000,000 shares. Par value, **Capitalization** $1.00. In treasury January 1, 1900, 500,000 shares.

Owns the Park View, 8 1-3 acres; the Horse Shoe, 10 acres; the **Property** Grover Cleveland, 10 acres, and the Admiral Duncan, 10 acres, all situated in the N. W. 1-4 of section 6, on Rhyolite mountain. The Park View, Horse Shoe and the Grover Cleveland are patented. The Admiral Duncan is held by location.

Log cabin and shaft house. The Park View has a shaft 115 feet **Development** deep and 260 feet of drifting. The Horse Shoe has a 70-foot shaft. The Grover Cleveland has a shaft 55 feet deep. The Admiral Duncan has a shaft 65 feet deep and 25 feet open cut. The greater part of the development work is being done on the Park View claim.

Ore has assayed from $2 to $50 per ton. The Park View showed **History** good assays at a depth of 60 feet. On the summit of Rhyolite mountain a phonolite dyke was cut, but the quantity of water encountered prohibited further extension. At the bottom of the 64-foot shaft on the Admiral Duncan are two distinct veins between porphyry and phonolite walls, assaying from $8 to $36. In an open cut at a depth of 12 feet $24 assays have been obtained. The Horse Shoe and the Grover Cleveland are covered with some very fine timber.

Highest price for stock during 1899, 2 3-4 cents; lowest price for stock during 1899, 1 1-4 cents.

The Pay Car Gold Mining and Leasing Company.

Incorporated February 1, 1896.

Directors

C. W. McReynolds, president; A. H. Garnett, vice-president; Chas. M. Kirk, secretary; John Killen, treasurer; J. L. Semmes, assistant secretary; Frank H. Kirk.

Main Office—Nos. 24 and 25 Exchange Bank building, Colorado Springs, Colorado.

Capitalization

1,500,000 shares. Par value, $1.00. In treasury January 1, 1900, 40,000 shares.

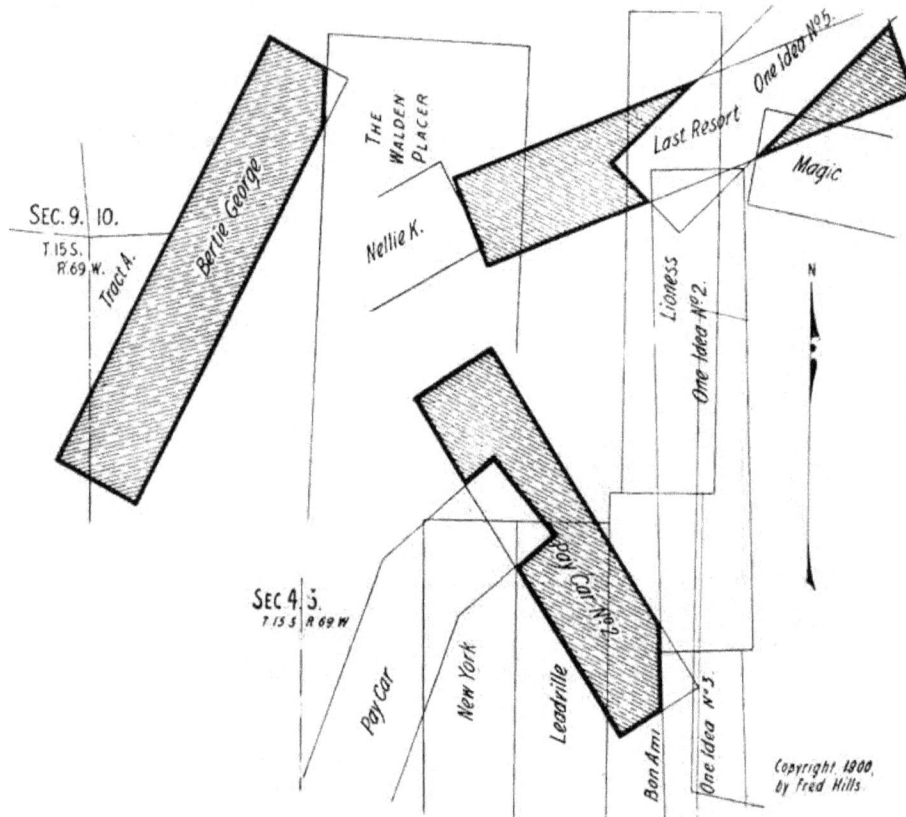

Property

Owns the Bertie George, 10.159 acres, in the W. 1-2 section 10; the Pay Car No. 2, 7 1-4 acres, in the S. W. 1-4 section 3; the Last Resort, 6 acres, in the N. W. 1-4 section 3—all in township 15 south, range 69 west, on Trachyte mountain. The Bertie George and the Pay Car No. 2 are patented. The Last Resort is in process of patenting.

Development

Pay Car No. 2 has a shaft house. Bertie George has shaft house, blacksmith shop, whim, tracks, car, etc., and about 160 feet of shafts and 500 feet of drifts. The Pay Car No. 2 has about 80 feet of shafts, and 100 feet of drifts. On the Last Resort there are 40 feet of shafts, and 60 feet of drifting. Large bodies of ore have been discovered but of too low grade to pay for treatment. The greater part of the development work is being done on the Bertie George. This claim is being developed by six of the largest stockholders of the company, with the understanding that they will be reimbursed for expenses, when paying ore is discovered. Up to January 1, 1900, about $4,500 has been expended and work is still being prosecuted.

Highest price for stock during 1899, 2 1-2 cents.

The Pelican Gold Mining Company.

Incorporated August 9, 1899.

W. B. Storer............................President

C. Edsall..........................Vice-President

N. H. Partridge.............Secretary and Treasurer

E. C. Fletcher.   J. J. Key.

Main Office—10, 11 and 12 Giddings block, Colorado Springs, Colorado.

1,250,000 shares. Par value, $1.00. In treasury January 1, 1900, 175,000 shares; in treasury January 1, 1900, $500.00 cash.

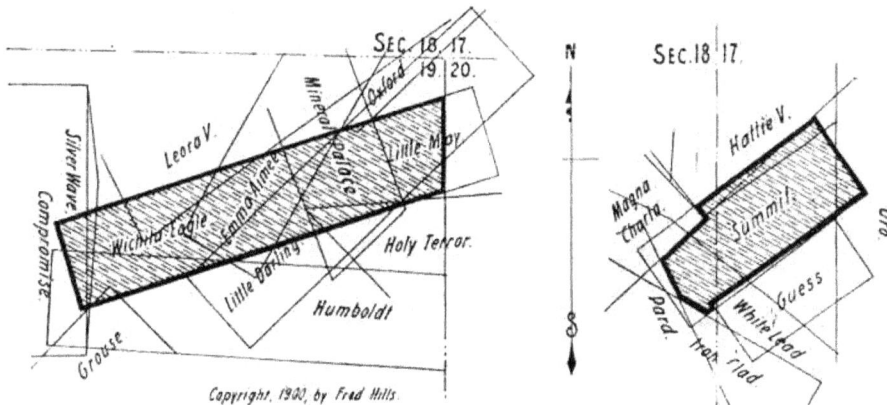

Copyright, 1900, by Fred Hills.

Owns the Summit, 4 acres, in the S. W. 1-4 section 17, patented. Also the Little May, 8 acres, in the N. E. 1-4 section 19. Receiver's receipt. All on Ironclad hill.

Some surface development on the property with shallow shafts less than 100 feet deep.

Highest price for stock during 1899, 5 1-2 cents; lowest price for stock during 1899, 2 1-2 cents.

The Pennsylvania Gold Mining Company.

Incorporated November, 1895.

Directors — Stephen W. Keene, president; A. L. Steinmeyer, vice-president; H. L. Beard, secretary and treasurer; F. I. Smith, assistant secretary; H. E. Stewart, J. B. Meyers.

Main Office—P. O. building, Colorado Springs, Colorado.

Capitalization — 1,000,000 shares. Par value, $1.00. In treasury January 1, 1900, 220,000 shares.

Property — Owns the Pennsylvania Gold Tunnel and tunnel site, located on the north slope of Tenderfoot hill near the northwest corner of the S. E. 1-4 of section 7, and extending in a southeasterly direction 3,000 feet from that point through Tenderfoot and Ironclad hills.

Development — The tunnel, size 6x7, has been driven 300 feet; timbered, tracked and drained. Blacksmith's shop, tools and cars.

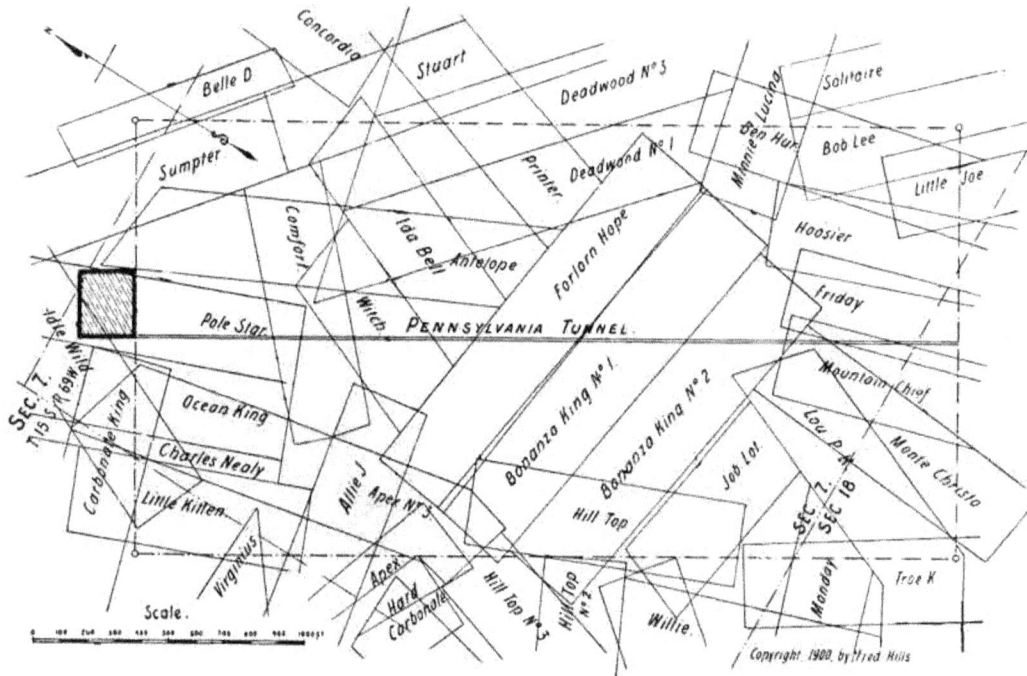

History and Object — The tunnel was located September 3, 1895, and the company formed for the purpose of working, mining and developing lodes, leads and veins, and locating the same under the laws governing tunnels and tunnel sites, and to do and perform all other acts and duties incidental to conducting and carrying on a general mining business; for the discovery of mines, lodes and leads and deposits of valuable mineral, and develop the same by excavating, running cross-cuts or such other workings as may be necessary for drainage and transportation purposes, through property owned by the company or other corporations or individuals, for the purpose of draining such property and transporting water, ores, rock and waste from said mines, or carrying men and supplies to said mines or mining claims, and when such transportation is done for other corporations or individuals by this company, to demand and receive compensation therefor.

The company has contracts for the purposes above named with the owners of the following named properties through which the tunnel will pass: The Pole Star, the Little Kitten, the Ocean King, the Charles Nealy, and the Forlorn Hope; also has leases on the ground of the first four named claims.

Stock not listed. All sales privately negotiated at 10 cents.

The People's Mining and Milling Company.

Incorporated July 14, 1892.

John P. Brockway, president; Geo. M. Mitchell, secretary; H. P. Directors
Bristol, treasurer.

Main Office—No. 308 Mining Exchange, Denver, Colorado.

500,000 shares. Par value, $1.00. In treasury January 1, 1900, Capitalization
$57.40 cash.

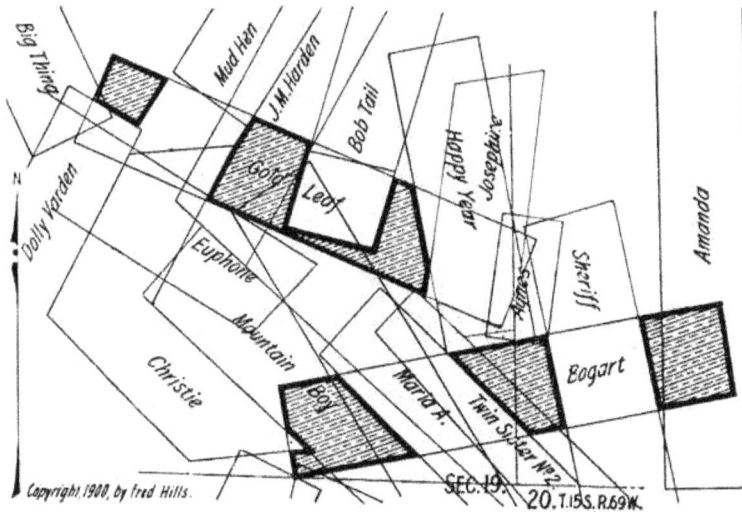

Owns the Bogart, 4.78 acres, and the Gold Leaf, 5 acres, both on Property
Raven hill, on a line between sections 19 and 20. A part of the Gold Leaf
is at present in litigation. Both claims patented.

Shaft house and steam hoist on the Bogart. There are several shafts, Development
the deepest being 284 feet; also some trenching. The greater part of the
development work is being done on the Bogart claim. The east end of this
property is leased and bonded.

All of the company's property is mortgaged to secure a debt of $6,-
700.00, contracted July, 1899, to run one year.

Gross production to January 1, 1900, about $20,000.00. Production

The Pharmacist Consolidated Mining Company.

Incorporated September 7, 1899.

Directors

James F. Burns, president; W. J. Chambers, vice-president; Chas. N. Miller, secretary; Larry Maroney, treasurer; A. D. Jones.

Main Office—271 Bennett avenue, Cripple Creek, Colorado.

Capitalization

1,500,000 shares. Par value, $1.00. In treasury January 1, 1900, 59,000 shares.

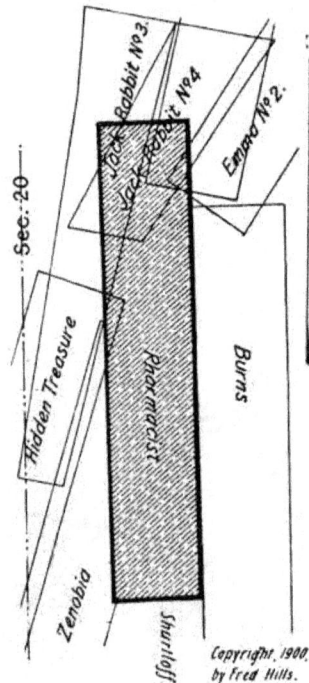

Copyright, 1900, by Fred Hills.

Property

Owns the Pharmacist claim, situated in the N. E. 1-4 of section 20, containing 10 1-3 acres, on Bull hill, adjoining the property of the Acacia Company. Patented.

Development

Two shaft houses, the old one, size 30x75, with a plant of machinery; the new one, size 75x100, with a splendid large plant of machinery, capable of doing the necessary work with a shaft sunk to a depth of 1,500 feet. The old incline shaft is 4x8 in the clear, 650 feet deep, on the incline, with levels at every 50 feet; 1,000 feet in drifts, cross-cuts, etc. The new straight shaft is 5x9 in the clear, 550 feet deep, levels every 100 feet, 1,000 feet in drifts, cross-cuts, etc.

Production and Dividend

Gross production to January 1, 1900, about $650,000.00. The old Pharmacist Company paid $84,000 in dividends in the past. The new company has not yet paid dividends.

Highest price for stock during 1899, 19 1-2 cents; lowest price for stock during 1899, 12 1-4 cents.

The Pike's Peak and Bull Hill Gold Mining Company.

Incorporated December 16, 1895.

A. H. Garnett..........................President
O. W. Spicer.........................Vice-President
F. B. Tiffany..............Secretary and Treasurer
S. T. Hamilton.   C. J. Tiffany.

Directors

Main Office—Nos. 9 and 10 Giddings building, Colorado Springs, Colorado.

1,500,000 shares. Par value, $1.00. In treasury January 1, 1900, 671,000 shares; in treasury January 1, 1900, $100.00 cash. Capitalization

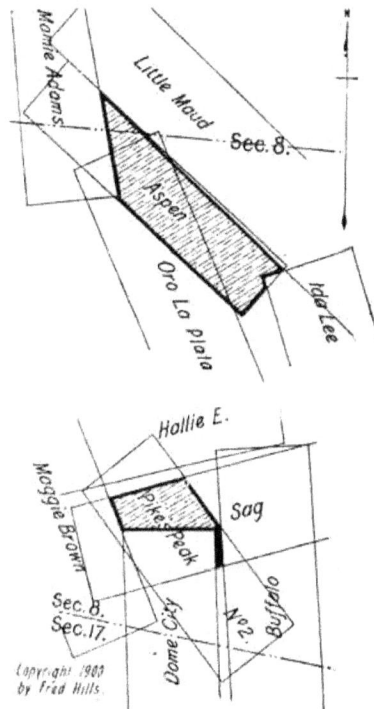

Owns the Pike's Peak No. 2 and the Aspen, about 4 acres, in the east 1-2 of section 8, on Galena hill. Receiver's receipt. Property

On the Pike's Peak No. 2 there is a shaft 50 feet deep, and also on the Aspen there is a shaft the same depth. No sales of stock are reported. Development

The Pilgrim Consolidated Gold Mining Company.

Incorporated January, 1896.

Directors

H. M. Blackmer, president; J. P. Curtis, vice-president; Sherwood Aldrich, secretary; E. P. Shove, treasurer; W. R. Barnes; W. H. Gumm.

Main Office—No. 9 S. Tejon street, Colorado Springs, Colo.

Capitalization

1,250,000 shares. Par value, $1.00. In treasury January 1, 1900, 180,000 shares; in treasury January 1, 1900, $900.00 cash.

Property

Owns the Maud Helena, containing 7.53 acres, and the Maud Helena No. 2, containing 4.992 acres, both situated on Bull hill in the E. 1-2 section 20. Patented.

Development

One electric hoisting plant.

The Maud Helena has a shaft 200 feet deep, with 100 feet of drifting; also several other shafts (probably six), ranging from 40 feet to 90 feet deep. On the Maud Helena No. 2 are a number of shafts ranging from 40 feet to 100 feet deep, a considerable amount of surface trenching and two tunnels, each of about 150 feet. All the property is leased. Three leases now being operated.

Production

Gross production to January 1, 1900, about $5,000.00.

Highest price for stock during 1899, 12 cents; lowest price for stock during 1899, 5 cents.

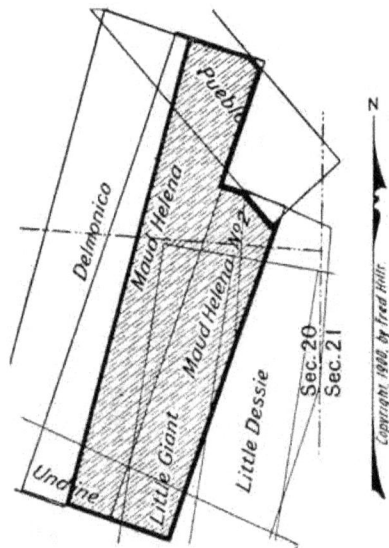

The Pilgrim Gold Mining Company.

Incorporated January 28, 1896.

F. B. Davis..............................President
C. A. Thompson....................Vice-President
Chas. H. Peters........................Secretary
F. A. Bailey...........................Treasurer

J. Weatherbee.   Van R. Kent.

E. W. Smith.

Directors

Main Office—307 Mining Exchange, Denver, Colorado.

1,500,000 shares. Par value, $1.00. In treasury January 1, 1900, 150,000 shares; in treasury January 1, 1900, $100.00 cash.

Capitalization

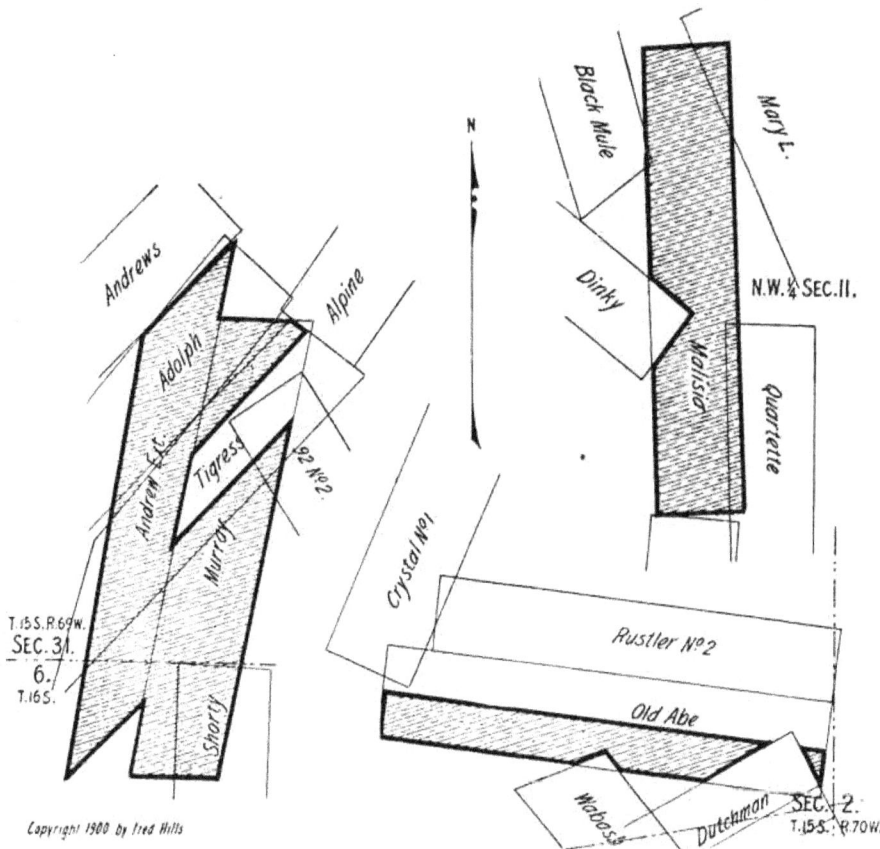

Copyright 1900 by Fred Hills

Owns the Murray, 8 1-4 acres, in the S. W. 1-4 of section 31; the Andrew Extension, 7 1-2 acres, in the S. W. 1-4 of section 31, on the southeast slope of Grouse mountain; the Malisia, 10 acres, in the N. W. 1-4 of section 11, township 15 south, range 70 west, and the south one-half of the Old Abe, 4 1-2 acres, in the center of section 2, on Copper mountain. All patented.

Property

About $3,500 worth of development work has been done to date. The company has no debts. The Malisia claim is leased.

Highest price for stock during 1899, 1 3-4 cents; lowest price for stock during 1899, 3-4 of a cent.

The Pinnacle Gold Mining Company.

Little Bonanza Placer

John R. Watt. Sec. 1.
Sec. 2.

Ben N. A. Miles

"77"

Ellis

Cornucopia

Margaret

Flying Cloud

Sarah Ann McDonald

Last Chance

Black Tail

Elk

Deer

Moose

Whistler

Royal Age

Ione

Horse Shoe

Mitchell

Black Tail

McClver

Bonanza

Ferguson

Free Coinage Grant

Concord

Timber

Lansing

First Chance

Emma No 2

Brandschild

Mollie M.

Sec. 17.

N

Hub

Iron King

Grouse

Josie S.

Colo. Boss No 2

Last Chance

Big Thing

Overlooked

White Elephant

Gold Leaf

Free Colorado Boss No 1

Mining

Euphone

Sec. 19.

Callie

Dolly R.

Little Clara

Copyright 1900 by Fred Hills

The Pinnacle Gold Mining Company.

Incorporated December, 1895.

Chas. Farnsworth..........President and Treasurer

Franklin E. Brooks.................Vice-President

Samuel L. Caldwell....................Secretary

F. M. Woods.   Frederick Farnsworth.

Directors

Main Office—Colorado Springs, Colorado.

Capitalization 2,000,000 shares. Par value, $1.00. In treasury January 1, 1900, 4,000 shares.

Property Owns the Mitchell, the McCluer, the Lansing, the Bonanza, the Girard, the Black Tail, the Brindsmaid, the Horse Shoe, the Olympia, the Mollie W. and a one-half interest in the Sarah Ann McDonald and the Flying Cloud, situated in the S. E. 1-4 section 17, on the northeast slope of Bull hill. Exclusive of the last two named, the total acreage of these claims is about 30 acres. Also the Overlooked, the Dolly R. and the White Elephant, containing in all about 4 acres, in the N. E. 1-4 section 19, on Gold hill, and the Gen. N. A. Miles claim, containing about 8 acres, situated between sections 1 and 12, on Mineral hill. All the above property is patented with the exception of the Horse Shoe, Olympia and Mollie W., which are in process.

Development Shaft house, gallows frame and ore house on the Lansing. Two gallows frames and shaft houses on the Mitchell. On the Lansing the main shaft is 250 feet deep, an inclined shaft of 96 feet, and about 800 feet of drifting. The Mitchell has a shaft 80 feet deep, a drift of 100 feet at the 56-foot level, and a winze 60 feet deep below said level. The greater part of the development work has been done on the Lansing and the McCluer.

Production The gross production of these claims up to January 1, 1900, has been not less than $200,000. About $30,000 in royalties has been received by the company.

History The present company, at the time of its formation, owned the following claims: The Dolly R., Overlooked, White Elephant, Gen. N. A. Miles, Mitchell, McCluer and Lansing. The remainder have since been acquired.

Highest price for stock during 1899, 39 cents; lowest price for stock during 1899, 3 1-2 cents.

The Plymouth Rock.

Not Incorporated.

Property   This group, which is owned by Mr. W. S. Stratton, of Colorado Springs, is situated on Ironclad mountain, between sections 17 and 18,

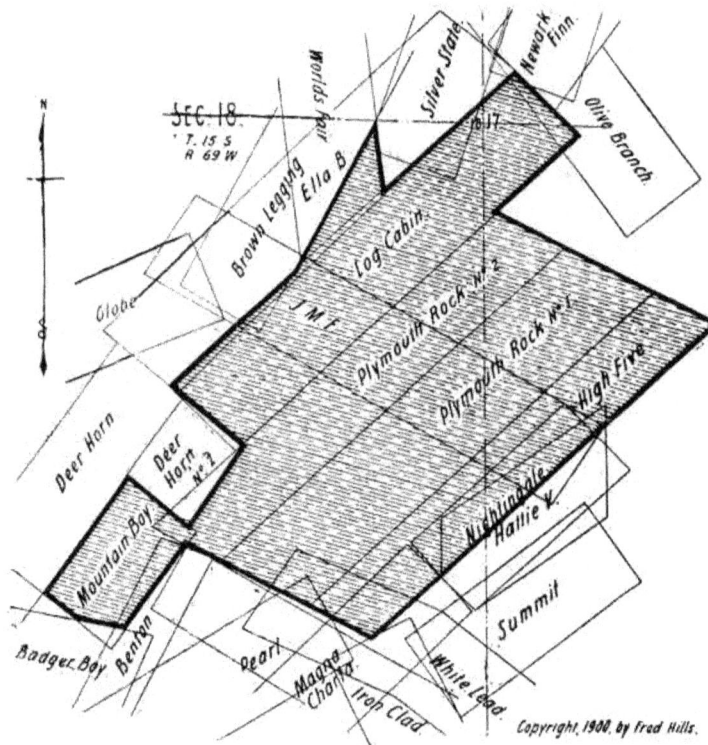

and consists of the following claims: The Log Cabin, Plymouth Rock No. 1, Plymouth Rock No. 2, Mountain Boy and the High Five, containing in all about 35 acres. Patented.

The Pointer Gold Mining Company.

Incorporated November 13, 1899.

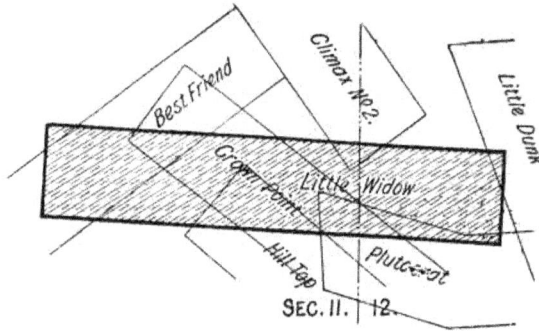

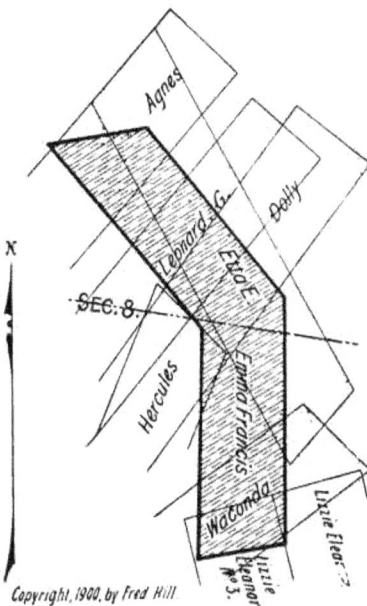

Copyright, 1900, by Fred Hill

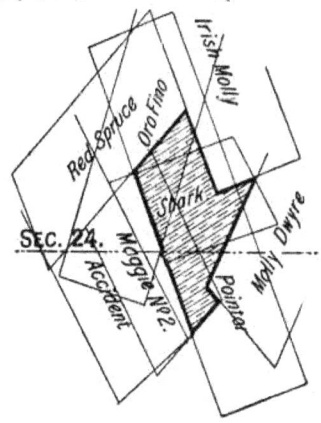

W. R. Foley..............President Directors
D. Weyand...........Vice-President
S. J. Mattocks. Secretary and Treasurer

Main Office—No. 104 Pike's Peak avenue, Colorado Springs, Colorado.

1,250,000 shares. Par value, $1.00. In treasury January 1, 1900, 250,000 shares; in treasury January 1, 1900, $2,000.00 cash. Capitalization

Owns the Pointer and Shark claims, in the E. 1-2 section 24, containing 3 1-2 acres, on Gold hill; the Emma Francis, in the W. 1-2 section 8, containing 10 acres, on Tenderfoot hill; and the Little Widow, in the W. 1-2 section 12, containing 10 acres, on Mineral hill. Property

Property is all patented.

On the Pointer claim there is a full plant of machinery; a good substantial shaft house and boiler room. The improvements cost about $5,000.00. Development

There is a shaft 250 feet deep, and 300 feet of drifting on the Pointer claim. The greater part of the development work is being done on this claim.

The Emma Francis has a shaft 100 feet deep.

About 300 tons of ore, at $20 a ton, were shipped from the Pointer claim about three years ago. From this claim some very high assays have been obtained. Production

On the Pointer claim there is good shipping ore in sight at the present time, and development work is being vigorously pushed. The rock is highly mineralized.

Highest price for stock during 1899, 5 cents; lowest price for stock during 1899, 4 1-2 cents.

The Portland Gold Mining Company.

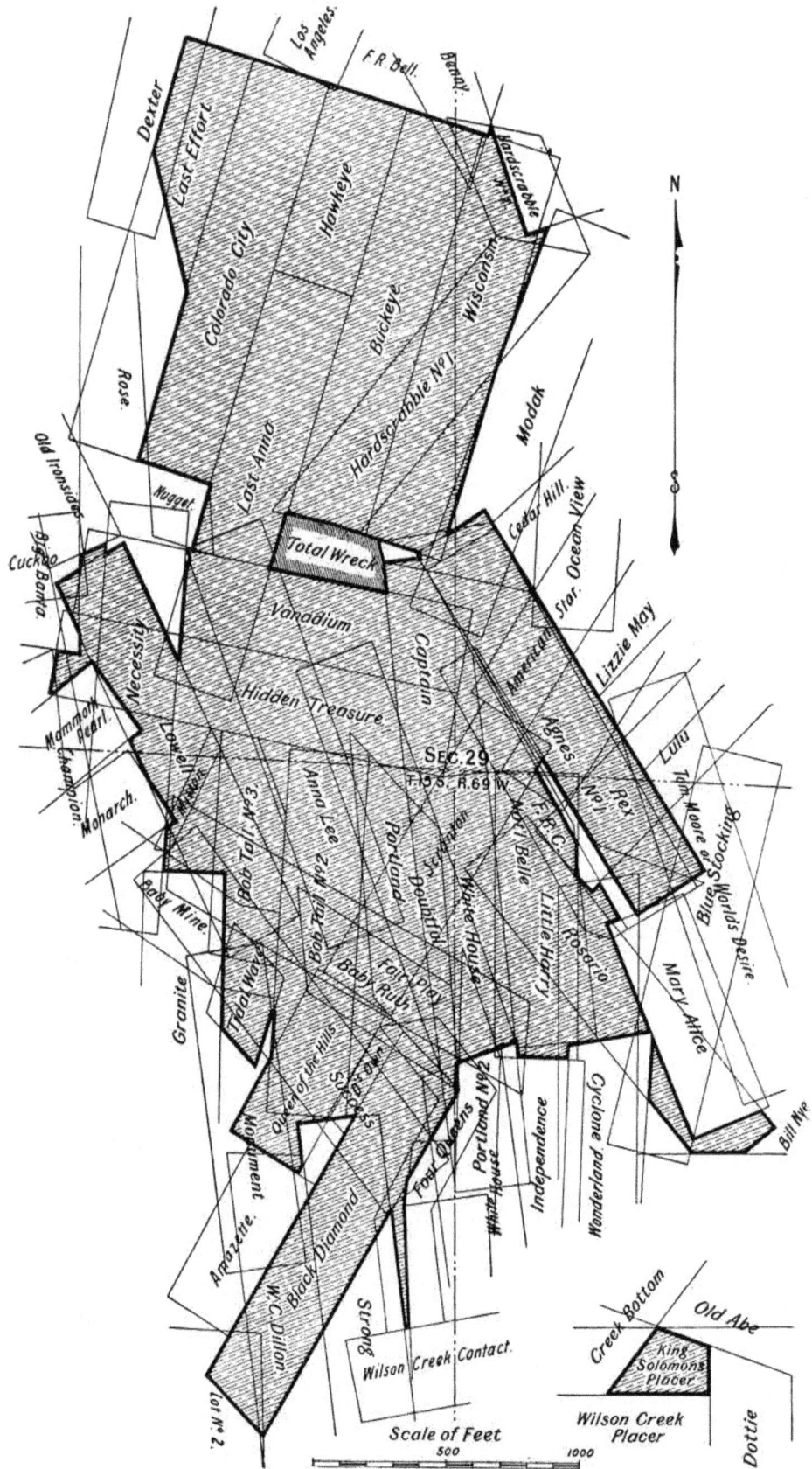

Dexter
Last Effort.
Los Angeles.
F.R. Bell.
Bendy.
Hardscrabble
W.E.
Colorado City
Hawkeye
Buckeye
Wisconsin
Rose.
Modak
Hardscrabble No 1.
Old Ironsides
Lost Annie
Cedar Hill.
Ocean View
Nugget.
Bravo
Bank
Cuckoo
Total Wreck
Star.
Vanadium
Lizzie May
Captain
American
Necessity
Hidden Treasure
Agnes No 1.
Lulu
Mammoth Pearl
Sec. 29
T.15 S. R.69 W.
Rex
Tom
Moore or
Blue Stocking
Champion.
Lovell
Anna Lee
Portland
F.R.C.
World's Desire
Monarch.
Bob Tail No 3.
Stephan
White House
Little Harry
Baby Mine.
Bob Tail No 2.
Doubtful
Granite
Tidal Wave
Baby Ruth
Fair Play
Rosario
Mary Alice
Queen of the Hills
Spokane
S.S.
Bill Nye
Monument.
Black Diamond
Four Queens
White House
Portland No 2
Independence
Cyclone
Wonderland No 2
Argazette.
W.C. Dillon
Strong
Wilson Creek Contact.
Lot No 2.
Creek Bottom
Old Abe
King Solomons Placer
Wilson Creek Placer
Dottie
Scale of Feet
500
1000

The Portland Gold Mining Company.

Incorporated 1894.

James F. Burns......................President
Irving Howbert....................Vice-President
Frank G. Peck..............Secretary and Treasurer
Chas. J. Moore...............Consulting Engineer
A. T. Gunnell....................General Counsel
    W. S. Stratton.   John Harnan.

Directors

Main Office—First National Bank block, Colorado Springs, Colorado.

Capitalization

3,000,000 shares. Par value, $1.00. In treasury January 1, 1900, $602,672.10 cash.

Property

Owns the Portland, Bobtail No. 2, Doubtful, Anna Lee, White House, Hidden Treasure, Vanadium, Captain, Queen of the Hills, Scranton, Baby Ruth, National Belle, Bobtail No. 3, Four Queens, King Solomon placer, Success, Fair Play, Cyclone, Lowell, Rosario, Black Diamond, Tidal Wave, Lost Anna, Milton, Little Harry, Devil's Own, and a one-fourth interest in the Blue Stocking, in all, about 183 acres, situated in section 29, on Battle mountain. In April, 1900, the company also purchased 52 1-2 per cent. of the Total Wreck, which property is shown in the center of the plat on the opposite page. All of the above property is covered by United States patent, excepting the Fair Play claim, containing about one-twentieth of an acre, which is held by location.

Development

The surface improvements and machinery consist of a dwelling house for officers of the company, dwelling house for general manager, mine office, surveyor's office, with fire-proof vault; assay office, sixty-ton railroad scale and ten-ton wagon scale; also the following shaft houses: The Burns, the Bobtail No. 2, the Scranton and the Lowell. One pump station on the King Solomon placer, with a pump of 300 gallons capacity, boiler, and a 55,000 and two 60,000-gallon reservoirs, used for fire purposes as well. The main working shaft house on the Burns is 40x130, with wing for boiler, 30x75, also drying room, 11x30, with wing for dynamo and carpenter shop. Large blacksmith shop, with five forges. A 600-horse power hoisting engine, which is the largest in the camp. The Burns shaft is a three-compartment shaft and the cages are double deckers. There is an immense compressor plant for supplying drills. Three compressors, one driving eight drills, one driving 16 drills, and one driving 24 drills. Four boilers. There are also two underground hoists, three large pumps; one of 800 gallons capacity at a depth of 900 feet; one of 1,000 gallons capacity at 1,000 feet; one of 1,200 gallons capacity at 1,000 feet lift. The principal development work has been done on the old portion of the Anna Lee and Doubtful claims. Shaft No. 2 is now being sunk on the Captain lode, and No. 3 on the Buckeye. The Portland Gold Mining Company stands alone in the fact that none of its treasury stock has ever been offered for sale; it has all been used in the purchase of new property. All litigation has now been practically settled and the mine stands as the largest producer in the district.

Production and Dividend

The total production of ore from April 1, 1894, to January 1, 1900, is 148,139.96 tons, of a value of $8,378,732.88. The amount paid in dividends from the commencement until January 1, 1900, $2,557,080.00. A regular dividend was paid in January, 1900, consisting of 2 cents per share and 1 cent extra, amounting to $90,000.00. The last regular dividend up to the time this Manual went to press was paid April 15, 1900, $60,000. Total dividends to April 15, 1900, $2,827,080.

The lowest price on record for Portland stock is 26 1-2 cents; highest price for stock during 1899, $2.60 per share; lowest price for stock during 1899, $1.45 7-8 per share; price of stock May 19, 1900, $3.00.

The Poverty Gulch Gold Mining Company.

Incorporated.

Directors

Wm. Lennox..............................President
John Lennox.......................Vice-President
C. H. Dudley.....Asst. Secretary and Asst. Treasurer

Main Office—14 N. Nevada avenue, Colorado Springs, Colorado.

Capitalization

3,000,000 shares.   Par value, $1.00.

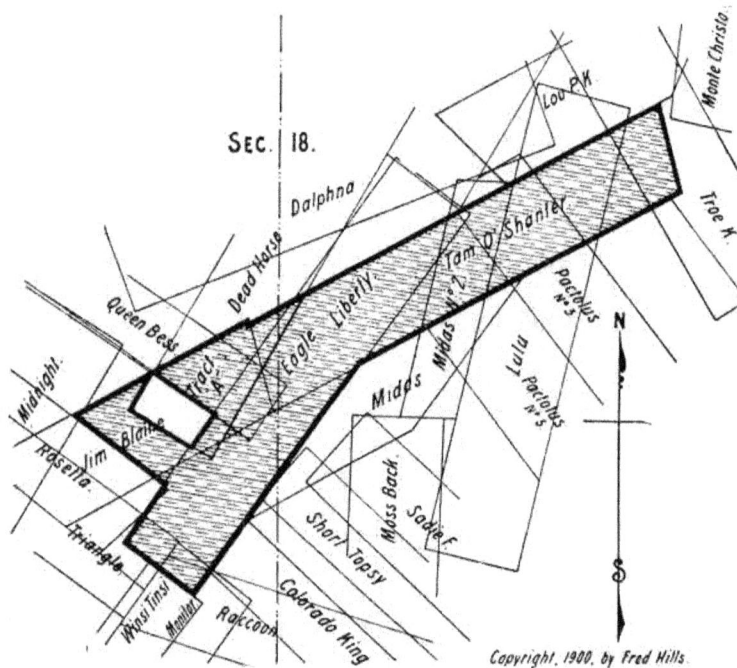

Copyright. 1900. by Fred Hills.

Property

Owns the Tam O'Shanter, Eagle and part of the Jim Blaine, 16 acres, in the N. E. 1-4 of section 18. Patented. A conflict, marked tract A, with the Dead Horse claim of the C. C. Consolidated Company, of about 2 acres, is in controversy.

Development

There has been very little development work done on this property, and, being a close corporation, no sales of stock have been reported.

The Prince Albert Mining Company, Limited.

Incorporated October, 1895.

R. J. Preston, president; K. R. Babbitt, vice-president and treasurer; Directors R. C. Thayer, secretary; L. S. Thompson, H. K. Devereux, Wm. A. Otis, A. E. Carlton.

Main Office—No. 5 Giddings building, Colorado Springs, Colorado.

3,000,000 shares. Par value, $1.00. In treasury January 1, 1900, Capitalization 485,000 shares; in treasury January 1, 1900, $2,000.00 cash.

Owns the Prince Albert, survey No. 8,883, containing 9.85 acres; the Property Beacon, 8.06 acres, and the Eureka, survey No. 8,161, 8.51 acres—all situated in the S. E. 1-4 section 25, on Beacon hill. All patented.

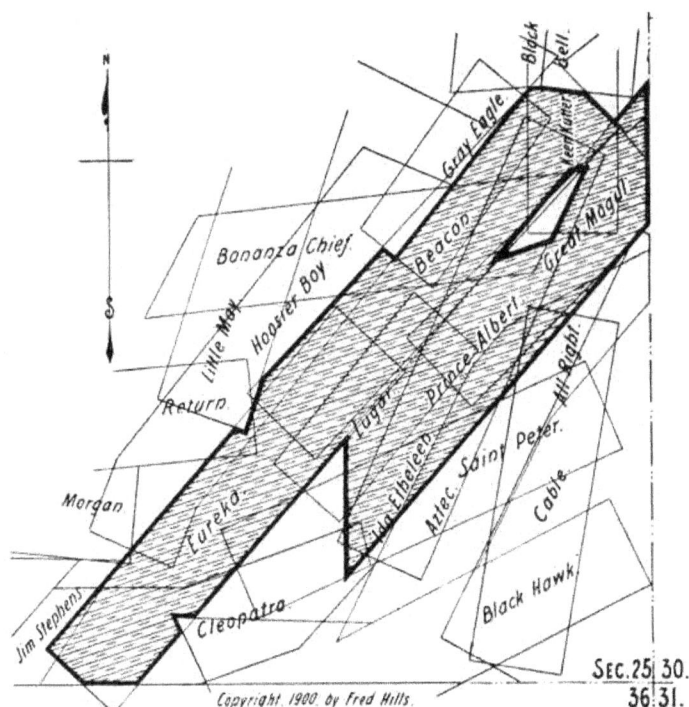

Copyright, 1900, by Fred Hills.

One large shaft house and steam hoist on block 12 of the Prince Al- Development bert. The property is divided into blocks of 300 feet square. Three blocks of the Prince Albert and three of the Beacon claim are leased. The development consists of 3,860 feet in drifts and cross-cuts, and about 740 feet in shafts, not including upraises and winzes. The greater part of work is being done on the Prince Albert and the Beacon.

Gross production to January 1, 1900, $500,000.00. The net value of Production ore mined during the year 1899 cannot be given, as all ore shipped has been mined by the lessees, and the profits have been expended in developing the underground workings.

The present holdings of the company resulted in a coalition of the History Ida Etheleen, Beacon, and Prince Albert Gold Mining Companies. The present condition of the mine is good. The lessees are shipping ore.

Highest price for stock during 1899, 7 3-4 cents; lowest price for stock during 1899, 4 3-4 cents.

The Princess Alice Gold Mining Company.

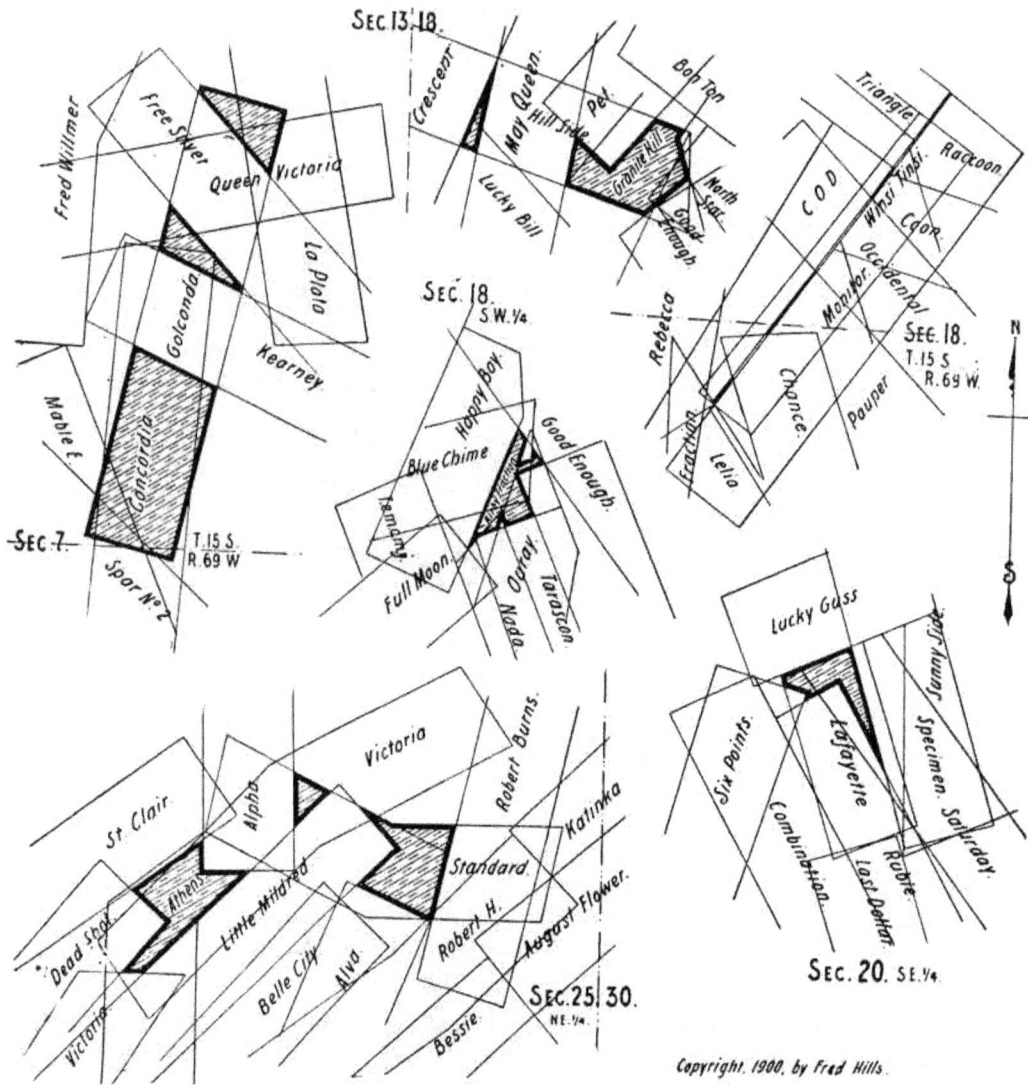

Copyright. 1900, by Fred Hills.

The Princess Alice Gold Mining Company.

Incorporated.

E. C. Lufken, president; Charlotte S. Williams, vice-president; Anna S. Douglas, secretary and treasurer; Jennie Rumrill; Eugene A. George; J. M. Hawkins, manager. **Directors**

Main Office—Buffalo, New York; transfer office, Colorado Springs, Colorado.

1,250,000 shares. Par value, $1.00. In treasury February 5, 1900, **Capitalization** 283,460 shares.

Owns the Standard and Athens, containing 6 1-2 acres, in the N. E. **Property** 1-4 section 25, on Guyot hill; the Kirby Fraction, 3-4 acre, and the Hill Side, 1 3-4 acres, in the S. W. 1-4 section 18, on Gold hill; the Golconda, 8 acres, in the N. E. 1-4 section 7, on Tenderfoot hill; the Linus, 10 acres, in the S. W. 1-4 section 8, township 15 south, range 70 west; the Havilah, 10 acres, in the W. 1-2 section 32, township 14 south, range 70 west; the Snow Flake, Yukas Nos. 1 and 2, Defender, Peurl, and North Star, containing in all about 50 acres; the Winsi Tinsi, containing a fraction of an acre, in the S. W. 1-4 section 18; the Lafayette, containing 3 acres, in the S. E. 1-4 section 20, on Bull hill. The company also holds bond and lease on the Rubie claim, containing 6 acres, and adjoining the Lafayette. Total acreage owned by the company is over 90 acres. The Lafayette, Hill Side, Defender, Linus, and the Havilah are patented. Remainder in process or held by receiver's receipt. Linus, Havilah and the Snow Flake et al. group are outside properties and are not shown on plat.

Two hoisting plants, with engine house and ore house on the Rubie **Development** and the Lafayette. The Lafayette has a shaft 640 feet deep, with levels of about 50 feet. The Rubie has a shaft 925 feet deep with levels of from 50 to 100 feet; also drifts of from 50 to 400 feet. The greater part of the work has been done on the Lafayette and the Rubie.

Highest price for stock during 1899, 50 cents; lowest price for stock during 1899, 10 cents.

The Princess Gold Mining Company.

Incorporated May 16, 1892.

Directors — Verner Z. Reed, president; A. S. Holbrook, vice-president; O. H. Shoup, secretary and treasurer; Sam Strong, W. H. Sutherland, W. A. Perkins, C. C. Hamlin.

Main Office—Bank building, Colorado Springs, Colorado, with the Reed & Hamlin Investment Co.

Capitalization — 1,000,000 shares. Par value, $1.00. In treasury January 1, 1900, $1,500.00 cash. No stock in treasury.

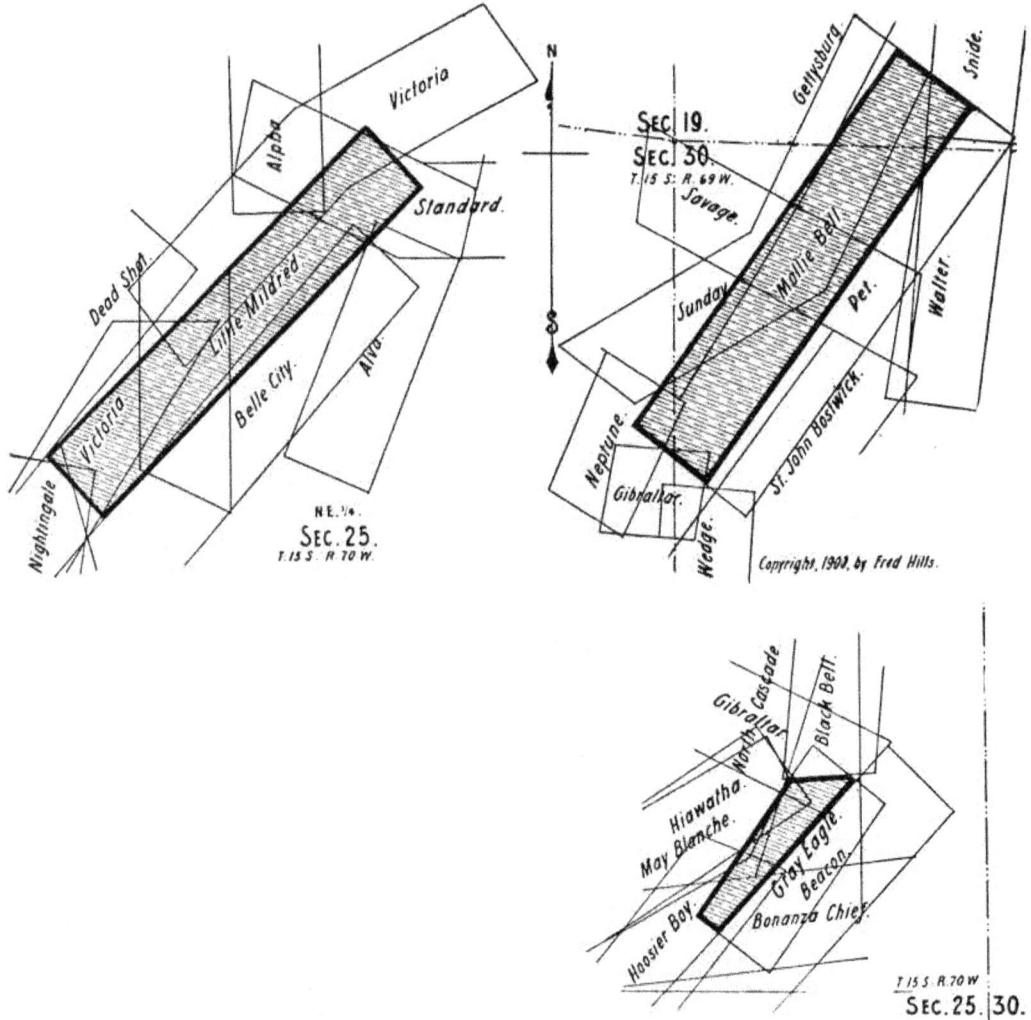

Property — Owns Mollie Bell, consisting of 10 1-3 acres, located in section 30, on Raven hill; also the Gray Eagle and Little Mildred, containing 2 and 10 acres, respectively, located on Beacon hill, in section 25. All patented.

Development — No considerable amount of development work has been done on any one claim, in addition to patent work, except Mollie Bell has one shaft 159 feet deep and a cross-cut 500 feet long. The location of the Mollie Bell claim is most excellent, and adjoins such properties as the Elkton, Tornado and Gould companies. The Beacon hill properties are excellent prospective claims. A lease is in operation on the Gray Eagle and looks favorable.

Highest price for stock during 1899, 6 1-2 cents; lowest price for stock during 1899, 5 1-2 cents.

390

The Progress Gold Mining Company.

Incorporated 1899.

Max Straus, president; W. F. Crosby, vice-president; L. R. Ehrich, _{Directors} secretary and treasurer; W. H. Spurgeon.

Main Office—No. 64 Hagerman building, Colorado Springs, Colorado.

1,500,000 shares. Par value, $1.00. In treasury January 1, 1900, _{Capitalization} 300,000 shares; in treasury January 1, 1900, $8,000.00 cash.

Owns the Gold King, containing 4 acres, on Gold hill, in the S. W. _{Property} 1-4 section 19; the Gold Cup and the Tin Cup, containing 19.729 acres, on Rhyolite mountain, in the S. W. 1-4 section 31, township 14 south, range 69 west; the Becky Sharp and the Dauntless, containing about 2 acres, on Raven hill, in the S. W. 1-4 section 20; the Spider and the Trinidad, containing 8.628 acres, on Beacon hill, in the S. E. 1-4 section 26, township 15 south, range 70 west. Property is all patented.

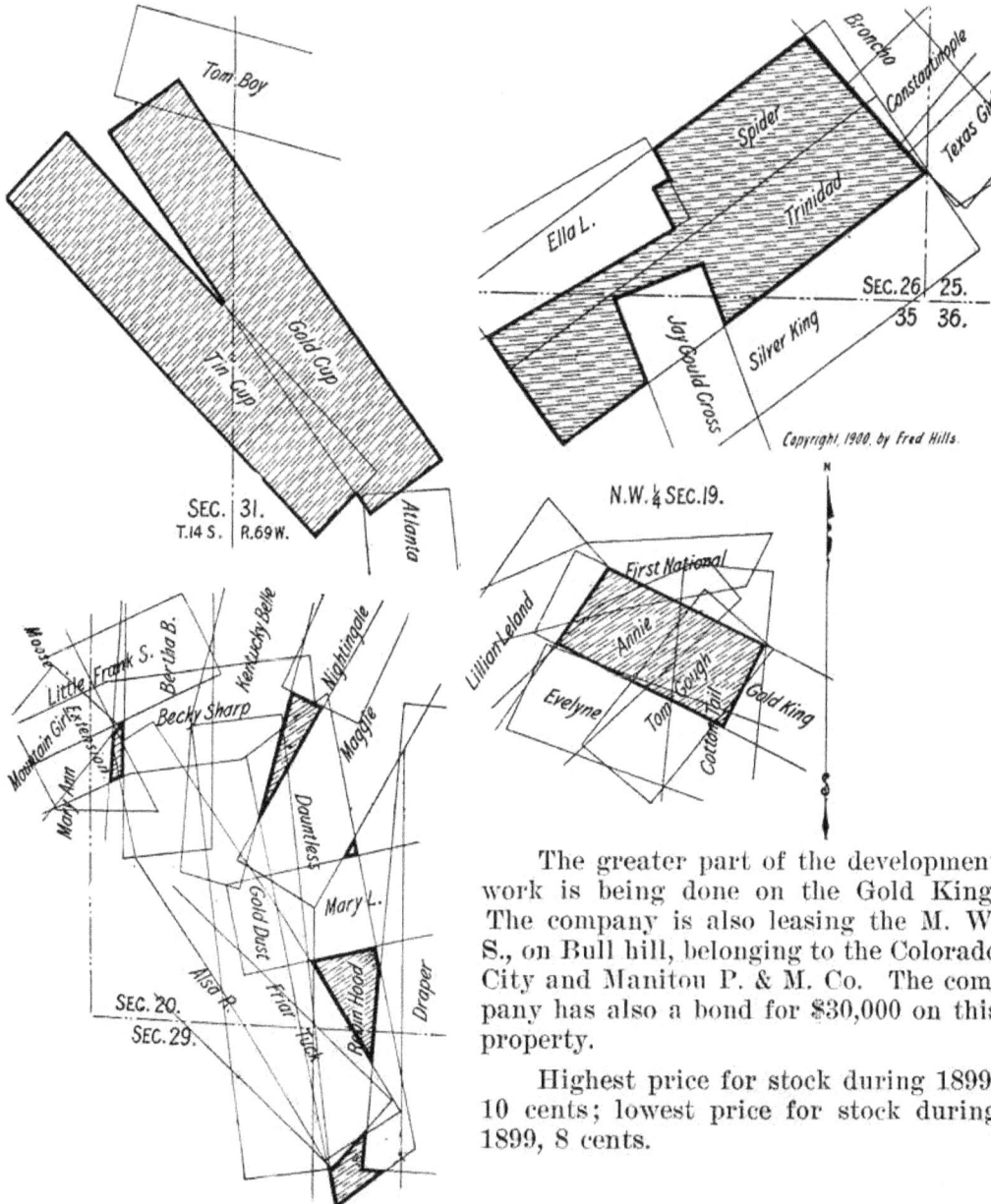

Copyright, 1900, by Fred Hills.

The greater part of the development _{Development} work is being done on the Gold King. The company is also leasing the M. W. S., on Bull hill, belonging to the Colorado City and Manitou P. & M. Co. The company has also a bond for $30,000 on this property.

Highest price for stock during 1899, 10 cents; lowest price for stock during 1899, 8 cents.

The Proposition Gold Mining Company.

Incorporated.

Main Office—No. 53 First National Bank building, Colorado Springs, Colorado.

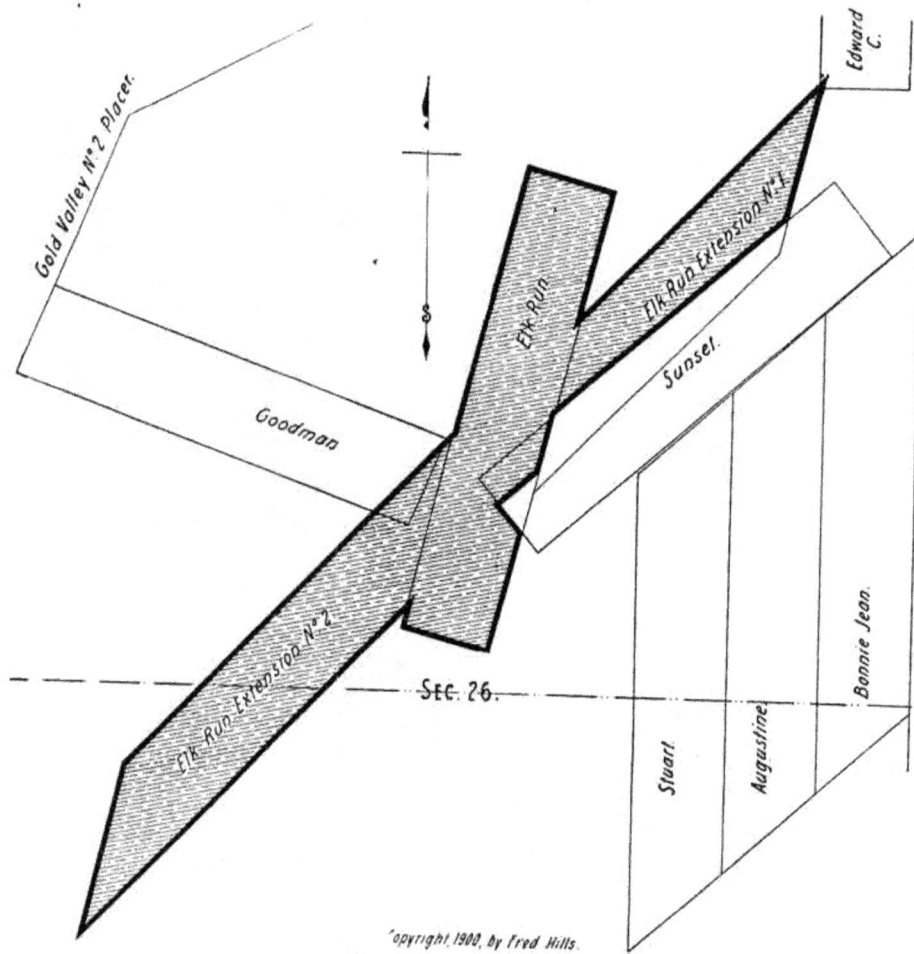

Copyright, 1900, by Fred Hills.

Capitalization

1,250,000 shares. Par value, $1.00. In treasury January 1. 1900, 83,000 shares; in treasury January 1, 1900, $321.00 cash.

Property

Owns the Elk Run, the Elk Run Extension No. 1 and No. 2, containing 26 1-2 acres, on Signal hill, in N. W. 1-4 section 26, township 15 south, range 70 west. All patented.

The Protection Gold Mining Company.

Incorporated 1896.

Phil. Strubel...........................President

Chas. E. Leibold.................Acting Secretary

Directors

Main Office—12 South Tejon street, Colorado Springs, Colorado.

1,000,000 shares. Par value, $1.00.

Capitalization

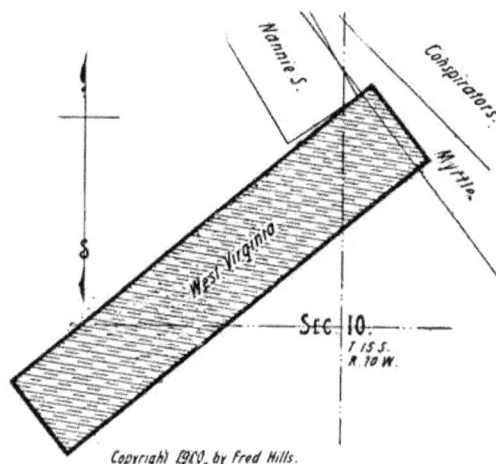

Copyright 1900, by Fred Hills.

Owns the West Virginia claim, 10 acres, in the N. W. 1-4 section 10, township 15 south, range 70 west, on Iron mountain. Patented.

Property

Development work only has been done on the property to obtain patent. The company is in debt about $400.00.

Highest price for stock during 1899, $4.00 per 1,000; lowest price for stock during 1899, $2.50 per 1,000.

The Provident Gold Mining Company.

Incorporated September 6, 1894.

Directors   E. A. Colburn...........................President

C. B. Seldomridge...................Vice-President

Horace H. Mitchell.........Secretary and Treasurer

Main Office—45 Bank building, Colorado Springs, Colorado.

Capitalization   1,500,000 shares. Par value, $1.00. In treasury January 1, 1900, about 150,000 shares.

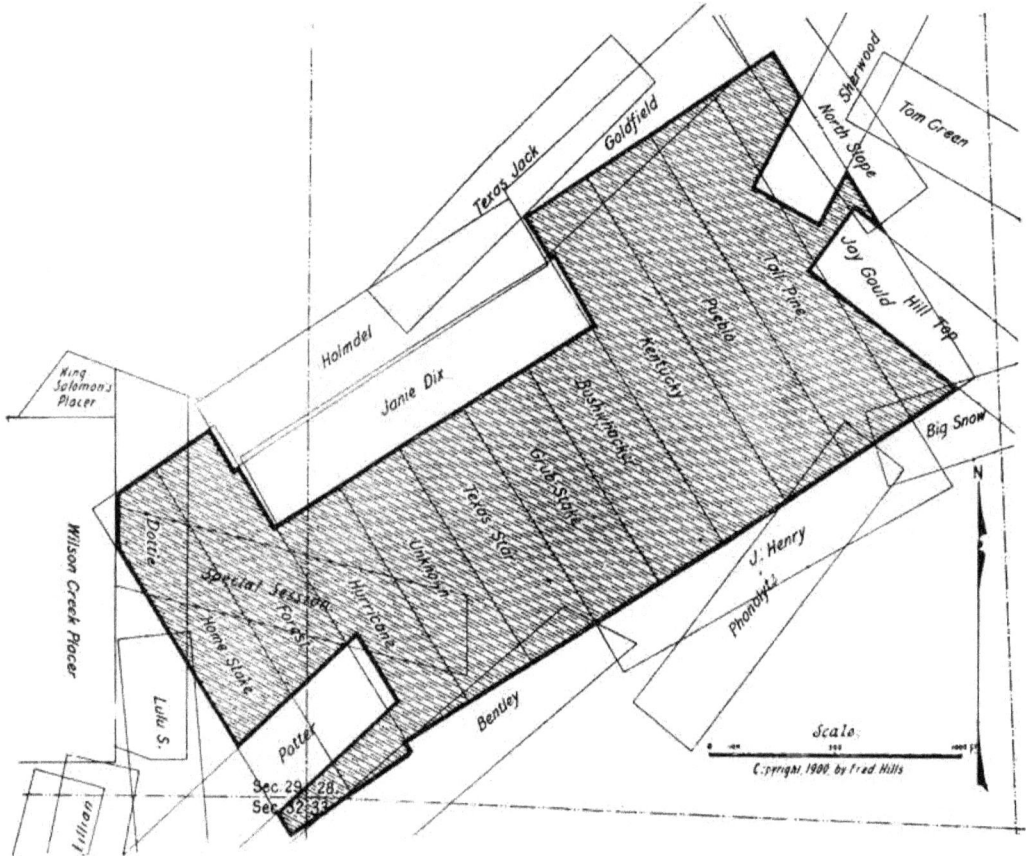

Property   Owns the Jay Gould, Tall Pine, Pueblo, Kentucky, Bushwhacker, Grub Stake, Texas Star, Unknown, Hurricane, Forest and Home Stake, 86 1-2 acres, in the S. W. 1-4 section 28, on the S. W. slope of Big Bull hill, adjoining Goldfield and Stratton's Independence Company's property. All patented.

Very little work has been done up to the present time. The policy of the company will be to develop the property by the leasing system.

No stock has been offered in public.

The Puritan Gold Mining Company.

Incorporated January 10, 1896.

Erastus W. Smith, President; F. B. Davis, Vice-President; Chas. H. Peters, Secretary; F. A. Bailey, Treasurer; C. A. Thompson, Van R. Rent, F. C. Vickers. Directors

Main Office—No. 307 Mining Exchange building, Denver, Colorado.

1,500,000 shares. Par value, $1.00. Capitalization In treasury January 1, 1900, 65,000 shares; in treasury January 1, 1900, about $100.00 cash.

Owns the Maggie Lynch, in the Property S. W. 1-4 of section 24 and the N. W. 1-4 of section 25, containing 6 1-2 acres; also the Holy Terror, on Bull hill, in the N. E. 1-4 of section 19 and N. W. 1-4 of section 20, containing 2 acres. All patented.

On these properties about $3,500 of development work has been done. The Holy Terror is leased for two years from November 14, 1899.

Highest price for stock during 1899, 3 1-8 cents; lowest price for stock during 1899, 1 cent.

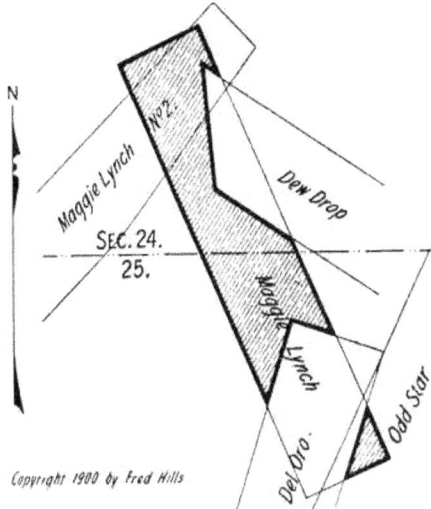

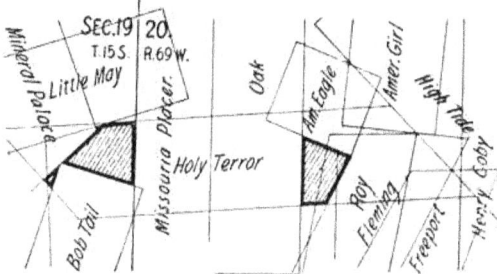

The Pythias Gold Mining Company.

Incorporated February 7, 1899.

Jas. F. Burns, President; Frank Smith, Vice-President; R. P. Davie, Secretary; C. M. MacNeill, Treasurer; A. T. Gunnell. Directors

Main Office—25 E. Pike's Peak avenue, Colorado Springs, Colorado.

1,250,000 shares. Par value, $1.00. In Capitalization treasury January 1, 1900, 90,000 shares.

Owns the Last Chance, 9.6 acres, and the Property Cornucopia, 5.3 acres, on Bull hill, in the S. E. 1-4 of section 17. Property all patented.

Steam and electric hoists. Property is Development being developed by shafts. Five lessees. The property is all near or adjacent to large producers.

Highest price paid for stock during 1899, 10 cents; lowest price paid for stock during 1899, 4 1-2 cents.

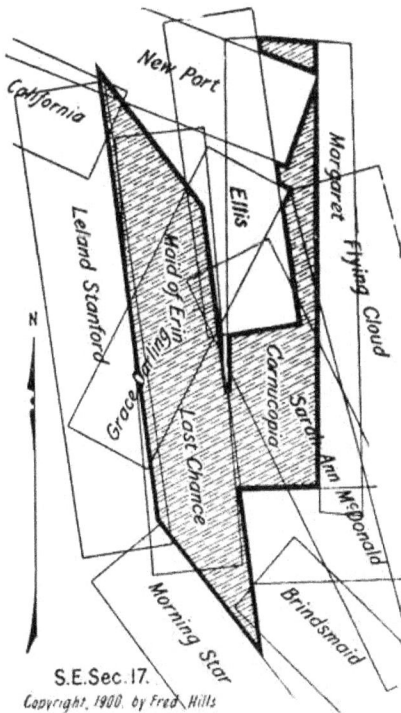

The Quartette Gold Mining Company.

Incorporated December 13, 1895.

T. C. Harbison.........................President

E. S. Joslyn........................Vice-President

P. Wollesen...............Secretary and Treasurer

A. F. Broege.   R. C. Day.

Main Office—Colorado Springs, Colorado.

Capitalization

1,000,000 shares. Par value, $1.00. In treasury January 1, 1900, 9,500 shares.

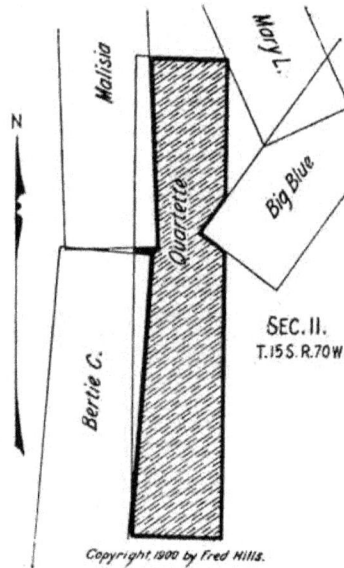

Copyright 1900 by Fred Hills.

Property

Owns the Quartette, containing 9.028 acres, situated on Mineral hill, in the N. W. 1-4 of section 11, township 15 south, range 70 west. Patented. There is a 50-foot shaft on the property, but at present no work is being done.

Highest price for stock during 1899, $2.50 per 1,000; lowest price for stock during 1899, $2.00.

The Queen of the Cripples Gold Mining Company.

Incorporated 1896.

John MacMillan........................President

Geo. D. Johnstone.......... Secretary and Treasurer

Directors

Main Office—Equitable building, Denver, Colorado.

100,000 shares.   Par value, $1.00.

Capitalization

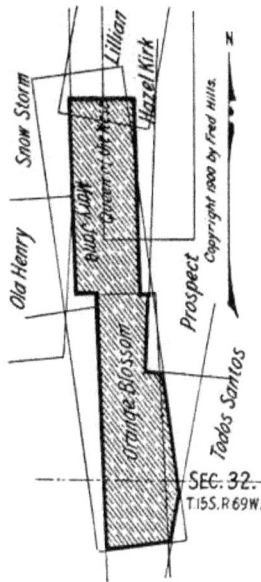

Owns the Queen of the West, 3 1-3 acres, and an undivided one-half interest in the ground in common between the Mary Jane and Berkey lodes, containing about 1 2-3 acres; all in section 32, township 15 south, range 69 west, on Battle mountain.   All the above property is patented. Sufficient development work has been done to secure a patent.   The company is now considering applications for leases.   A good vein of $16 ore has been disclosed.   The company purpose increasing their capitalization to $1,000,000 in the near future, and to add to their holdings other properties in the Cripple Creek district.   Stock not listed.

Property

The Ramona Gold Mining Company.

Incorporated 1898.

Directors   A. S. Holbrook, president; W. S. Jackson, vice-president; E. S. Johnson, secretary and treasurer; Arthur Perkins.

Main Office—Postoffice building, Colorado Springs, Colorado.

Capitalization   1,000,000 shares. Par value, $1.00. In treasury January 1, 1900, 15,000 shares; in treasury January 1, 1900, $1,330.00 cash.

Copyright 1900 by Fred Hills.

Property   Owns the Inez A., situated in the N. E. 1-4 of section 30, and containing 5 1-4 acres; the Ramona, in the N. E. 1-4 of section 30, containing .729 acre; the Phoenix, in the N. W. 1-4 of section 29, containing two acres, and the Lizzie S., in the N. W. 1-4 of section 29, containing 2 1-2 acres, and deeds from conflicting claims, making, in all, 15 acres—all on Raven hill.

Development   One shaft on the Inez A., vertical, 4 1-2x8 in the clear, 300 feet deep. The Phoenix has a shaft 75 feet deep, and a tunnel connecting with the Moose. On the Ramona is a shaft about 75 feet deep, and on the Lizzie S. one of 85 feet.

Highest price for stock during 1899, 16 cents; lowest price for stock during 1899, 4 cents.

The Rattler Gold Mining Company.

Incorporated January 10, 1896.

Hon. John Campbell.....................President
Franklin E. Brooks..................Vice-President
S. L. Caldwell..........................Secretary
W. H. McIntyre.........................Treasurer

S. S. Bernard.   Geo. Bernard.

Directors

Main Office—56 Bank building, Colorado Springs, Colorado.

1,250,000 shares.   Par value, $1.00.

Capitalization

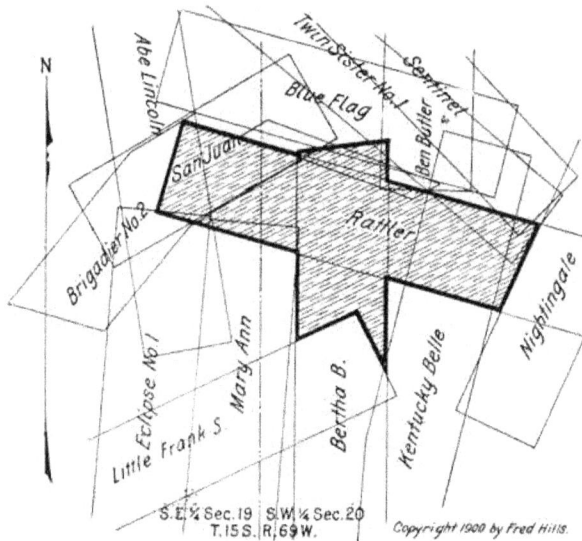

Owns the Rattler, and 2-3 interest in that part of Bertha B. lying north of Little Frank S., containing 10.5 acres (patented), situated in the S. W. 1-4 of section 20, Raven hill. A portion of the property is leased.

Property

Improvements consist of two shafts 50 feet deep, one shaft 100 feet deep, and one shaft 125 feet deep with drifting. One shaft 165 feet deep with drifts, all on the Rattler. There are two shafts, each 100 feet deep, on the Bertha B. claim.

Development

Gross production to January 1, 1900, about 30 tons of ore. This portion of the property is at present being worked under lease.

Production

Highest price for stock during 1899, 7 1-2 cents; lowest price for stock during 1899, 3 cents.

The Raven Gold Mining Company.

Incorporated April 11, 1894.

Directors

E. M. De LaVergne, president; E. R. Stark, vice-president and treasurer; Thomas Stark, secretary; C. E. Noble; M. F. Stark.

Main Office—Nos. 5 and 7 Barnes building, Colorado Springs, Colorado.

Capitalization

1,500,000 shares. Par value, $1.00. In treasury January 1, 1900, 500,000 shares.

Property

Owns the Raven and Princess E., 18.267 acres; the Snowy Range, 2.659 acres; the Maid of Erin, 2.12 acres; the Raven Tunnel, 0.046 acre—all in the S. E. 1-4 section 19, on Raven hill; also the Mill Site, 6 acres, and the Rainbow Placer, adjoining, 24 acres, in the N. W. 1-4 section 10, township 15 south, range 69 west. In April, 1900, the consolidation of this company and the Tornado with the Elkton was agreed upon, as will be seen by referring to the Elkton Cons. M. & M. Co., on another page.

Development

On the property are three ore houses, one shaft house, one engine house, two hoists, one six-drill compressor, one mile of track, two blacksmith shops, one office building, one powder house. Developments consist of 20,000 feet of tunnels, shafts, drifts, upraises, and winzes.

The Raven Hill property was located in May and June, 1892. The production has been steadily turned back into development until a pay mine was made of the property. There are some small short-dated leases on unimportant portions of the Raven Hill property. No bonds.

Production and Dividend

Gross production to January 1, 1900, about $750,000.00. Total amount of dividends to April 20, 1900, $99,500, the last being $10,000.00, on April 20, 1900.

Highest price for stock during 1899, about 90 cents; lowest price for stock during 1899, about 40 cents.

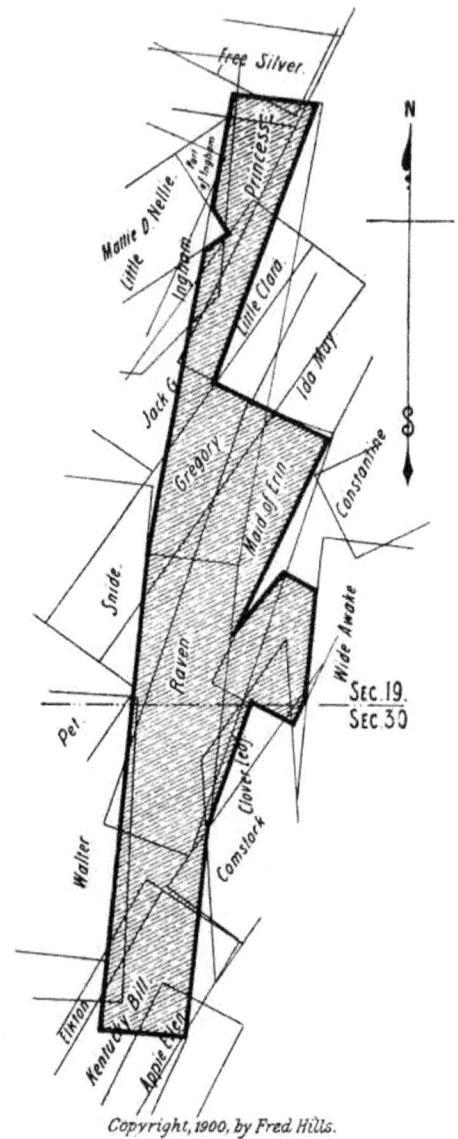

Copyright, 1900, by Fred Hills.

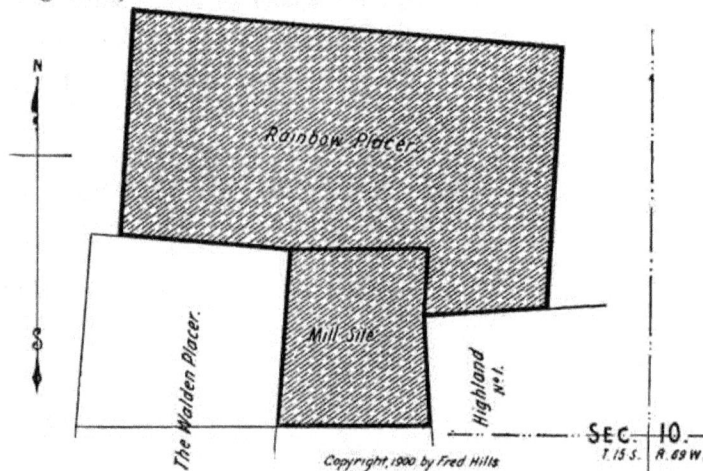

Copyright, 1900 by Fred Hills

The Rebecca Gold Mining Company, Limited.

Incorporated May 24, 1895.

Earl B. Coe......................President
W. N. McBird..............Secretary and Treasurer
Geo. W. Mayhew.   Henry Higgins.

Directors

Main Office—Suite No. 811-814 Ernest & Cranmer building, Denver, Colorado.

200,000 shares.   Par value $5.00.

Capitalization

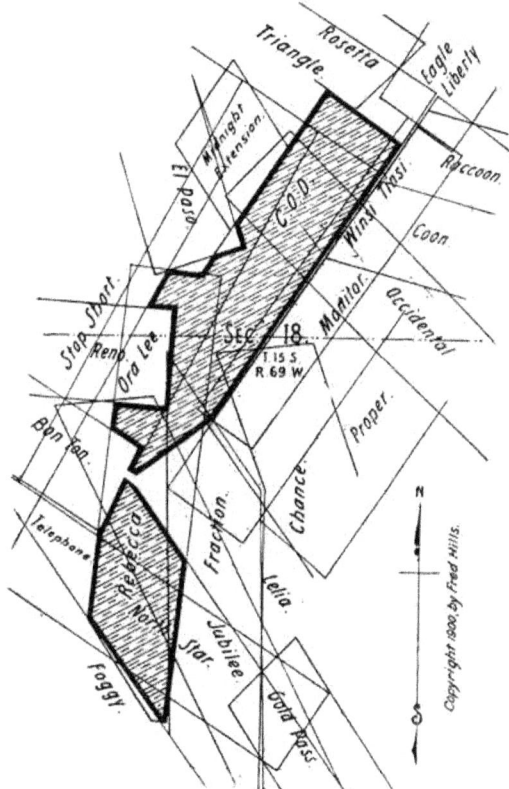

Property Owns the Rebecca, survey No. 8,790, and the C. O. D., survey No. 7,523—in all 14 acres in W. 1-2 section 18, Poverty gulch. All patented except that fraction in conflict between the Rebecca and the C. O. D., which is being settled at present.

Development There are two shaft houses, one large and one small, a blacksmith shop, a carpenter shop, ore house and bins, engine and boiler house, office building and lodging place. In underground machinery there is a Snow triple expansion pump, capacity of 1,000 gallons, at a depth of 1,000 feet. Three station pumps, sinkers, etc. A shaft of 3 compartments has been sunk about 700 feet, and there is about 2,000 feet of drifting. The greater part of the development work has been done on the C. O. D.

Production and Dividend Gross production to January 1, 1900, about $300,000. Dividends to January 1, 1900, about $150,000. Last dividend of 2 cents was paid February, 1896.

The Red Umbrella Mining Company.

Incorporated October 8, 1894.

Directors

H. P. Lillibridge........................President

E. F. Smith...........Vice-President and Treasurer

W. S. Reynolds..........................Secretary

Louis R. Ehrich.   Chas. L. Smith.

Main Office—51 and 52 Hagerman building, Colorado Springs, Colorado.

Capitalization

500,000 shares. Par value, $1.00.

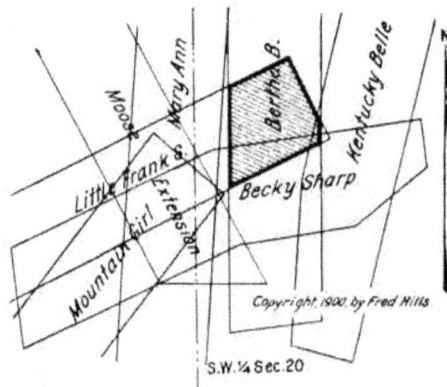

Copyright, 1900, by Fred Hills

S.W. ¼ Sec. 20

Property

Owns that portion of the Bertha B. and Frank S. conflict known as the Red Umbrella, containing two acres, in section 20, on Raven hill. Patented.

Development

There is a steam hoist and commodious shaft house located at the Murray shaft. There are three shafts, of which the deepest is 235 feet, with drifts and cross-cuts.

Production

Production to January 1, 1900, $4,994.67.

Highest price for stock during 1899, 10 cents; lowest price for stock during 1899, 1 1-2 cents.

The Reno Mining and Milling Company.

Incorporated April 13, 1892.

W. R. Foley...............................President

S. J. Mattocks.............Secretary and Treasurer

A. L. Houck........................Vice-President

D. C. Sindlinger.   W. M. Dutton.

Main Office—104 East Pike's Peak avenue, Colorado Springs, Colorado.

1,000,000 shares. Par value, $1.00. Increased April 10, 1900, to
1,250,000 shares.

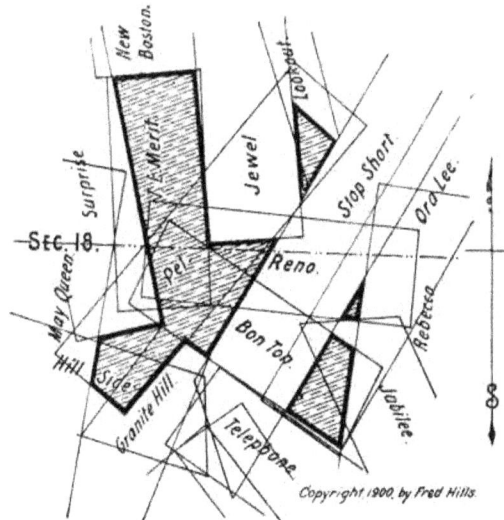

Copyright 1900, by Fred Hills.

Owns the Reno, Bon Ton, T. E. Merit and Pet claims, containing Property
nearly 10 acres, located in section 18, in Poverty gulch.

There is no machinery on the property as yet. A shaft has been sunk Development
100 feet and a tunnel run about 500 feet. The property adjoins the Gold
King and is surrounded by shipping mines; there is an 8-foot vein opened
up on the property and at a depth of 50 feet shows an average of about
$12.00 per ton. The directors expect to have the property leased soon
and development will be pushed as fast as possible.

Highest price for stock during 1899, 6 3-4 cents; lowest price for
stock during 1899, 3 cents.

The Republic Gold Mining Company.

Incorporated.

Directors

J. R. McKinnie.......................President

H. S. Hawkes......................Vice-President

M. F. Stark................Secretary and Treasurer

C. F. Potter.   D. G. C. MacNeill.

Main Office—Room No. 8, Barnes building, Colorado Springs, Colorado.

Capitalization

1,250,000 shares.   Par value, $1.00.   In treasury January 1, 1900, $750.00 cash.

Property

Owns the J. I. C., containing 5 1-4 acres, on Battle mountain, in the N. E. 1-4 section 29; Sweepstake, containing 1-2 acre, on Battle mountain, in the N. E. 1-4 section 29; the Janet W., the Laura M., and the Lester W., on Beacon hill, in the S. 1-2 section 25, containing about 9 acres.

Highest price for stock during 1899, 9 1-8 cents; lowest price for stock during 1899, 2 cents.

The Requa Savage Gold Mining Company.

Incorporated May, 1894.

Frank Finegan.........................President

Wm. Goshen.........................Vice-President

R. B. Taylor..........................Secretary

John Pedersen.   Pat Jones.

Directors

Main Office—No. 25 Midland block, Colorado Springs, Colorado.

500,000 shares.   Par value, $1.00.   In treasury January 1, 1900, 45,000 shares.   Present indebtedness of the company, $90.00.

Capitalization

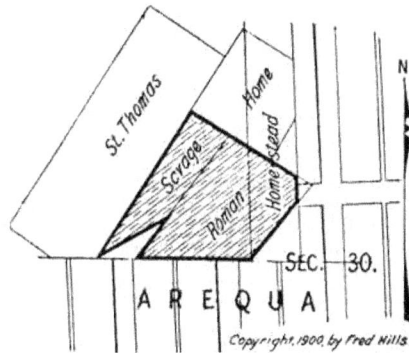

Copyright, 1900, by Fred Hills

Owns the Savage and the Roman, in the N. E. 1-4 section 30, containing 4.058 acres; also the Trojan, containing about 2 1-4 acres, in section 30, not shown on map. All on Beacon hill. The Savage and the Roman are patented. The property of this company adjoins the Mabel M., the St. Thomas, the Commonwealth and the Little Nell.

Property

About $8,500.00 have been expended in developing these claims. Thirteen shafts have been sunk, ranging in depth from 65 to 110 feet. The greater part of the development work has been done on the Savage.

Development

Highest price for stock during 1899, 4 cents; lowest price for stock during 1899, 2 1-2 cents.

The Reserve Gold Mining Company.

Incorporated 1897.

Directors

Franklin E. Brooks......................President

A. B. Heath........................Vice-President

Chas. F. Potter............Secretary and Treasurer

C. C. Brown.   L. W. Davis.

John L. Armit.

Main Office—Colorado Springs, Colorado.

Capitalization   300,000 shares. Par value, $1.00.

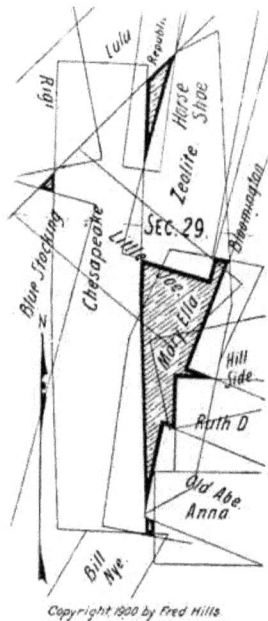

Copyright 1900 by Fred Hills

Property   Owns the Mary Ella, Survey No. 10,091; the Horse Shoe, Survey No. 11,277; the Little Joe, Survey No. 9,863; a part of the Hill Side lode, Survey No. 9,841; a part of the Republic lode, Survey No. 11,830, containing, in all, 3 acres, in the E. 1-2 of section 29, on the east slope of Battle mountain. All the above property is patented.

History   This property, adjoining the Portland and Independence properties, has great prospective value. The stock of the corporation is very closely held and has never been put on the market for sale by the original subscribers. It is the intention of this company in the near future to increase its capitalization for the purposes of acquiring additional territory.

The Rex Gold Mining and Milling Company.

Incorporated November, 1895.

Directors

S. D. McCracken....................Vice-President

W. A. Davis..............Secretary and Treasurer

Wm. Clark.   A. M. Selfridge.

J. J. Meyer.

Main Office—55 Bank building, Colorado Springs, Colorado.

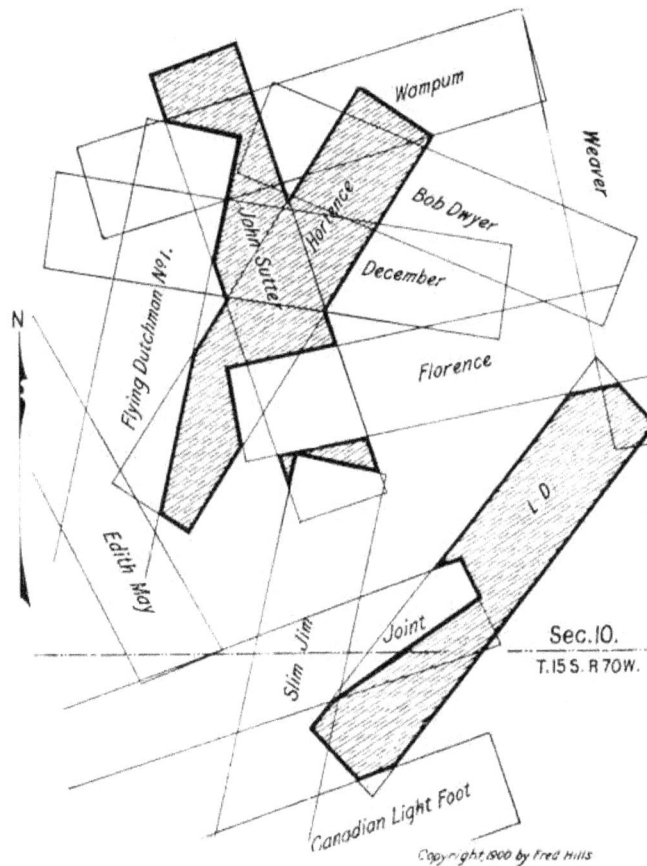

Copyright 1900 by Fred Hills

1,000,000 shares.   Par value, $1.00.

Capitalization

Owns the L. D., 7 1-2 acres, patented; also the John Sutter and Hortence, 15 acres, receiver's receipt; all in the N. E. 1-4 of section 10, township 15 south, range 70 west. There is about 250 feet of shaft work on the property.

Property

Highest price for stock during 1899, $4.50 per thousand; lowest price for stock during 1899, $1.50 per thousand.

The Rigi Group Gold Mining Company, Limited.

Incorporated.

Directors

The Earl of Essex..........President and Chairman
Geo. R. Saunders.....................Secretary
W. H. Cullen.   W. Keen.
W. T. E. Fosbery.

Main Office—3 Copthall buildings, London, E. C., England.

Capitalization

130,000 shares. Par value, one pound sterling.

Property

    Owns the Rigi, Lizzie May, Yucatan and the Lulu, containing, in all, 16 acres, all patented, in the N. E. 1-4 of section 29, on Battle mountain, adjoining the Portland property on the east.

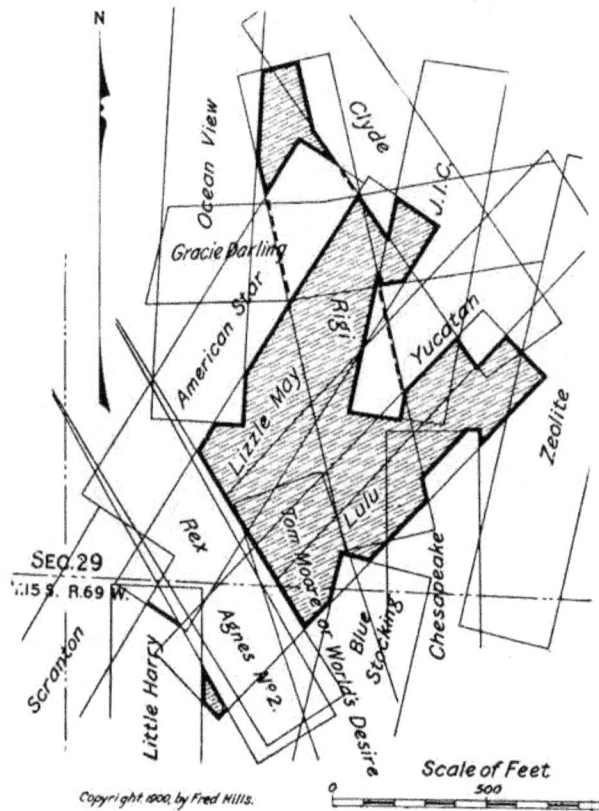

Copyright 1900, by Fred Hills.

Development

    About 4 acres of the property containing the main workings are under lease. There are two shaft houses and ore bins, two blacksmith shops, three hoisting frames, two 6x8 double hoisting engines. There are five shafts, as follows: One, 9x4 1-2 feet, 500 feet deep; one, 8x4, 143 feet deep; one, 6x3, 106 feet deep; one, 7x3 1-2, 112 feet deep; one, 9x4 1-2, 60 feet deep; five levels from the main shaft aggregating 1,350 feet; cross-cuts aggregating 940 feet; stoping 100 feet from second shaft levels, aggregating 610 feet; cross-cuts 230 feet; from third shaft, crosscuts 125 feet long; numerous surface trenches, etc. This property is leased by the North American Gold Mining and Development Company until March 1, 1902. No dividends have been paid, as all money received on ore shipments has been put into permanent improvements and developments.

The Rio Grande Gold Mining Company.

Incorporated March 20, 1896.

A. L. Lawton, President; M. A. Leddy, Vice-President; J. H. **Directors**
Thedinga, Secretary and Treasurer; Chas. Dickens, F. F. Horn.

Main Office—17 E. Pike's Peak avenue, Colorado Springs, Colorado.

1,000,000 shares. Par value, $1.00. In treasury January 1, 1900, **Capitalization**
400,000 shares; in treasury January 1, 1900, $250.00 cash.

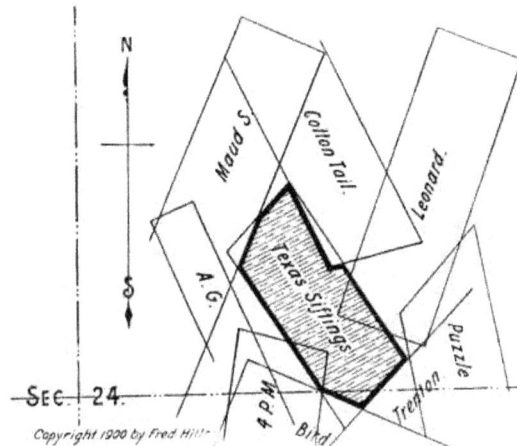

Owns the Texas Siftings, containing 4.042 acres, located in E. 1-2 **Property**
of section 24, on Gold hill. Patented. There is one shaft on the property which has been sunk to a depth of 85 feet.

Highest price for stock during 1899, 1 1-2 cents; lowest price for stock during 1899, 1 cent.

The Rittenhouse Gold Mining Company.

Incorporated.

Silas W. Pettit, President; W. V. Pettit, Vice-President; C. H. **Directors**
Bryan, Secretary and Treasurer.

Main Office—No. 10 Hagerman building, Colorado Springs, Colorado.

1,250,000 shares. Par value, $1.00. In treasury January 1, 1900, **Capitalization**
250,000 shares.

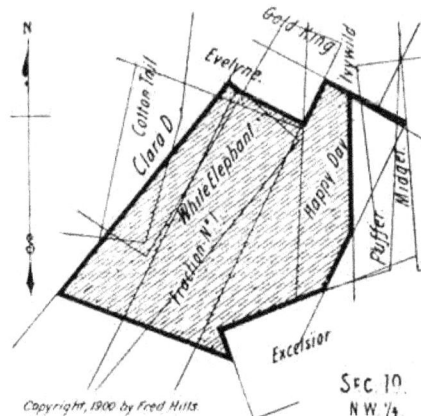

Owns the upper portions of White Elephant, Happy Day, and Frac- **Property**
tion No. 1 claims, 11 acres in the N. W. 1-4 section 19, situated on Gold
hill. All the property is patented.

Two shafts have been sunk, one of them 200 feet deep, the other 175 **Development**
feet deep, with 150 feet of drifting.

Stock not listed.

The Robert Burns Mining Company.

Incorporated August 14, 1899.

Directors

Verner Z. Reed..........................President
C. C. Hamlin.......................Vice-President
O. H. Shoup...............Secretary and Treasurer
J. S. Tucker.   J. R. McKinnie.

Main Office—With the Reed & Hamlin Investment Co., Bank building, Colorado Springs, Colorado.

Capitalization

1,500,000 shares. Par value, $1.00. In treasury January 1, 1900, 100,000 shares; in treasury January 1, 1900, about $3,500.00 cash.

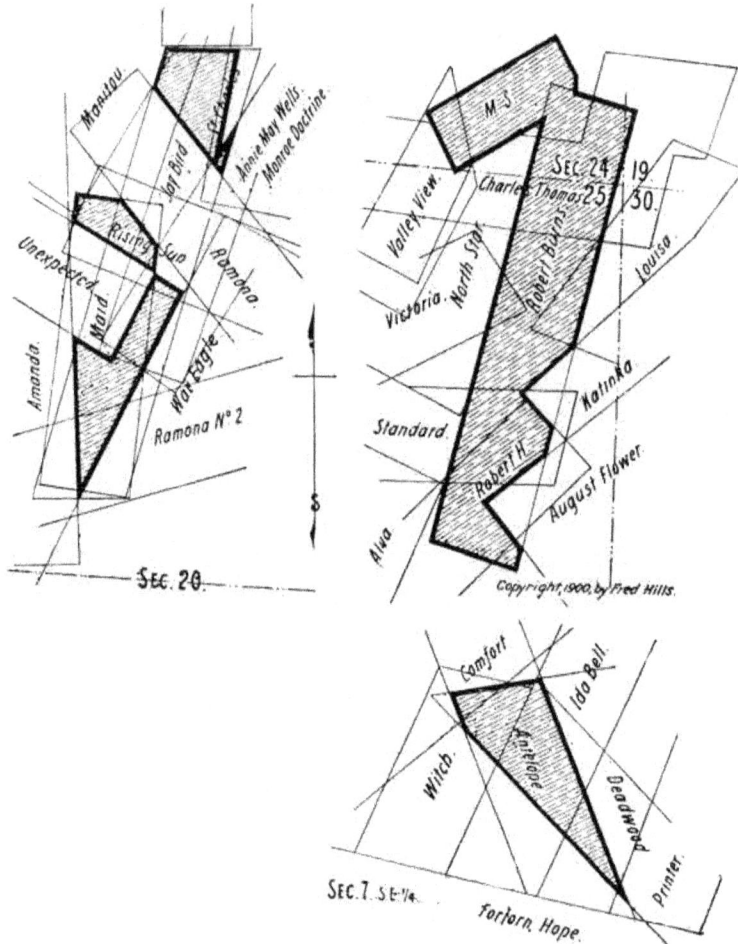

Property

Owns the Robert Burns lode and Millsite on Guyot hill, containing 11 3-4 acres, in the N. E. 1-4 section 25; the Jay Bird and the Maid, about 4 acres, on Bull hill, in the N. W. 1-4 section 20; also a part of the Antelope, 3.10 acres, in the S. E. 1-4 section 7, on Tenderfoot hill. All the above property is patented.

Development

The company is working a part of the Robert Burns claim, and recent development has disclosed a vein of good ore which has been followed downward for about 50 feet. The formation is such that the company feels that the systematic development of this claim will be most satisfactory. A lease is in operation on the Jay Bird and, from adjoining properties, the company believe that this property will have an extension of a good ore body.

Highest price for stock during 1899, 6 1-2 cents; lowest price for stock during 1899, 4 1-2 cents.

The Robinson Victor Mines Company.

Incorporated June 16, 1896.

F. G. Peck.............................President

Jesse F. McDonald.................Vice-President

F. M. Woods...............Secretary and Treasurer

H. E. Woods.   Warren Woods.

Directors

Main Office—Giddings building, Colorado Springs, Colorado; branch offices, Victor, Robinson, and Leadville, Colorado.

Capitalization

1,500,000 shares. Par value, $1.00. In treasury January 1, 1900, 287,000 shares; in treasury January 1, 1900, $11,500.00 cash.

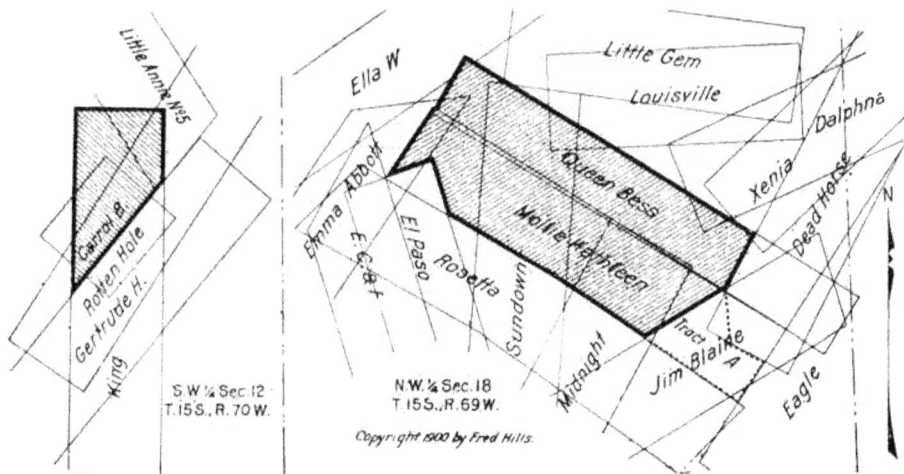

Property

Owns the King, 2.82 acres; also bond and lease on the Mollie Kathleen, 6 acres, and the Queen Bess, 5.75 acres; also owns the following property at Robinson, Colorado: The Lulu, 4.75 acres; the Black Diamond, 3.33 acres; the Silver Tip, 3 acres; the Midget, the Idalia, the San Francisco, and the Bradford Belle; also an undivided 1-8 interest in the Robinson mine, together with other holdings in Teller and Summit counties, Colorado. (Tract A on the map is claimed by this company, by an agreement to deed.)

Development

The main working shaft on the Mollie Kathleen is 700 feet deep, with 1,300 feet of underground work.

Highest price for stock during 1899, 35 cents; lowest price for stock during 1899, 17 cents.

The Rocky Mountain Gold Mining and Milling Company.

Incorporated.

Directors

T. P. Airheart..............................President
W. P. K. Hedrick....................Vice-President
S. T. Miller.............................Secretary
E. P. Arthur...............................Treasurer
E. D. Marr...................Assistant Secretary

Main Office—No. 257 Bennett avenue, Cripple Creek, Colorado; transfer office, Colorado Springs, Colorado.

Capitalization

1,500,000 shares. Par value, $1.00. In treasury January 1, 1900, 195,000 shares.

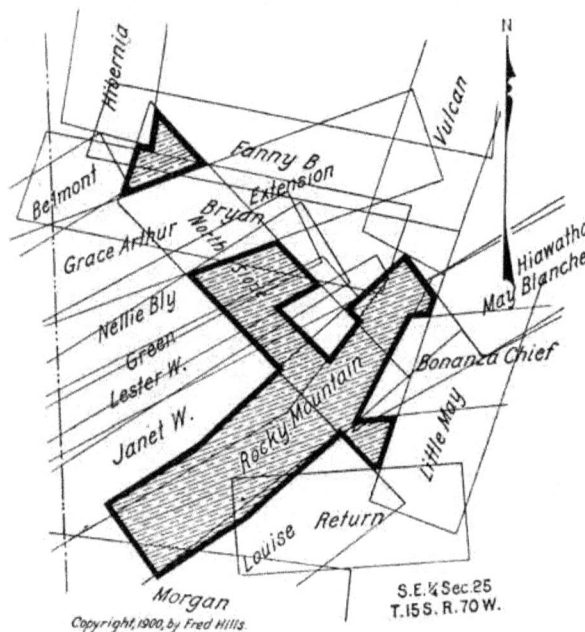

S.E. ¼ Sec. 25
T. 15 S. R. 70 W.

Copyright, 1900, by Fred Hills.

Property

Owns the Rocky Mountain and the North Slope, 9.55 acres, patented, in the S. E. 1-4 section 25, on Beacon hill; the Oregon and Texas, 20 acres, patented, in the North Cripple Creek District; the American Nos. 1 and 2, 13.584 acres, in process of patenting, in the N. E. 1-4 section 26, on Buck mountain; also the Mogul and the Daisy Bell, not patented, on the west slope of Cow mountain. Only the Beacon hill property is shown on plat.

Development

Shaft house and ore bins; 22 H. P. gasoline hoist; blacksmith shop. The deepest shaft on the Rocky Mountain and the North Slope is 200 feet deep, and about 150 feet of drifting. There are 14 other shafts, ranging in depth from 50 to 125 feet; also some drifting. The greater part of the development work is being done on the Rocky Mountain and the North Slope.

Production

Gross production to January 1, 1900, $6,000 to $8,000.

Highest price for stock during 1899, 9 1-2 cents; lowest price for stock during 1899, 4 cents.

The Rose Maud Gold Mining Company.

Incorporated September 12, 1899.

F. M. Woods, President; J. Arthur Connell, Vice-President; Irving W. Bonbright, Secretary and Treasurer; Asa T. Jones, Wm. A. Otis, Eugene P. Shove, Fred L. Ballard. **Directors**

Main Office—Colorado Springs, Colorado.

1,250,000 shares. Par value, $1.00. In treasury January 1, 1900, 95,000 shares; in treasury January 1, 1900, $4,062.92 cash. **Capitalization**

Owns one claim, the Rose Maud, containing 10 acres, Survey No. 7,903, in the S. W. 1-4 of section 19, adjoining the properties of the Work Company, on Raven hill. Patented. The property is leased to Mr. F. M. Woods until October 2, 1902. **Property**

Highest price for stock during 1899, 9 1-4 cents.

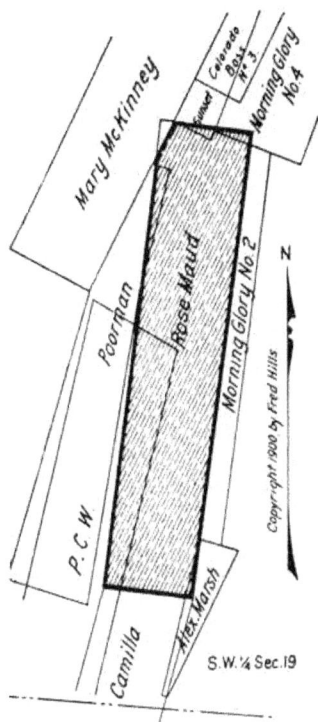

The Rose Nicol Gold Mining Company.

Incorporated September 30, 1899.

Verner Z. Reed, President; Clarence Edsall, Vice-President; O. H. Shoup, Secretary and Treasurer; Duncan Chisholm, W. B. Jenkins. **Directors**

Main Office—With the Reed & Hamlin Investment Company, Bank Block, Colorado Springs, Colorado.

1,500,000 shares. Par value, $1.00. In treasury January 1, 1900, 200,000 shares; in treasury January 1, 1900, $25,000. **Capitalization**

Owns the Rose and the Gurley claims, located in the N. W. 1-4 of section 29, on Battle mountain and Bull hill. Both the claims are patented. **Property**

There is a steam hoist on the property, and the company is at present engaged in sinking a shaft and in working through tunnel. **Development**

This company was incorporated late last year, and as its side line is a side line of a portion of the Portland company's property, its location is remarkably good. Work is steadily progressing and the company intends to use its cash in the treasury for the systematic development of the property. From the surface indications, there is every reason to suppose that with sufficient development work, the Rose Nicol territory has all the indications of a good mine. A shaft has now been sunk 80 feet in depth. Recent assays have been from $12 to $15 per ton.

Highest price for stock during 1899, 14 cents; lowest price for stock during 1899, 11 1-4 cents.

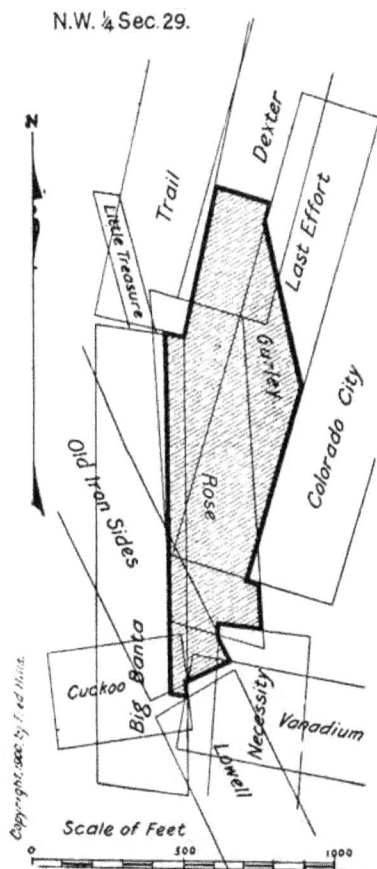

The Royal Oak Gold Mining Company.

Incorporated December 23, 1895.

Main Office—827 Equitable building, Denver, Colorado.

Capitalization

1,000,000 shares. Par value, $1.00. In treasury January 1, 1900, 150,000 shares; in treasury January 1, 1900, $1,000.00 cash.

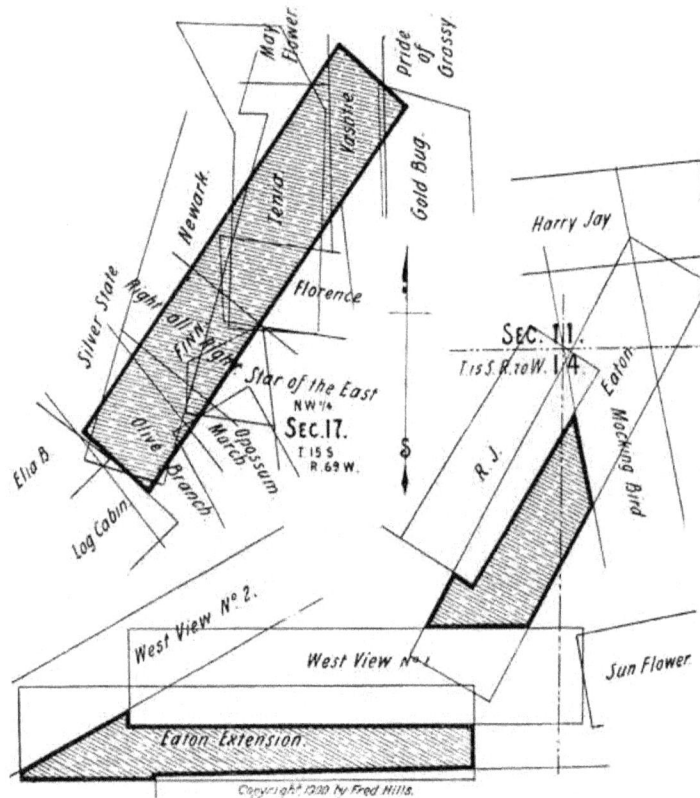

Copyright 1900 by Fred Hills.

Property

Owns the Finn lode, 5 acres, Survey No. 9,037, in section 17, on Ironclad hill, west of Grassy; the Eaton Extension lode, 5 acres, in the N. W. 1-4 of section 14, Survey No. 10,941; also an undivided one-half interest in the Eaton lode, 3 acres, Survey No. 10,941, in the N. W. 1-4 of section 14. All patented.

Development

Large double compartment shaft on the Finn lode, over 200 feet deep, a square set shaft, 4 1-2x9 in the clear. The Eaton properties have each received $500 worth of assessment work. The greater part of the development work is being done on the Finn lode. Outside of location work, the first permanent work done on the property was when the sinking of the main shaft was begun in November, 1899.

Highest price for stock during 1899, 7 cents; lowest price for stock during 1899, 7 cents.

VIEW OF RAVEN HILL.

SHOWING ELKTON CONSOLIDATED AND OTHER MINES

The Rubicon Gold Mining Company.

Incorporated 1894.

Directors

O. C. Townsend........................President

E. S. Woolley.......................Vice-President

H. V. Wandell.............Secretary and Treasurer

W. P. Wight (deceased).   J. M. Harden.

Main Office—No. 112 1-2 East Pike's Peak avenue, Colorado Springs, Colorado.

Capitalization   1,000,000 shares. Par value, $1.00. In treasury January 1, 1900, 80,000 shares; in treasury January 1, 1900, $150.00 cash.

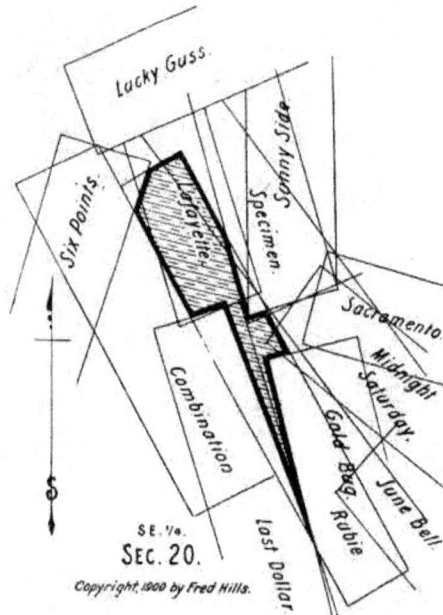

SE.¼.
SEC. 20.
Copyright 1900 by Fred Hills.

Property   Owns the Rubie, containing about 4 acres, situated on Bull hill, in the S. 1-2 section 20. Patented.

Development   Shaft house, 60-horse power boiler and hoist, ore bins, screens, etc., underground machinery, air pipes and machine drills. A shaft has been sunk 850 feet, and there are drifts at intervals of from 50 to 100 feet. This work has all been done on the Last Dollar vein. At the present time nothing but development work is being done. The Rubie is bonded to the Princess Alice Company.

Production   Gross production to January 1, 1900, about $50,000.00.

The stock is closely held. None is changing hands.

The Ruth D. Mining Company.

Incorporated 1894.

D. G. C. MacNeill........................President   Directors

J. W. Gathright.....................Vice-President

R. A. Dana................Secretary and Treasurer

L. C. Dana.

Main Office—No. 4 and 5 Hagerman block, Colorado Springs, Colorado.

100,000 shares. Par value, $1.00. In treasury January 1, 1900,   Capitalization
23,186 shares.

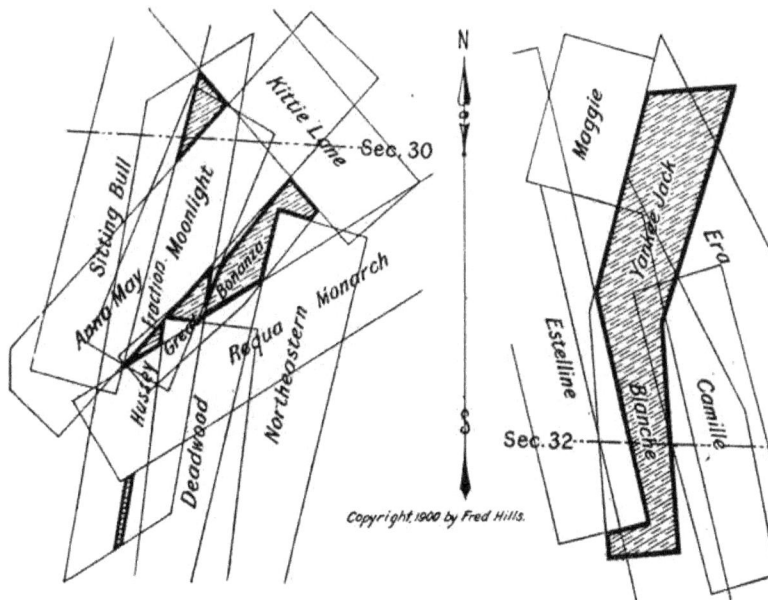

Copyright 1900 by Fred Hills.

Owns the Yankee Jack, containing 7.249 acres, located in the N. E.   Property
1-4 section 32, on Battle mountain. Patented. Also owns the Great
Bonanza and an undivided one-fourth interest in the Hussey Fraction,
both located in section 30, on Raven hill, containing in all 2.652 acres.
Receiver's receipt for the Great Bonanza. The Hussey Fraction is patented.

About $6,500.00 have been expended on the Yankee Jack claim in   Development
development work, and $500.00 on the Great Bonanza. Owing to some
litigation in relation to the title, this company has done very little development
work, but recently everything has been settled and work will
commence almost immediately.

Highest price for stock during 1899, 10 cents bid; none sold.

The Sacremento Gold Mining Company.

Incorporated April 24, 1893.

Directors

Sherwood Aldrich.........................President

E. P. Shove............Vice-President and Treasurer

C. G. White...............................Secretary

John Harnan.   F. G. Peck.

Main Office—No. 9 South Tejon street, Colorado Springs, Colorado.

Capitalization   1,000,000 shares. Par value, $1.00. In treasury January 1, 1900, 6,000 shares; in treasury January 1, 1900, $400.00 cash.

Property   Owns the Sacremento and the Midnight, situated on Bull hill, in section 20, and containing about 7.950 acres. All patented.

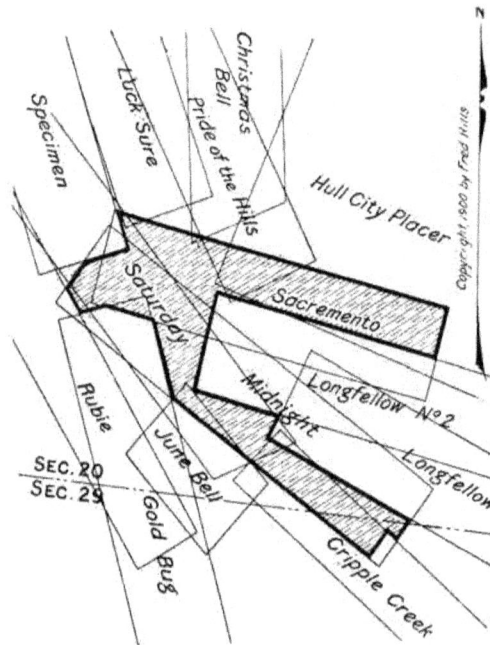

Development   Electric hoist and shaft house on the Midnight, and a shaft house on the Sacremento. There are three shafts, one of 140 feet, one of 165 feet, and one of 250 feet. There are 750 feet of drifting on these shafts, and numerous other trenches and cuts.

History   Since the formation of the company the property has been worked entirely by lessees, save during the first six months.

Production   Gross production to January 1, 1900, about $10,000. Net value to the company of ore mined during the year 1899, $7,000.00.

Highest price for stock during 1899, 14 1-2 cents; lowest price for stock during 1899, 2 cents.

The Safety Gold Mining Company.

Incorporated April 9, 1894.

Geo. B. Sherwood.......................President Directors

Charles F. Potter...........Secretary and Treasurer

A. Danford.

Main Office—No. 619 Ernest & Cranmer building, Denver, Colorado.

1,000,000 shares.   Par value, $1.00.   In treasury January 1, 1900, Capitalization about $2,500.00 cash.

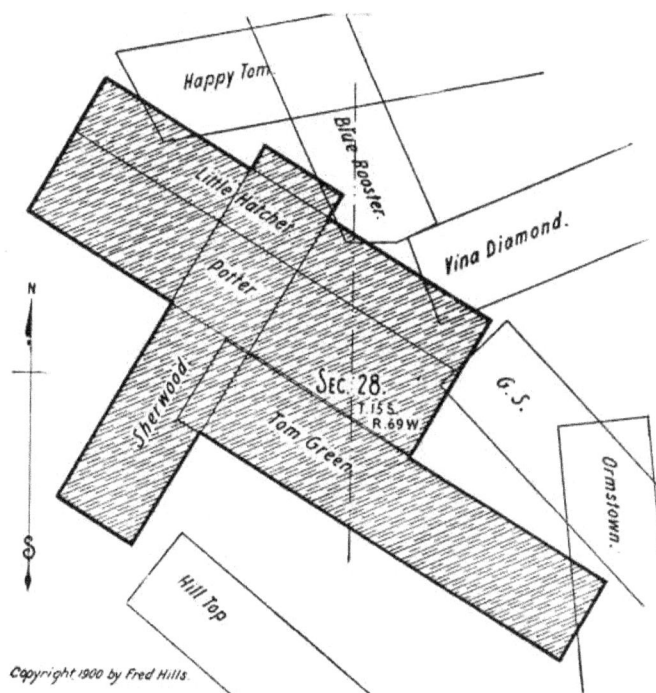

Copyright 1900 by Fred Hills.

Owns the Little Hatchet lode, the Sherwood lode, the Tom Green Property lode and the Potter lode, all in one group, containing in all 31.433 acres, on Big Bull mountain, in section 28, township 15 south, range 69 west. All the above property is patented.

Shaft house, 8-horse power hoist, plant of machinery.   One shaft Development has been sunk 150 feet.   In shafts and levels about 1,000 feet of development work has been done.   The greater portion of the development work has been done on the Little Hatchet lode.   The property is now being worked by the company.   The stock is closely held by eight stockholders, who own nine-tenths of the capitalization.

The Santa Fe Gold Mining Company.

Incorporated 1892.

Directors

H. S. Hawks.............................President
Phil. S. Delany.....................Vice-President
T. C. Delany..............Secretary and Treasurer
Wm. Banning.

Main Office—35 and 37 Hagerman building, Colorado Springs, Colorado.

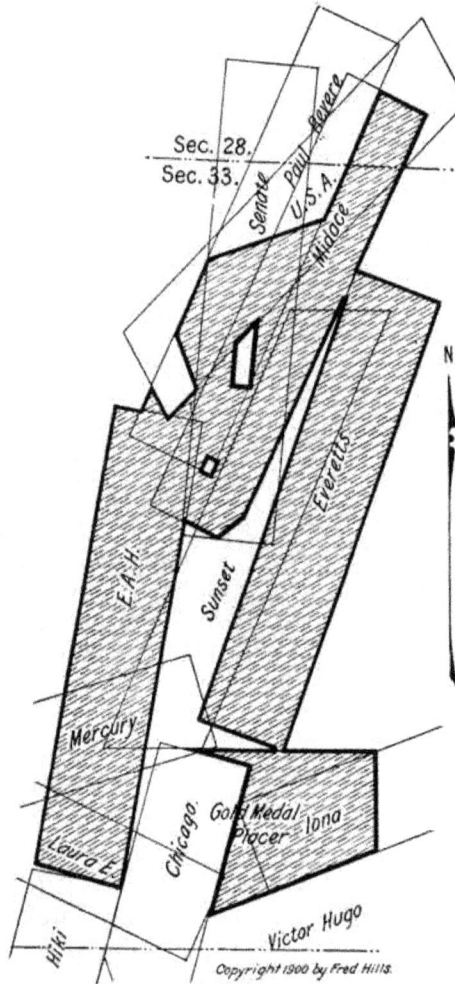

Copyright 1900 by Fred Hills.

Capitalization   1,250,000 shares.   Par value, $1.00.

Property   Owns the E. A. H., 10 acres, in the N. W. 1-4 section 33, on Big Bull mountain.   Patented.   Also the Paul Revere, Midace, Everetts and Gold Medal placer, 20 acres in all, in the N. W. 1-4 section 33, one-half mile east of Victor, ready for patenting.   No adverses.

Development $4,000.00 has been expended in development in the way of shafts and trenches.

Price for stock during 1899, 1-2 cent.

The Santa Rita Gold Mining Company.

Incorporated 1895.

Main Office—Victor, Colorado.

1,000,000 shares. Par value, $1.00. In treasury January 1, 1900, Capitalization
over $2,000.00 cash.

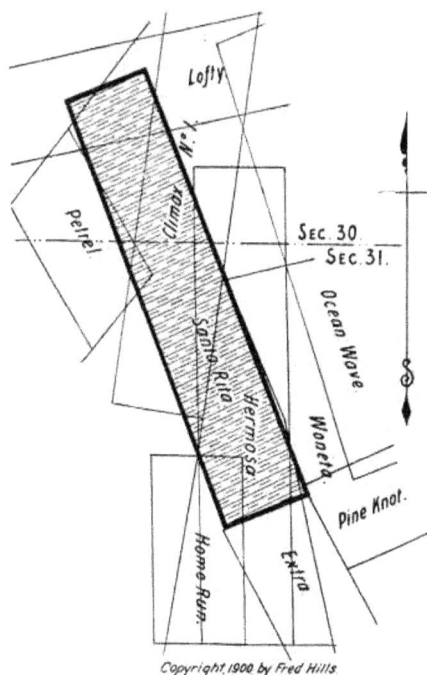

Copyright, 1900 by Fred Hills

Owns the Santa Rita, about 10 acres, in the N. E. 1-4 of section 31, Property
on Squaw mountain. Patented.

Improvements consist of a shaft house, hoist and ore bins. There Development
is one shaft 500 feet deep, with 450 feet of drifting. The property is
bonded and leased to the Atlantis Mines Corporation for $100,000.00.
Lease expires October, 1901. This property has always been worked under lease.

Gross production to January 1, 1900, about $70,000.00. Production

The stock is held by the parties who incorporated the company, and
has never been offered on the open market.

The Santa Rosa Gold Mining Company.

Incorporated November 4, 1895.

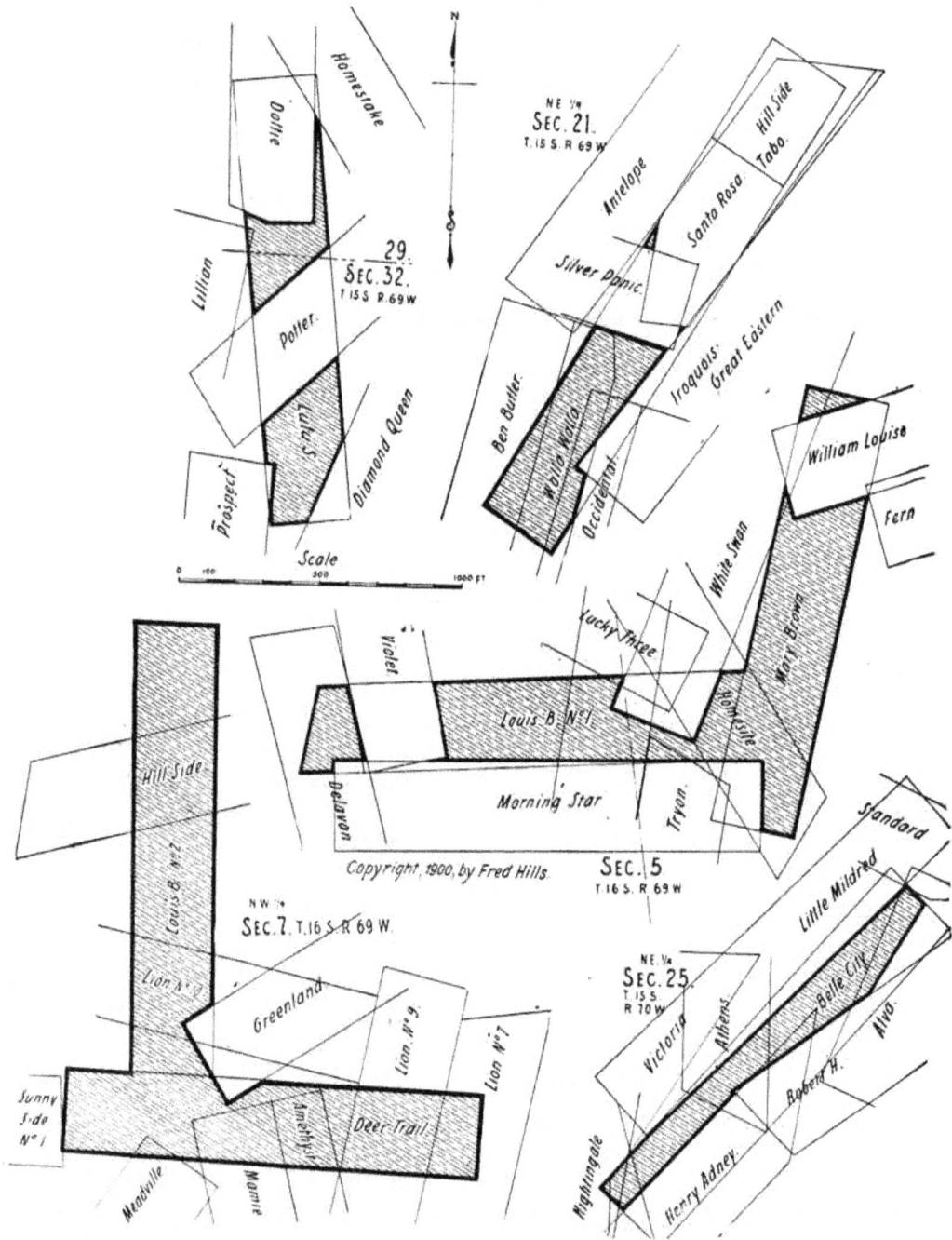

Copyright, 1900, by Fred Hills.

Directors — R. P. Davie, president; L. A. Keys, vice-president; F. H. Pettingell, secretary and treasurer; J. C. Williams, C. A. Stein.

Main Office—11 Bank block, Colorado Springs, Colorado.

Capitalization — 1,250,000 shares. Par value, $1.00. In treasury January 1, 1900, 527 shares.

Property — Owns the Santa Rosa and the Hill Side, 6 1-2 acres, Survey No. 10,734, in the N. E. 1-4 of section 21; the Mary Brown and the Louis B. No. 1, Survey No. 10,687, containing 13.735 acres, in the N. 1-2 of section 5, township 16 south, range 69 west; the Deer Trail and the Louis B. No.

(CONTINUED ON PAGE 423.)

2, 19.922 acres, Survey No. 10,877, in the N. W. 1-4 of section 7, township 16 south, range 69 west; the Belle City, 4.240 acres, on Beacon hill, Survey No. 10,371, in the N. E. 1-4 of section 25; the Lulu S., 5.046 acres, Survey No. 10,413, one-half mile east of Victor, in the S. E. 1-4 of section 29 and the N. E. 1-4 of section 32. All the above property is patented. Only enough work to obtain patent has been done.

The Savage Gold and Copper Mining Company.

Incorporated 1899.

J. R. McKinnie, president; M. W. Savage, vice-president; L. L. Aitken, secretary and treasurer; J. L. Lindsay, E. C. Sharer. *Directors*

Main Office—25 E. Pike's Peak avenue, Colorado Springs, Colorado.

1,500,000 shares. Par value, $1.00. In treasury January 1, 1900, 350,000 shares; in treasury January 1, 1900, $7,500.00 cash. *Capitalization*

Property Owns that part of the Gold King lode heretofore owned by the Gold & Globe H. M. Co., Survey No. 7,669, 2 1-5 acres, on Gold hill, in the N. W. 1-4 of section 19; the San Jose lode, Survey No. 9,288, 5 1-4 acres, on Gold hill, in the S. E. 1-4 of section 24; the Savage claim, Survey No. 8,191, 1 3-4 acres, on Raven hill, in the N. W. 1-4 of section 30; the Pearl Macy, Survey No. 8,815, 3 acres, on Squaw mountain, in the S. E. 1-4 of section 30; also own in Grant county, New Mexico, the Hanover Annex lode, the Briggs and Doyle group, and the Rattler mine, in all about 114 acres. These latter are copper mines, and, being outside the Cripple Creek district, are not shown on the map. The Gold King, San Jose and Savage are patented. Receiver's receipt held for the Pearl Macy.

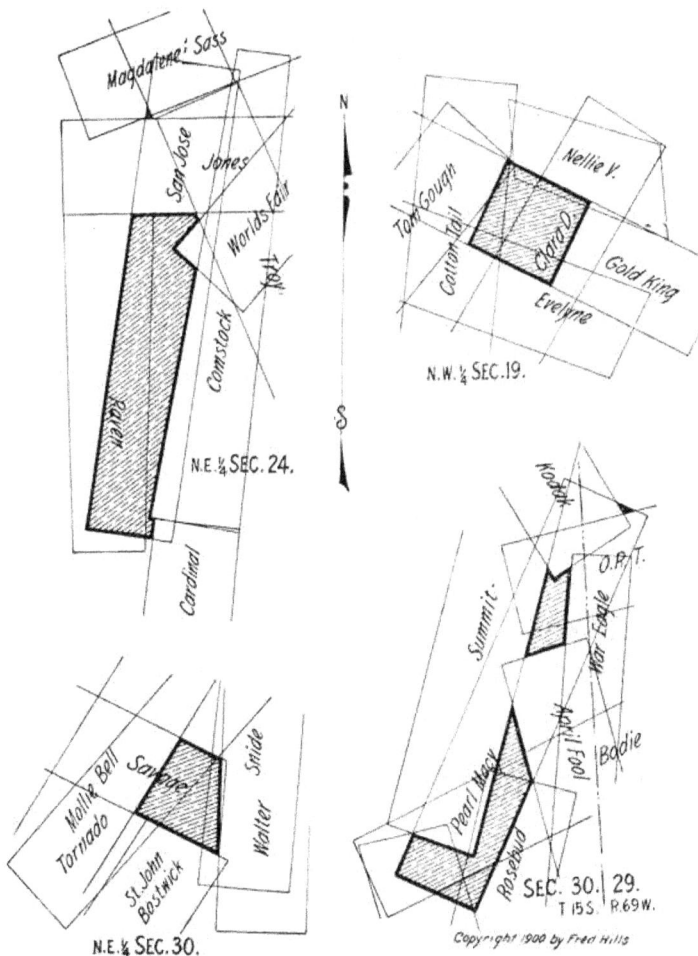

N.E. 1/4 SEC. 24.

N.W. 1/4 SEC. 19.

N.E. 1/4 SEC. 30.

SEC. 30. | 29.
T 15S. R.69W.

Copyright 1900 by Fred Hills

Development $5,000 plant on the Gold King. Prospecting outfit on the Hanover Annex and the Briggs and Doyle group. The Gold King has a shaft 425 feet deep. The Hanover Annex has a shaft 125 feet deep and other prospecting shafts and trenches. The Briggs and Doyle group have three tunnels. The greater part of the development work is being done on the Hanover Annex.

Stock is not listed at the present time.

The Sedan Gold Mining Company.

Incorporated.

Main Office—Wm. A. Otis & Co., Colorado Springs, Colorado.

Capitalization   1,250,000 shares. Par value, $1.00. In treasury January 1, 1900, $50.00 cash; in treasury January 1, 1900, 27,000 shares.

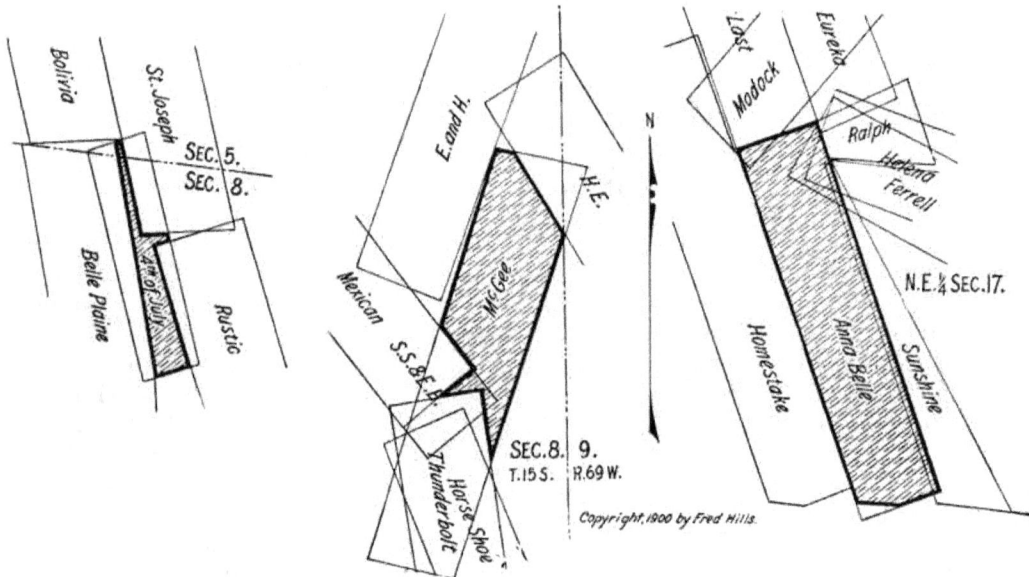

Copyright, 1900 by Fred Hills.

Property   Owns the Anna Belle, containing 7 3-4 acres, north of Cameron, in the N. E. 1-4 section 17, township 15 south, range 69 west; the Fourth of July, containing 1.50 acres, in the N. E. 1-4 section 8, township 15 south, range 69 west; the McGee, containing 4.441 acres, in the S. E. 1-4 section 8, township 15 south, range 69 west. The Anna Belle is patented. The McGee and the Fourth of July are in process. The total holdings of this company comprise 13.738 acres. The Anna Belle has just been leased and active development is now progressing.

The greater part of the work is being done on the Anna Belle. The Sedan has a shaft 200 feet deep and 160 feet of drifting.

Highest price for stock during 1899, 3 5-8 cents; lowest price for stock during 1899, $7.00 per 1,000.

The Shannon Gold Mining Company.

Incorporated.

Geo. A. Cockburn........................President
Julius Gump.......................Vice-President
W. J. Hendrickson..........Secretary and Treasurer
C. P. Bently...................Assistant Secretary
H. H. Mitchell.   John J. Key.

Directors

Main Office—No. 53 First National Bank building, Colorado Springs, Colorado.

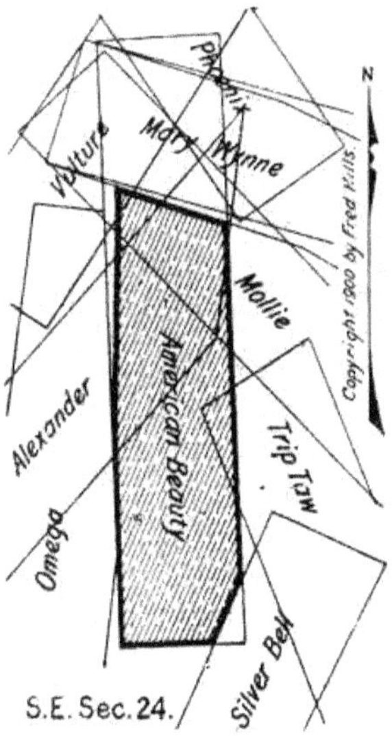

1,750,000 shares. Par value, $1.00. In treasury January 1, 1900, 132,000 shares; in treasury January 1, 1900, $1,600.00 cash.

Capitalization

Owns the American Beauty claim, containing 7 1-2 acres, situated on the S. W. slope of Gold hill, in S. E. 1-4 section 24. Patented.

Property

There are 250 feet of drifts and shafts. The property is leased.

Highest price for stock during 1899, 2 1-8 cents; lowest price for stock during 1899, 1 7-8 cents.

The Shasta Mining, Milling and Prospecting Company.

Incorporated 1892.

Directors

W. W. Williamson.........President and Treasurer
A. F. Woodward......................Secretary
R. D. Munson.   H. T. Cooper.
D. C. Davis.

Main Office—25 1-2 N. Tejon street, Colorado Springs, Colorado.

Capitalization

1,250,000 shares. Par value, $1.00. In treasury January 1, 1900, 56,921 shares. Present indebtedness of the company, $2,000.00.

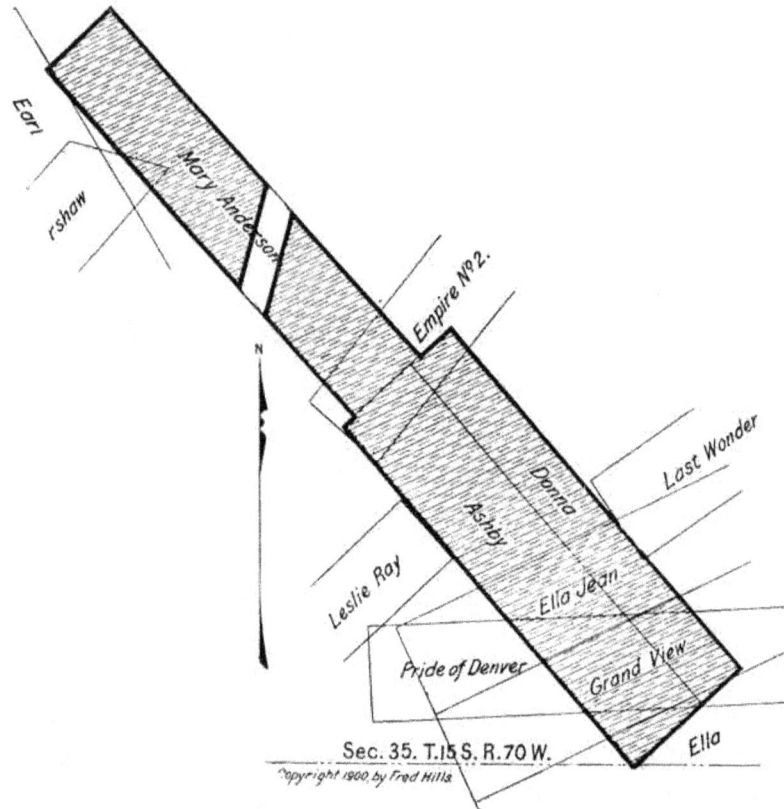

Property

Owns the Mary Anderson, Ashby, and the Donna, containing 25.31 acres, in the S. E. 1-4 section 35, on the S. W. slope of Grouse mountain. Patented. The company also owns the Joe Johnson, in sections 9 and 10, township 15 south, range 70 west, containing 9.72 acres. Plat of this last-named property does not show on map.

Development

The Joe Johnson claim has a 70-foot incline, with true fissure vein of low grade ore. The work on the property has heretofore been limited, there practically being only enough done to secure patent.

Highest price for stock during 1899, $8.00 per 1,000; lowest price for stock during 1899, $2.00 per 1,000.

The Sheriff Gold Mining Company.

Incorporated December 24, 1895.

Matt France (deceased) . . . . President   Directors
E. A. Colburn. . . . . Vice-President
Mort Parsons. . . . . Secretary
Leonard Jackson. . . . . Treasurer
  Francis Wright.   Wm. B. Jenkins.

Main Office—No. 34 Giddings building, Colorado Springs, Colorado.

1,250,000 shares. Par value, $1.00. In treasury January 1, 1900,   Capitalization
250,000 shares; in treasury January 1, 1900, $150.00 cash.

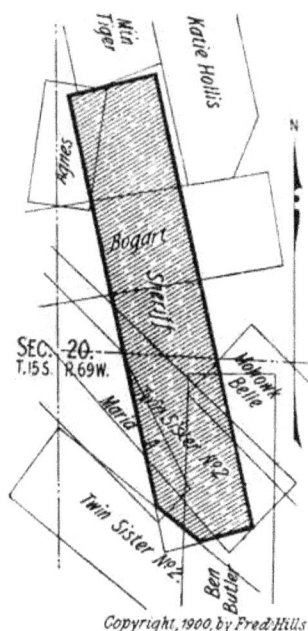

Copyright, 1900, by Fred Hills

Owns the Sheriff lode, survey No. 7,508, containing 10 acres, patented,   Property
in the E. 1-2 section 20, on the saddle between Raven and Bull hills.

One steam plant on the Jenkins and Wright lease. The deepest work-   Development
ing shaft on the Jenkins and Wright lease is nearly 300 feet deep. There
are a number of other shafts ranging in depth from 10 to 200 feet; also
some 400 feet of drifting, cross-cutting, trenching, etc.

Ore of paying grade has been shipped from nearly every working on
the property, but no large continuous chute has been located.

No stock on the market.

The Silver Panic Gold Mining and Tunnel Company.

Incorporated December 11, 1895.

Directors

V. Sargent..........................Vice-President

Frank Cotten...............Secretary and Treasurer

E. D. Lowe.

Main Office—23 1-2 North Tejon street, Colorado Springs, Colorado.

Capitalization

1,250,000 shares. Par value, $1.00. In treasury January 1, 1900, 30,000 shares.

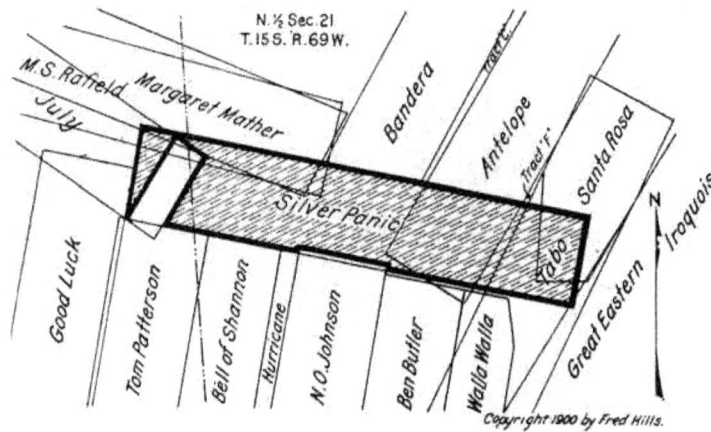

Property

Owns the Silver Panic, 9.592 acres, in the N. 1-2 section 21, on the northeast slope of Bull hill, about 1,500 feet from the Isabella group. Patented. There are several shafts on the property.

Highest price for stock during 1899, 3 cents per share; lowest price for stock during 1899, $4.00 per 1,000.

The Silver State Consolidated Gold Mining Company.

Incorporated March 3, 1898.

John J. Key..............................President   Directors
Clarence Edsall.....................Vice-President
F. B. Tiffany............................Secretary
B. N. Beal.............................Treasurer
S. P. Beal.

Main Office—Nos. 10 and 12, Giddings building, Colorado Springs, Colorado.

1,250,000 shares. Par value, $1.00. In treasury January 1, 1900, Capitalization
87,000 shares; in treasury January 1, 1900, $100 cash.

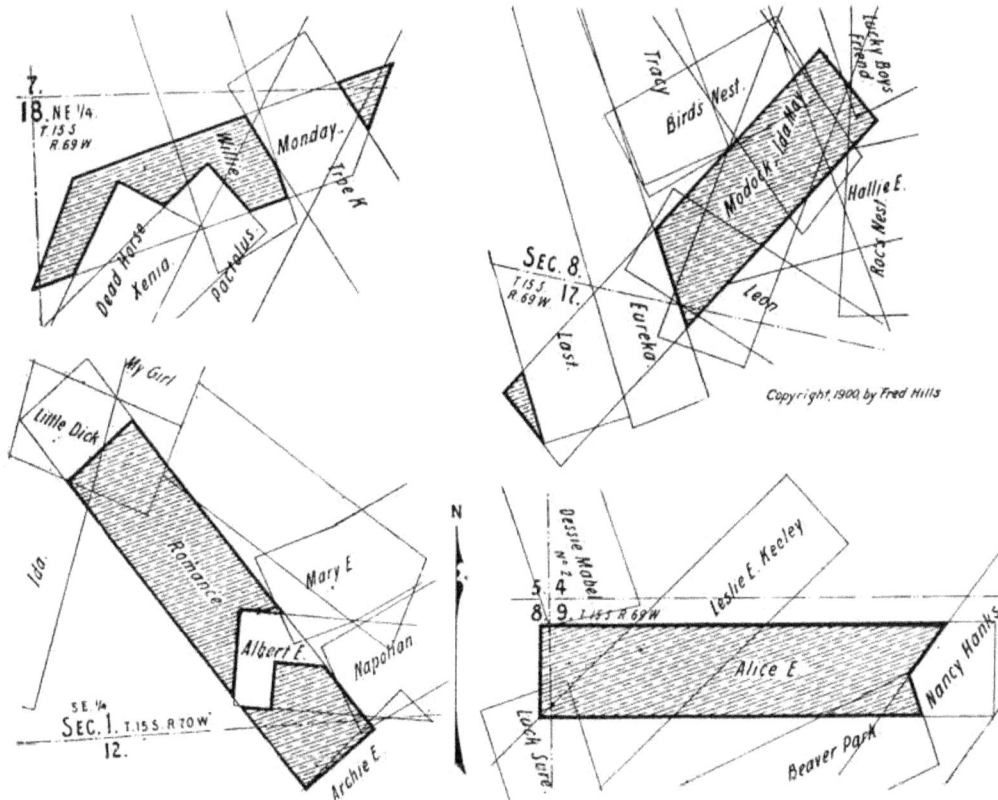

Copyright, 1900, by Fred Hills

Owns the Monday, comprising 7 3-4 acres, on Tenderfoot hill, sit- Property
uated in the N. E. 1-4 of section 18; the Modock, 8 acres, on Galena hill,
in the S. E. 1-4 of section 8; the Alice E., 10 acres, on Trachyte mountain,
in the N. W. 1-4 of section 9; and the Romance, 7 acres, on Rhyolite
mountain, in the S. E. 1-4 of section 1. Receiver's receipt held for the
Alice E. and the Romance. Monday and Modock patented. On April 30,
1900, the Alice E. claim was sold to the Hobart Gold Mining Company.

With the exception of the Monday but little work has been done on Development
these claims. The Monday has a tunnel of 200 feet, one shaft of 75 feet,
and several shafts of from 20 to 40 feet.

Highest price for stock during 1899, 3 3-4 cents; lowest price for stock
during 1899, 1 cent.

The Sitting Bull Gold Mining Company.

Incorporated January, 1898.

Directors

E. C. Glenn............................President

W. E. Hughes.....................Vice-President

A. Riesenecker...........................Secretary

T. G. McCarthy.........................Treasurer

Geo. K. Kirchner.

Main Office—Rooms Nos. 220-221, Central block, Pueblo, Colorado.

Capitalization

750 shares. Par value, $100.00 each. In treasury January 1, 1900, 95 1-4 shares.

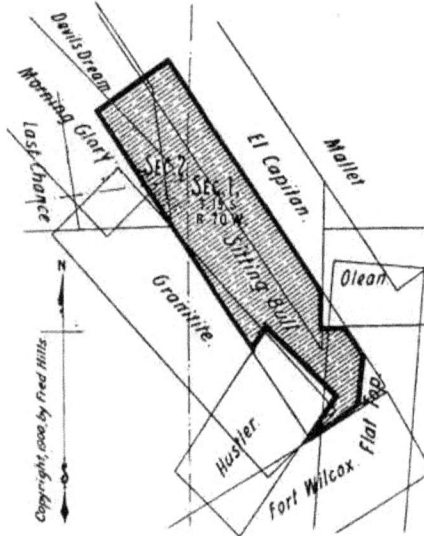

Property

Owns the Sitting Bull, containing about 8 acres, on Copper mountain, in the W. 1-2 section 1 and the E. 1-2 section 2, township 15 south, range 70 west. The property is patented.

Development

Two small shaft houses. Three shafts have been sunk, ranging in depth from 60 to 80 feet each. The property was located by Mr. W. J. McGee, on March 17, 1892. It was patented in December, 1898.

Highest price for stock during 1899, $25.00; lowest price for stock during 1899, $20.00.

The Six Points Gold Mining Company.

Incorporated 1896.

J. A. Cummings........................President
Robert Hillhouse....................Vice-President
C. H. Dudley..............Secretary and Treasurer
E. A. Colburn.   A. E. Carlton.

Directors

Main Office—14 North Nevada avenue, Colorado Springs, Colorado.

1,250,000 shares. Par value, $1.00. In treasury January 1, 1900, 250,000 shares.

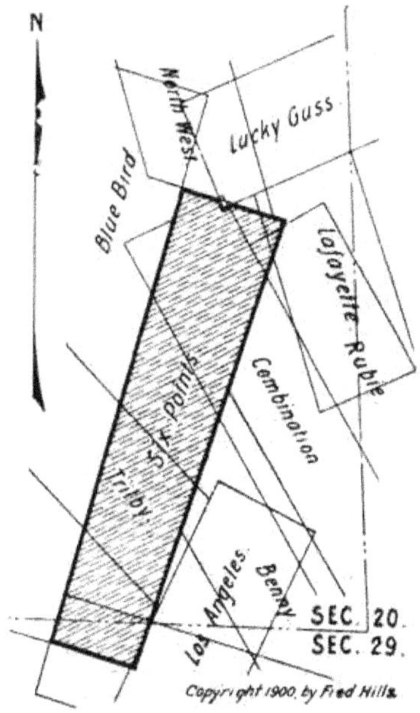

Capitalization

Copyright 1900, by Fred Hills.

Owns the Six Points, 8 acres, in the S. W. 1-4 section 20, on Bull hill. Patented.

Property

There is a shaft on the north end 170 feet deep, with from 100 to 200 feet of drifting, and one on the south end 210 feet deep, with from 300 to 400 feet of drifting. The property has been worked heretofore by lessees. It is a close corporation. No stock changes hands.

Development

The Slim Jim Gold Mining and Milling Company.

Incorporated 1896.

Directors   Phil. Strubel............................President
A. F. Woodward...................Vice-President
Chas. E. Leibold.........................Secretary

Main Office—No. 12 South Tejon street, Colorado Springs, Colorado.

Capitalization   1,000,000 shares. Par value, $1.00. In treasury January 1, 1900, about 300,000 shares; in treasury January 1, 1900, about $75.00 cash.

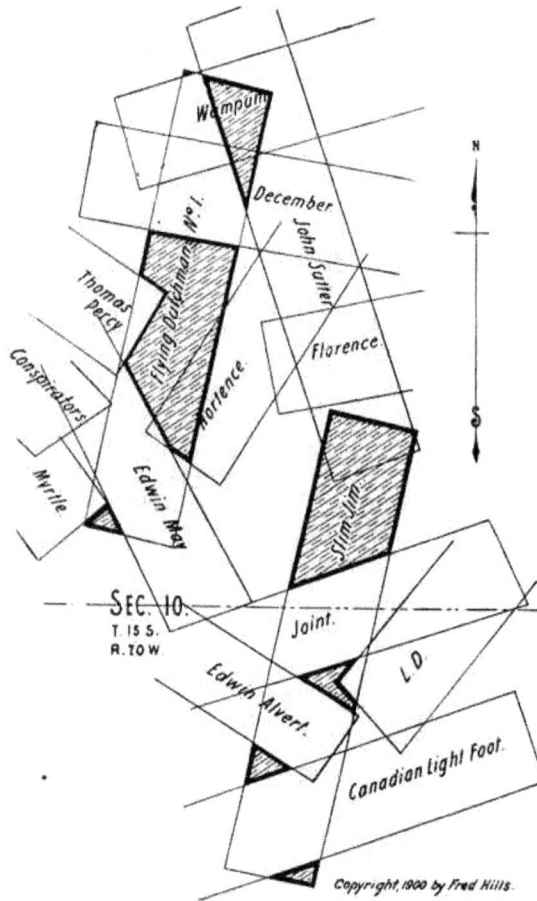

Copyright, 1900 by Fred Hills.

Property   Owns the Slim Jim and the Flying Dutchman No. 1, containing 5 acres, situated in the N. E. 1-4 section 10, township 15 south, range 70 west, on Iron mountain. Receiver's receipt for the Slim Jim. The Flying Dutchman is in process of patenting.

Development   Sufficient work has been done to obtain a patent. Good veins of low grade ore have been discovered in both shafts.

Highest price for stock during 1899, $4.00 per M; lowest price for stock during 1899, $1.00 per M.

The Sliver Gold Mining Company.

Incorporated November 23, 1895.

Clarence Edsall.......................President

John J. Key........................Vice-President

Bertram N. Beal...........Secretary and Treasurer

J. McK. Ferriday.   Geo. R. Buckman.

Directors

Main Office—Hagerman building, Colorado Springs, Colorado.

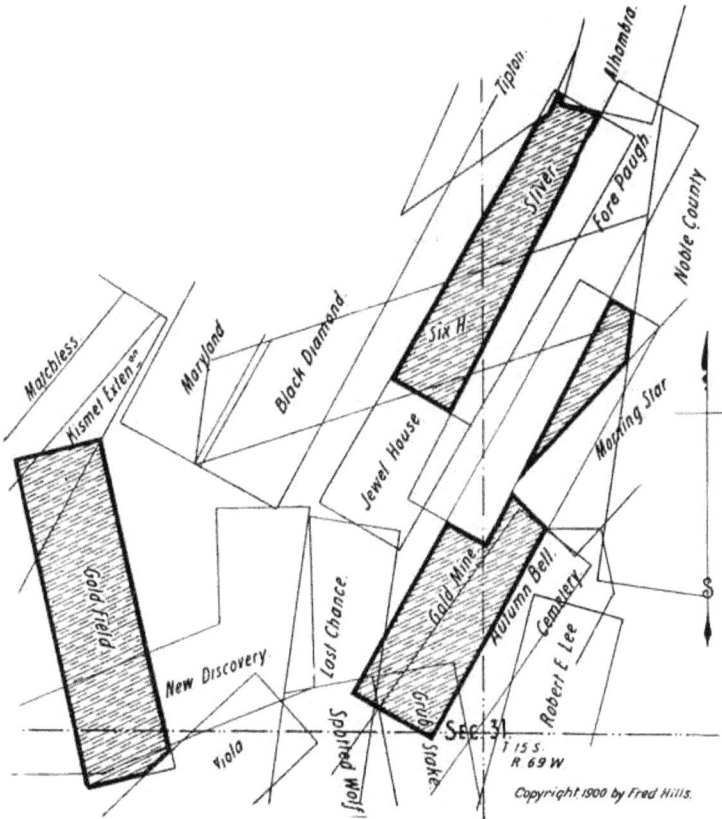

Copyright 1900 by Fred Hills.

1,000,000 shares. Par value, $1.00. In treasury January 1, 1900, Capitalization 26,000 shares.

Owns the Sliver, 4 acres, and the Gold Mine, 6 acres, in the N. 1-2 Property of section 31; also the Gold Field, 7 acres, in the N. E. 1-4 of section 31, all on Squaw mountain. Patented. The property is being developed by lessees.

Highest price for stock during 1899, 2 1-2 cents; lowest price for stock during 1899, 1 cent.

The Solitaire Gold Mining Company.

Incorporated January 16, 1900.

Directors   W. A. B. MacDonald, president; T. C. MacDonald, vice-president;
H. H. Barbee, secretary; J. F. De Berry, treasurer; A. B. Risk.

Main Office—Postoffice block, Colorado Springs, Colorado.

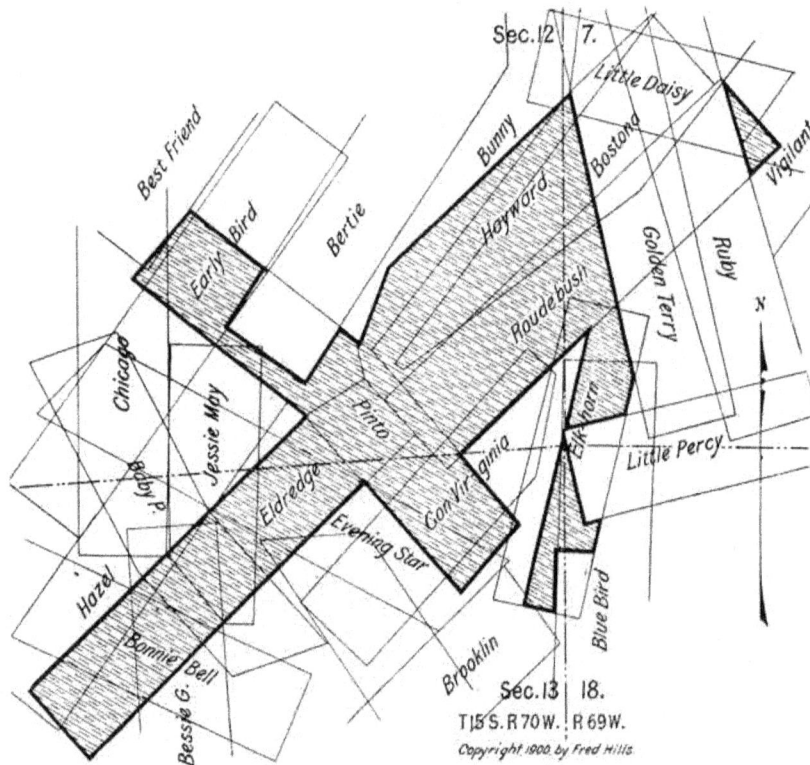

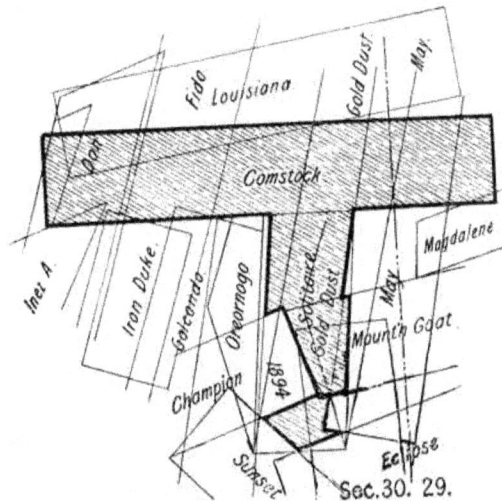

Capitalization

    1,500,000 shares. Par value, $1.00. In treasury January 1, 1900,
200,000 shares; in treasury January 1, 1900, $10,000.00 cash.

Property

    Owns the Solitaire, about 5 acres, in the N. E. 1-4 of section 30, on
Raven hill and Battle mountain; the Roudebush, the Hayward, the Eld-
redge, the Pinto, about 30 acres in all, situated in the S. E. 1-4 of section
12, about 1,000 feet from the Gold King mine, on Carbonate hill. The

(CONTINUED ON PAGE 435.)

company also owns the Elkhorn, adjoining the above property, on Carbonate hill, and a three-fourths interest in the Comstock, 10 acres, on Raven hill. All the above property is patented.

The greater part of the development work is being done on the Elkhorn claim, which has a shaft 450 feet deep and considerable drifting and stoping. The company holds, by lease, the Columbia, owned by the Cripple Creek Columbia G. M. Company. *Development*

The Elkhorn is a shipping mine and has produced some very rich ore, running from 5 to 20 ounces. There are now a number of applications for leases which are under consideration and will soon be granted. The Elkhorn promises to be a good mine. *History*

The stock was not on the market in 1899. Since then has been placed at 10 cents.

The Southern Boy Gold Mining Company.

Incorporated 1899.

R. P. Davie, president; H. L. Shepherd, vice-president; A. J. Bendle, secretary and treasurer; Newton S. Gandy, Henry M. Blackmer. *Directors*

Main Office—25 E. Pike's Peak avenue, Colorado Springs, Colorado.

1,250,000 shares. Par value, $1.00. In treasury January 1, 1900, 550,000 shares. *Capitalization*

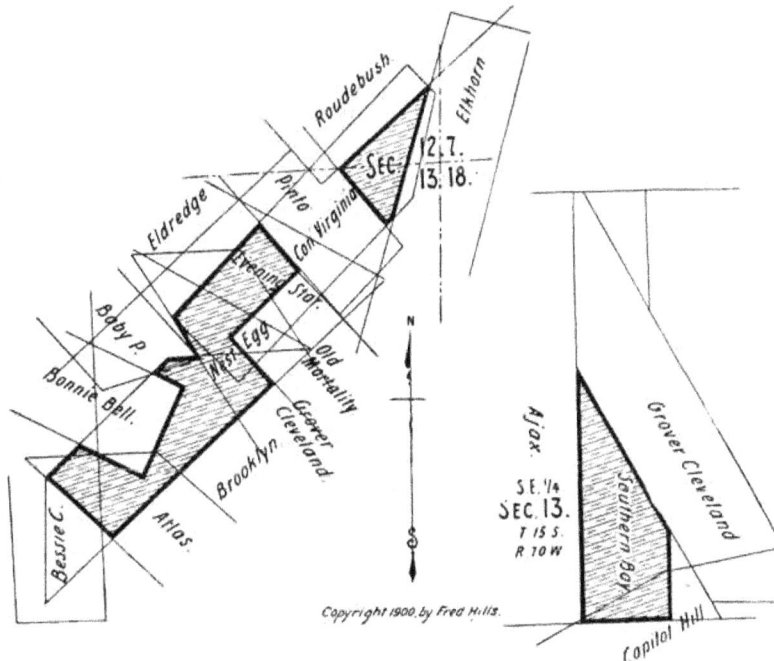

Copyright 1900 by Fred Hills.

Owns the Southern Boy, 4 acres, adjoining the Oriole and M. J. T. property, on Gold hill, in the S. E. 1-4 of section 13. Patented. Also owns the Consolidated Virginia and the Nest Egg, a total of 6 acres, on Mineral hill, in the N. E. 1-4 of section 13, adjoining the Hayden placer townsite on the northeast. *Property*

Just as this Manual goes to press, May, 1900, the company has sold the Nest Egg and Consolidated Virginia claims, for $11,250 cash, to the Key West Gold Mining Company. On the folding map in pocket on back cover of this book this latter property is shown as Key West ground.

The Specimen Gold Mining and Milling Company.

Incorporated March, 1897.

Main Office—No. 19 and 20 Hagerman block, Colorado Springs, Colorado.

Capitalization

1,200,000 shares. Par value, $1.00. In treasury January 1, 1900, 84,000 shares; in treasury January 1, 1900, $1,200 cash.

Property

Owns the Specimen, situated on Bull hill, in S. 1-2 section 20, and containing 10 acres. Patented.

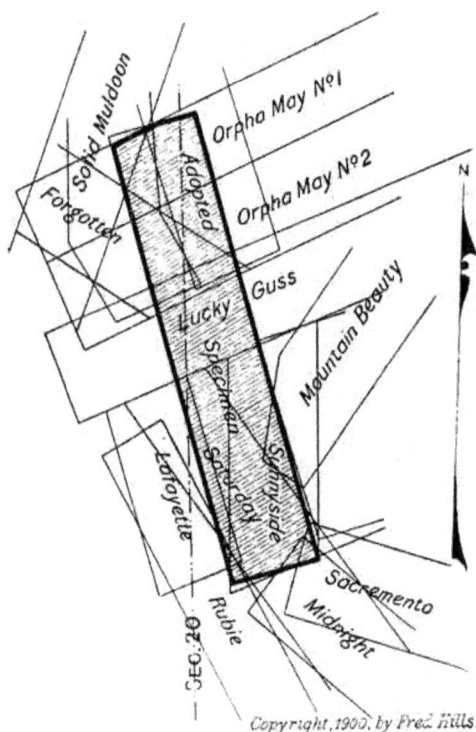

Copyright, 1900, by Fred Hills.

Development

There are several electric and steam hoists owned by the lessees. The company owns one 30 H. P. engine and boiler. There are also two fair-sized ore houses and shaft houses. Four shafts have been sunk, ranging in depth from 200 to 450 feet deep. About 1,500 feet of cross-cutting and leads have been followed. The property is all leased.

History

The claim was first worked in 1892. Since then work by the company and by lessees has been nearly continuous. Some very rich ore has been discovered, but the shafts have not been sunk sufficiently deep to encounter permanent ore bodies. In May, 1900, this property was sold to J. A. Sill for $156,000 cash, which will be distributed in dividends.

Production and Dividend

Gross production to January 1, 1900, $125,000; $6,000 in royalties paid the company during the past year.

Highest price for stock during 1899, 15 cents; lowest price for stock during 1899, 7 1-2 cents.

The Spicer Mining and Milling Company.

Incorporated April 30, 1894.

O. W. Spicer.........................President   Directors

W. W. Maybury.....................Vice-President

R. H. Atchison............Secretary and Treasurer

J. C. Spicer.   Thos. Spicer.

Main Office—Colorado Springs, Colorado; branch office, Victor, Colorado.

300,000 shares.   Par value, $1.00.   In treasury January 1, 1900, $2,-   Capitalization
000.00 cash.

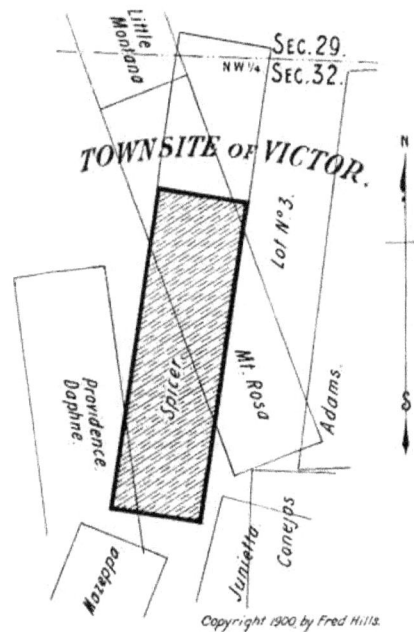

Owns the Spicer lode mining claim, survey No. 8,682, containing 7   Property
acres, in the N. W. 1-4 section 32, at the base of Battle mountain.   Patented.

The Spicer Shaft No. 2 is an incline shaft 400 feet deep, and the   Development
Spicer Shaft No. 3, where work is being done at present, is a vertical shaft
200 feet deep.   Three well-defined mineral veins run through this entire
property.   A porphyry dyke and phonolite dyke cross these veins on the
property.   The greater part of the development work has been done on
the north end, or Spicer Shaft No. 2.   The property is under lease and is
being worked through Shaft No. 3.

Total dividends up to January 1, 1900, about $35,000.00.   Production

The Spicer lode claim was located in 1891 and patented early in 1894.   History
This is a close corporation, no changes having taken place in its officials,
and no sales of its shares having been made since its organization.

The Spring Creek Deep Mining and Drainage Tunnel Company.

Incorporated 1896.

Directors
Stephen W. Keene, president; William E. Jones, vice-president; Sidney R. Bartlett, secretary and treasurer; Eugene G. Weidner, James L. Lindsay.

Main Office—Postoffice building, Colorado Springs, Colorado.

Capitalization
2,000,000 shares. Par value, $1.00. In treasury January 1, 1900, 780,000 shares.

Property
Owns the St. Louis lode mining claim, now being patented; the Spring Creek Deep Mining and Drainage Tunnel and tunnel site, located on the north slope of Tenderfoot hill.

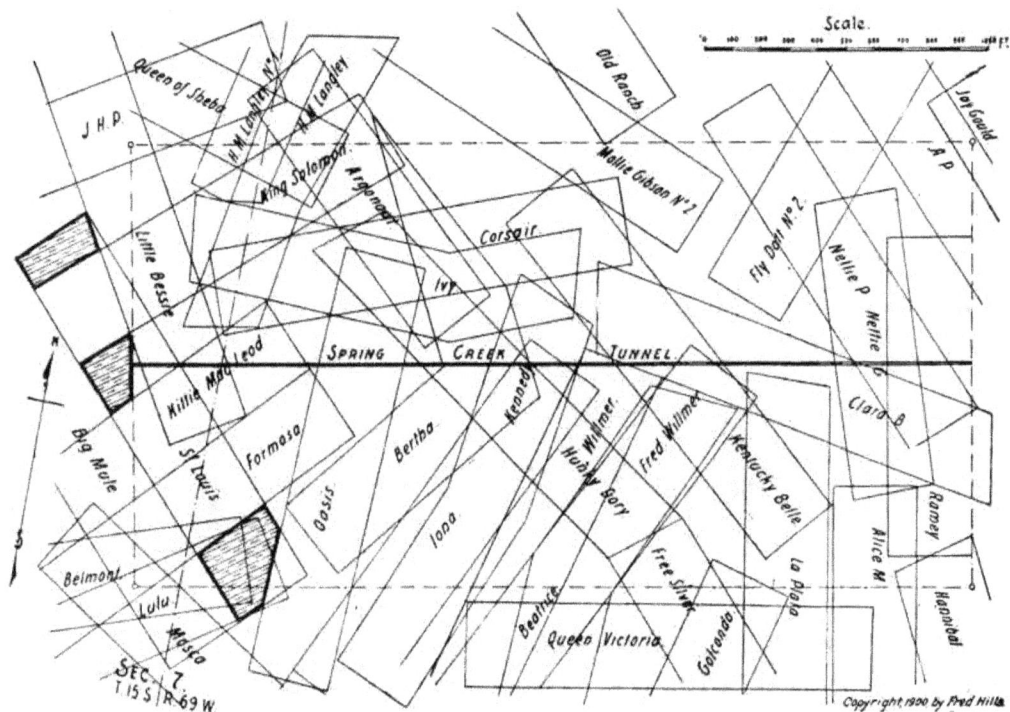

Development
The surface improvements consist of building containing boiler, air drills, tools and supplies; blacksmith's shop; office; boarding house. Seven hundred dollars have been expended on the St. Louis lode claim for labor and to secure patent. The principal work has been done in driving the tunnel 425 feet. Course of tunnel is north 77° 38′ east 3,000 feet. Size of the tunnel is 7 feet wide by 8 feet high. Drainage tracks and timbering.

History and Object
The tunnel was located November 16, 1895, for the discovery of mines and development of lodes, leads, and deposits of valuable mineral, and the company organized to work, operate and develop the same; to excavate and run cross-cuts or such other workings as may be necessary for drainage and transportation purposes, through property owned by the corporation, or other corporations or individuals, for the purpose of draining such property and transporting water, ores, rock and waste from said mines or carrying men and supplies to said mines or mining claims, and when such transportation is done for other corporations or individuals by this com-

(CONTINUED ON PAGE 439.)

pany, to demand and receive compensation therefor. The company has contracts for the purposes above named with the owners of the following properties through which the tunnel will pass: The Kittie MacLeod, Oasis, Hunky Dory, Clara B., the Nettie G., and the Ramey lode mining claims, at prices ranging from 50 cents to $1.00 per ton on pay ore, and 25 cents per car on waste, rock, etc. Drainage to be settled according to quantity of flow. These contracts secure to the company a revenue independent from their own properties.

The stock has not been listed and all sales have been privately negotiated at 10 cents per share.

The Squaw Mountain Mining Company.

Incorporated 1893.

J. R. McKinnie...........................President
A. May..............................Vice-President
L. L. Aitken...............Secretary and Treasurer
A. D. Aitken.   E. C. Sharer.

Directors

Main Office—25 E. Pike's Peak avenue, Colorado Springs, Colorado.

1,000,000 shares. Par value, $1.00. In treasury January 1, 1900, 250,000 shares.

Capitalization

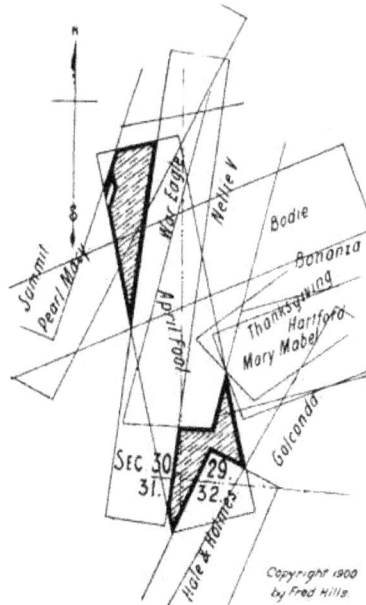

Owns the April Fool lode mining claim, containing about 3 acres, situated between sections 29 and 30, near the Nellie V., on Squaw mountain. Patented.

Property

The property is bonded and leased for $30,000, expiring July 15, 1900. Royalty of 20 per cent.

Development

Highest price for stock during 1899, 1 cent; lowest price for stock during 1899, $4.00 per thousand.

The St. Paul Gold Mining and Tunnel Company.

Incorporated 1895.

Directors

J. R. Hankey..............................President

J. O. Hardwick.....................Vice-President

E. E. McMahan............................Secretary

L. E. Sherman...........................Treasurer

A. G. Lewis.

Main Office—52 P. O. building, Colorado Springs, Colorado.

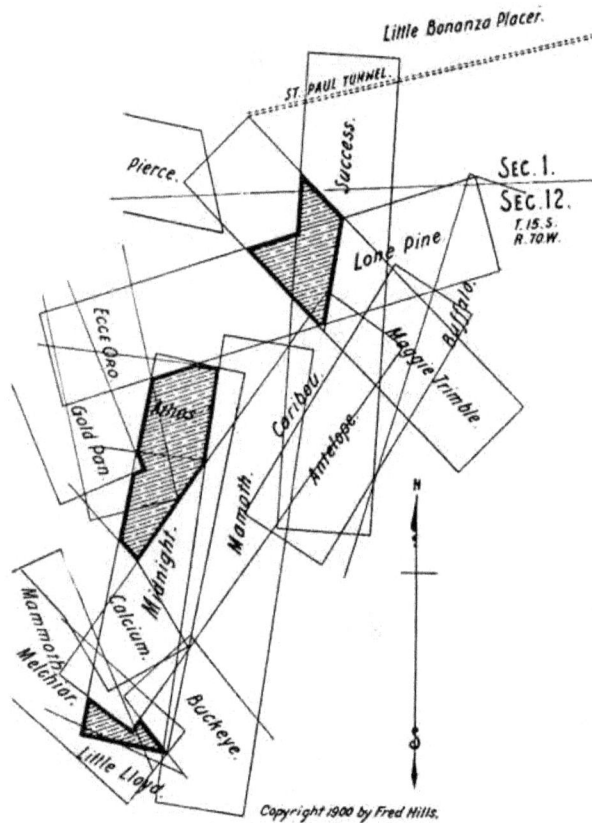

Copyright 1900 by Fred Hills.

Capitalization   1,000,000 shares. Par value, $1.00. In treasury January 1, 1900, 7,000 shares.

Property   Owns the Maggie Trimble, containing 2 acres, in the N. W. 1-4 of section 12, the Midnight, containing 3 acres, in the N. W. 1-4 of section 12, and the St. Paul tunnel. All above property situated on north slope of Mineral hill and is patented.

Development   There is a cabin 15x30 feet. About 600 feet of development work has been done, a tunnel of about 400 feet and several small shafts. The larger part of the work has been done on the Maggie Trimble and the St. Paul tunnel.

The Starr King Gold Mining Company.

(Formerly the Anaconda Extension Gold Mining Company.)

Incorporated July, 1892.

P. S. Baily............................President
F. H. Pettingell....................Vice-President
Peter McCourt.............Secretary and Treasurer
Jos. Monnig.   W. F. De Groot.

Directors

Main Office—813 Seventeenth street, Denver, Colorado.

1,000,000 shares. Par value, $1.00.

Capitalization

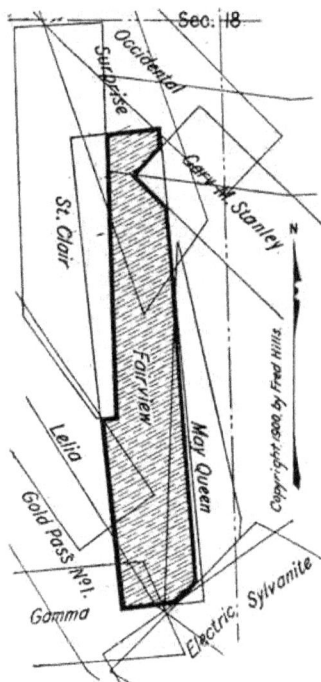

Owns the Fairview claim, 7.610 acres, in the S. W. 1-4 of section 18, township 15 south, range 69 west, on Gold hill. Patented.

Property

Cabin, blacksmith shop, etc., on the property. One shaft 80 feet deep, with 120 feet of drifting. There are numerous other shafts and trenches. The Chicago & Cripple Creek tunnel runs to the middle of the claim, at about 500 feet depth. The property is being worked by lessees.

Development

Stratton's Group.

Copyright, 1900 by Fred Hills.

SCALE.

Stratton's Group.

Consisting of—

The Arcadia Consolidated Mining Company.

Companies

The Franklin-Roby Gold Mining Company.

The Marinette Gold Mining Company.

The Mars Consolidated Gold Mining Company.

And two claims formerly owned by the Cripple Creek Consolidated Gold Mining Company.

The Mars Consolidated: This company, it is reported, has sold its Property controlling interest to Mr. W. S. Stratton.

The Franklin-Roby sold its belongings, consisting of the Lillie claim, to Mr. W. S. Stratton.

The Marinette Company is also owned by Mr. Stratton.

The Arcadia Company is owned and controlled by Mr. Stratton.

The Key West Company sold the eastern portion of the Key West claim to Mr. Stratton.

The May Queen and Geneva lodes, as shown, formerly belonged to the Cripple Creek Consolidated G. M. Company and were sold to Mr. Stratton.

The Temomj Mining and Milling Company control was recently purchased by Mr. Stratton. This comprises the following claims: Temomj, Home Fraction, Baby McKee and the Clayton E., which claims are more fully described and shown under the "Temomj Mining and Milling Company."

Stratton's Independence, Limited.

Copyright, 1900, by Fred Hills.

Scale of Feet

444

Stratton's Independence, Limited.

Incorporated in England, 1899.

W. F. Orriss (Chairman) ....London — Directors
The Earl of Chesterfield....................London
F. W. Baker.....................Wimbledon, Eng.
W. S. Stratton.........Colorado Springs, Colorado
T. A. Rickard, Consulting Engineer.. Denver, Colorado

Main Office—No. 3 Princes street, London, E. C. Local Office—McPhee building, Denver, Colorado.

1,100,000 shares. Par value, £1. In treasury January 1, 1900, 100,000 shares set aside for working capital. — Capitalization

Owns 14 mining claims, consolidated into a compact group covering 112 acres, on the south slope of Battle mountain. All patented. — Property

The mine is well equipped. The steam plant consists of three Heine water tube boilers, each of 300-horse power. The machinery consists of a 225-horse power hoisting engine, capable of working the mine to a depth of 1,500 feet; one Ingersoll-Sargent air compressor, equal to three drills, and two Norwalk compressors, equal to twelve drills. — Surface Improvements

The pumping plant consists of two compound Snow pumps, each of a capacity of 1,000 gallons per minute. These are stationed at the 900-foot level. — Underground Machinery

The workings consist of two shafts, known as the No. 1 and the No. 2, respectively. The No. 1 is 920 feet deep; No. 2 is 625 feet deep. The workings have a total aggregate length of 35,000 feet, or a length of seven miles. The largest part of the production is derived from three principal veins, known as the Independence, the Bobtail and the Emerson. — Development

From its inception to the end of 1899, the amount produced was 41,694 tons of ore, having a gross value of $3,837,360.00, which gave a net profit of $2,402,164.00 after deducting every expense. The total production to June 1, 1900, will be 58,000 tons, having a value of, approximately, $7,000,000.00, yielding a profit of $4,500,000.00. Production for April, 1900, $310,000.00. — Production and Dividend

This mine was discovered by Mr. W. S. Stratton on the Fourth of July, 1891, he being one of the first prospectors who visited the camp, and, as he staked out the first two claims on that day, he named them the Independence and the Washington. This mine has been one of the greatest producers in the district, and the amount of ore blocked out is a guarantee that it will remain so for many years to come. Since the incorporation of the new company four quarterly dividends have been declared, equal to $488,000.00 for each dividend, the last being payable on the 14th of June, 1900. — History

The Strong Gold Mining Company.

Incorporated 1892.

Directors

Wm. Lennox.........................President
N. B. Scott.......................Vice-President
E. A. Colburn.........................Secretary
E. W. Giddings, Jr....................Treasurer

Main Office—Giddings building, Colorado Springs, Colorado.

Capitalization

500,000 shares. Par value, $1.00. In treasury January 1, 1900, large cash surplus.

Property

Owns the Strong claim, containing 7 1-2 acres, situated in the S. 1-2 section 29, on Battle mountain. All patented.

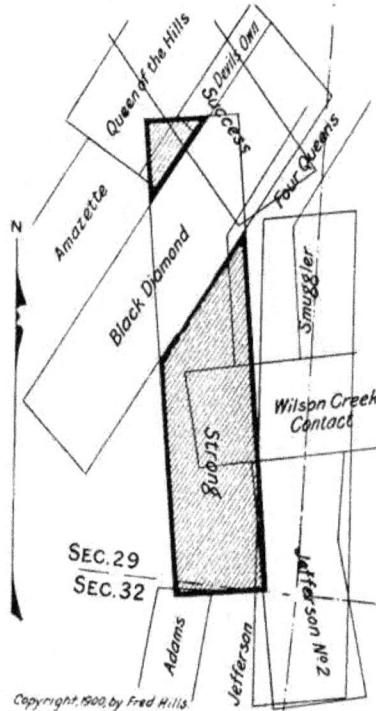

Development

One of the most commodious plants, consisting of a large shaft house, ore houses, blacksmith shop, engine and boiler house. There is a double-compartment shaft, 800 feet deep; a two-compressor plant, with six machine drills. Underground machinery: Large duplex pump, with a capacity of 500 gallons per minute, at a depth of 1,000 feet. There are about 10,000 feet of drifts and cross-cuts.

Production and Dividend

In dividends considerably over $500,000 have been paid. The mine is now producing 1,000 tons of ore per month. This is a very close corporation and information is hard to obtain.

The stock is held entirely by the directorate.

The Sumpter Gold Mining Company.

Incorporated January 29, 1896.

S. B. Carrington, president; Samuel Lindsey, vice-president; Chas. Directors
E. Snider, secretary; Thos. F. Creighton, treasurer; Robert Schwartz.

Main Office—Colorado Springs, Colorado.

1,500,000 shares. Par value, $1.00. In treasury January 1, 1900, Capitalization
138,500 shares; in treasury January 1, 1900, $150.00 cash.

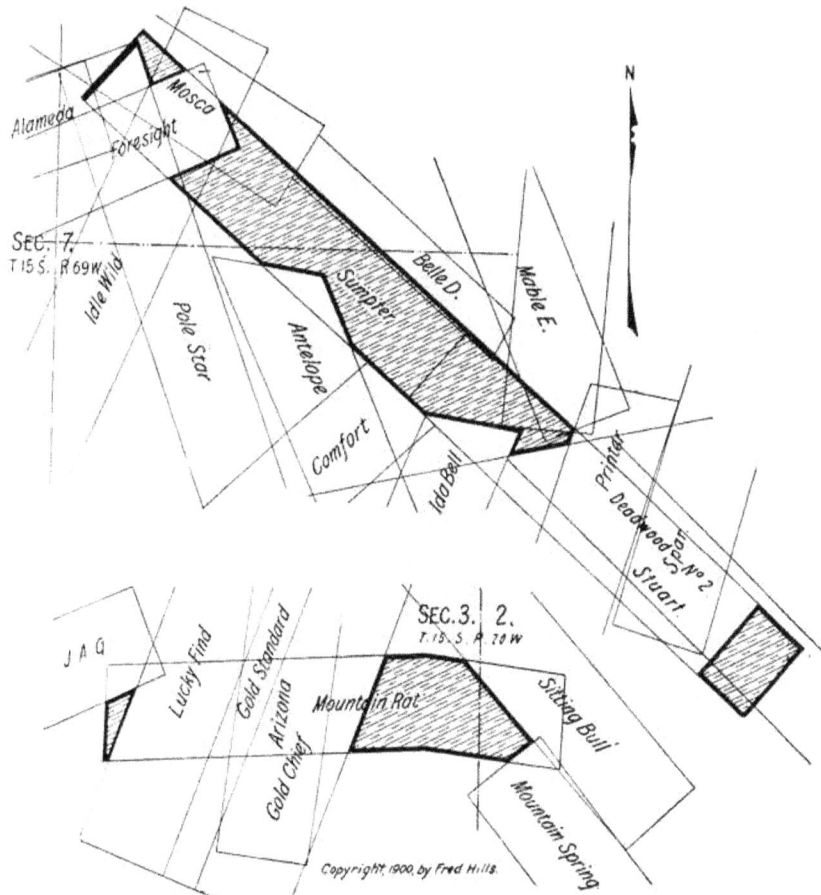

Owns the Sumpter and Stuart claims, comprising nearly 9 acres, in Property
the E. 1-2 of section 7, on the north end of Tenderfoot hill, in the vicinity
of Deadwoods Nos. 1, 2, 3 and 4, and north of the Forlorn Hope and
Antelope claims; also the Mountain Rat claim, comprising 3 1-2 acres, in
the S. E. 1-4 of section 3, township 15 south, range 70 west, on Red mountain, northwest of Cripple Creek. Mountain Rat is patented. Receiver's
receipts are held for balance of property.

There are three shafts on the Sumpter and Stuart claims, one of 105 Development
feet, with about 20 feet cross-cut; one of 60 feet, with a 15-foot cross-cut,
and one of 84 feet, with a 30 or 40-foot cross-cut.

These claims were located in 1892 and purchased by S. B. Carrington, History
Thos. Creighton and Chas. E. Snider in '92 or '93. The present company
was organized in 1896.

Highest price for stock during 1899, 4 1-2 cents.

The Sun Consolidated Gold Mines Company.

Incorporated February 1, 1899.

Directors

Dan Weyand, president; J. R. McKinnie, vice-president; L. L. Aitken, secretary and treasurer; M. S. Herring; Chas. F. Potter.

Main Office—No. 25 E. Pike's Peak avenue, Colorado Springs, Colorado.

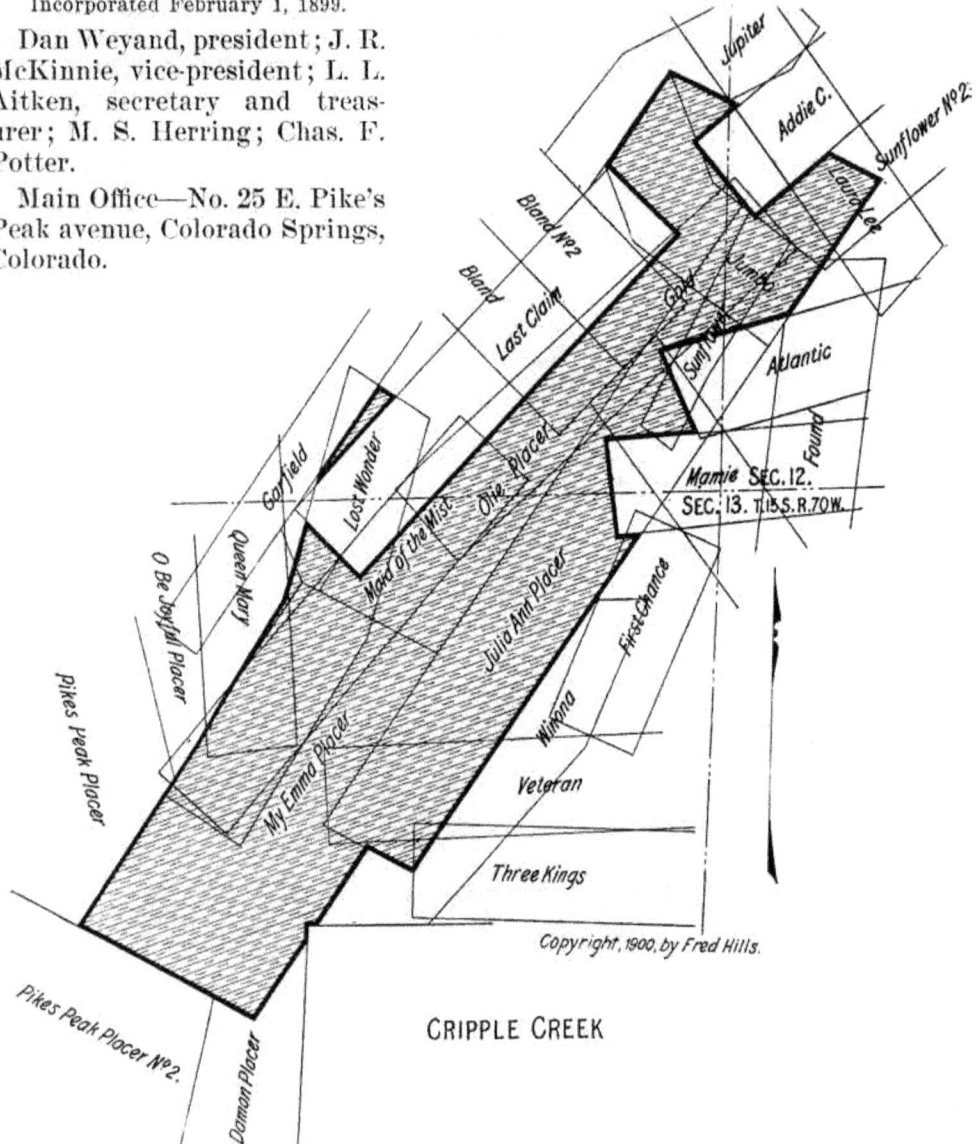

Copyright, 1900, by Fred Hills.

CRIPPLE CREEK

Capitalization

1,000,000 shares. Par value, $1.00. In treasury January 1, 1900, 196,995 shares.

Property

Owns the Olie Placer, the Julia Ann Placer, My Emma Placer, Lost Wonder, Maid of the Mist, containing in all about 50 acres, situated on south slope of Mineral hill, in the N. W. 1-4 section 13 and in the S. W. 1-4 section 12. In process of patenting. The company also has control of the Bradford and Cripple Creek Gold Mining Company, owning the Sunflower, Jumbo and Gold lodes. The company also has vein rights on the Atlantic and the Mamie.

Development

One dwelling house, one blacksmith shop with bellows and anvil, one whim, 200 feet of steel wire cable with two iron whim buckets, one iron dump car, one truck for bucket, 600 feet of T rail. On the Sunflower claim are two shafts, each 100 feet deep; also another shaft 75 feet deep, together with 800 feet of drifting. The greater part of the development work is being done on this claim.

Highest price for stock during 1899, 8 1-2 cents; lowest price for stock during 1899, 6 cents.

The Sunflower Gold Mining Company.

Incorporated December 11, 1896.

W. H. Bacon..............President and Treasurer
John Pedersen ....Vice-President
Taylor J. Downer.......................Secretary

Directors

Main Office—104 1-2 East Pike's Peak avenue, Colorado Springs, Colorado.

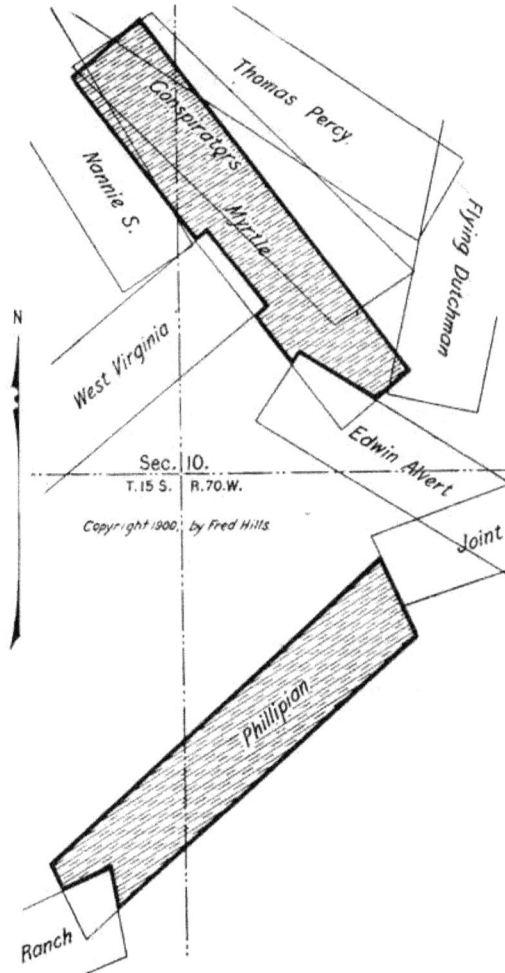

Capitalization 1,000,000 shares. Par value, $1.00. In treasury January 1, 1900, 243,000 shares; in treasury January 1, 1900, $100.00 cash. No debts.

Property Owns the Phillipian, 8 1-4 acres, in the S. E. 1-4 section 10, township 15 south, range 70 west; also the Myrtle, 8 1-2 acres, in the N. E. 1-4 section 10, township 15 south, range 70 west, on Iron mountain. Patented.

Development The Myrtle claim has a shaft 45 feet deep, with a vein 12 inches wide, which gave a mill test of $17.00 per ton. The Phillipian has a shaft 35 feet deep and several open cuts. Prospects are very good.

Highest price for stock during 1899, 1 1-2 cents; lowest price for stock during 1899, 1 cent.

The Sunset Consolidated Gold Mining Company.

Incorporated November 5, 1896.

Directors   Geo. L. Keener, president; W. B. Pullen, vice-president; W. S. Nichols, secretary and treasurer; M. S. Herring; Robert Denton.

Main Office—No. 16 N. Nevada avenue, Colorado Springs, Colorado.

Capitalization   2,000,000 shares. Par value, $1.00. In treasury January 1, 1900, 420,000 shares; in treasury January 1, 1900, $500.00 cash.

Property   Owns the Herbert Allen, the Baby June, the Justice, the Kodak, the Topsy, the Sunset, the Summit, and the vein rights crossing the Big Theatre, containing about 30 acres, situated in E. 1-2 section 30 in the saddle between Squaw and Battle mountains.

The Summit is patented.

Development   About $8,000.00 has been expended in development work. The greater part of the development work is being done on the Sunset and the Topsy.

The property of this company is leased.

Highest price for stock during 1899, 12 cents; lowest price for stock during 1899, 5 cents.

The Surprise Gold Mining Company.

Incorporated December 21, 1895.

Frank H. Hall............................President

A. T. Gunnell......................Vice-President

C. P. Bently...............Secretary and Treasurer

L. Hall.   J. W. D. Stovell.

Directors

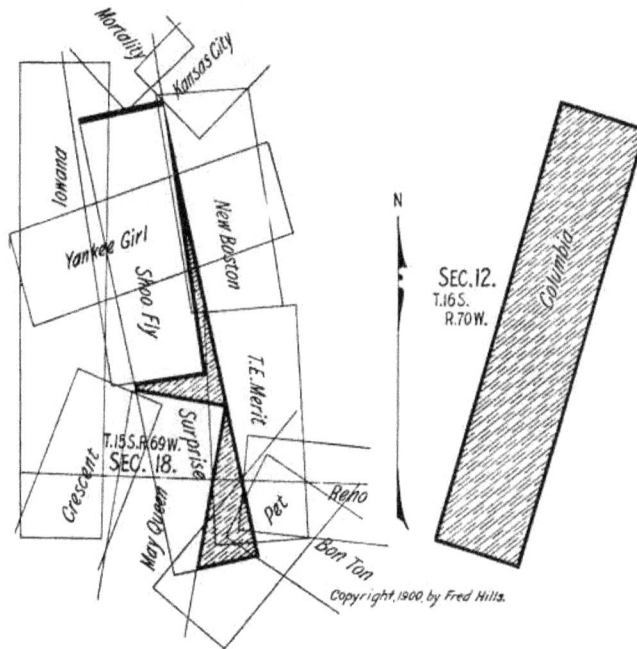

Main Office—No. 53 First National Bank building, Colorado Springs, Colorado.

600,000 shares. Par value, $1.00. In treasury January 1, 1900, 100,000 shares. Capitalization

Owns the Surprise, survey No. 8,371, 2 1-2 acres, in the N. W. 1-4 section 18, on Womack hill, and the Columbia, survey No. 9,457, in the N. E. 1-4 section 12, township 16 south, range 70 west on the S. W. slope of Straub mountain, containing 10 acres. All patented. Property

The T. Bone Gold Mining Company.

Incorporated.

Directors

Chas. N. Miller, President; S. M. Perry, Vice - President; W. L. Clark, Secretary and Treasurer.

Main Office—Cripple Creek, Colorado.

Capitalization

2,000,000 shares. Par value, $1.00. In treasury January 1, 1900, 20,000 shares; in treasury January 1, 1900, $565.00 cash.

Property

Owns the E. J., containing 10 1-3 acres, patented; the Thelma, the Chrystal, the Mayflower, and the T. Bone, containing in all 20 acres, all in process of patenting. These claims are situated in the S. E. 1-4 of section 11, and in the N. E. 1-4 of section 14, on the southwest slope of Mineral hill.

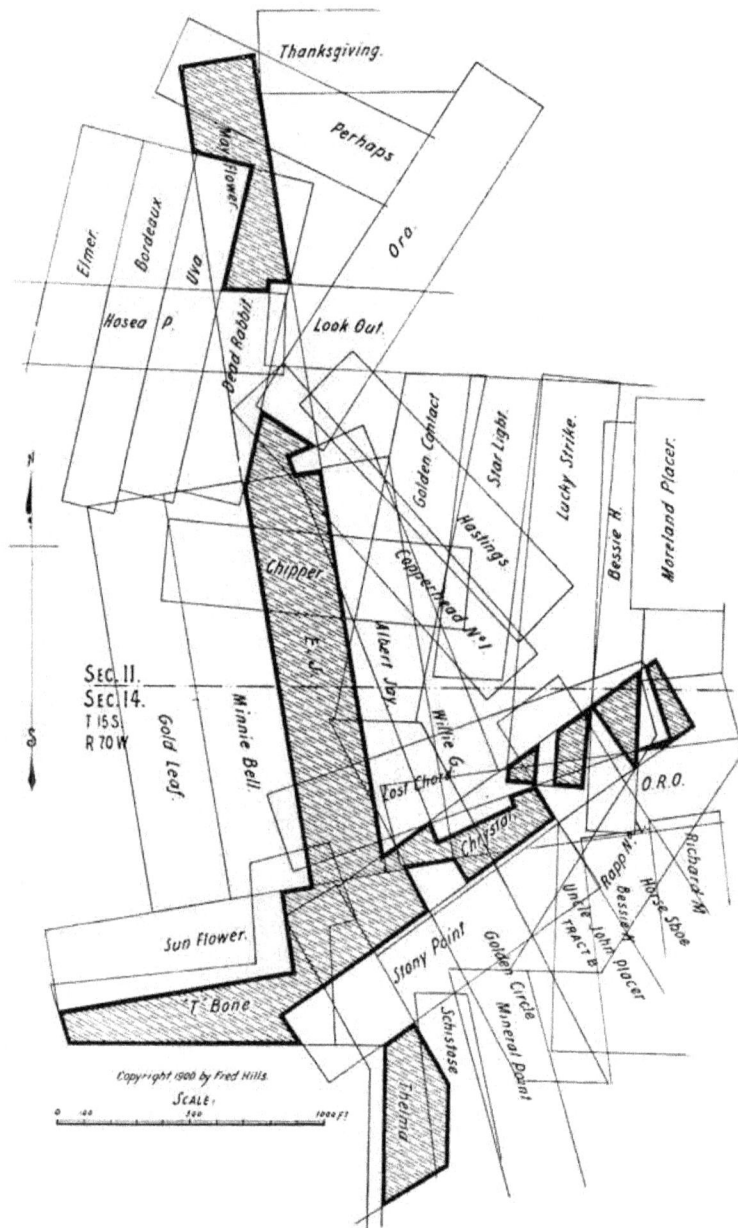

Copyright 1900 by Fred Hills.

SCALE.

Development

Shaft house 16x20 feet. The Thelma has a shaft 175 feet deep and 250 feet bottom drifting, with 100 feet of drifting at the first level. The Chrystal has two shafts, one of 80 feet depth and one of 60 feet depth, with 70 feet of drifting. On the E. J. are two shafts, one of 125 feet depth with 30 feet of drifting, and one of 75 feet depth with 50 feet of drifting. There are also 5 other shafts, 40, 20, 35, 60, and 20 feet deep, respectively. The greater part of the development work is being done on the Thelma.

Highest price for stock during 1899, 1 1-2 cents; lowest price for stock during 1899, $5.00 per 1,000.

TOWN OF ANACONDA AND MARY McKINNEY MINE.

COLD HILL IN BACKGROUND.

The Tejon Mining Company.

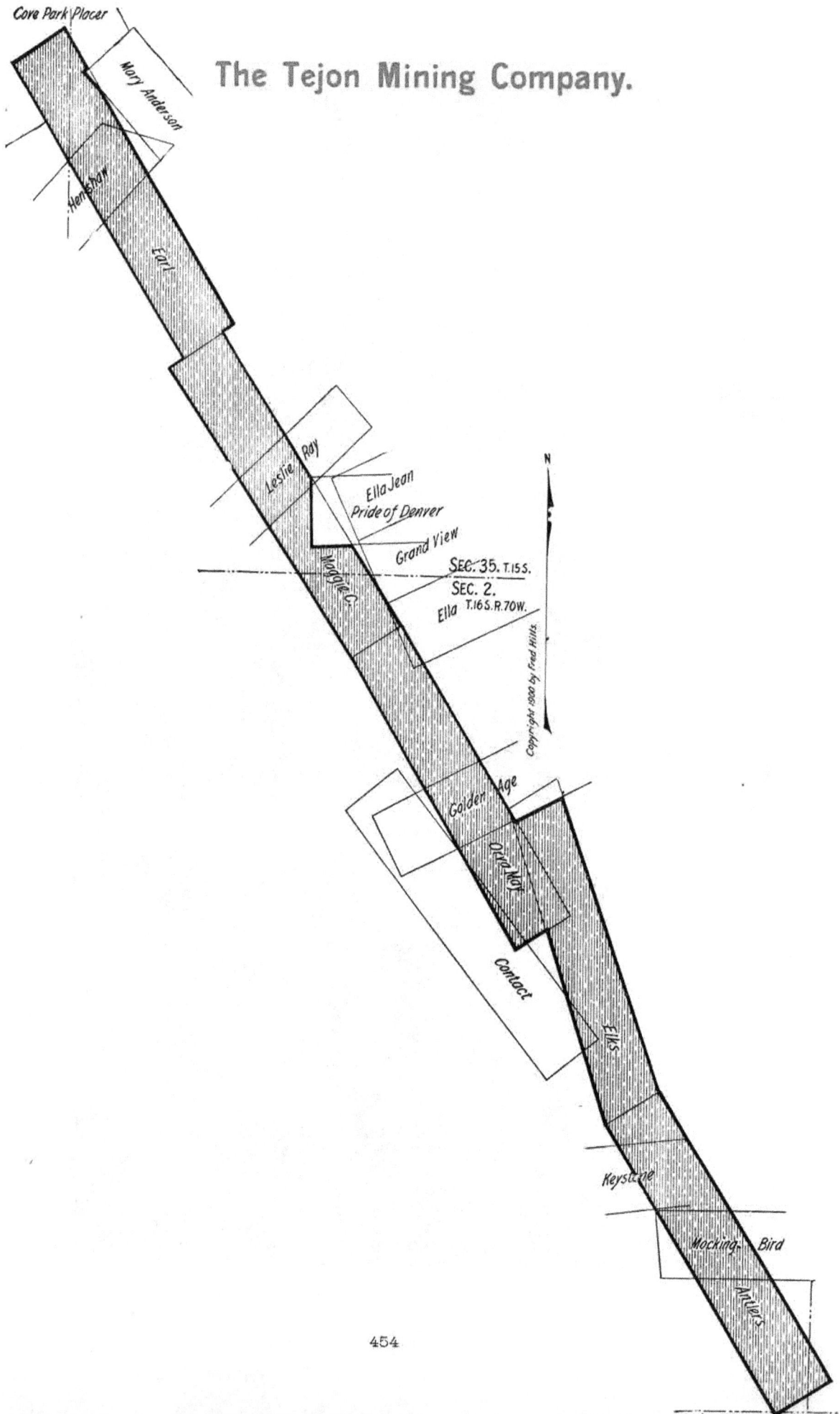

Cove Park Placer

Mary Anderson

Henshaw

Earl

Leslie Ray

Ella Jean
Pride of Denver

Grand View

Maggie C.

SEC. 35. T.15 S.
SEC. 2.
Ella T.16 S. R.70 W.

N

Copyright 1900 by Fred Hills.

Golden Age

Ora May

Contact

Elks

Keystone

Mocking Bird

Antlers

The Tejon Mining Company.

Incorporated under the title of The Grouse Mt. Mining and Milling Company, July 12, 1892. Changed to present title, December 2, 1897.

Theoph. Harrison, president; J. C. Plumb, vice-president; P. C. Dockstader, secretary and treasurer; E. Barnett, Thos. J. Fisher, M. V. Andre. **Directors**

Main Office—Bank Building, Colorado Springs, Colorado.

1,000,000 shares. Par value, $1.00. In treasury January 1, 1900, **Capitalization** 63,000 shares.

Owns the Earl, Orva May, Maggie C., Elks, and the Antlers, em- **Property** bracing in all between 47 and 48 acres, situated on Grouse mountain in the N. E. 1-4 section 2, township 16 south, range 70 west, and in the S. E. 1-4 of section 35, township 15 south, range 70 west, survey No. 10,439. All patented.

The Earl has a 210-foot tunnel; the Maggie C., a 20-foot shaft and **Development** 40-foot open cut; the Orva May, a 40-foot shaft and three trenches; the Elks, an 85-foot shaft; the Antlers, a 20-foot shaft and two trenches. The greater part of the development work is being done on the Earl.

Highest price for stock during 1899, 1 1-4 cents; lowest price for stock during 1899, 1 cent.

The Temomj Mining and Milling Company.

Incorporated.

1,000,000 shares. Par value, $1.00. **Capitalization**

Owns the T. E. M. O. M. J., the Home Fraction, the Baby McKee, and **Property** the Clayton E., containing in all about 27 acres, situated on Gold hill, in section 18. All patented.

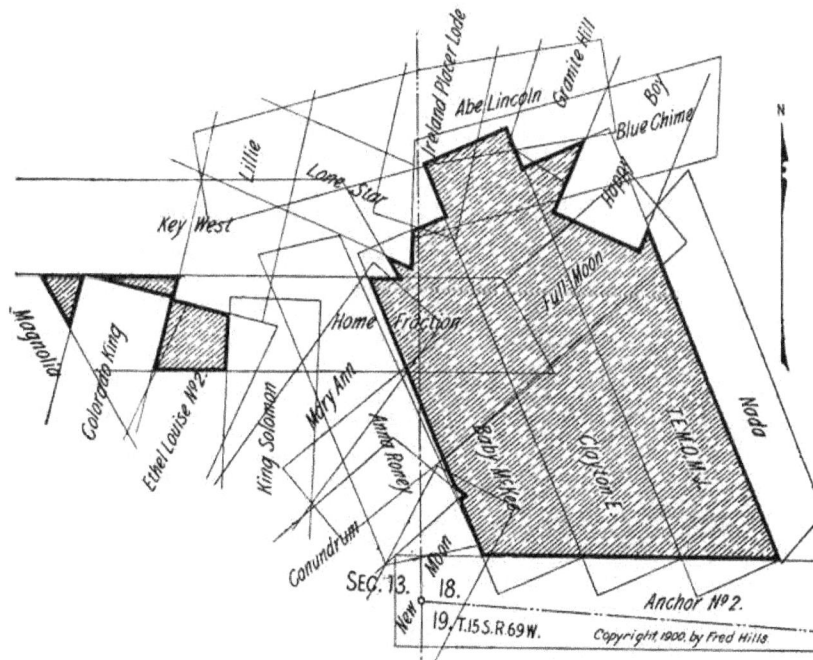

This property has produced about $25,000.00. **Production**

In March, 1900, the company passed into the hands of Mr. W. S. **History** Stratton of Colorado Springs, who now practically owns the property, as well as the stock of the company.

The Tenderfoot Hill Consolidated Mining Company.

Incorporated February, 1900.

Directors

William A. Otis.........................President
C. C. Hamlin......................Vice-President
Oliver H. Shoup...........Secretary and Treasurer
Verner Z. Reed.   Philip B. Stewart.
E. D. Wetmore.

Main Office—The Reed & Hamlin Investment Company, Bank building, Colorado Springs, Colorado.

Capitalization

1,500,000 shares. Par value, $1.00. In treasury March 1, 1900, $50,000.00 cash.

Property

Owns the Lawrence Worden, the Sarah Bell, the Hill Top and the Hill Top Nos. 2 and 3, a total, in all, of 37 acres, formerly being the property of the Alta Mont G. M. Company. Also owns the Deadwood Nos. 1, 2, 3, and 4, containing 23 acres. The Deadwood No. 4 has vein rights through the Peacock. All the above property, a total of over 60 acres, is situated in the S. E. 1-4 of section 7, on Tenderfoot hill. The Alta Mont group is patented. The Deadwood group has receiver's receipt.

Development

On the Alta Mont group there is one shaft 100 feet deep. The company are now prospecting this territory to determine the location of a new shaft. The Deadwood has a shaft 200 feet deep and from 200 to 300 feet of drifting. A deep shaft is now being sunk on the Deadwood No. 3 and already, at a depth of 100 feet, shows a strong phonolite dyke over 10 feet wide. The greater part of the development work is being done on this claim.

On March 1, 1900, the market price of this stock was 15 cents.

The Tenderfoot Hill Consolidated Mining Company.

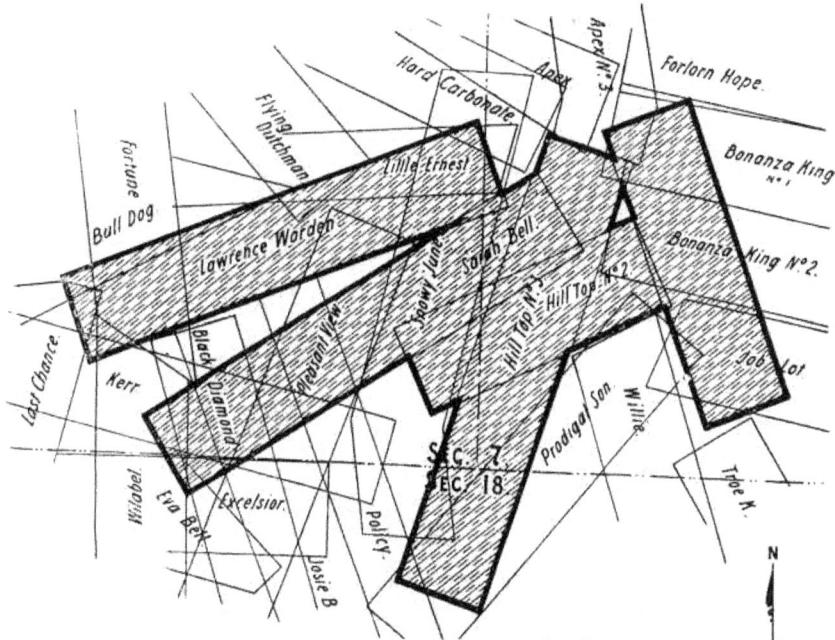

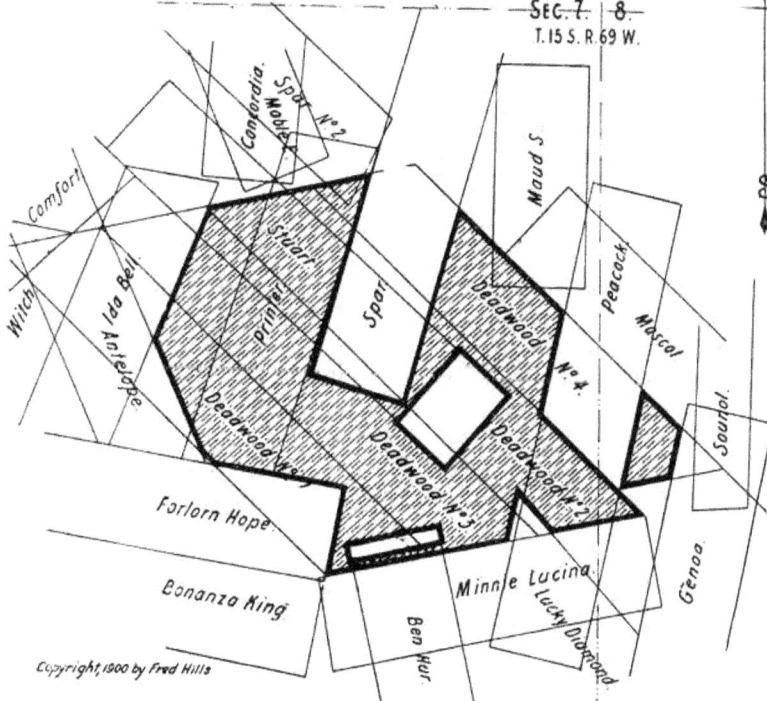

The Teutonic Consolidated Gold Mines Company.

Incorporated 1899.

Directors M. E. O'Bryan, president; A. G. Young, vice-president; Chas. N. Miller, secretary and treasurer; H. E. Hoyt, Zeno Felder.

Main Office—271 Bennett avenue, Cripple Creek, Colorado.

Capitalization 1,250,000 shares. Par value, $1.00. In treasury January 1, 1900, 126,760 shares.

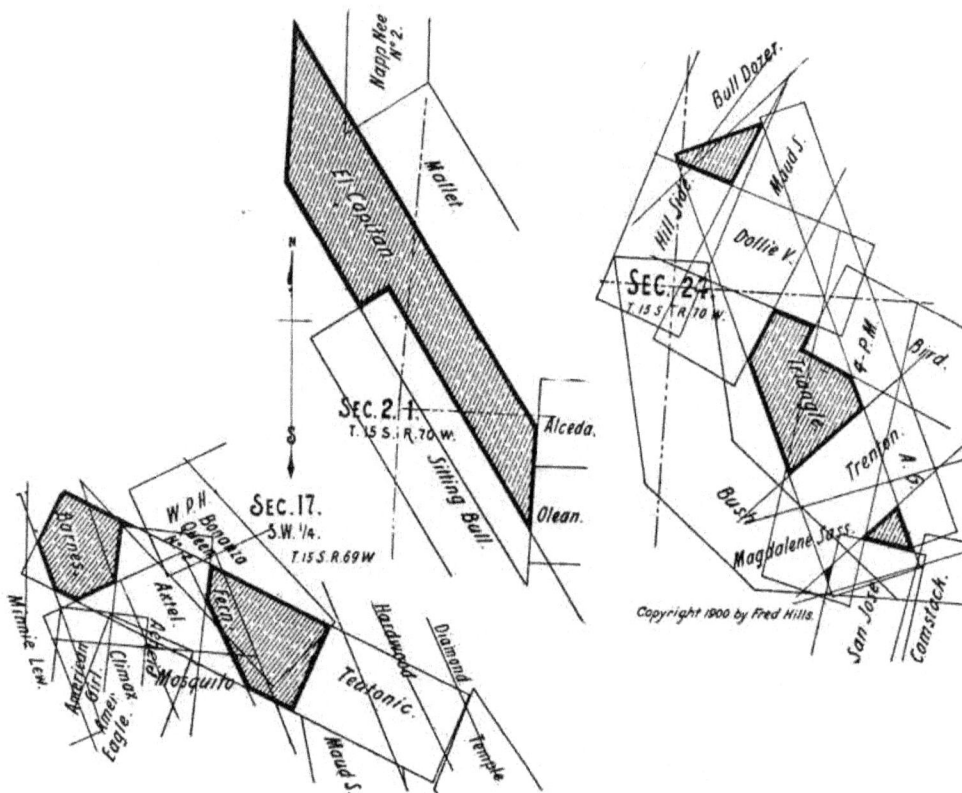

Copyright 1900 by Fred Hills.

Property Owns the Teutonic, in the S. W. 1-4 of section 17, containing 4 1-2 acres, on Bull hill, adjoining the Damon; the Triangle, in the S. E. 1-4 of section 24, on Gold hill, adjoining the El Reno, and containing 4 acres; the Conflict, in the S. E. 1-4 of section 24, containing 7 acres, on Gold hill, adjoining the El Reno; and the El Capitan, in the W. 1-2 of section 1, on Copper mountain, and near the Fluorine, containing 8 acres. The Conflict is not shown on plat. Triangle and El Capitan are patented; Teutonic held by receiver's receipt; Conflict by purchase.

Development On the Teutonic is a shaft house with whim. This claim has three shafts, the main one being 110 feet deep, the others 50 feet and 40 feet deep, respectively. The Triangle has a shaft about 50 feet deep. El Capitan is developed by tunnel. The Teutonic has received the greater portion of the development work.

History The present company is an organization of 1899 and the directors, in view of their well located properties, consider that a shipping mine ought to be opened up on any one of the above claims.

Highest price for stock during 1899, 7 3-4 cents; lowest price for stock during 1899, 5 cents.

The Texas Girl Gold Mining Company.

incorporated October 10, 1899.

J. M. Auld............................President Directors

L. L. Aitken........................Vice-President

E. C. Sharer...............Secretary and Treasurer

    J. R. McKinnie.   W. W. Williamson.

Main Office—25 E. Pike's Peak avenue, Colorado Springs, Colorado.

1,500,000 shares. Par value, $1.00. In treasury January 1, 1900, **Capitalization**
$6,500 cash.

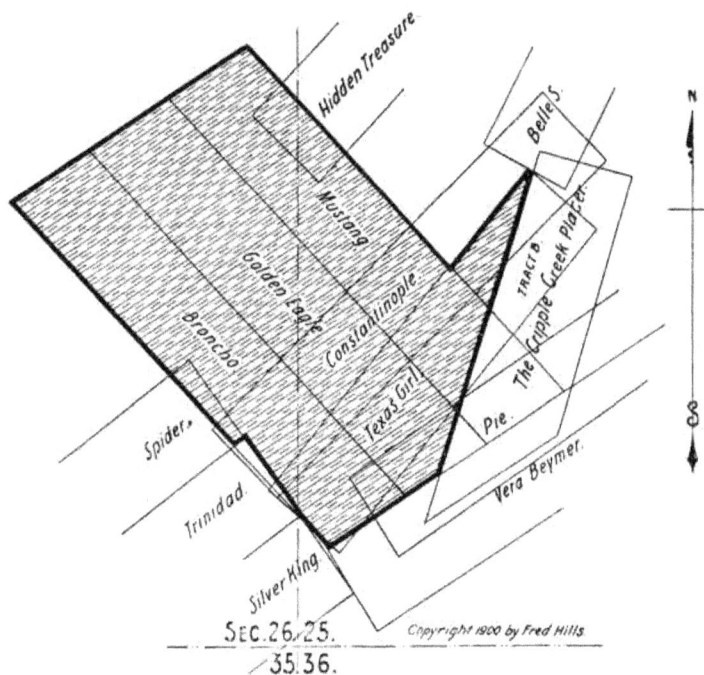

Owns the Texas Girl, 6 acres; the Golden Eagle, 6 1-2 acres; the Mus- **Property**
tang, 4 acres, and the Broncho, 6 1-2 acres, a total, in all, of 23 acres, situated in the S. W. 1-4 of section 25, on Beacon hill. Patented.

Ninety feet of tunnel work has been done on the Texas Girl. Other **Development**
claims in the group have location work only. The property is now being
opened up from a joint shaft that is now being sunk by the Texas Girl
and the Banner Gold Companies and is down 140 feet.

Highest price for stock during 1899, 2 7-8 cents; lowest price for
stock during 1899, 1 5-8 cents.

The Theresa Gold Mining Company.

Incorporated 1895.

Directors E. A. Colburn, president; M. B. Irvine, vice-president; C. H. Dudley, secretary and treasurer; W. R. Barnes, Frank Cotten.

Main Office—14 N. Nevada avenue, Colorado Springs, Colorado.

Capitalization 1,250,000 shares. Par value, $1.00. In treasury January 1, 1900, 155,000 shares.

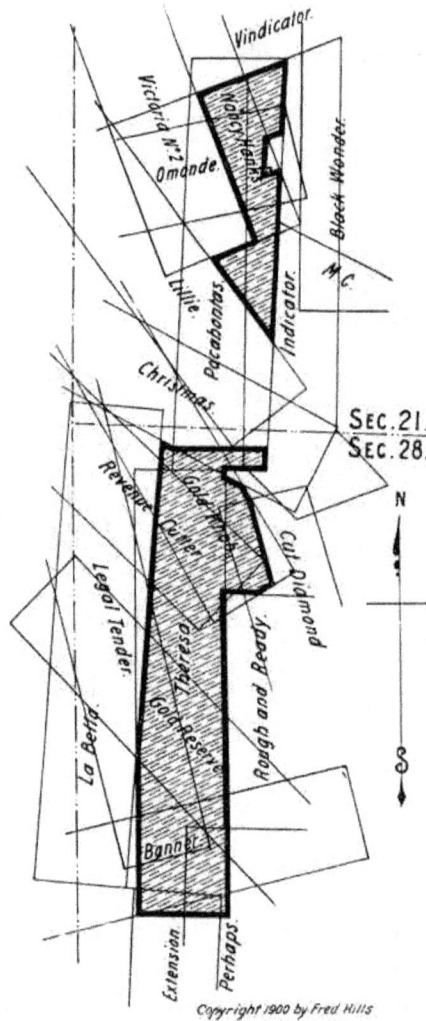

Copyright 1900 by Fred Hills

Property Owns the Theresa and part of Gold Knob, in the N. W. 1-4 section 28, and S. E. 1-4 section 20, and the Pocahontas, in the S. W. 1-4 section 21, all on Bull hill. Patented. About 15 acres near the Legal Tender mine.

Development One blacksmith shop; one 100 H. P. boiler; one compressor; one 10x12 hoist and double-compartment shaft with cage. The main shaft is 400 feet deep; there is also a shaft 150 feet deep, one 100 feet deep, and one 50 feet deep. The company has been shipping more or less for five years past; closed down January 1, 1900. Part of the property has been leased.

Highest price for stock during 1899, 9 1-4 cents; lowest price for stock during 1899, 7 3-4 cents.

The Three H. Gold Mines Company.

Incorporated January, 1896.

Dennis Murto, president; S. A. Osborn, vice-president; A. W. Vandeman, secretary and treasurer; E. L. Shannon, H. L. Ritter. **Directors**

Main Office—No. 607 Ernest & Cranmer building, Denver, Colorado.

2,000,000 shares. Par value, $1.00. In treasury January 1, 1900, **Capitalization**
120,000 shares.

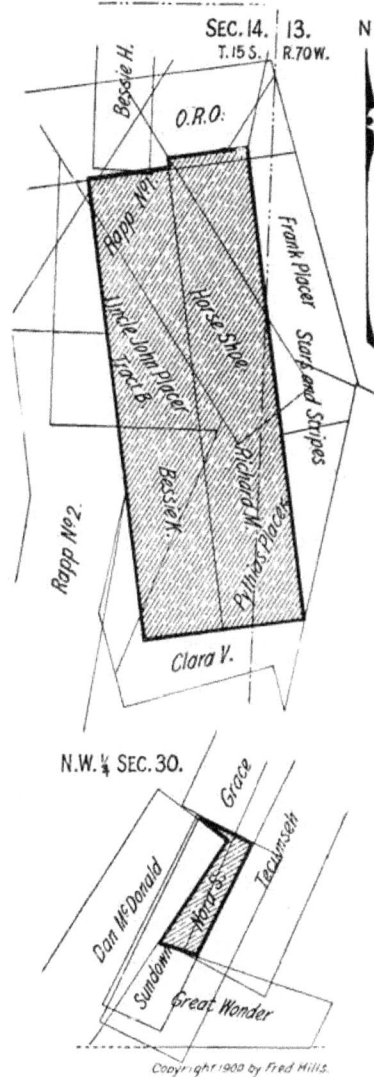

Copyright 1900 by Fred Hills.

Owns the Bessie K. and the Richard M. lodes, containing 15 acres, situated on Mineral hill in the N. E. 1-4 section 14; also the Sundown lode, containing two acres, situated on Raven hill, in the N. W. 1-4 section 30. All patented. **Property**

On the Sundown there is a shaft 200 feet deep, and drifting has been commenced. On this claim, which is leased to the Sundown Development Company, the greater part of the development work is being done. The Bessie K. and the Richard M. lodes, being within the corporate limits of Cripple Creek, the surface of same has been platted into town lots, from which the company expects to realize $10,000 or more, all mineral rights being reserved when lots are sold. **Development**

Highest price for stock during 1899, 2 1-8 cents; lowest price for stock during 1899, 1 5-8 cents.

461

The Titan Consolidated Gold Mining Company.

Incorporated July 15, 1896.

Directors

H. V. Wandell, president; C. E. Minier, vice - president; James Earengey, secretary and treasurer.

Main Office—No. 64 Hagerman Building, Colorado Springs, Colorado.

Capitalization

1,500,000 shares. Par value, $1.00. In treasury January 1, 1900, 400,000 shares.

Property

Owns the Amethyst, Mamie, Sunnyside, Great Western, Great Western No. 2, Bolivar, and the Eclipse, in all 46.897 acres, in the W. 1-2 section 7, township 16 south, range 69 west, in Pot gulch.

Receiver's receipt held for all the above property.

Sufficient development work to obtain a patent has been done.

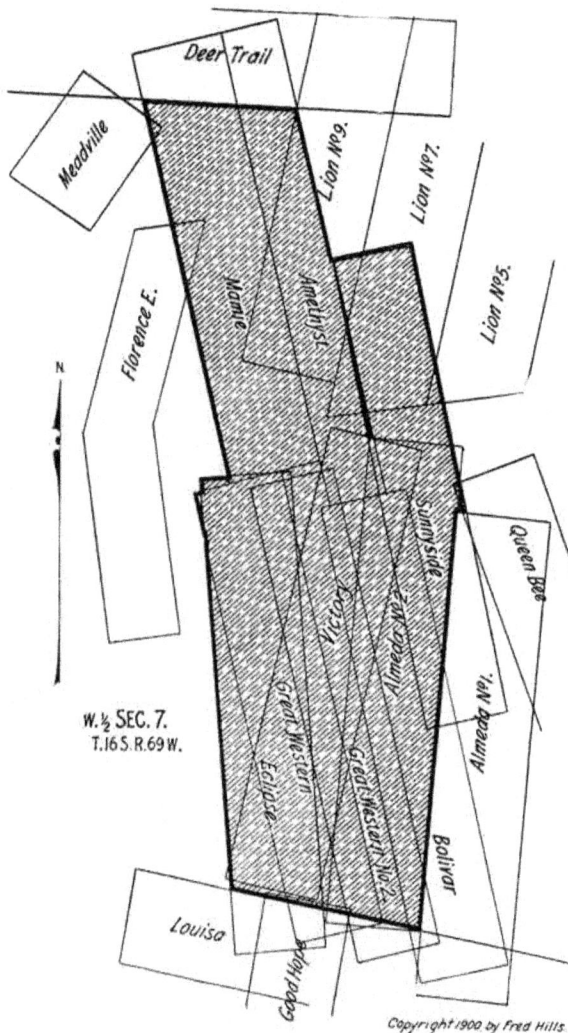

W. ½ SEC. 7.
T. 16 S. R. 69 W.

Copyright 1900 by Fred Hills.

The Tornado Gold Mining Company.

Incorporated April 1, 1895.

J. W. Graham, President; Sherwood Aldrich, Vice-President; E. P. Shove, Secretary and Treasurer; C. J. Cooper.

Directors

Main Office—No. 9 S. Tejon street, Colorado Springs, Colorado.

1,250,000 shares. Par value, $1.00. In treasury January 1, 1900, 100,000 shares; in treasury January 1, 1900, $30,000.00 cash.

Capitalization

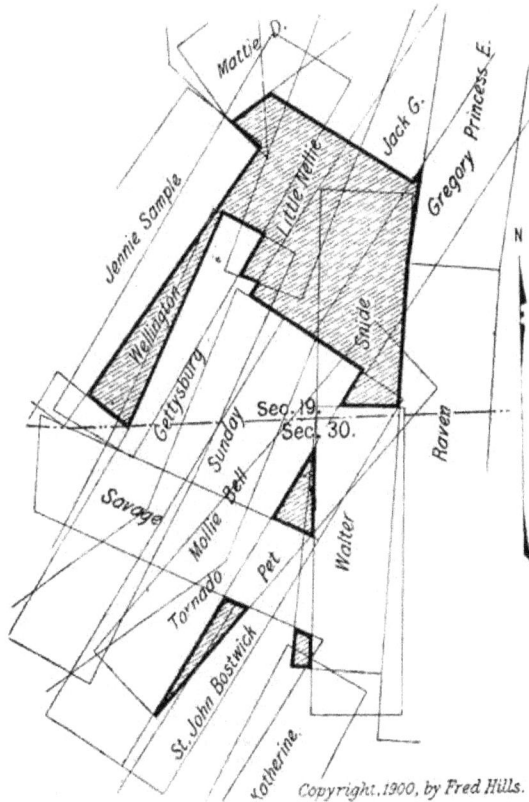

Copyright, 1900, by Fred Hills.

Property

Owns parts of the Tornado, Snide, Pet, Gregory, Jack G., Little Nellie, and the Wellington claims, all situated in sections 30 and 19, on Raven hill.

The total acreage is about 14 acres. All patented. In April, 1900, the consolidation of this company and the Raven with the Elkton was agreed upon, as will be seen by referring to the Elkton Cons. M. & M. Co., on another page.

Development

There is a large shaft house with engines and hoisting plant, good for 1,500 feet, a Norwalk air-compressor for six drills, ore bins, and full surface equipment. The main shaft is 800 feet deep with levels at about 100 foot intervals; on all the claims over 1,800 feet of drifting, discovery shafts, and shallow workings. The greater part of the development work is being done on the Snide.

History

The property was operated entirely by the company to a depth of over 500 feet. It showed scattered ore bodies of good grade, and but a small extent below that depth, ore bodies of liberal size and very high grade were encountered. Shipments, averaging over $150 per ton, began September, 1899.

Production and Dividend

Gross production to January 1, 1900, $75,000.00; net value of ore mined during 1899, $60,000.00; net profits on ore mined during 1899, $30,000.00.

Highest price for stock during 1899, 51 cents; lowest price for stock during 1899, 20 cents.

The Touraine Gold Mining Company.

Incorporated 1899.

Main Office—Care Wm. P. Bonbright & Company, Colorado Springs, Colorado.

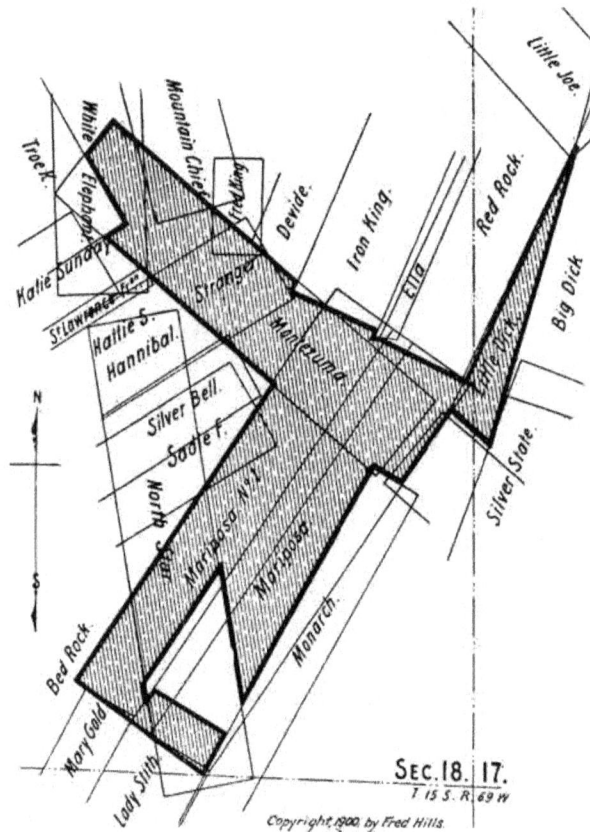

SEC. 18. | 17.
T 15 S. R. 69 W
Copyright 1900 by Fred Hills.

Capitalization

1,250,000 shares. Par value, $1.00.

History

Up to March of this present year this company owned the group of property shown herewith, in section 18, but at that time sold all its holdings to C. W. Kurie, trustee, and is distributing the net proceeds in dividends among the stockholders.

The Trachyte Gold Mining and Milling Company.

Incorporated August 10, 1893.

Warren Woods..........................President   Directors

F. E. Brooks........................Vice-President

F. M. Woods........Secretary and General Manager

H. E. Woods..........................Treasurer

W. C. Frost.................Assistant Secretary

Main Office—Giddings building, Colorado Springs, Colorado.

1,500,000 shares. Par value, $1.00. In treasury January 1, 1900,   Capitalization
$852.89 cash.

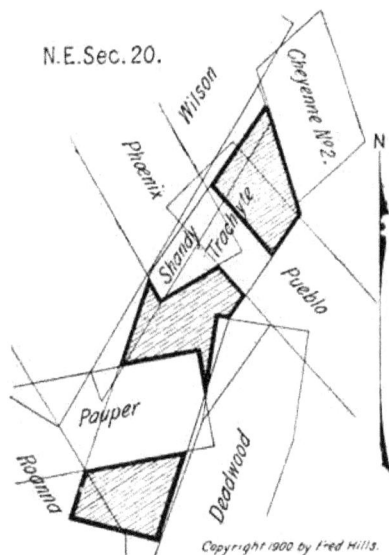

N.E.Sec.20.

Copyright 1900 by Fred Hills.

Owns the Trachyte, on Bull hill, in the N. E. 1-4 of section 20, con-   Property
taining 5 acres, patented. The company also owns 60 per cent. of the
stock of the Compromise G. M. Company, which latter company was
formed on the basis of the settlement of conflicting claims between the
Trachyte and the New Zealand Gold Mining Companies.

The Trachyte property is being worked entirely by lessees and is sur-   Development
rounded by mining companies which are producing ore. Great expectations are held that it will soon become a producer.

Highest price for stock during 1899, 9 cents; lowest price for stock
during 1899, 4 cents.

The Trenton Gold Mining Company.

Incorporated December 16, 1895

Directors

Clarence Edsall.........................President
Geo. R. Buckman....................Vice-President
T. B. Tiffany...............Secretary and Treasurer
O. C. Townsend.   B. N. Beal.

Main Office—31 and 32 Giddings building, Colorado Springs, Colorado.

Capitalization   1,500,000 shares. Par value, $1.00. In treasury January 1, 1900, 256,500 shares.

Property   Owns the Trenton and Bird claims, consisting of 9.5 acres, situated in section 24, on Gold hill. The property is practically patented. There has been $1,000.00 expended on the property in shafts and cuts.

Highest price for stock during 1899, 2 cents; lowest price for stock during 1899, 1 cent.

The Twin Sisters Mining and Milling Company.

Incorporated October 9, 1899.

A. P. Mackey, president; Chas. Farnsworth, vice-president; Wm. P. Sargeant, secretary and treasurer; C. E. Titus, assistant secretary; B. B. Brown, J. C. Connor.

Directors

Main Office—Wm. A. Otis & Company, Giddings block, Colorado Springs, Colorado.

1,500,000 shares. Par value, $1.00. In treasury January 1, 1900, 200,000 shares. **Capitalization**

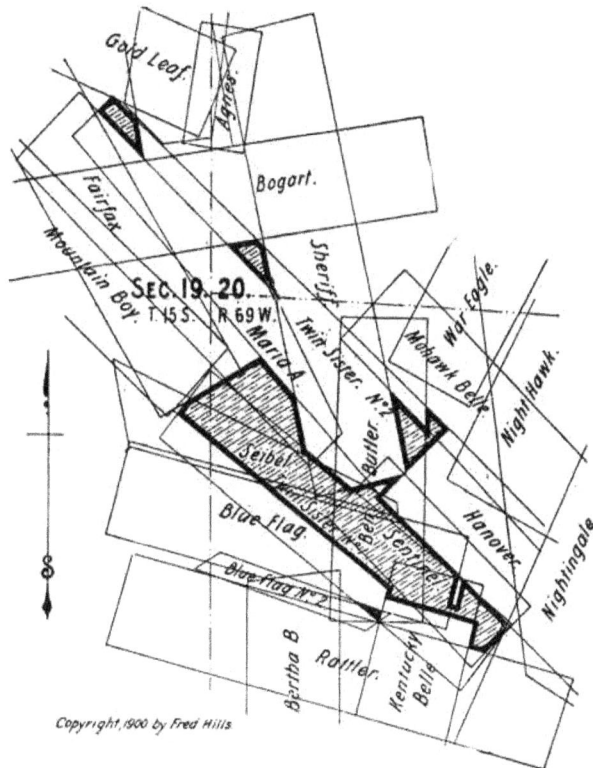

Copyright, 1900 by Fred Hills

Owns the Twin Sister Nos. 1 and 2, and the Sentinel, containing 8 acres in all, situated on Bull hill, in the S. W. 1-4 of section 20, Survey No. 7,917. This includes the following property, acquired by deed from the respective companies: .883 acre of the Kentucky Belle, .207 acre of the Sheriff, .593 acre of the Maria A., and the vein rights from the Seibel and the Blue Flag. Property is all patented. **Property**

There are two shafts on the Twin Sisters, one of 60 feet, and one of 65 feet; also 150 feet of drifting. At present, drifting is being done on a vein from 2 to 3 feet wide, recently struck. Assays average from 2 to 6 ounces. The mine gives indications of soon being a regular shipper. **Development**

Gross production to May 1, 1900, 48 tons, of a value of $2,250.00. **Production**

Highest price for stock to May 1, 1900, 9 cents; lowest price for stock to May 1, 1900, 6 cents.

The Tycoon Gold Mining Company.

Incorporated January, 1896.

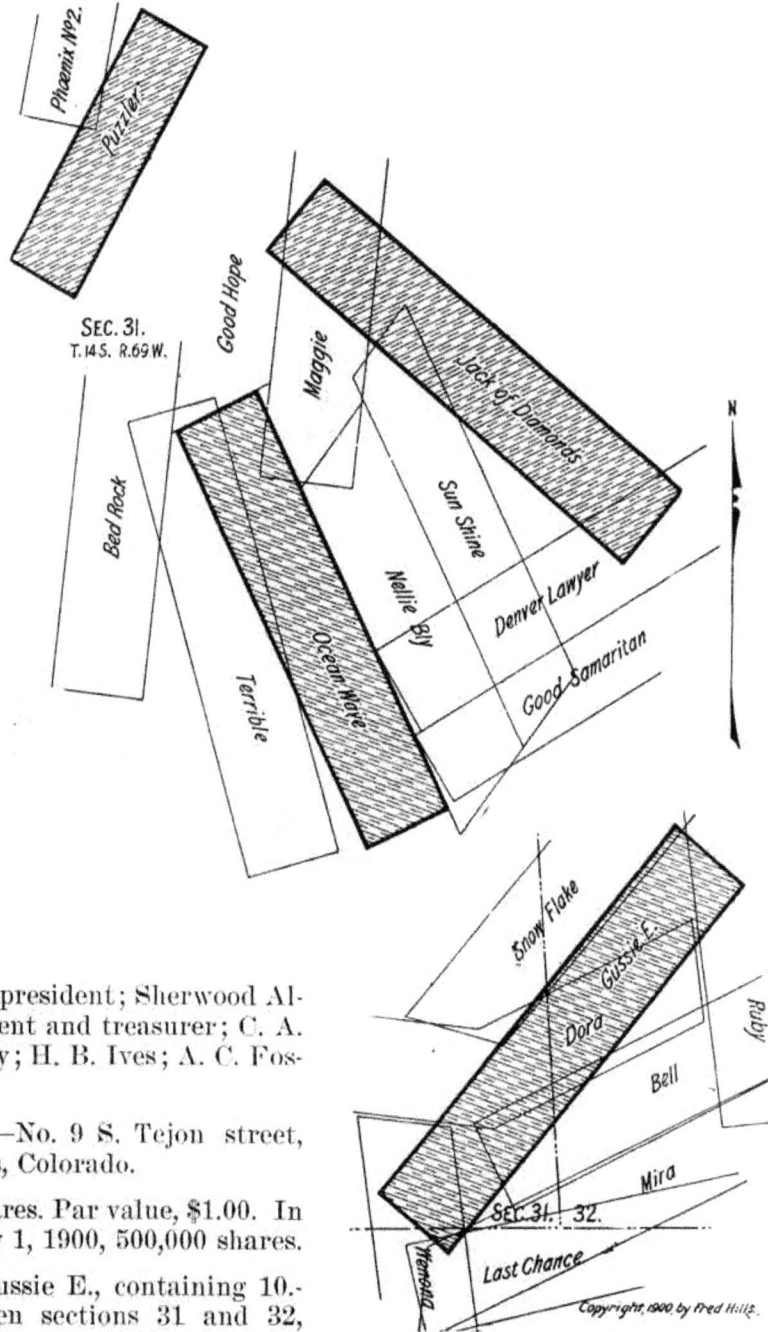

SEC. 31.
T. 14 S. R. 69 W.

Copyright, 1900, by Fred Hills.

Directors   E. P. Shove, president; Sherwood Aldrich, vice-president and treasurer; C. A. Ralston, secretary; H. B. Ives; A. C. Foster.

Main Office—No. 9 S. Tejon street, Colorado Springs, Colorado.

Capitalization   1,500,000 shares. Par value, $1.00. In treasury January 1, 1900, 500,000 shares.

Property   Owns the Gussie E., containing 10.-330 acres between sections 31 and 32, township 14 south, range 69 west; the Ocean Wave, containing 9.761 acres, in the S. E. 1-4 of section 31, township 14 south, range 69 west; the Jack of Diamonds, containing 10.331 acres, in the N. E. 1-4 of section 31, township 14 south, range 69 west, and the Puzzler, containing 5.165 acres, in the N. W. 1-4 of section 31, township 14 south, range 69 west. Total acreage owned by the company is 35.587 acres. The property is all patented.

Development   On each of the above claims there are shafts ranging in depth from 50 to 100 feet. At present active work has ceased. No sales of stock have been reported.

The Uncle Sam Mining and Milling Company.

Incorporated October 10, 1892.

J. Bateman, president; J. W. Ogden, vice-president; C. S. Bancroft, secretary; A. England, treasurer; E. S. Bach, K. Macdermid, Robert Beers.

Main Office—26 Midland block, Colorado Springs, Colorado.

1,000,000 shares. Par value, $1.00. In treasury January 1, 1900, 67,095 shares; in treasury January 1, 1900, $125.00 cash.

Owns the Little Leota and the Alta, containing 15,750 acres, situated on Bull hill, in the N. W. 1-4 of section 21, township 15 south, range 69 west, also the Priscilla, containing 8.500 acres in the N. W. 1-4 of section 5, township 15 south, range 69 west, on Lincoln hill. All patented.

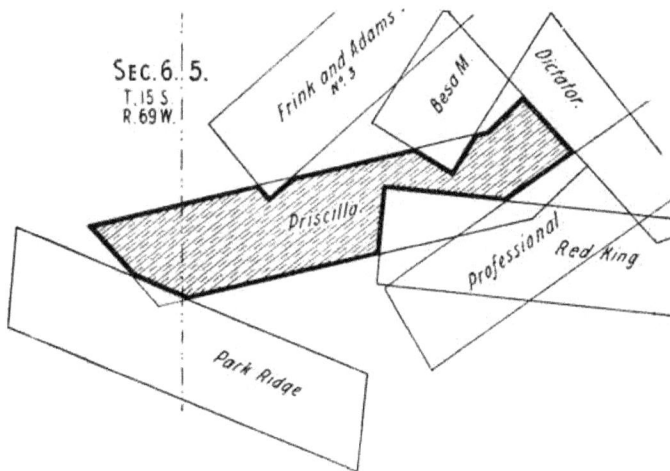

One 12-horse power gasoline hoist, engine house and blacksmith shop, gallows frame and all necessary tools for working the property. Drifting was first done from the Alta shaft to a distance of 40 feet. On a cross-cut a run was struck and 80 feet more of drifting was done. The shaft on this claim is down 165 feet and active work is now being done to sink it to a depth of 200 feet. The Little Leota shaft is 100 feet deep and is being sunk deeper. On this claim there is about 250 feet of cross-cutting. These two claims are receiving the greater part of the development work. When the sinking of the shafts of both claims is completed the company is confident the mine will produce good shipping ore.

Highest price for stock during 1899, 7 1-2 cents; lowest price for stock during 1899, 3 cents.

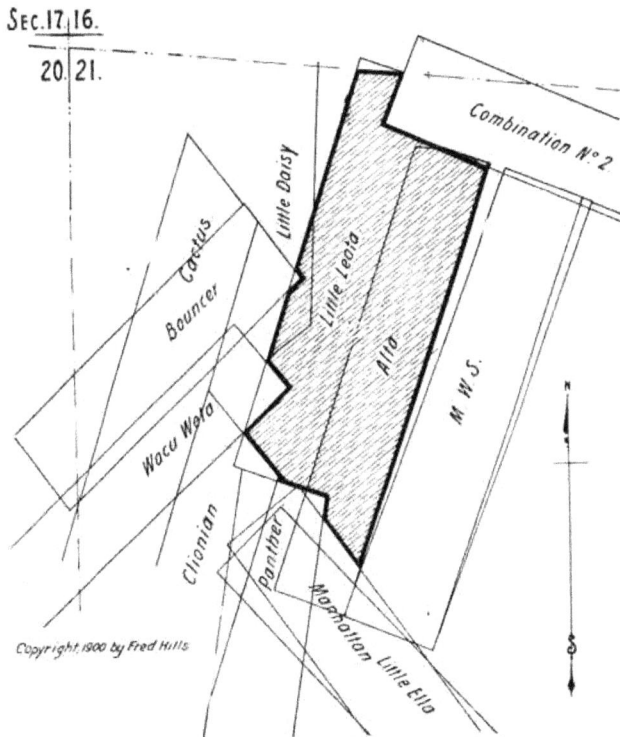

Copyright 1900 by Fred Hills

The Union Belle Gold Mining Company.

Incorporated January 2, 1896.

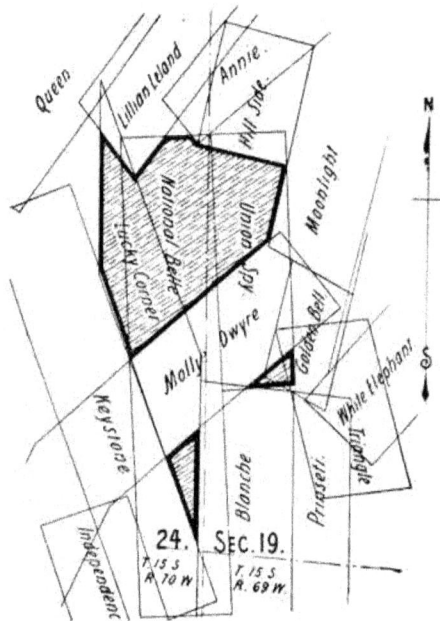

Main Office—Colorado Springs, Colorado.

Capitalization   1,500,000 shares. Par value, $1.00. In treasury January 1, 1900, 27,854 shares.

Property   Owns the National Belle, the Union Spy and the Lucky Corner, 9 acres in all, situated between sections 19 and 24, on Gold hill. All the above property is patented.

No stock has been placed on the market.

The Union Gold Mining Company.

Incorporated June 16, 1892.

J. A. Sill, president; A. S. Holbrook, vice-president; C. H. Morse, secretary and treasurer; John Dern, E. R. Stark, C. H. White, J. F. Burns, R. Clough.   Directors

Main Office—Colorado Springs, Colorado.

1,250,000 shares. Par value, $1.00. In treasury January 1, 1900,   Capitalization
$12,000.00 cash. No stock in treasury.

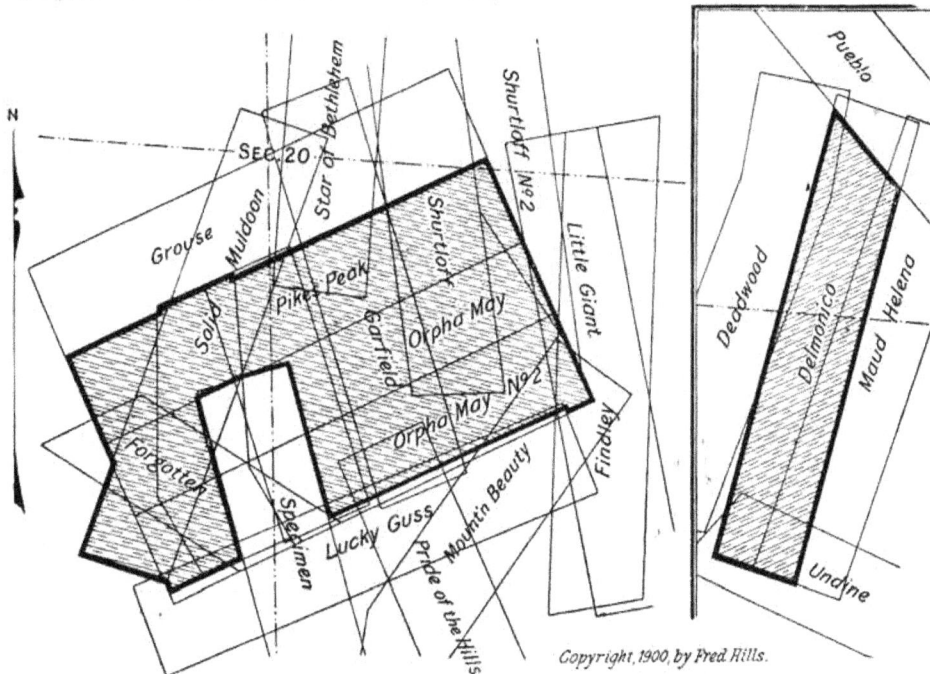

Copyright, 1900, by Fred Hills.

Owns Orpha May Nos. 1 and 2, Delmonico, Pike's Peak, Solid Mul-   Property
doon, Forgotten, and Adopted, containing 34 1-2 acres in all, located in
section 20, south slope of Bull hill. All patented except Solid Muldoon,
Forgotten, and Adopted. In April, 1900, the control of this company was
sold to John Dern of Mercur, Utah, for $350,000.

The main workings of Orpha May are equipped with a 75 H. P.   Development
hoist, also with air drills, and lighted by electricity. On the east end of
the Pike's Peak there is a 25 H. P. plant, and on the west end of this property there is another plant of the same capacity. The workings on the east
end of the Orpha May No. 2 are equipped with a 30 H. P. electric hoist.
Orpha May No. 2 has one shaft 900 feet; Pike's Peak shaft, 400 feet; Por-
cupine shaft, 300 feet; Seahorn shaft, 425 feet; Delmonico, 150 feet.
There are several other shafts on the property, ranging from 50 to 150
feet, also several thousand feet of drifting and cross-cutting.

Production to January 1, 1900, $1,000,000.00; dividends to January
1, 1900, $82,744.00; last dividend paid June 25, 1896, $11,669.07.

Highest price for stock during 1899, 37 3-4 cents; lowest price for
stock during 1899, 19 3-4 cents.

The Venus Gold Mining Company.

Incorporated November, 1895.

Directors

Chas. F. Potter........................President

D. W. Greene......................Vice-President

Horace H. Mitchell.........Secretary and Treasurer

Main Office—No. 45 Bank block, Colorado Springs, Colorado.

Capitalization

1,250,000 shares. Par value, $1.00. In treasury January 1, 1900, 40,000 shares.

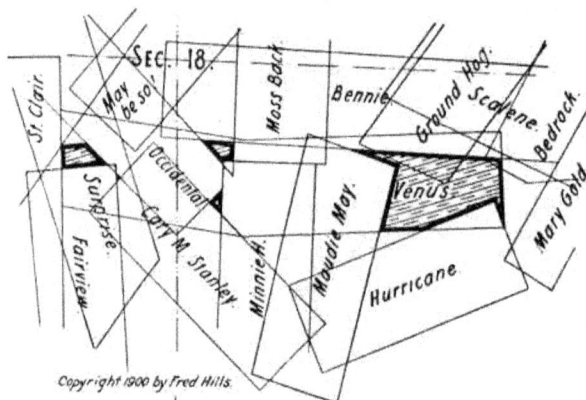

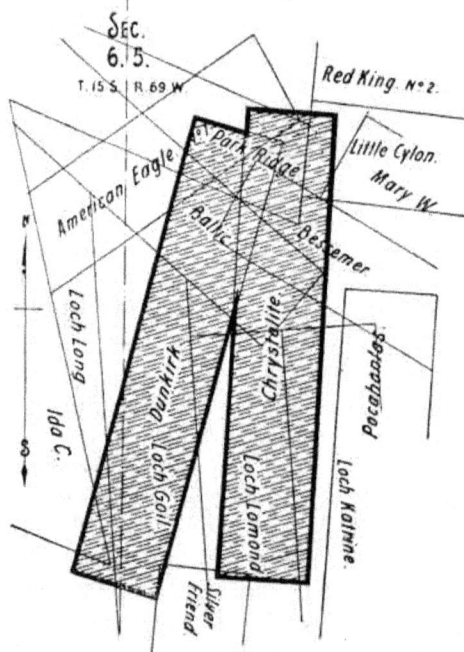

Copyright 1900 by Fred Hills.

Property

Owns a 3-4 interest in the Venus, containing 3 1-2 acres, on Globe hill, in center of section 18; the Dunkirk and Chrystolite, about 18 1-2 acres, on Tenderfoot hill, W. 1-2 section 5. All patented.

The Victor Gold Mining Company.

Incorporated February 16, 1893.

W. H. Brevoot.........................President

Eben Smith ....Vice-President

R. H. Reid...........................Secretary

G. E. Ross-Lewin......................Treasurer

Jas. A. McClurg.

Directors

Main Office—No. 822 Equitable building, Denver, Colorado.

200,000 shares.   Par value, $5.00.

Capitalization

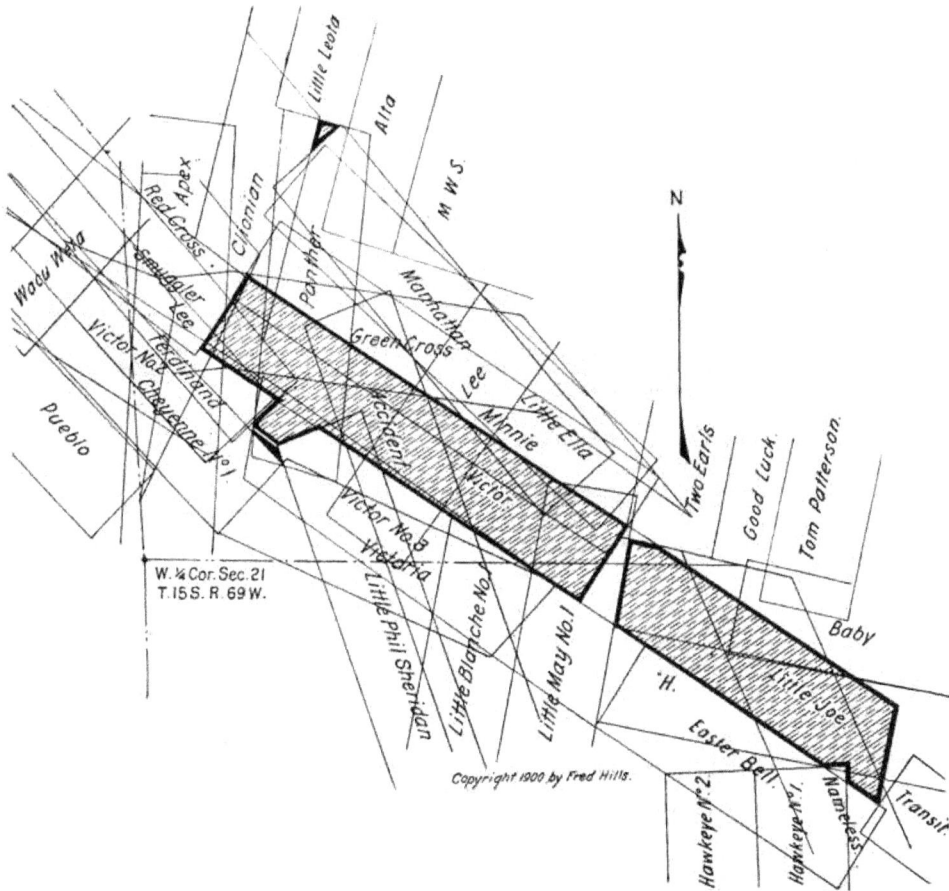

Owns the Victor, 10 acres; the Little Joe, 7 acres; the Panther, 1-2 acre, and the Victor No. 2, 1-5 acre—all situated on Bull hill in the N. W. 1-4 section 21. All patented.

Property

There is a complete equipment for hoisting 300 tons of ore per day. Several veins have been disclosed; all are being worked.

Development

Gross production to January 1, 1900, $2,216,670.95; amount of dividends to January 1, 1900, $1,155,000.00; last dividend, paid December 1, 1898, was for $100,000.00.

Production and Dividend

The Victor Mines and Land Company.

Incorporated.

Directors
Warren Woods.........................President
H. E. Woods........................Vice-President
F. M. Woods...............Secretary and Treasurer
C. L. Arzeno.   J. M. Allen.

Main Office—Victor, Colorado.

Capitalization   600,000 shares.   Par value, $1.00.   In treasury January 1, 1900, $5,000.00 cash.

Property   Owns a 1-4 interest in the New Port, and a 3-4 interest in the Menona; also owns the Long John, and a 2-5 interest in the Eva L.   All situated in the S. 1-2 section 17.

Development   The Menona has a tunnel 400 feet long.   On the Long John is a shaft 50 feet deep and an incline of 40 feet from the bottom.

There has been small production of ore by the lessees.

Highest price for stock during 1899, 20 cents; lowest price for stock during 1899, 12 1-2 cents.

The Vindicator Consolidated Gold Mining Company.

Incorporated December, 1896.

F. L. Sigel, president; G. S. Wood, vice-president; F. J. Campbell, Directors secretary and general manager; A. J. Zang, treasurer; P. J. Friederich.

Main Office—1424 Sixteenth street, Denver, Colorado.

1,500,000 shares. Par value, $1.00. In treasury January 1, 1900, Capitalization 435,000 shares; in treasury January 1, 1900, $86,957.42 cash.

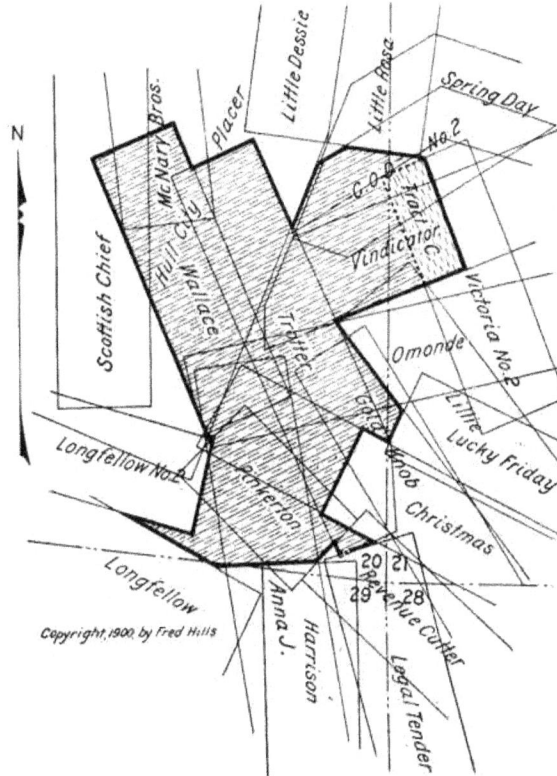

Copyright, 1900, by Fred Hills

Owns the west 692 feet of the Vindicator and the C. O. D. No. 2; Property the west fraction of the Gold Knob; the Wallace, Trotter, Omonde; the west 450 feet of the Christmas; and all that part of the Pinkerton, Propolite and Anna J. north of the north end line of the Harrison claim, in sections 20 and 28, on Bull hill; all patented except the Anna J., Pinkerton and Propolite, for which receiver's receipt is held. Total acreage is about 28 acres. Tract "C" is in controversy with the Keystone Mining and Milling Company.

The improvements consist of shaft No. 1, 820 feet deep, and shaft Development No. 2, 407 feet deep, besides several others worked by lessees. The property is equipped with complete plants costing $75,000. One 13x25 and 7x24 compound condensing Janesville station pump, geared sinking hoist 6x8, and sinking pumps. There are approximately five miles of underground workings. On the No. 1 shaft a new ore house has just been completed at a cost of $12,000.00, besides water tanks and an electric light plant. A new 264-horse power Babcock and Wilcox boiler has been added, and an order has been placed for a 20x48 Corliss hoisting engine. The company has also an interest in mill for treating ore.

(CONTINUED ON PAGE 476.)

Production
and Dividend

Production to January 1, 1900, $2,000,000.00; production for the year 1899, $607,778.08; total dividends to January 1, 1900, $357,750.50; dividend paid January 25, 1900, $53,250.00; last dividend paid April 25, 1900, $53,250.00; total dividends paid to April 25, 1900, $411,000.00.

Highest price for stock during 1899, $1.74 1-2; lowest price for stock during 1899, 92 cents.

The Virginia M. Consolidated Mining Company.

Incorporated February 20, 1892.

Directors

E. M. De La Vergne.....................President
Chas. E. Noble......................Vice-President
S. N. Nye...............................Treasurer
H. F. Noble...........................Secretary
    M. F. Stark.   W. G. Newman.

Main Office—No. 49 Hagerman building, Colorado Springs, Colorado.

Capitalization

1,000,000 shares. Par value, $1.00. In treasury January 1, 1900, $900.00 cash.

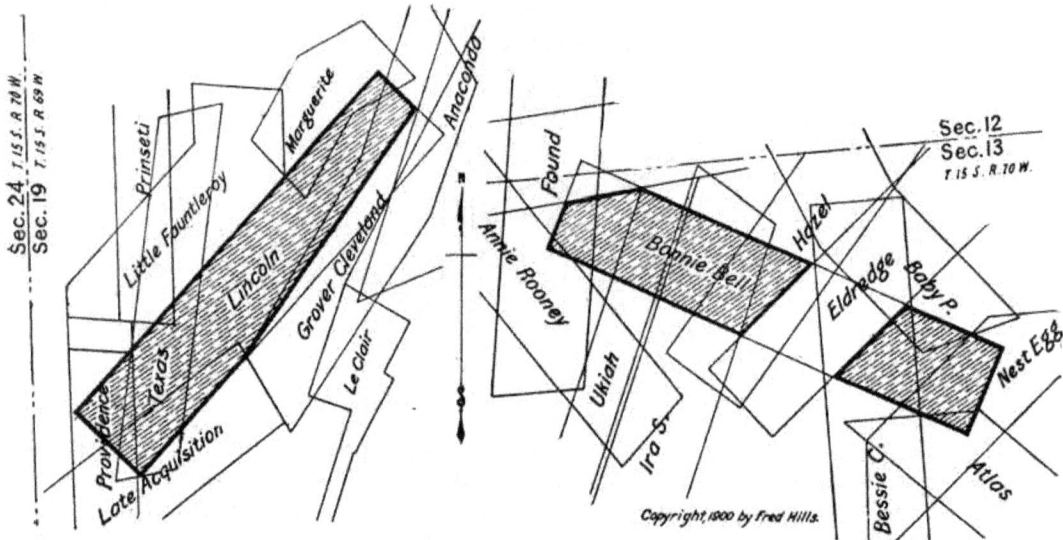

Property

Owns the Lincoln, 9 1-2 acres, situated in the S. E. 1-4 section 19, on Gold hill; and a one-half interest in the Bonnie Bell, 8 acres, situated in the N. E. 1-4 section 13, just north of the town of Cripple Creek. The above property is patented.

Development

There is a shaft house and hoist on the property. One shaft 225 feet deep, with about 1,000 feet of drifts; one two-compartment shaft is now down 60 feet; it is to be sunk 400 feet. Several prospect shafts representing about 400 feet of work in all. The property has been leased practically the whole time. At the present time the work is being pushed by the lessees.

Highest price for stock during 1899, 7 cents; lowest price for stock during 1899, 2 1-2 cents.

The Volcano Gold Mining Company.

Incorporated September 11, 1899.

W. R. Foley.............................President

Dan Weyand......................Vice-President

S. J. Mattocks............Secretary and Treasurer

Directors

Main Office—No. 104 Pike's Peak avenue, Colorado Springs, Colorado.

1,250,000 shares. Par value, $1.00. In treasury January 1, 1900, 200,000 shares; in treasury January 1, 1900, $1,000.00 cash. Capitalization

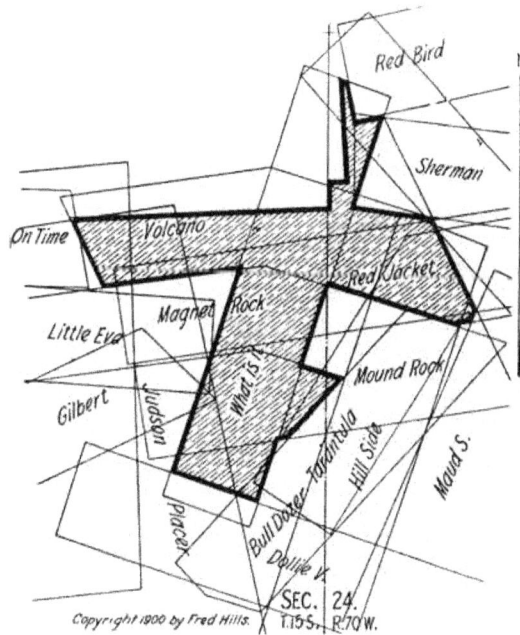

Owns the Mound Rock, the Red Jacket, the Volcano and the What Is It, all in one group, on Gold hill, in N. 1-2 of section 24, a total of 15 acres. Receiver's receipt issued for Volcano and the What Is It claims. Mound Rock and Red Jacket in process of patenting. Property

Full plant of machinery, steam hoist, good buildings, air drills. Machinery cost about $8,000.00. Underground work operated by air drills. The Volcano has a shaft 200 feet deep. The greater part of the development work is being done on this claim. The main shaft shows good values at the first level at a depth of 100 feet. The vein uniformly assays from $50 to $75, and varies in width from 8 inches to 3 feet. The same vein opened up in the second level, shows slightly better values. The vein is being drifted upon and improves in value as depth is gained. Ore could be shipped that would run two ounces. No ore has been shipped. About 25 tons were ready for shipment in April, 1900. Development

Highest price for stock during 1899, 13 1-4 cents; lowest price for stock during 1899, 5 cents.

The Wabash Gold Mining Company.

Incorporated 1896.

Directors

Wesley Gourley........................President
W. H. McGuire.....................Vice-President
W. C. Calhoun.............Secretary and Treasurer

Main Office—No. 633 Cooper building, Denver, Colorado.

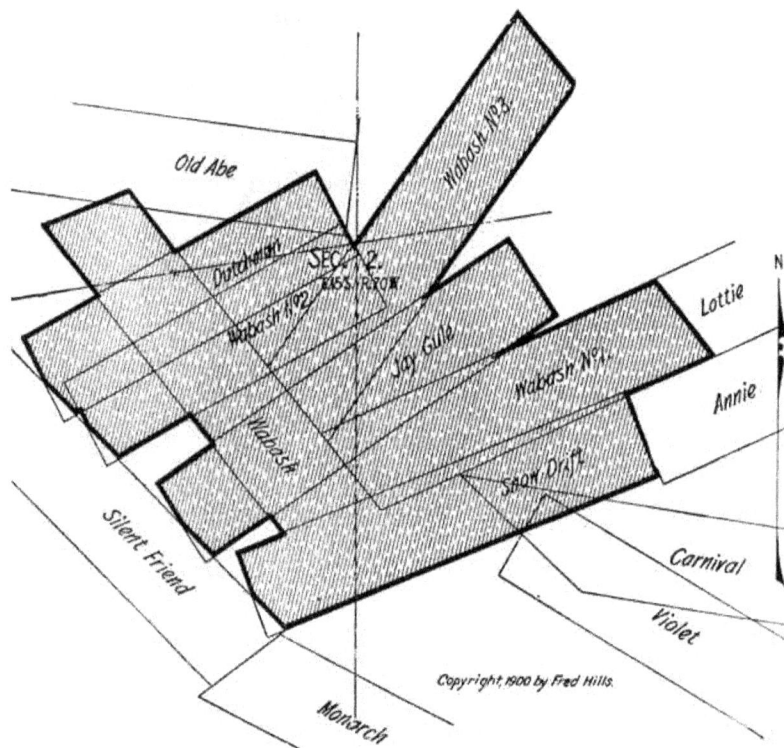

Copyright, 1900 by Fred Hills.

Capitalization   1,500,000 shares. Par value, $1.00. In treasury January 1, 1900, 500,000 shares; in treasury January 1, 1900, $1,500.00 cash.

Property   Owns the Wabash, the Wabash Nos. 1, 2 and 3, the Dutchman, the Jay Gule and the Snow Drift, containing about 50 acres, in the center of section 2, on the north slope of Copper mountain. All patented.

Development   Over $10,000 has been expended in developing this property. The Snow Drift is receiving the greater part of the development work.

The Wanda Gold Mining Company.

Incorporated September 26, 1896.

Henry I. Seemann......................President
Chas. S. Ellis.......................Vice-President
John MacMillan............Secretary and Treasurer

Directors

Main Office—227 and 228 Equitable building, Denver, Colorado.

1,000,000 shares. Par value, $1.00. In treasury January 1, 1900, 190,000 shares; in treasury January 1, 1900, $2,500.00 cash. Capitalization

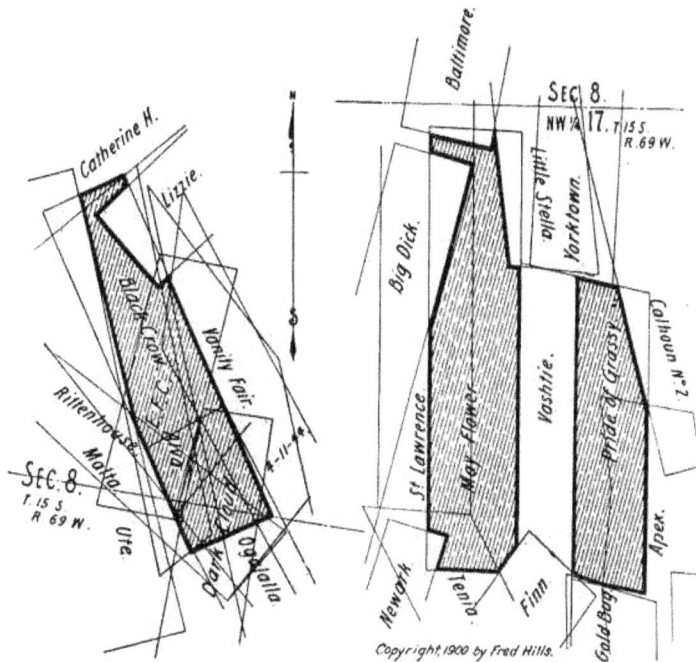

Copyright, 1900 by Fred Hills.

Owns the May Flower, Survey No. 10,788, containing 9 1-2 acres, situated in the N. W. 1-4 of section 17; the Pride of Grassy, containing 5 acres, in the N. W. 1-4 of section 17, both on the north slope of Ironclad hill; the Black Crow, containing 5 1-2 acres, in the N. E. 1-4 of section 8, Survey No. 11,530, on Galena hill. May Flower and Pride of Grassy are patented. The Black Crow is in process of patenting. Property

The May Flower has two shafts, one of 75 feet depth and one of 50 feet depth, with about 100 feet of drifting. There are three shafts on the Pride of Grassy, one of 40 feet depth, one of 50 feet and one of 100 feet, with 95 feet of cross-cutting. The Black Crow has four shafts of 20, 60, 75 and 130 feet, respectively, with 75 feet of cross-cutting. Two blocks of the May Flower and two blocks of the Pride of Grassy are leased for 18 months, at a 25 per cent. royalty. Development

This stock is not listed.

The Waverly Gold Mining and Milling Company.

Incorporated January 31, 1896.

Main Office—252 Equitable building, Denver, Colorado.

Capitalization   1,500,000 shares.   Par value, $1.00.   In treasury January 1, 1900, about 200,000 shares.

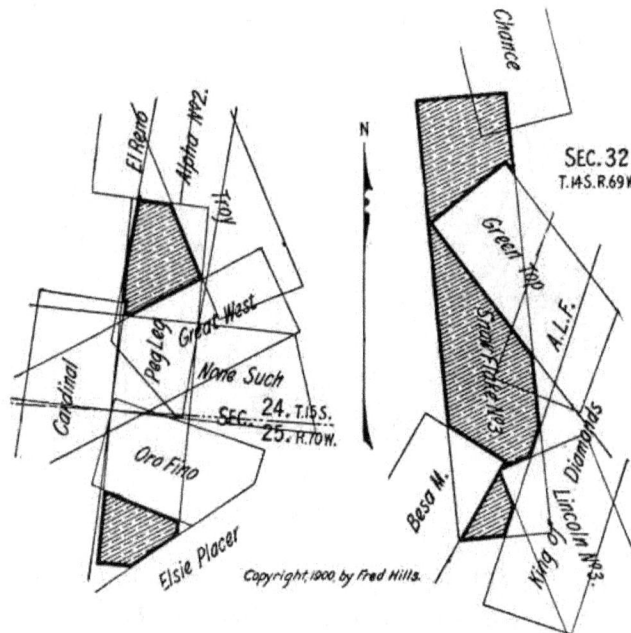

Copyright, 1900, by Fred Hills.

Property   Owns the Peg Leg, in the N. E. 1-4 of section 25, containing 2 1-3 acres, on Gold hill, and the Snow Flake No. 3, in the S. W. 1-4 of section 32, township 14 south, range 69 west, containing 6 acres, on Lincoln hill. All patented. Also holds school lease on Rhyolite mountain.

Development   Shaft houses and whims on the Peg Leg. There is an 80-foot shaft on Peg Leg and a shaft about 50 feet deep on the Snow Flake No. 3. Up to date about $2,000 worth of development work has been done in all. The greater part of the development work is being done on the Peg Leg, which is leased.

Highest price for stock during 1899, 3 cents; lowest price for stock during 1899, 1 5-8 cents.

The Wells Gold Mining Company.

Incorporated February 12, 1896.

James A. Orr, president; B. A. Hughes, vice-president; C. L. McKesson, secretary and treasurer; F. J. Webber, S. P. Newman.

Main Office—21 and 22 Giddings block, Colorado Springs, Colorado.

1,000,000 shares. Par value, $1.00. In treasury January 1, 1900, 224,000 shares; in treasury January 1, 1900, $20.00 cash.

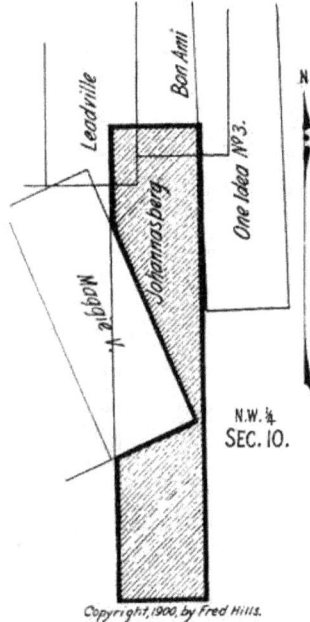

Copyright, 1900, by Fred Hills.

Owns the Ida May lode, about 2 3-4 acres, adjoining the northeast corner of the town of Lawrence, and the southeast corner of the town of Victor; the I. X. L. lode, about 6 3-4 acres, on Tenderfoot hill; an undivided one-third interest in the Johannasberg, about 9 acres, in the N. W. 1-4 of section 10, township 15 south, range 69 west; an undivided one-third interest in the Kimberly of Diamonds, about 10 acres, on Trachyte mountain; also owns the Red Mountain Chief and the Apex lodes, 18 acres, in the Lincoln mining district, Pitkin county, Colorado. The company also have a lease on the Ruby No. 4 lode in Pitkin county. The Johannasberg is patented and is the only property shown on plat. Remainder of the property of this company in the Cripple Creek district is held by location.

The Ida May has a shaft 40 feet deep; the I. X. L. has one 15 feet deep. The Johannasberg has a shaft 60 feet deep and a 340-foot tunnel. The Kimberly of Diamonds has a 40-foot shaft. Considerable development has also been done on the Pitkin county property. The greater part of the development work is being done on the Johannasberg, the Ida May and the Ruby No. 4.

Gross production to January 1, 1900, $466 from the Ruby No. 4. Net profit on ore for 1899, $466, 3 1-2 tons.

The Wells Gold Mining Company passed under a new management in September, 1899, since which date all the properties now owned by the company have been acquired.

Highest price for stock during 1899, 3 cents; lowest price for stock during 1899, 1 cent.

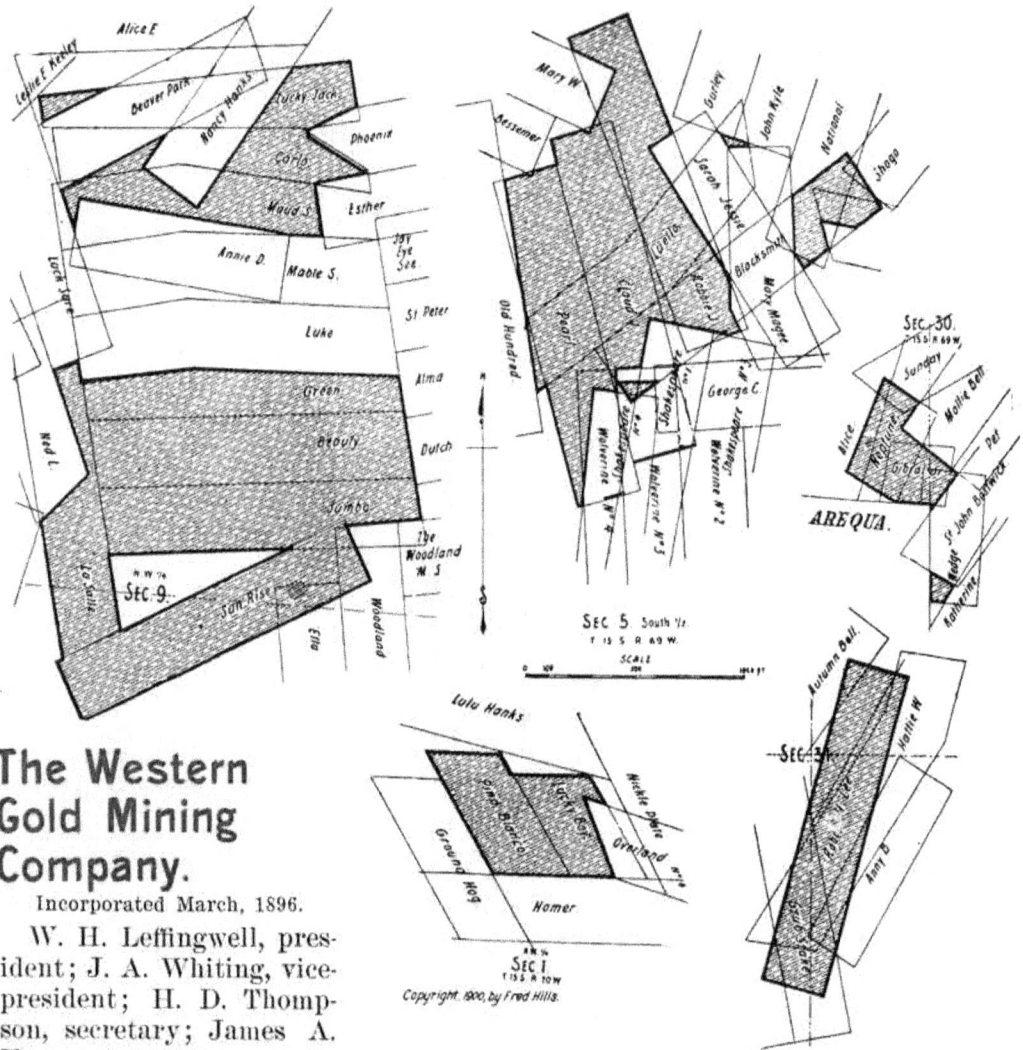

Copyright. 1900, by Fred Hills.

The Western Gold Mining Company.

Incorporated March, 1896.

Directors

W. H. Leffingwell, president; J. A. Whiting, vice-president; H. D. Thompson, secretary; James A. Howze, treasurer.

Main Office—Mining Exchange, Cripple Creek, Colorado.

Capitalization

1,250,000 shares. Par value, $1.00. In treasury March 1, 1900, 350,000 shares; in treasury March 1, 1900, $1,700.00 cash.

Property

Owns the Neptune, the Gibraltar, the Wedge, containing, in all, 4.50 acres, in the N. W. 1-4 of section 30; the Robert E. Lee, containing 10.331 acres, on Squaw mountain, in the S. E. 1-4 of section 31; the Pearl, Claude J., Robbie J., Luella and the Blacksmith, containing 21.50 acres, in the center of section 5, township 15 south, range 69 west; the Sun Rise, La Salle, Jumbo, Green, Beauty, Maude S., Mabel S., Carlo and the Lucky Jack, containing 60 acres, in the N. W. 1-4 of section 9; the Pino Blanco and the Lucky Boy, containing 8 acres, in the N. W. 1-4 of section 1. All patented except the Lucky Boy and Pino Blanco.

Development

The Neptune has a 140-foot shaft, the Gibraltar a 145-foot shaft, the Wedge a 135-foot shaft, the Lee a shaft 60 feet deep and an equal amount of drifting. On the other claims sufficient work to obtain a patent has been done. The greater part of the development work is being done on the Neptune, Gibraltar, Wedge and the Lee. These claims are leased.

History

In October of 1899 the company was revived and reorganized. At the present time it does not owe any debts and is prepared to push development work vigorously. This company is conservatively and carefully managed and the directors feel confident that these claims will show great results in the future.

Highest price for stock during 1899, 5 cents; lowest price for stock during 1899, 4 cents.

482

The Western Union Gold Mining Company.

Incorporated.

S. R. Bartlett, president; F. Gilpin, vice-president; Wm. A. Otis & Directors Company, secretary and treasurer; C. E. Titus, assistant secretary; L. C. Hall, F. H. Frankenberg.

Main Office—Wm. A. Otis & Company, Colorado Springs, Colorado.

2,000,000 shares. Par value, $1.00. In treasury January 1, 1900, Capitalization 665,000 shares.

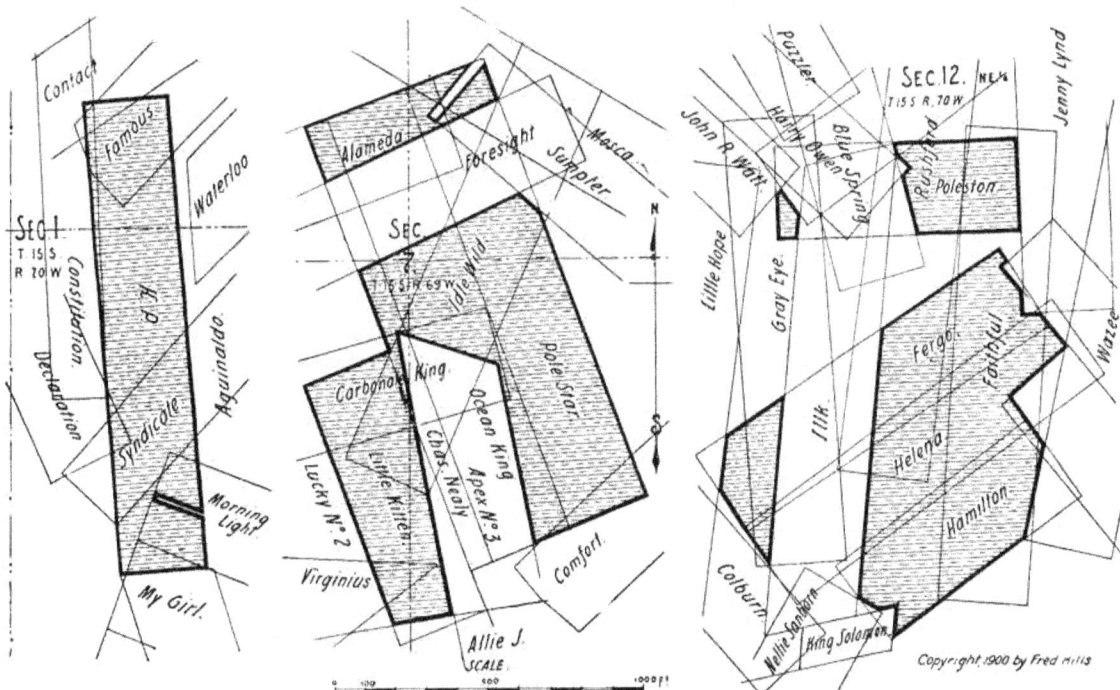

Owns the Poleston, 2.768 acres; the Hamilton, 4.104 acres; the Property Helena, 4.133 acres, and the Fergo, 5.331 acres, on the north slope of Carbonate hill, in the N. E. 1-4 of section 12; the Little Kitten, 5.446 acres; the Ocean King, 7.349 acres; the Pole Star, 8.672 acres; the Charles Nealy, 2.024 acres, on Tenderfoot hill in the center of section 7, and the K. P., 10.214 acres, on the south slope of Copper mountain, in the S. E. 1-4 of section 1, township 15 south, range 70 west. Part of this property is patented, the balance held by receiver's receipts.

All work necessary to obtain a patent has been done. Development

The West Virginia Gold Mining and Milling Company.

Incorporated November, 1892.

Directors

T. W. Fleming.................President
C. B. Fleming.................Secretary
A. H. Fleming.   John Sweeney.
J. M. Harden.

Main Office—Fairmont, West Virginia.

Capitalization

1,000,000 shares. Par value, $1.00. In treasury January 1, 1900, 50,000 shares.

Property

Owns the Omonde, 3-4 of an acre, and 1 1-2 acres mineral rights, patented, in the S. W. 1-4 of section 21, on Bull hill, near the Hull City. Also the Little Lulu and Guadaloup, near Marigold City, 16.28 acres, in the S. E. 1-4 of section 9, township 16 south, range 70 west. Patented. Not shown on plat.

Development

There is a shaft house on each of these claims. The Omonde claim has a shaft 200 feet deep; the Guadaloup a shaft 100 feet deep and the Little Lulu a shaft 50 feet deep. The Omonde is a fractional claim with end lines 1,500 feet apart, adjoining the Vindicator. The Guadaloup and Little Lulu lay together; $5,000.00 has been expended on development work, with a good vein assaying $20.00 to $140.00 to the ton.

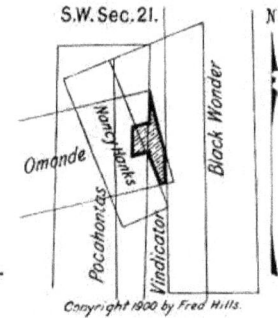

The Wheel of Fortune Consolidated Gold Mining Company.

Incorporated October 6, 1896.

Directors

Wm. Clark.............................President
Wm. Helm..........................Vice-President
E. S. Cohen...............Secretary and Treasurer
R. Hillhouse.   C. A. McLain.
John Pedersen.   Chas. E. Liebold.

Main Office—3 N. Tejon street, Colorado Springs, Colorado.

Capitalization

1,500,000 shares. Par value, $1.00.

Property

Owns the Wheel of Fortune, 5.93 acres; the Green Top and Ed. Wolcott, 12.778 acres, all in N. 1-2 of section 5; the Voumoltree, 8.703 acres, on Rhyolite mountain, in the S. E. 1-4 of section 6; the B. & M. and the Missouri Boy, 14.386 acres, in the S. E. 1-4 of section 6; the Bob and Big 400, 16.884 acres, in the S. 1-2 of section 35, township 15 south, range 70 west, and the N. 1-2 of section 2, township 16 south, range 70 west; the Gold Charm, 2.867 acres, on Tenderfoot hill, in the S. W. 1-4 of section 8; the Falcon, 4.835 acres, and the Yankee, 8.71 acres, on Rhyolite mountain, in S. 1-2 of section 6. United States patents and receiver's receipts on all the above except the Bob and the Big 400, which are in process of patenting.

Highest price for stock during 1899, 2 1-4 cents; lowest price for stock during 1899, $2.50 per thousand.

The Wheel of Fortune Consolidated Gold Mining Company.

Elbow Placer

Saddle

Big 400.

Bobb.

T. 15 S. R. 70 W.
SEC. 35.
SEC. 2.

Leslie Ray.

Frink & Adams Nº 1

Hagerman

Gurley

John Kyle

Conmearwehith

Wheel of Fortune

Nº ½
SEC. 5.
T. 15 S. R. 69 W.

National

Champion Nº 2 A. C. Gillem

Calhoun Nº 4

Lizzie Cranor Nº 5

Lizzie Eleanor, Gold Chord.

Tacky Gertie.

Gold Leaf

Dark

Little Maggie

Copper Signal

Annie C.

Emma D. Nº 3.

SW ¼
SEC. 8
T. 15 S. R 69 W.

Chance

Ed Wolcott

Little Dolly

A. C. F.

Hudson

Snow Flake Nº ½

SEC. 32.
SEC. 5.

Mines of Breckenridge

T. 14 S. R 69 W.

Owl

Green Top

Mains

B. H. Bryant.

SEC. 6.
T. 15 S. R. 69 W.

Keeley

Redondo.

Sandstone

Yountree

Lord Lyman

Yankee.

Little Alice

Uncle Tom.

Mojave

Mocking Bird

Chimney Rock

B. & M.

Falcon

Missouri Boy

Fly Ott

Old Ranch

Seven out.

National Debt

Rosanna

Algiers

Mollie Gibson, Nº 2

Copyright 1900, by Fred Hills.

Scale
0   100   500   1000 ft

485

The Wide Awake Gold Mining Company.

Incorporated November, 1899.

Directors

Walter F. Crosby...................President

M. I. Appel......................Vice-President

Louis R. Ehrich............Secretary and Treasurer

J. W. Wright.   G. Warrh.

Main Office—64 Hagerman building, Colorado Springs, Colorado.

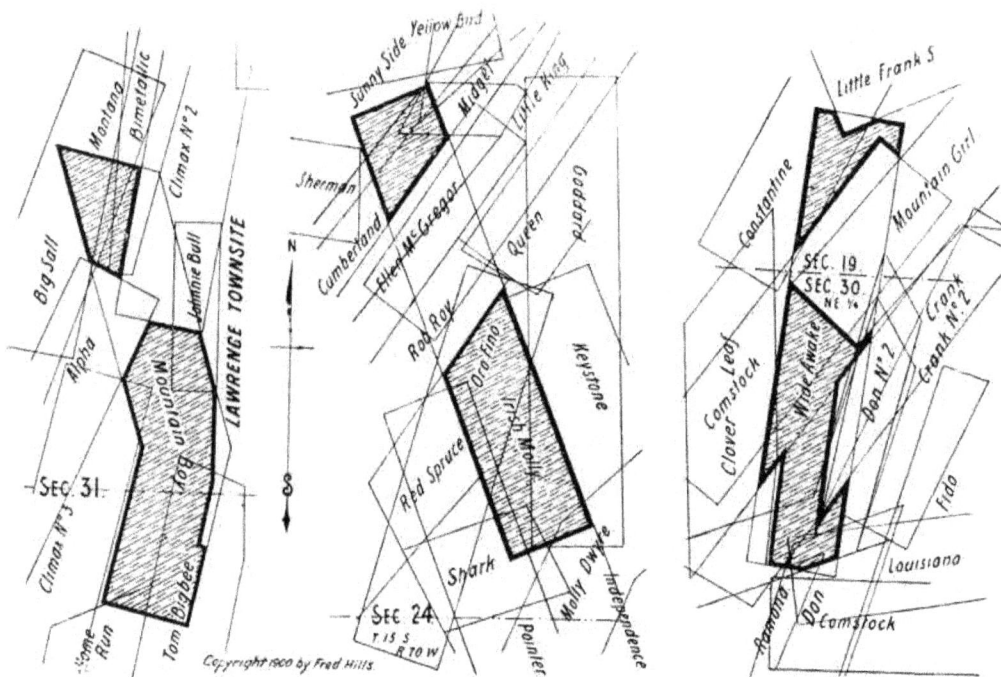

Copyright 1900 by Fred Hills

Capitalization   1,500,000 shares. Par value, $1.00. In treasury January 1, 1900, 300,000 shares.

Property   Owns the Wide Awake, 5 acres, in the N. E. 1-4 of section 30, on Raven hill; the Mountain Boy, 7.6 acres, in the E. 1-2 of section 31, on the south slope of Squaw mountain, and the Irish Molly (held under bond and lease), 7 acres, in the N. E. 1-4 of section 24, on Gold hill. All patented. This property will be vigorously developed by lessees. Gross production to January 1, 1900, about $30,000.00.

The Wilson Creek Consolidated Gold Mining Company.

Incorporated.

W. P. Dunham.........................President
W. P. Sargent.....................Vice-President
R. H. Reid................Secretary and Treasurer
Geo. C. Wallace.   J. M. Bonney.

Directors

Main Office—827 Equitable building, Denver, Colorado.

1,350,000 shares.   Par value, $1.00.

Capitalization

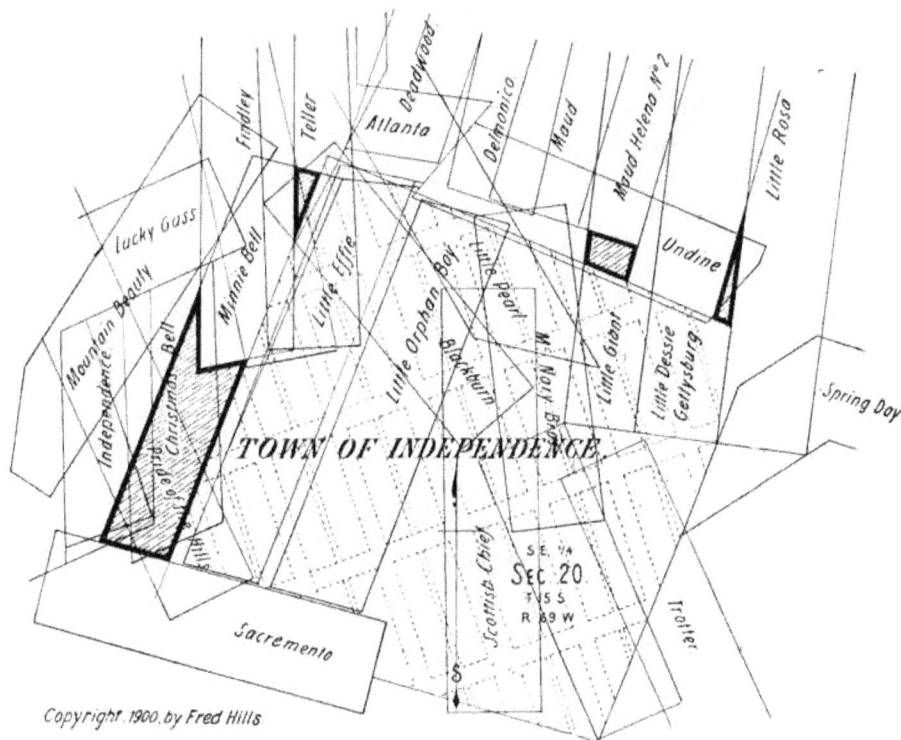

Copyright. 1900, by Fred Hills

Owns the Minnie Bell, Little Effie, Little Dessie, Little Giant, and the Little Orphan Boy, situated in the S. E. 1-4 of section 20; also north one-half conflict with Gettysburg and of the Maud Helena No. 2, containing, in all, about seven acres.

Property

As is known, this company has been in litigation since July, 1899, with the Independence Town and Mining Company. The theory upon which the various suits were instituted was that the lode claims were valid subsisting lode locations at the date of the application for the placer patent and as such did not pass to the patentees under the placer patent. Litigation was still pending in June, 1900.

History

Highest price for stock during 1899, 24 cents; lowest price for stock during 1899, 17 cents.

The Wire Gold Mining and Milling Company.

Incorporated.

F. B. Davis, president; Chas. H. Peters, secretary; F. A. Bailey, treasurer; H. S. Shaw, O. L. Linch.

Main Office—No. 307 Mining Exchange building, Denver, Colorado.

1,500,000 shares. Par value, $1.00. In treasury January 1, 1900, 525,000 shares.

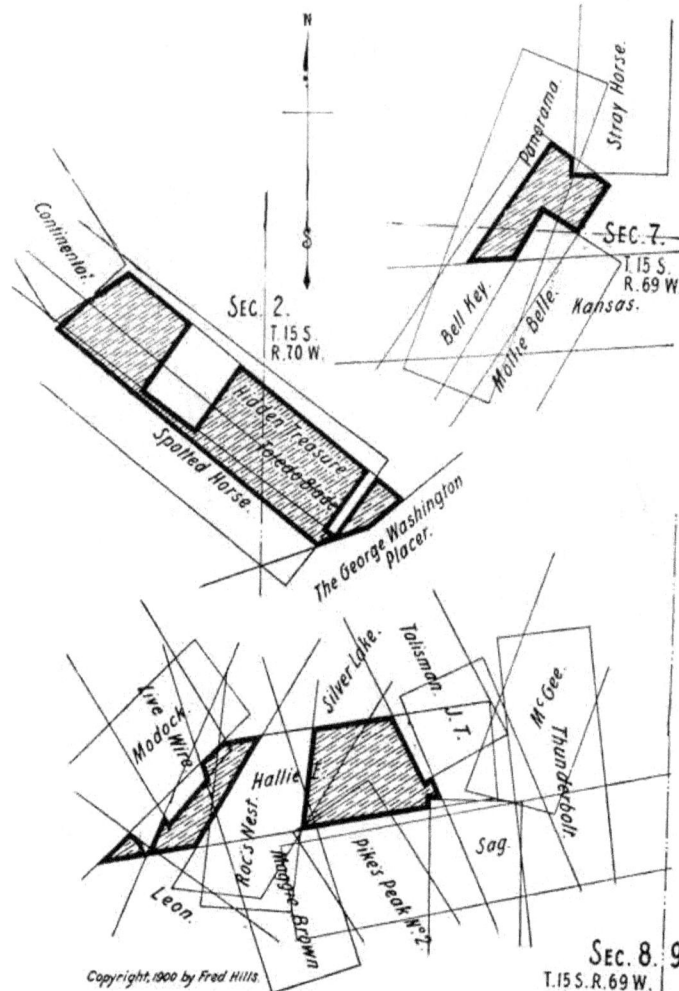

Owns the Hallie E., situated on Galena hill, in the S. E. 1-4 section 8, containing 4.32 acres; the Toledo Blade, containing 7.935 acres, survey No. 10,291, in the S. 1-2 section 2, township 15 south, range 70 west, on Copper mountain; the Bell Key, containing 3.801 acres, survey No. 11,241, in the S. W. 1-4 section 7, township 15 south, range 69 west, on Tenderfoot hill. The Toledo Blade and the Bell Key are patented. Receiver's receipt held for the Hallie E. The company also owns the Wire Gold, the Yankee Hill, the Mamie L., the Badge and the Plainview, all patented, in Gilpin county. The company also expects to purchase in the near future about 17 acres situated on Copper mountain. Active development work is in progress.

Highest price for stock during 1899, 1 3-4 cents; lowest price for stock during 1899, 1 cent.

The Woman's Gold Mining Company.

Incorporated May, 1893.

Jacob Bishoff..........................President

Henry McAllister, Jr........Secretary and Treasurer

W. J. Nesbit.   A. Huyser.

J. H. Avery.

Main Office—33 Giddings building, Colorado Springs, Colorado.

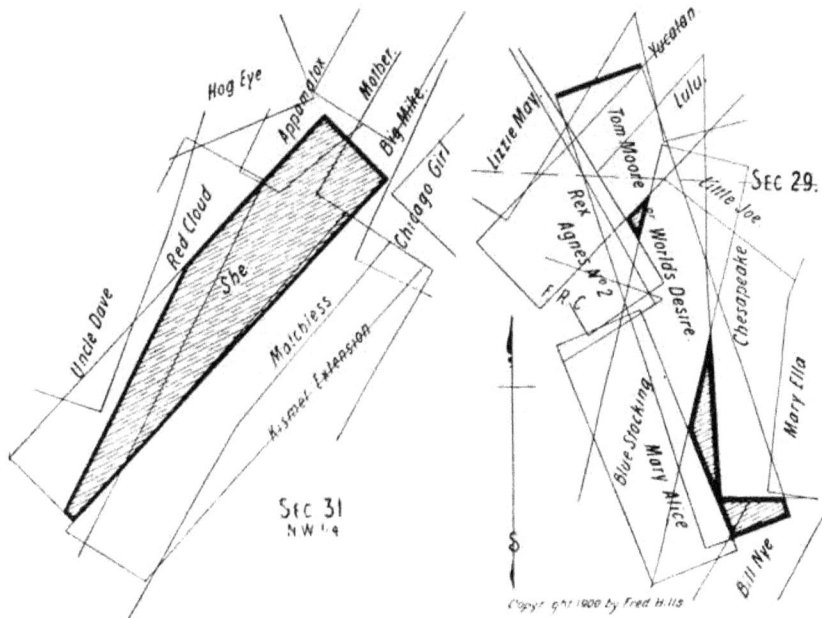

800,000 shares. Par value, $1.00.

Owns the She lode, 7.5 acres, in the N. W. 1-4 section 31, on Squaw mountain, and the World's Desire (or Tom Moore), 1.47 acres, in the S. E. 1-4 section 29, on Battle mountain. All patented.

About $500.00 worth of work has been done on each of the claims.

Highest price for stock during 1899, 2 cents; lowest price for stock during 1899, under 1 cent.

The Worcester Gold Mining Company.

Incorporated January 6, 1896.

Directors

D. W. Walsh..............President and Treasurer

Sylvian Levy....Vice-President and General Manager

Thos. A. Hart.........................Secretary

W. W. Smith.   Wm. H. Gallagher.

Main Office—1325 Lincoln avenue, P. O. box 861, Colorado Springs, Colorado.

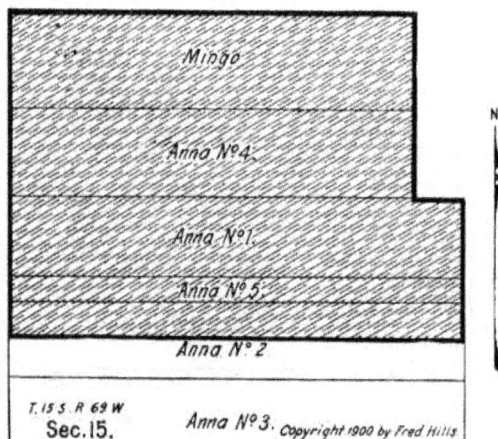

Capitalization   1,500,000 shares. Par value, $1.00. In treasury January 1, 1900, 342,000 shares.

Property   Owns the Anna Nos. 4, 1, 5 and part of 2, containing 25 acres in all, in the N. W. 1-4 section 15, on Calf mountain. Patented. Also the Mingo claim, 10 acres, in process, in the N. W. 1-4 section 15.

Development   A tunnel on property extending 180 feet and shaft 65 feet, with a 35-foot drift, and several other shafts from 20 to 40 feet deep. No lessees working at present, but the company is negotiating for two sets to commence work soon.

The stock being principally held by the directors, there have been no quotations.

The Work Mining and Milling Company.

Incorporated 1892.

Irving W. Bonbright, president; J. Arthur Connell, vice-president and general manager; W. P. Wight (deceased), treasurer; D. D. Lord, secretary; George Rex Buckman. **Directors**

Main Office—68 Postoffice building, Colorado Springs, Colorado.

1,500,000 shares. Par value, $1.00. In treasury January 1, 1900, **Capitalization** about $10,000.00 cash.

Copyright, 1900, by Fred Hills

Owns the Morning Glory, 10.331 acres; Morning Glory No. 2, 2 acres; **Property** Morning Glory No. 4, 9 acres; Poorman, 5.55 acres, and Ida B., 2 acres— all in the S. W. 1-4 section 19, on Raven hill, and the Little Clara, 9.4 acres, in the N. E. 1-4 section 19, Gold hill. All patented.

Several shaft houses on property, but no machinery. The Poorman, **Development** Ida B., Morning Glory, and Morning Glory Nos. 2 and 4 are being worked under a two-years lease by the Morning Glory Mining and Leasing Company, a large corporation described on another page of this work.

Highest price for stock during 1899, 33 1-8 cents; lowest price for stock during 1899, 18 1-4 cents.

491

The World's Fair Mining Company.

Incorporated April 23, 1892.

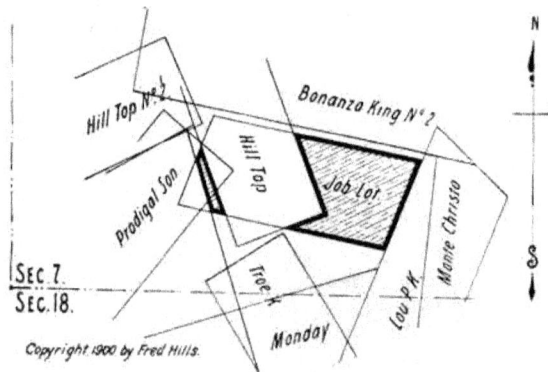

Main Office—112 E. Pike's Peak avenue, Colorado Springs, Colorado.

Capitalization

1,000,000 shares. Par value, $1.00. In treasury January 1, 1900, 148,250 shares.

Property

Owns the Job Lot, a little over 2 acres, in the S. E. 1-4 section 7, on Tenderfoot hill. Patented.

Development

The property is being developed by lessees, and up to the present time only such work has been done as is required to secure patent.

Highest price for stock during 1899, $6.50 per M.; lowest price for stock during 1899, $2.00 per M.

The Yorktown Gold Mining Company.

Incorporated January, 1896.

C. F. Rickey..............................President
J. H. Ryan........................Vice-President
G. D. Kennedy.........................Secretary
H. C. Shimp..........................Treasurer
John McConaghy.

Directors

Main Offices—Colorado Springs, Colorado.

1,500,000 shares. Par value, $1.00. In treasury January 1, 1900, Capitalization
165,000 shares; in treasury January 1, 1900, $527.00 cash.

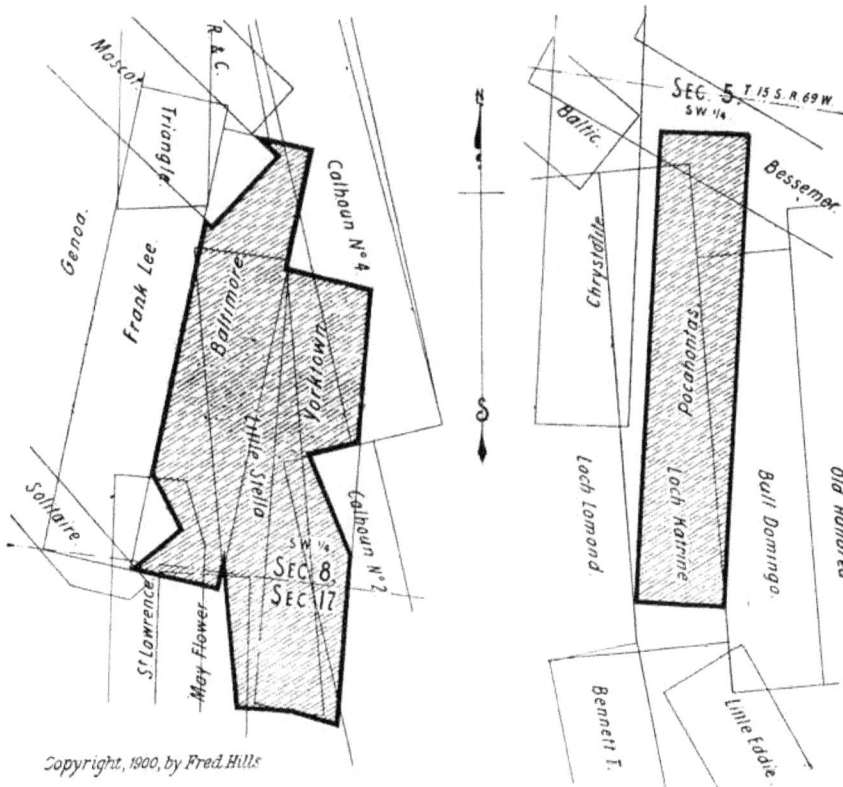

Copyright, 1900, by Fred Hills.

Owns the Pocahontas, containing 9 1-2 acres, in the S. W. 1-4 section Property
5, and the Yorktown, Little Stella, and the Baltimore, containing in all
19.67 acres, on Tenderfoot hill, in the S. W. 1-4 section 8. All patented.

The company has a shaft down about 100 feet, and when down 150 Development
feet a station will be made and cross-cutting commenced to tap a basalt
dyke which traverses the property in a northeasterly and southwesterly
direction. On the S. end of Yorktown there are some leases to parties
who are now down 80 feet, on a contact between schist and breccia.

Highest price for stock during 1899, 7 cents; lowest price for stock
during 1899, 5 cents.

The Zenobia Gold Mining Company.

Incorporated July 31, 1893.

Directors

Edgar H. Brennan.....................President

D. H. Rice.........................Vice-President

Walter H. Baldwin.........Secretary and Treasurer

Main Office—119 Pike's Peak avenue, Colorado Springs, Colorado.

Capitalization

1,000,000 shares. Par value, $1.00. In treasury January 1, 1900, 106,525 shares; in treasury January 1, 1900, $4,583.44 cash.

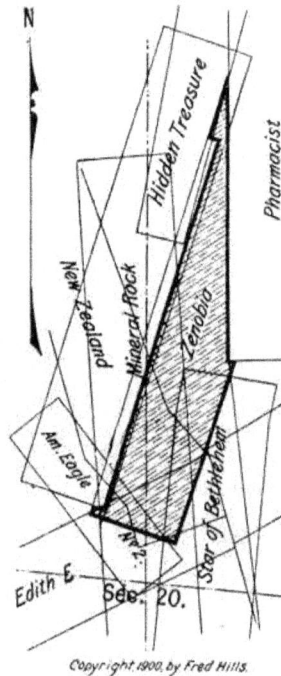

Copyright 1900, by Fred Hills.

Property

Owns the Zenobia, containing 5 1-2 acres, in the N. E. 1-4 of section 20, on Bull hill. Patented. The company also owns the Christopher Columbus No. 5, in section 24, township 14 south, range 69 west. This latter property, being outside the limits of the district shown on map, is not placed thereon.

Development

There is a shaft house, ore house and steam hoist on the Zenobia claim. There are two shafts, one being 500 and the other 600 feet deep.

Production and Dividend

Production to January 1, 1900, $140,863.00. Dividends to January 1, 1900, $10,500.00, which were paid in 1893. On May 10, 1900, the control of this stock, comprising 670,000 shares, was sold for $143,000 cash.

Highest price for stock during 1899, 23 cents; lowest price for stock during 1899, 16 1-4 cents.

The Zeus Gold Mining and Milling Company.

Incorporated February 6, 1896.

Main Office—P. O. box No. 123, Denver, Colorado.

1,000,000 shares. Par value, $1.00.

Capitalization

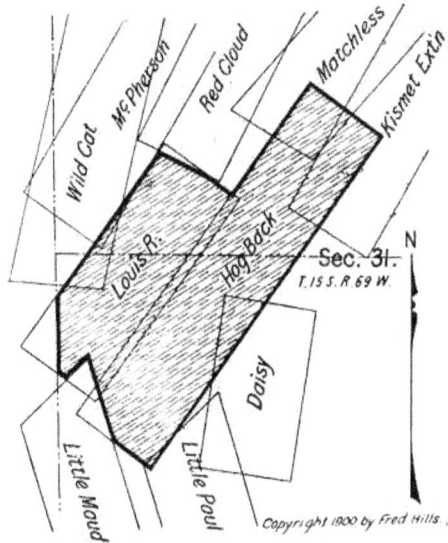

Copyright 1900 by Fred Hills.

Property — Owns the Hog Back claim and the Louis R. claim, both in one body, patent No. 9,000, containing 12.965 acres, in section 31, township 15 south, range 69 west, on the east slope of Grouse mountain. Both patented.

Development — The machinery owned by the company has been removed, owing to its insufficient power. On the Hog Back there is a discovery shaft 72 feet deep, with 30 feet of cross-cuts and drifts and a working shaft 150 feet deep, with 175 feet of cross-cuts and drifts. The Louis R. has a shaft of 50 feet depth and 50 feet of cross-cutting and drifting. Thus far no ore has been shipped.

History — Work was being prosecuted in the main working shaft when the proceeds of the sale of treasury stock was exhausted, making it impossible for this company to carry on the development work. Hence, although there seemed promise of pay ore not far distant, work ceased.

Highest price for stock during 1899, 5 cents; lowest price for stock during 1899, 3 cents.

www.ingramcontent.com/pod-product-compliance
Lightning Source LLC
Chambersburg PA
CBHW051114200326
41518CB00016B/2503